# Recent Advances in Analytical Chemistry

# Recent Advances in Analytical Chemistry

Editor: Johana Meyer

MURPHY & MOORE
www.murphy-moorepublishing.com

Published by Murphy & Moore Publishing
1 Rockefeller Plaza,
New York City, NY 10020, USA
www.murphy-moorepublishing.com

Recent Advances in Analytical Chemistry
Edited by Johana Meyer



International Standard Book Number: 978-1-63987-473-6 (Hardback)



**Cataloging-in-Publication Data**

Recent advances in analytical chemistry / edited by Johana Meyer.
p. cm.
Includes bibliographical references and index.
ISBN 978-1-63987-473-6
1. Chemistry, Analytic. 2. Chemistry. I. Meyer, Johana.
QD75.22 .R43 2022
543--dc23

# Contents


**Preface** .......... IX

Chapter 1 **Efficiency of Polymeric Membrane Graphene Oxide-$TiO_2$ for Removal of Azo Dye** .......... 1
Elahe Dadvar, Roshanak Rezaei Kalantary, Homayon Ahmad Panahi and Majid Peyravi

Chapter 2 **Synthesis, Characterization, and Evaluation of Antimicrobial and Antifertility Efficacy of Heterobimetallic Complexes of Copper (II)** .......... 14
Ashu Chaudhary and Anshul Singh

Chapter 3 **Online Solid-Phase Extraction LC-MS/MS: A Rapid and Valid Method for the Determination of Perfluorinated Compounds at Sub ng·$L^{-1}$ Level in Natural Water** .......... 23
Salvatore Barreca, Maddalena Busetto, Matteo Vitelli, Luisa Colzani, Laura Clerici and Pierluisa Dellavedova

Chapter 4 **Adsorptive Removal of Copper by using Surfactant Modified Laterite Soil** .......... 32
Tien Duc Pham, Hoang Hiep Nguyen, Ngoc Viet Nguyen, Thanh Tu Vu, Thi Ngoc Mai Pham, Thi Hai Yen Doan, Manh Ha Nguyen and Thi Mai Viet Ngo

Chapter 5 **Optimization of the Extraction of the Volatile Fraction from Honey Samples by SPME-GC-MS, Experimental Design, and Multivariate Target Functions** .......... 42
Elisa Robotti, Federica Campo, Marta Riviello, Marco Bobba, Marcello Manfredi, Eleonora Mazzucco, Fabio Gosetti, Giorgio Calabrese, Emanuele Sangiorgi and Emilio Marengo

Chapter 6 **Development of a Methodology for the Determination of Pesticide Residues in Cajá-Manga Pulp (*Spondias dulcis* L.) Using Solid-Liquid Extraction with Low-Temperature Partitioning** .......... 56
Escarlet T. D. Santana, Dayane F. Soares and Anizio M. Faria

Chapter 7 **Three-Dimensional Excitation and Emission Fluorescence-Based Method for Evaluation of Maillard Reaction Products in Food Waste Treatment** .......... 66
Jiaze Liu, Jun Yin, Xiaozheng He, Ting Chen and Dongsheng Shen

Chapter 8 **A Stability-Indicating RP-HPLC-UV Method for Determination and Chemical Hydrolysis Study of a Novel Naproxen Prodrug** .......... 77
Mohamed H. M. Hamid and Tilal Elsaman

Chapter 9 **Determination of the Phenolic Composition and Antioxidant Activity of Pear Extracts** .......... 87
Mindaugas Liaudanskas, Kristina Zymonė, Jonas Viškelis, Almantas Klevinskas and Valdimaras Janulis

Chapter 10 **Comparison of an Electronic Nose based on Ultrafast Gas Chromatography, Comprehensive Two-Dimensional Gas Chromatography, and Sensory Evaluation for an Analysis of Type of Whisky** ........ 96
Paulina Wiśniewska, Magdalena Śliwińska, Tomasz Dymerski, Waldemar Wardencki and Jacek Namieśnik

Chapter 11 **Analytical Methods for the Determination of the Distribution of Total Petroleum Hydrocarbons in the Water and Sediment of Aquatic Systems** ........ 109
A. O. Adeniji, O. O. Okoh and A. I. Okoh

Chapter 12 **Flavonoids in Different Parts of *Lysimachia clethroides* Duby Extracted by Ionic Liquid: Analysis by HPLC and Antioxidant Activity Assay** ........ 122
Jin-feng Wei, Zhi-juan Zhang, Li-li Cui and Wen-yi Kang

Chapter 13 **Equilibrium Measurements of the $NH_3$-$CO_2$-$H_2O$ System: Speciation based on Raman Spectroscopy and Multivariate Modeling** ........ 132
Maths Halstensen, Henrik Jilvero, Wathsala N. Jinadasa and Klaus-J. Jens

Chapter 14 **Adsorptive Removal of Methylene Blue and Crystal Violet onto Micro-Mesoporous $Zr_3O$/Activated Carbon Composite: A Joint Experimental and Statistical Modeling Considerations** ........ 145
H. Ait Ahsaine, Z. Anfar, M. Zbair, M. Ezahri and N. El Alem

Chapter 15 **Fabrication of Functionalized MOFs Incorporated Mixed Matrix Hollow Fiber Membrane for Gas Separation** ........ 159
Haitao Zhu, Xingming Jie and Yiming Cao

Chapter 16 **A New Approach for Accurate Prediction of Liquid Loading of Directional Gas Wells in Transition Flow or Turbulent Flow** ........ 168
Ruiqing Ming and Huiqun He

Chapter 17 **Vibrational Spectroscopies and Chemometry for Nondestructive Identification and Differentiation of Painting Binders** ........ 177
Serena Carlesi, Maurizio Becucci and Marilena Ricci

Chapter 18 **3-Mercaptopropyltrimethoxysilane Modified Diatomite: Preparation and Application for Voltammetric Determination of Lead (II) and Cadmium (II)** ........ 187
Dinh Quang Khieu, Bui Hai Dang Son, Vo Thi Thanh Chau, Pham Dinh Du, Nguyen Hai Phong and Nguyen Thi Diem Chau

Chapter 19 **Application of Central Composite Design in the Adsorption of Ca(II) on Metakaolin Zeolite** ........ 197
Upenyu Guyo, Lycenter Yard Phiri and Fidelis Chigondo

Chapter 20 **Determination of 2-Propenal using Headspace Solid-Phase Microextraction Coupled to Gas Chromatography–Time-of-Flight Mass Spectrometry as a Marker for Authentication of Unrefined Sesame Oil** ........ 207
Ahmad Rois Mansur, Tae Gyu Nam, Hae Won Jang, Yong-Sun Cho, Miyoung Yoo, Dongwon Seo and Jaeho Ha

Chapter 21 **Hetarylazopyrazolone Dyes based on Benzothiazole and Benzimidazole Ring Systems: Synthesis, Spectroscopic Investigation, and Computational Study** ..... 219
Ebru Aktan and Tahsin Uyar

Chapter 22 **LIBS Analysis of Geomaterials: Comparative Study of Basalt Plasma Induced by TEA $CO_2$ and Nd:YAG Laser in Air at Atmospheric Pressure** ..... 229
Jelena Savovic, Milos Momcilovic, Sanja Zivkovic, Andrei Stancalie, Milan Trtica and Miroslav Kuzmanovic

**Permissions**

**List of Contributors**

**Index**

# Preface

Over the recent decade, advancements and applications have progressed exponentially. This has led to the increased interest in this field and projects are being conducted to enhance knowledge. The main objective of this book is to present some of the critical challenges and provide insights into possible solutions. This book will answer the varied questions that arise in the field and also provide an increased scope for furthering studies.

Analytical chemistry is the branch of chemistry that encompasses the study and uses different instruments and methods for identifying, separating and quantifying matter. Qualitative and quantitative analysis are two methods. They identify analytes and determine the numerical count of concentration, respectively. This field includes classical, modern, wet chemical and instrumental methods. Analytical chemistry is focused on the improvement and advancement of experimental design, the creation of new measurement tools and chemometrics. It has significant applications in the diverse areas of bioanalysis, nanotechnology, clinical analysis, forensic screening, environmental analysis and materials analysis. This book discusses the fundamentals as well as modern approaches of analytical chemistry. It is compiled in such a manner, that it will provide in-depth knowledge about the theory and practice of this field. In this book, using case studies and examples, constant effort has been made to make the understanding of the difficult concepts of analytical chemistry as easy and informative as possible, for the readers.

I hope that this book, with its visionary approach, will be a valuable addition and will promote interest among readers. Each of the authors has provided their extraordinary competence in their specific fields by providing different perspectives as they come from diverse nations and regions. I thank them for their contributions.

**Editor**

# 1

# Efficiency of Polymeric Membrane Graphene Oxide-$TiO_2$ for Removal of Azo Dye

**Elahe Dadvar,[1] Roshanak Rezaei Kalantary,[2] Homayon Ahmad Panahi,[3] and Majid Peyravi[4]**

[1]*Department of Environmental Engineering, Faculty of Environment and Energy, Islamic Azad University, Science and Research Branch, Tehran, Iran*
[2]*Department of Environmental Health Engineering, School of Public Health, Iran University of Medical Sciences, P.O. Box 15875-4199, Tehran, Iran*
[3]*Chemistry Department, Islamic Azad University, Central Tehran Branch, Tehran, Iran*
[4]*Nanotechnology Research Institute, Chemical Engineering Department, Babol University of Technology, Babol, Iran*

Correspondence should be addressed to Roshanak Rezaei Kalantary; rezaei.r@iums.ac.ir

Academic Editor: Zuhui Zhang

Achieving the desired standard of drinking water quality has been one of the concerns across water treatment plants in the developing countries. Processes such as grid chamber, coagulation, sedimentation, clarification, filtration, and disinfection are typically used in water purification plants. Among these methods, unit filtration which employs polymers is one of the new technologies. There have been many studies about the use of semiconductive $TiO_2$ with graphene oxide (GO) on the base of different polymeric membranes for the removal of azo dyes, especially methylene blue (MB). Polymeric GO-$TiO_2$ membranes have high photocatalytic, antifouling property and permeate the flux removal of organic pollutants. The aim of this study was to investigate the characteristics of different polymeric membranes such as anionic perfluorinated polymer (Nafion), cellulose acetate, polycarbonate (PC), polysulfone fluoride (PSF), and polyvinylidene fluoride (PVDF). The result of this study showed that the GO-$TiO_2$ membrane can be used in the field of water treatment and will be used for the removal of polycyclic aromatic hydrocarbons (PAHs) from wastewater.

## 1. Introduction

With the development of heavy industries in recent decades, obtaining healthy drinking water has become a global concern [1]. Organic substances, industrial dyes, microorganisms, and heavy metals (heavy metals include cobalt, cadmium, mercury, chromium, and lead) are among the water pollutants [2–7]. In recent decades, there has been increasing interest in the use of clean water [8–10]. With high efficiency and low energy consumption, filtration is one of the most appropriate technologies for decreasing pollution [11–15]. The formation of a good membrane is an important and practical step toward increasing the efficiency of water treatment; this film could be made from materials such as polymer, fiber, ceramic, and carbon nanotubes [16–19].

In recent decades, ceramic and polymeric membranes have been increasingly used due to their strong mechanical properties, chemical stability, high efficiency in minimizing pollutants, high photocatalytic power, and odor reduction ability [20, 21]. The photocatalytic process of separating organic pollutants from the membrane, under visible light and UV radiation, can occur without any energy or chemical consumption. This process is especially important for the removal of different types of environmental pollutants [22]. The photocatalytic process of isolating organic pollutants from membrane, under visible light and UV radiation, can demonstrate higher efficiency [23]. Titanium dioxide is one of the important semiconductors, having properties such as low toxicity, low cost, and highly efficient removal of pollutants [21, 24, 25].

$TiO_2$ has the ability to degrade organic materials (organic dye, oil, and toxic pollutants) into $H_2O$ and $CO_2$ [24, 25]. As a semiconductor, nanoparticles of anatase titanium dioxide have higher photocatalytic power than rutile and brookite crystals and exhibit absorption at 390 nm when exposed to UV radiation [26]. The present study includes a review of the semiconductive properties of $TiO_2$ semiconductor and its photocatalytic power. Different methods, such as doping and surface chemical modification, were used to increase the photocatalytic power of $TiO_2$ [27]. It is worth mentioning that, in different industries, such as the textile industry, $TiO_2$ graphene membranes are increasingly used to improve the efficiency of water treatment.

In recent decades, there have been many reports on the mechanical and chemical properties of graphene (G) and graphene oxide (GO) [28]. In studies concerning water treatment and purification, nanostructured carbon has been used in the form of carbon nanotube and graphene. These structures have high absorption capacity and the ability to absorb organic materials in aqueous solutions [29]. After studies on the absorption of organic aromatic pollutants by graphene, carbon nanotube, and granular activated carbon, Onu et al. discovered that graphene possessed higher efficiency in the absorption of natural organic materials.

The presence of natural organic matter (NOM) reduced the absorption of synthetic organic compound (SOC) in carbon nanotube, granular activated carbon, and, finally, graphene [30]. After comparing the absorption of methylene blue and atrazine pollutants by super fine powdered activated carbon (SPAC), carbon nanotubes (CNTs), and graphene, Ellerie et al. found that SPAC had the highest efficiency, while multiwalled carbon nanotube (MWCNT) was the least efficient in absorbing pollutants [31].

Compared to membranes without absorbents, the direct use of absorbents in membrane structure improves the efficiency of absorbed pollutants. GO consists of graphene sheets having functional groups (R-O-H, epoxy, COOH, and C=O). Therefore, the GO-$TiO_2$ nanocomposite, under UV radiation and visible light, has the capability of photocatalytic degradation, high absorption capacity, and the capability of being more exposed to water pollutants as a result of its huge surface area of 2360 $m^2$/gr [32]. Liu et al. studied the nanorod of $TiO_2$ on the surface of graphene oxide sheets interface water and toluene. They produced a high efficiency nanorod (GO-$TiO_2$ NRCs) composite. The $TiO_2$ nanorod was stabilized with oleic acid, low temperature, and hydrolysis approach. The photocatalytic activity (GO-$TiO_2$ NRCs) desired in the degradation of methylene blue (MB) under UV radiation was higher than GO-25 and the other original $TiO_2$ nanorod states [33].

Rao et al. reported efficient removal of 4-chlorophene, 2,4-dichlorophene, and 2,4,6-trichlorophene by $ZrO_2$ graphene composite and, herein, removal efficiency of 4-chlorophene was by far better than the others, and the absorbability decreased with the increase in pH [34].

According to the report of Zhang et al., $TiO_2$ nanowire membrane was able to remove TOC and HA pollution with a removal efficiency of 93.6 and 100%, respectively. This membrane was able to degrade organic materials and produce $CO_2$ [35].

Mele et al. carried out a characterization of polycrystalline bare $TiO_2$ for the degradation of 4-nitrophenol (4-NP) in aqueous suspension using TRMC, EPR, and XPS. The results of this study showed that propyline impregnated on the surface of $TiO_2$ improved the photocatalytic properties compared to bare $TiO_2$ [35].

Afzal et al. synthesized anatase $TiO_2$/TCPP (meso-tetra(4-carboxyphenyl)porphyrin) coated with cotton. $TiO_2$/TCPP coated with cotton for the purpose of making comparison with bare $TiO_2$ was found to be better. There was complete degradation of methylene blue in 110 min and removal of coffee and red wine in 16 h by visible light [36].

There are many studies regarding the $TiO_2$ semiconductor and its photocatalytic features in removing pollutants, especially industrial dyes and heavy metals; however, to date, there has been no comprehensive study regarding the photocatalytic features, flow permeability, and antifouling property. Fouling resistance in GO-$TiO_2$ membrane, with regard to various polymer bases, has been used in this article.

TABLE 1: The structural properties of $TiO_2$ crystals [43–45].

| Properties | Rutile | Anatase | Brookite |
|---|---|---|---|
| Crystal structure | Tetragonal | Tetragonal | Orthorhombic |
| Lattice constant (Å) | $a = 4.5936$<br>$c = 2.9587$ | $a = 3.784$<br>$c = 9.515$ | $a = 9.184$<br>$b = 5.447$<br>$c = 5.154$ |
| Space group | $p4_2/mnm$ | $I4_1/amd$ | pbca |
| Molecule (cell) | 2 | 2 | 4 |
| Volume/molecule ($Å^3$) | 31/2160 | 34.061 | 32/172 |
| Density ($g\,cm^{-3}$) | 4.13 | 3.79 | 3.99 |
| Ti-O band length (Å) | 1.949 (4) | 1.937 (4) | 1.87–2.04 |
| O-Ti-O band | 81.2°<br>90.0° | 77.7°<br>92.6° | 77.0°–105° |

*1.1. Titanium Dioxide.* Titanium(IV) oxide is the oxide of titanium, with a molecular weight of 79.87 g/mol and chemical formula of $TiO_2$. When used as a pigment, it is called white titanium or pigment white. Titanium dioxide is used in industries as a mineral pigment. It occurs in nature in three forms: brookite, rutile, and anatase. White titanium dioxide is 200 to 300 nm in diameter and has a particular geometric form and structure. Table 1 shows the properties of different kinds of titanium. The first production of $TiO_2$ was manufactured in Norway, USA, and Germany in 1918 [39–42].

*1.2. Photocatalysts.* For several years in industrialized countries, photocatalysts have been used for the removal of pollutants which are not removed by bioprocesses. Most photocatalysts are semiconductor solid oxides which, under radiation, are activated with sufficient energy [47]. Chlorophyll in plants acts in a similar manner to photocatalysts. When compared with photosynthesis in which chlorophyll absorbs sunlight and produces oxygen and glucose by water and carbon dioxide, in the process of photocatalysis, organic materials are converted into water and carbon dioxide in the presence of light, water, and catalyst (Figure 1) [48].

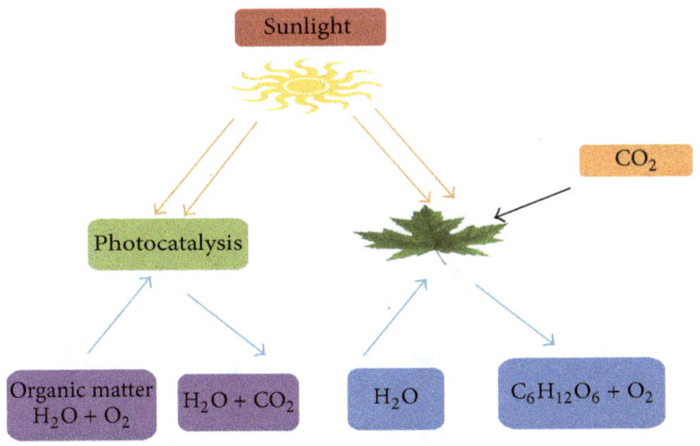


FIGURE 1: Comparison between the actions of a catalyst and chlorophyll.

TABLE 2: Band gap energy for some common semiconductor materials [46].

| Semiconductors | Band gap energy (eV) |
|---|---|
| Diamond | 5.4 |
| CdS | 2.42 |
| ZnS | 3.6 |
| ZnO | 3.436 |
| $TiO_2$ | 3.03 |
| CdS | 2.582 |
| $SnO_2$ | 3.54 |
| CdSe | 1.7 |
| $WO_3$ | 2.76 |
| Si | 1.17 |
| Ge | 0.744 |
| $Fe_2O_3$ | 2.3 |
| PbS | 0.286 |
| PbSe | 0.165 |
| $ZrO_2$ | 3.87 |
| $Cu_2O$ | 2.172 |

*1.3. $TiO_2$ Photocatalytic Degradation Mechanism.* One of the important properties of $TiO_2$ solid and inorganic nanomaterials is the photocatalytic activity. In many cases, this feature is used for antibacterial surfaces. Also, the photocatalytic activity of $TiO_2$ has been applied in a wide range of metal oxides and their sulfides [49, 50] including ZnO [51], $WO_3$ [52], $WS_2$ [53], $Fe_2O_3$ [54], $V_2O_5$ [55], $CeO_2$ [56], CdS [57], and ZnS [58]. The band gap energy values for some common semiconductor materials are presented in Table 2 [46].

$TiO_2$ was found to be the best semiconductor because of its chemical stability and nontoxicity. Also, it is cost-effective and has an excellent photocatalytic activity in the presence of UV irradiation [59–62].

One of the commercial uses of $TiO_2$ is $P_{25}$, which is often used as a photocatalytic material. It consists of 70 to 80% anatase and 20 to 30% rutile, with specific surface area of 50 $m^2/g$. There exists an energy gap between the valence band

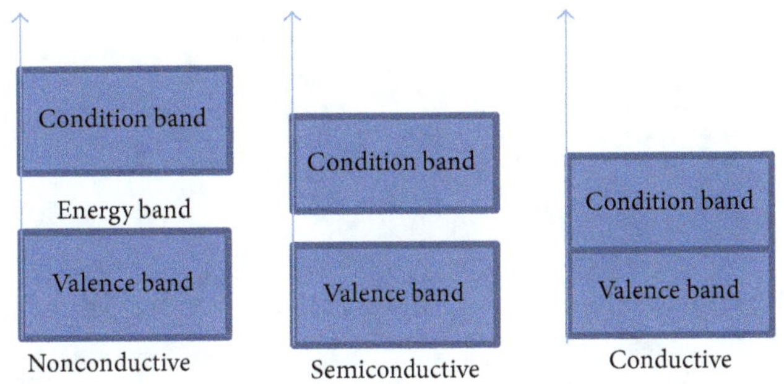


FIGURE 2: Energy gap in semiconductive, nonconductive, and conductive material.

and the conduction band in semiconductors (Figure 2). Contrary to nonconductors, it is small; therefore, after radiation of light to the photocatalysts of semiconductors, photons with energy equal to or higher than the gap energy are absorbed, and electrons can be excited from the valence band to the conduction band.

After receiving solar energy (photons), semiconductors gain more energy than usual, and, as a result, electrons travel from the valence band to the conduction band and create holes in the valence band ($h_{vb}^{+}$) (Figure 3).

$TiO_2$ produces this energy by receiving sunlight or UV radiation and acts as an important photocatalyst [63]. Moreover, the use of $TiO_2$ nanomaterials as a photocatalyst for the removal of pollutants has attracted much attention due to its physical and chemical properties, high efficiency, low cost, and low toxicity.

$$(1)\ TiO_2 + h\nu \rightarrow e_{(cb)}^{-} + h_{vb}^{+} \quad h\nu < 390\,\text{nm}$$

$h^+$ is suitable for producing $OH^{\bullet}$ hydroxyl radicals on the surface of $TiO_2$ and also allows excitation of electrons from the valence band to the conduction band [51] and excitation of conduction band electrons to reduce the oxygen molecule.

$$(2)\ e^- + O_2 \rightarrow O_{2(ads)}^{\bullet}$$

$$(3)\ e^- + H_{(ads)}^{+} \rightarrow H_{(ads)}^{\bullet}$$

$$(4)\ h^+ + OH_{(ads)}^{-} \rightarrow OH_{(ads)}\ \text{(in alkaline solution)}$$

$$(5)\ h_{(VB)}^{+} + H_2O_{(ads)} \rightarrow H^+ + OH_{(ads)}^{\bullet}\ \text{(in neutral solution)}$$

Anatase and rutile are the two main types of $TiO_2$ with energies of 3.2 and 3.1 eV, respectively, showing that the photocatalytic activity of anatase is much greater than that of rutile [64, 65].

The photocatalytic power of $TiO_2$ is dependent not only on the energy band but also on the surface characteristics. The presence of a high surface area in each mass increases the photocatalytic power. The degradation of azo dye, especially methylene blue, by $TiO_2$ film, depends on the surface of the photocatalyst. This film could be either in anatase crystal form, having different thicknesses and surfaces, or in low pressure and chemical deposition, from which vapor is produced [66].

$TiO_2$ nanomaterials can be used for the photocatalytic cleaning of different types of organic compounds, such as

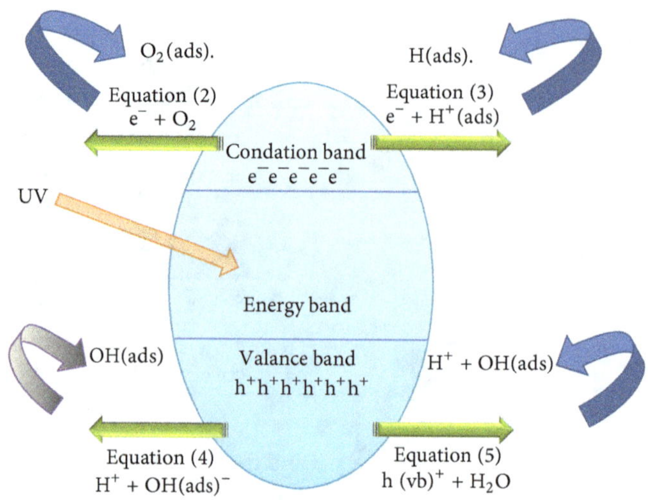


FIGURE 3: The mechanism of photocatalysis on $TiO_2$ nanomaterials [37].

chlorinated hydrocarbons (like organic dye, insecticides, surfactants, $CCl_4$, $CHCl_4$, $C_2HCl_3$, and phenol), decreasing heavy metals such as $Pt^{4+}$, $Pb^{2+}$, $Au^{3+}$, $Rh^{3+}$, and $Cr^{3+}$, and are also capable of the high efficiency removal of bacteria and viruses [65–69].

The pH of the solution has a great influence on the degradation of dye and $TiO_2$ surface charge. The surface of $TiO_2$ is positively charged in acidic pH and negatively charged in alkaline condition. The anionic dyes have strong adsorption in acidic pH while the cationic dyes have weak adsorption in alkaline pH [70, 71].

## 2. Background of Azo Dye

Dyes of textile industries and other kinds of dyes are dangerous to humans and the environment. During coloring, 1 to 20% of dyes are expelled from the wastewater of textile industries [72–75].

The discharge of wastewater from textile industries into the environment is regarded as a major source of pollution and main cause of eutrophication in rivers and lakes. Also, the oxidation, hydrolysis, and chemical activities in the discharged wastewater cause potential dangers [76–78].

The application of old processes (such as absorption, activated carbon, ultrafiltration, reverse osmosis, coagulation, and ion exchange) for the removal of dyes shows high efficiency. The transmission of organic materials from water to other phases, regeneration of absorbents and wastewater pretreatment, is quite expensive [79–82].

Due to high levels of aromatic compounds in dyes and their stability, the biological treatment takes a longer time and the crystals produced are not suitable for degradation [83–86]. Chlorination and ozonation are practical for the removal of some kinds of dyes. Both methods are expensive since they consume chemical materials, consume power, have low efficiency and limitations in the removal of carbon, and are not suitable for degradation [87–90].

In recent decades, advanced oxidation processes (AOPs) have become widespread. This process is based on the production of super active species like hydroxyl ($^{\bullet}OH$), which is capable of oxidizing a wide range of materials, but its action is nonselective. Fenton and photo-Fenton, photocatalytic activities [91–95], $UV/H_2O_2$ [96, 97], and modified photocatalytic activity of $TiO_2$ [98–101] are some of the advanced oxidation processes (AOPs). In AOPs, the use of $TiO_2$ as a photocatalyst has been recognized as a suitable technology [102–107].

Finally, the use of $TiO_2$ as a photocatalyst under visible light has attracted much attention because it is economical. The use of a photocatalyst in the presence of sunlight can help in reducing dyes, especially methylene blue and methyl orange, and also their mineralization [108–110]. Many studies have considered the use of the desired photocatalyst for cleaning of textile industrial wastewater [111–114].

*2.1. Azo Dye.* The main problem of pollution in textile industries is water pollution. Methylene blue and methyl orange are among the commonly used dye compounds in the industry, and they are often known as azo dyes because they contain the functional group R-N=N-R, in which N=N is known as azo. The structures of methyl orange and methylene blue compounds are shown in Figure 4. Azo dyes are divided into three groups: mono, di, and tri, and this classification is dependent on the number of azo groups. R and $R'$ can be aryl and alkyl groups, and as we know it is quite difficult to break down benzene ring; hence, the simplest benzene ring is $C_6H_5$ in the aryl group.

Studies have shown that the presence of UV radiation helps $TiO_2$ in removing organic dyes [50, 51].

## 3. Strategies for Improved Power of $TiO_2$

*3.1. Doping.* Doping titanium dioxide with metal elements such as Cr, Co, Cu, V, Mo, V, Ag, Au, Pt, Nb, and Ru and nonmetal elements such as C, N, S, P, I, F, and B increases the charge transfer reaction and thermal stability of the photocatalyst [27]. For example, X. Z. Li and F. B. Li reported the photodegradation of methylene blue in aqueous solutions

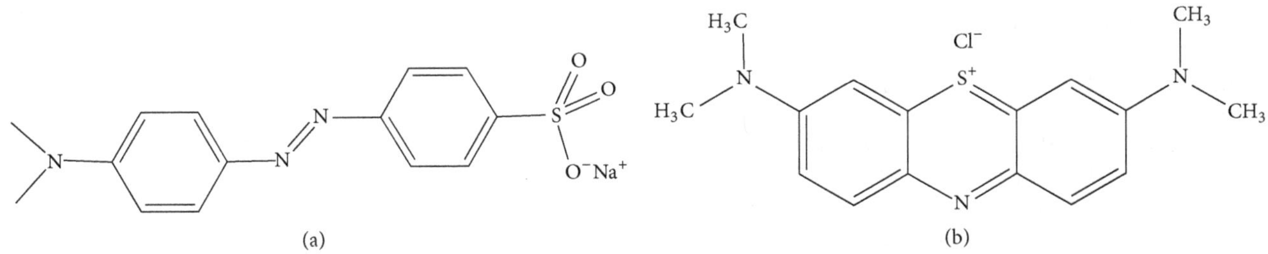


FIGURE 4: Structure of (a) methyl orange and (b) methylene blue.

under visible light over $AU^{3+}$ with modified $TiO_2$ power [115].

*3.2. Surface Chemical Modification.* Surface chemical modification is essential for increasing the efficiency of photocatalysts, thus preventing the recombination charge (electron and hole ($h^+$)) and their further separation, which is performed through two methods: sensitization and coupling. Sensitization, where different groups have used narrow band gap semiconductors to enhance optical absorption properties of $TiO_2$ nanoparticles in the visible light region, can be used to sensitize $TiO_2$ materials. Coupling different semiconductors with different energy systems provides another way to improve charge-carrier separation.

*3.2.1. Synthesis of Types of GO-TiO$_2$ Membrane.* The use of graphene, especially in the water treatment industry, has increased significantly in recent years due to its perfect mechanical and electrical features as well as high surface area. In the water treatment industry, graphene has been so useful in the removal of organic pollution, bacteria, heavy metals, and dyes, especially azo dyes [116], MB dyes and MO dyes [117, 118]. These dyes are produced, as mentioned before, in textile factories and have to be closely monitored to have standard concentration before being released into the environment [116]. In recent years, the synthesis of GO-$TiO_2$ membrane for the reduction of dye pollutants, especially azo dyes, has been studied in different cases in the water treatment industry to improve $TiO_2$.

The GO-$TiO_2$ membrane can be placed on any texture or on any membrane surface of the water filter. The photocatalytic process is an introduction to membrane properties, such as hydrophilicity, water penetration, and degradation of pollutants, and, as a result, there is a decrease in the odor produced from degradation of organic materials [117].

Studies were performed to achieve two major objectives: (1) studying the absorption ability of the membrane without considering the photocatalytic properties [118] and (2) studying the levels of degradation of organic dyes under UV radiation and considering the photocatalytic properties of the membrane [34, 117–119].

The GO-$TiO_2$ sheet can be placed on the following:

(1) Anionic perfluorinated polymer (Nafion) [120]

(2) Cellulose acetate [118]

(3) Polycarbonate (PC) [117]

(4) Polysulfone fluoride (PSF) [116]

(5) Polyvinylidene fluoride (PVDF) [120]

The polymeric membranes possess excellent chemical resistance, thermal stability, and good membrane formation [121, 122].

The photocatalytic effects of the produced membrane on the removal of organic dyes, especially for removed RB [117], MB [116], and MO dyes [117], which are considered as aromatic compounds, have been studied.

A comparison of hybrid membranes and standard nanomembranes showed that hybrid membranes act 5 times faster than standard nanomembranes in removing pollution [123]. Also, the energy consumption of hybrid membranes is half that of standard nanomembranes. Due to their mechanical and chemical features, these filters have stable photocatalytic activity in the presence of UV radiation [124, 125]. It is worth mentioning that GO is synthesized from natural graphite through Hommer' methods, similar to previous studies [126]. The $TiO_2$ microsphere was synthesized by the reported method, with some modifications [127–129].

The LbL (layer-by-layer) method has the ability to form a hybrid membrane. This experiment used a gold coated sensor to determine the mass of $TiO_2$ and GO during the LbL procedure. The sensor was coated with a base polymer membrane. The sensor was put in a desiccator for drying, after which it was put in DI water. A fixed concentration on the surface of the sensor was at 0 ng/cm$^2$ after 15 min, and the DI water was replaced with $TiO_2$ solution. $TiO_2$ nanoparticles were adsorbed on the base polymer and, after 2 h, the mass of the nanoparticle slowed down and became stabilized. After that, $TiO_2$ solution was replaced with DI water for the removal of packed nanoparticles. Next, the $TiO_2$ coated sensor was soaked in the GO solution [130].

According to Athanasekou et al., the reduced graphene oxide-$TiO_2$ (GO-$TiO_2$) was synthesized with GO suspension. GO was prepared according to previous studies, similar to the Hommer method, using LPD (liquid phase deposition) method at room temperature. Among all the membranes, GO-$TiO_2$ was the best membrane for removing MO under UV, and GO-T10 was the best membrane for removing MB. The hybrid membrane when compared with nanofiltration was better for the removal of pollutants and energy consumption [131].

*3.2.2. Reaction Schemes for Surface Modification GO-$TiO_2$.* By forming Ti-O band between $Ti^{4+}$ or functional group membrane and hydrogen band, $TiO_2$ nanomaterials are placed on the surface of the base membrane between $TiO_2$ groups and the functional group [132]. In the next phase, GO is placed on $TiO_2$ layers via Ti-O band or the band between hydrogen and $Ti^{4+}$ and carboxyl groups of GO, and, eventually, GO reduced and bonded to $TiO_2$ by ethanol under UV radiation [133, 134]. One of the advantages of the GO-$TiO_2$ membrane is flexibility and great force in passing flow; the resistance to GO-$TiO_2$ is obvious after drying GO-$TiO_2$ and the absence of cracks on the membrane. $TiO_2$ alone is not highly efficient in removing organic materials and pollutants [135, 136].

## 4. GO-$TiO_2$ Membrane

Membranes are typically made of inorganic (ceramic) materials and a polymeric membrane. Membrane separation is basically on three principles: adsorption, sieving, and electrostatics as shown in Figure 5 [38]. The pore size determines the different membrane types: microfiltration (MF), ultrafiltration (UF), nanofiltration (NF), and reverse osmosis (RO). Among the membrane processes, UF with polymeric membrane has the efficiency of removing contaminants such as organic pollutants, heavy metal, suspended solid, and different dyes. The polymeric membrane graphene oxide-$TiO_2$ is significantly efficient in the removal of organic pollutants and dye from solution. It has specific characteristics which completely explain some special traits.

*4.1. Analysis of the GO-$TiO_2$ Membrane*

(i) Membrane analysis through transmission electron microscopy (TEM) and EDX shows the location of $TiO_2$ anatase crystal particles on the surface of GO [137, 138].

(ii) The amount of $TiO_2$ on the surface of GO sheets is determined by thermogravimetric analysis (TGA) [139].

(iii) The amounts of Ti and GO have been studied in the following researches through CS and XPS analysis [140–143].

(iv) The absorption capacity of the GO-$TiO_2$ membrane is determined when different dye solutions are used in water [144, 145].

*4.1.1. Porosity of Membrane.* Atomic force microscopy (AFM) can record the roughness and surface morphology of polymeric membranes. The overall porosity of a membrane can be calculated using the gravimetric method as presented in [146]

$$\varepsilon = \frac{W_1 - W_2}{A \times L \times d_W}, \tag{1}$$

where $W_1$ and $W_2$ are the wet and dry weights of the membrane, $A$ is the effective area of the membrane ($m^2$), $L$ is the membrane thickness (m), and $d_W$ is the water of density 0.998 g/cm$^3$. Using pure water flux, the porosity of the membrane and the mean pore radius of the membrane ($r_m$) can be calculated using the Guerout–Elford–Ferry equation:

$$r_m = \sqrt{\frac{(2.9 - 1.75\varepsilon) \times 8\eta LQ}{\varepsilon \times A \times \Delta P}}, \tag{2}$$

where $\eta$ is the water viscosity ($8.9 \times 10^{-4}$ pas), $\Delta P$ is the operation pressure (3 MPa), $Q$ is the permeated pure water amount ($m^3$/s), $L$ is the membrane thickness (m), and $A$ is the surface area ($m^2$) [146, 147].

Atomic force microscopy, field emission scanning electron microscopy, and the Guerout–Elford–Ferry equation applied for pure water permeation rate were used for determination of average pore size. The pore size calculated from the method of Guerout–Elford–Ferry was the highest, and, based on the make-ready sample (FESEM), the pore size calculated was the least in size [148].

*4.2. GO-$TiO_2$ Photocatalytic Membrane for Dye Removal.* Photocatalytic properties of $TiO_2$ nanoparticles under UV radiation for removal of dye occur according to the following reactions [149]. In the photocatalytic oxidation, $TiO_2$ has to be irradiated and excited in near-UV energy to induce charge separation. On the other hand, dyes rather than $TiO_2$ are excited by visible light followed by electron injection onto the $TiO_2$ conduction band, which leads to photosensitized oxidation. Subsequently, this is followed by electron injection from the excited dye molecule onto the conduction band of the $TiO_2$ particles, whereas the dye is converted to the cationic dye radicals ($Dye^{\bullet+}$) that undergo degradation to yield products:

(6) $dye_{ads} + h\nu \rightarrow dye_{ads}^{*}$ excitation

(7) $dye_{ads}^{*} \rightarrow dye_{ads}^{\bullet+} + TiO_2(e_{cb}^{-})$ electron injection

(8) $dye_{ads}^{\bullet} + TiO_2(e_{cb}^{-}) \rightarrow dye_{ads}$ recombination

(9) $dye_{ads}^{\bullet+} \rightarrow$ degradation

(10) $O_2 + TiO_2(e_{cb}^{-}) \rightarrow O_2$ (electron scavenging)

Anatase crystals which form mesosphere $TiO_2$ can have high mass transfer flux and photocatalytic activity [150–152]. The mesosphere structure of GO-$TiO_2$ has a huge surface area which forms numerous channels for passing water flow and photocatalysts for photocatalytic degradation [153, 154]. It is absolutely obvious that a membrane with high cross flow, removal efficiency, and low cost is suitable for filtration. GO sheets increase the filtration efficiency [154].

According to studies performed in different researches, GO-$TiO_2$ on a base membrane and under UV radiation shows great photocatalytic power in degrading organic materials. Membrane pretreatment, using ethanol and UV radiation, helps in increasing the degradation of MB [117].

The photocatalytic properties of porphyrin $TiO_2$ and graphene $TiO_2$ under visible light and UV radiation have been studied. Porphyrin graphene $TiO_2$ shows 35% higher efficiency than pure $TiO_2$ [155].

GO-$TiO_2$ is used for the degradation of RhB and $AO_7$ dyes under UV radiation within 20 to 30 min. In the absence of UV radiation, GO-$TiO_2$ membrane shows low efficiency,

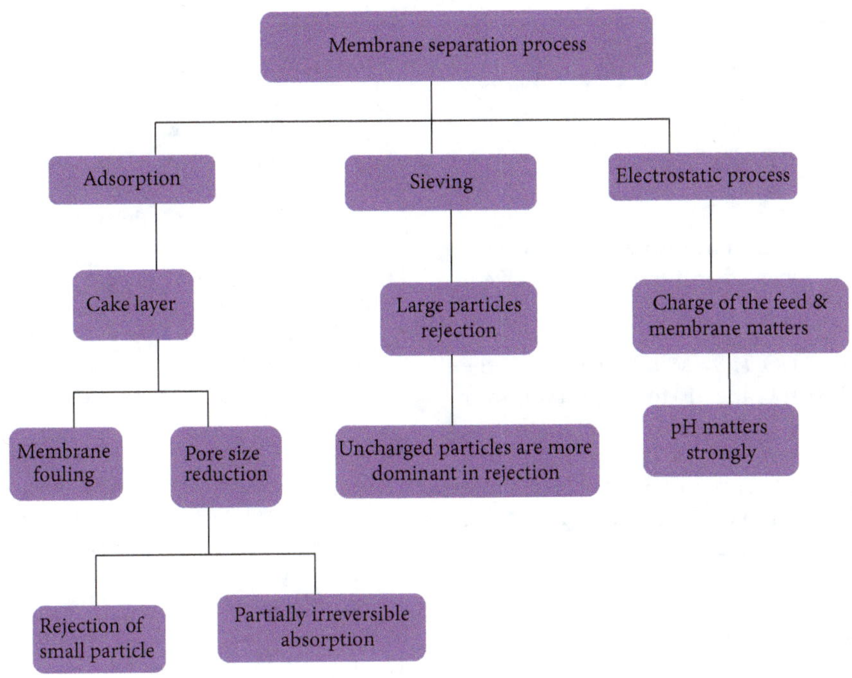


FIGURE 5: Schematic representing the basic principles involved in membrane separation [38].

lower than 15%, in the degradation of RhB and $AO_7$ dyes. The presence of UV radiation increases removal efficiency by 50% [118].

The presence of photocatalysts on a membrane prevents the accumulation of organic materials and macromolecules and therefore allows higher flow to pass through the membrane. Absorption of MO in darkness has minimal effects on the Nafion-GO SULF membrane, but under UV radiation the initial concentration is reduced by 90% [28, 118]. Pores size depends on the amount of GO-$TiO_2$ which is placed on the base membrane. GO-10 membrane (membrane having 10 nm pores) shows higher removal efficiency of dyes than other membranes [156].

The application of the method of liquid phase deposition (LPD) for synthesis of GO-$TiO_2$ composites and their stabilization on the integrated membranes is suitable for the removal of industrial dyes. Membranes with 1 m length and 30 monoliths consume less energy than nanomembranes [156].

*4.3. Permeation Flux through GO-$TiO_2$ Membrane.* Filtration tests on GO-$TiO_2$ membrane with 5%-5 mL/min flow rate, 0.6 nm constant membrane thickness, and 30 mg/L concentration cobalt ion show that the lesser the flow which is passed through, the more the cobalt removed. As the initial concentration of cobalt ion increases from 1 to 9 mg/L, the removal efficiency decreases from 89 to 21%. The high absorption capacity of nanosheets of graphene oxide ammonium for the removal of cobalt could be related to the ammonium functional groups. These absorbents are able to remove 30 mg/L concentration of cobalt ions in 5 min with 90% efficiency [119].

The presence of sulfonic functional group, along with graphene on Nafion membrane, allows more water to pass through it [28, 38, 118–157]. The movement of proton on Nafion membrane can help in separating ion compounds from the water molecule. These effects have been studied by adding the sulfonic functional group to the surface membrane of Nafion [38, 109–158].

The thickness of GO-$TiO_2$ membrane could be changed easily by changing the GO-$TiO_2$ mass which is placed on the base. Membrane thickness affects the flow passing through the membrane, meaning that a decrease in membrane thickness allows higher flow to pass through the membrane and decreases the removal efficiency [118].

*4.4. Fouling Resistant Membrane.* The total fouling resistance of the polymeric membrane ($R_T$) is calculated from $r_r$ and $r_{ir}$. $r_r$ is a reversible fouling ratio which describes the fouling caused by concentration polarization, and $r_{ir}$ is an irreversible fouling ratio which describes the fouling caused by adsorption or deposition of protein molecules on the membrane surface.

$$r_r = \frac{J_{W2} - J_P}{J_{W1}}$$
$$r_{ir} = \frac{J_{W1} - J_{W2}}{J_{W1}}. \quad (3)$$

$R_T$ is the sum or $r_r$ and $r_{ir}$, and $J_{W1}$ is the permeation water flux (kg/m$^2$ h):

$$J_{W1} = \frac{M}{At}, \quad (4)$$

where $M$ is the weight of collected permeation flux, $A$ is the membrane effective area, $t$ is the permeation time, $J_{W2}$ is the water flux cleaned membrane, and $J_P$ is the flux of the BSA solution [144]. Hence,

$$R_T = \left(1 - \frac{J_P}{J_{W1}}\right) \times 100. \quad (5)$$

Under special pressure, after the pure water test, the BSA solutions are immediately replaced in the filtration cell for 90 min [159].

*4.5. Antifouling Property of GO-TiO$_2$ Membrane.* Since the cost of cleaning normal membranes is too high and, over time, they produce odors because of the accumulation of organic materials, the application of GO-$TiO_2$ membrane in the water treatment industry seems to have become more widespread. Humic acid as a standard pollutant formed from natural organic materials and carcinogen disinfectants has been studied [160].

In the first phases of filtration, the accumulation of humic acid on the surface of membrane and closing of the pores are really obvious after 30 min [161]. After 2 h, with 7 $L/m^2 \cdot h$ flow rate, a layer of odorous HA cake becomes noticeable [162]. UV radiation reduces the produced odor significantly [118]. The reason is the degradation of humic acid and production of $CO_2$ and $H_2O$ under UV radiation; in this state, there is no reduction in the flow which passed through the membrane.

## 5. Future Work

GO-$TiO_2$ membrane, because of its significant applications such as high photocatalytic and antifouling property, flux permeation, and removal of organic pollutants, can be potentially used in the degradation of polycyclic aromatic hydrocarbon (PAH) as a low cost membrane in the future.

## 6. Conclusion

This study was carried out on the characteristics of different polymeric membranes of GO-$TiO_2$. According to previous studies, this membrane has lots of advantages like the capability of passing flow. Graphene oxide has a number of covalent attached oxygen containing groups such as epoxy, carbonyl, and carboxyl. The existence of these groups makes membrane possess good hydrophilicity. Another good property of this membrane is the existence of photocatalytic $TiO_2$ semiconductor, which can help in the degradation of the cake layer of the organic material and the reduction of odor. The object is very economical to reduce cost of washing membrane compared to the usual form. Fouling of this membrane because of GO-$TiO_2$ photocatalytic properties is also reduced from other forms of membranes. This result shows that the GO-$TiO_2$ polymer membrane could make considerable improvement in the field of water treatment and could have significant effects on the filtration efficiency.

## Competing Interests

The authors declare that they have no competing interests.

## References

[1] M. Elimelech and W. A. Phillip, "The future of seawater desalination: energy, technology, and the environment," *Science*, vol. 333, no. 6043, pp. 712–717, 2011.

[2] M. A. Shannon, P. W. Bohn, M. Elimelech, J. G. Georgiadis, B. J. Marĩas, and A. M. Mayes, "Science and technology for water purification in the coming decades," *Nature*, vol. 452, no. 7185, pp. 301–310, 2008.

[3] S. Kaur, R. Gopal, W. J. Ng, S. Ramakrishna, and T. Matsuura, "Next-generation fibrous media for water treatment," *MRS Bulletin*, vol. 33, no. 1, pp. 21–26, 2008.

[4] Z. Wu and D. Zhao, "Ordered mesoporous materials as adsorbents," *Chemical Communications*, vol. 47, no. 12, pp. 3332–3338, 2011.

[5] S. E. Bailey, T. J. Olin, R. M. Bricka, and D. D. Adrian, "A review of potentially low-cost sorbents for heavy metals," *Water Research*, vol. 33, no. 11, pp. 2469–2479, 1999.

[6] A. Netzer and D. E. Hughes, "Adsorption of copper, lead and cobalt by activated carbon," *Water Research*, vol. 18, no. 8, pp. 927–933, 1984.

[7] C. Gómez-Lahoz, F. Garcia-Herruzo, J. M. Rodriguez-Maroto, and J. J. Rodriguez, "Cobalt(II) removal from water by chemical reduction with sodium borohydride," *Water Research*, vol. 27, no. 6, pp. 985–992, 1993.

[8] M. M. Pendergast and E. M. V. Hoek, "A review of water treatment membrane nanotechnologies," *Energy and Environmental Science*, vol. 4, no. 6, pp. 1946–1971, 2011.

[9] H.-W. Liang, X. Cao, W.-J. Zhang et al., "Robust and highly efficient free-standing carbonaceous nanofiber membranes for water purification," *Advanced Functional Materials*, vol. 21, no. 20, pp. 3851–3858, 2011.

[10] Y. Liu, Z. Wu, X. Chen, Z. Shao, H. Wang, and D. Zhao, "A hierarchical adsorption material by incorporating mesoporous carbon into macroporous chitosan membranes," *Journal of Materials Chemistry*, vol. 22, no. 24, pp. 11908–11911, 2012.

[11] J. Cadotte, R. Forester, M. Kim, R. Petersen, and T. Stocker, "Nanofiltration membranes broaden the use of membrane separation technology," *Desalination*, vol. 70, no. 1–3, pp. 77–88, 1988.

[12] E. A. Jackson and M. A. Hillmyer, "Nanoporous membranes derived from block copolymers: from drug delivery to water filtration," *ACS Nano*, vol. 4, no. 7, pp. 3548–3553, 2010.

[13] H. P. Dijkstra, G. P. M. van Klink, and G. van Koten, "The use of ultra- and nanofiltration techniques in homogeneous catalyst recycling," *Accounts of Chemical Research*, vol. 35, no. 9, pp. 798–810, 2002.

[14] C. Stoquart, P. Servais, P. R. Bérubé, and B. Barbeau, "Hybrid membrane processes using activated carbon treatment for drinking water: a review," *Journal of Membrane Science*, vol. 411-412, pp. 1–12, 2012.

[15] P. Wang, J. Ma, F. Shi, Y. Ma, Z. Wang, and X. Zhao, "Behaviors and effects of differing dimensional nanomaterials in water filtration membranes through the classical phase inversion process: a review," *Industrial and Engineering Chemistry Research*, vol. 52, no. 31, pp. 10355–10363, 2013.

[16] B. W. Stanton, J. J. Harris, M. D. Miller, and M. L. Bruening, "Ultrathin, multilayered polyelectrolyte films as nanofiltration membranes," *Langmuir*, vol. 19, no. 17, pp. 7038–7042, 2003.

[17] S. El-Safty, A. Shahat, M. R. Awual, and M. Mekawy, "Large three-dimensional mesocage pores tailoring silica nanotubes

as membrane filters: nanofiltration and permeation flux of proteins," *Journal of Materials Chemistry*, vol. 21, no. 15, pp. 5593–5603, 2011.

[18] B. Zhao, L. Zhang, X.-Y. Wang et al., "Research progress in nanofiltration membrane based on carbon nanotubes," *New Carbon Materials*, vol. 26, no. 5, pp. 321–327, 2011.

[19] J. Sekulić, J. E. Ten Elshof, and D. H. A. Blank, "A microporous titania membrane for nanofiltration and pervaporation," *Advanced Materials*, vol. 16, no. 17, pp. 1546–1550, 2004.

[20] B. Van Der Bruggen, C. Vandecasteele, T. Van Gestel, W. Doyen, and R. Leysen, "A review of pressure-driven membrane processes in wastewater treatment and drinking water production," *Environmental Progress*, vol. 22, no. 1, pp. 46–56, 2003.

[21] X. Zhang, T. Zhang, J. Ng, and D. D. Sun, "High-performance multifunctional $TiO_2$ nanowire ultrafiltration membrane with a hierarchical layer structure for water treatment," *Advanced Functional Materials*, vol. 19, no. 23, pp. 3731–3736, 2009.

[22] M. N. Chong, B. Jin, C. W. K. Chow, and C. Saint, "Recent developments in photocatalytic water treatment technology: a review," *Water Research*, vol. 44, no. 10, pp. 2997–3027, 2010.

[23] J. Grzechulska-Damszel, M. Tomaszewska, and A. W. Morawski, "Integration of photocatalysis with membrane processes for purification of water contaminated with organic dyes," *Desalination*, vol. 241, no. 1–3, pp. 118–126, 2009.

[24] G. L. Liu, C. Han, M. Pelaez et al., "Enhanced visible light photo catalytic activity of C-N-codoped $TiO_2$ films for the degradation of microcystin-LR," *Journal of Molecular Catalysis A: Chemical*, vol. 372, pp. 58–65, 2013.

[25] G. L. Liu, C. Han, M. Pelaez et al., "Synthesis, characterization and photocatalytic evaluation of visible light activated C-doped $TiO_2$ nanoparticles," *Nanotechnology*, vol. 23, no. 29, Article ID 294003, 2012.

[26] X. Zhang, A. J. Du, P. Lee, D. D. Sun, and J. O. Leckie, "$TiO_2$ nanowire membrane for concurrent filtration and photocatalytic oxidation of humic acid in water," *Journal of Membrane Science*, vol. 313, no. 1-2, pp. 44–51, 2008.

[27] R. M. Mohamed, D. L. McKinney, and W. M. Sigmund, "Enhanced nanocatalysts," *Materials Science and Engineering: R: Reports*, vol. 73, no. 1, pp. 1–13, 2012.

[28] T.-F. Yeh, J. Cihlář, C.-Y. Chang, C. Cheng, and H. Teng, "Roles of graphene oxide in photocatalytic water splitting," *Materials Today*, vol. 16, no. 3, pp. 78–84, 2013.

[29] G. P. Rao, C. Lu, and F. Su, "Sorption of divalent metal ions from aqueous solution by carbon nanotubes: a review," *Separation and Purification Technology*, vol. 58, no. 1, pp. 224–231, 2007.

[30] A. G. Onu, W. Qil, Z. Yan, and K. Tan, "Adsorption of aromatic organic contaminants by graphene nanosheets: comparison with carbon nanotubes and activated carbon," *Water Research*, vol. 47, no. 4, pp. 1648–1654, 2013.

[31] J. R. Ellerie, O. G. Apul, T. Karanfil, and D. A. Ladner, "Comparing graphene, carbon nanotubes, and superfine powdered activated carbon as adsorptive coating materials for microfiltration membranes," *Journal of Hazardous Materials*, vol. 261, pp. 91–98, 2013.

[32] A. W. H. Mau, C. B. Huang, N. Kakuta et al., "Hydrogen photoproduction by Nafion/cadmium sulfide/platinum films in water/sulfide ion solutions," *Journal of the American Chemical Society*, vol. 106, no. 65, pp. 37–42, 1984.

[33] J. Liu, H. Bai, Y. Wang, Z. Liu, X. Zhang, and D. D. Sun, "Self-assembling $TiO_2$ nanorods on large graphene oxide sheets at a two-phase interface and their anti-recombination in photocatalytic applications," *Advanced Functional Materials*, vol. 20, no. 23, pp. 4175–4181, 2010.

[34] R. A. K. Rao, S. Singh, B. R. Singh, W. Khan, and A. H. Naqvi, "Synthesis and characterization of surface modified graphene-zirconium oxide nanocomposite and its possible use for the removal of chlorophenol from aqueous solution," *Journal of Environmental Chemical Engineering*, vol. 2, no. 1, pp. 199–210, 2014.

[35] G. Mele, R. D. Sole, G. Vasapollo et al., "TRMC, XPS, and EPR characterizations of polycrystalline $TiO_2$ porphyrin impregnated powders and their catalytic activity for 4-nitrophenol photodegradation in aqueous suspension," *The Journal of Physical Chemistry B*, vol. 109, no. 25, pp. 12347–12352, 2005.

[36] S. Afzal, W. A. Daoud, and S. J. Langford, "Self-cleaning cotton by porphyrin-sensitized visible-light photocatalysis," *Journal of Materials Chemistry*, vol. 22, no. 9, pp. 4083–4088, 2012.

[37] A. R. Khataee and M. B. Kasiri, "Photocatalytic degradation of organic dyes in the presence of nanostructured titanium dioxide: influence of the chemical structure of dyes," *Journal of Molecular Catalysis A: Chemical*, vol. 328, no. 1-2, pp. 8–26, 2010.

[38] M. Padaki, R. Surya Murali, M. S. Abdullah et al., "Membrane technology enhancement in oil-water separation. A review," *Desalination*, vol. 357, pp. 197–207, 2015.

[39] Kirk-Othmer, *Encyclopadia of Chemical Technology*, vol. 19, Wiley-Inter Science Publication, 4th edition, 1996.

[40] H. Takeda and O. Ishitani, "Development of efficient photocatalytic systems for $CO_2$ reduction using mononuclear and multinuclear metal complexes based on mechanistic studies," *Coordination Chemistry Reviews*, vol. 254, no. 3-4, pp. 346–354, 2010.

[41] R. Zallen and M. P. Moret, "The optical absorption edge of brookite $TiO_2$," *Solid State Communications*, vol. 137, no. 3, pp. 154–157, 2006.

[42] X. Chen, "Titanium dioxide nanomaterials and their energy applications," *Chinese Journal of Catalysis*, vol. 30, no. 8, pp. 839–851, 2009.

[43] D. T. Chomer and K. Herrington, "The structures of anatase and rutile," *Journal of the American Chemical Society*, vol. 77, no. 18, pp. 4708–4709, 1955.

[44] V. W. H. Baur, "Atomabstände und bindungswinkel im brookit, $TiO_2$," *Acta Crystallographica*, vol. 14, no. 3, pp. 214–216, 1961.

[45] S.-D. Mo and W. Y. Ching, "Electronic and optical properties of three phases of titanium dioxide: rutile, anatase, and brookite," *Physical Review B*, vol. 51, no. 19, pp. 13023–13032, 1995.

[46] R. Richards, *Surface and Nanomolecular Catalysis*, CRC Press/Taylor & Francis Group, 2006.

[47] M. Zhang, G. Sheng, J. Fu, T. An, X. Wang, and X. Hu, "Novel preparation of nanosized $ZnO–SnO_2$ with high photocatalytic activity by homogeneous co-precipitation method," *Materials Letters*, vol. 59, no. 28, pp. 3641–3644, 2005.

[48] M. Soltaninezhad and A. Aminifar, "Study nanostructures of semiconductor zinc oxide (ZnO) as a photocatalyst for the degradation of organic pollutants," *International Journal of Nano Dimension*, vol. 2, no. 2, pp. 137–145, 2011.

[49] A. Mills and S. Le Hunte, "An overview of semiconductor photocatalysis," *Journal of Photochemistry and Photobiology A: Chemistry*, vol. 108, no. 1, pp. 1–35, 1997.

[50] N. Daneshvar, D. Salari, and A. R. Khataee, "Photocatalytic degradation of azo dye acid red 14 in water: investigation of the effect of operational parameters," *Journal of Photochemistry and Photobiology A: Chemistry*, vol. 157, no. 1, pp. 111–116, 2003.

[51] N. Daneshvar, D. Salari, and A. R. Khataee, "Photocatalytic degradation of azo dye acid red 14 in water on ZnO as an alternative catalyst to $TiO_2$," *Journal of Photochemistry and Photobiology A: Chemistry*, vol. 162, no. 2-3, pp. 317–322, 2004.

[52] X. F. Cheng, W. H. Leng, D. P. Liu, J. Q. Zhang, and C. N. Cao, "Enhanced photoelectrocatalytic performance of Zn-doped $WO_3$ photocatalysts for nitrite ions degradation under visible light," *Chemosphere*, vol. 68, no. 10, pp. 1976–1984, 2007.

[53] D. Jing and L. Guo, "$WS_2$ sensitized mesoporous $TiO_2$ for efficient photocatalytic hydrogen production from water under visible light irradiation," *Catalysis Communications*, vol. 8, no. 5, pp. 795–799, 2007.

[54] J. Bandara, U. Klehm, and J. Kiwi, "Raschig rings-$Fe_2O_3$ composite photocatalyst activate in the degradation of 4-chlorophenol and Orange II under daylight irradiation," *Applied Catalysis B: Environmental*, vol. 76, no. 1-2, pp. 73–81, 2007.

[55] K. Teramura, T. Tanaka, M. Kani, T. Hosokawa, and T. Funabiki, "Selective photo-oxidation of neat cyclohexane in the liquid phase over $V_2O_5/Al_2O_3$," *Journal of Molecular Catalysis A: Chemical*, vol. 208, no. 1-2, pp. 299–305, 2004.

[56] Y. Zhai, S. Zhang, and H. Pang, "Preparation, characterization and photocatalytic activity of $CeO_2$ nanocrystalline using ammonium bicarbonate as precipitant," *Materials Letters*, vol. 61, no. 8-9, pp. 1863–1866, 2007.

[57] K. G. Kanade, J.-O. Baeg, U. P. Mulik, D. P. Amalnerkar, and B. B. Kale, "Nano-CdS by polymer-inorganic solid-state reaction: visible light pristine photocatalyst for hydrogen generation," *Materials Research Bulletin*, vol. 41, no. 12, pp. 2219–2225, 2006.

[58] C. L. Torres-Martínez, R. Kho, O. I. Mian, and R. K. Mehra, "Efficient photocatalytic degradation of environmental pollutants with mass-produced ZnS nanocrystals," *Journal of Colloid and Interface Science*, vol. 240, no. 2, pp. 525–532, 2001.

[59] M. R. Hoffmann, S. T. Martin, W. Y. Choi, and D. W. Bahnemann, "Environmental applications of semiconductor photocatalysis," *Chemical Reviews*, vol. 95, no. 1, pp. 69–96, 1995.

[60] A. L. Linsebigler, G. Q. Lu, and J. T. Yates, "Photocatalysis on $TiO_2$ surfaces: principles, mechanisms, and selected results," *Chemical Reviews*, vol. 95, no. 3, pp. 735–758, 1995.

[61] O. Zahraa, S. Maire, F. Evenou et al., "Treatment of wastewater dyeing agent by photocatalytic process in solar reactor," *International Journal of Photoenergy*, vol. 2006, Article ID 46961, 9 pages, 2006.

[62] A. Fujishima, T. N. Rao, and D. A. Tryk, "Titanium dioxide photocatalysis," *Journal of Photochemistry and Photobiology C: Photochemistry Reviews*, vol. 1, no. 1, pp. 1–21, 2000.

[63] N. Daneshvar, D. Salari, A. Niaei, M. H. Rasoulifard, and A. R. Khataee, "Immobilization of $TiO_2$ nanopowder on glass beads for the photocatalytic decolorization of an azo dye C.I. Direct Red 23," *Journal of Environmental Science and Health, Part A: Toxic*, vol. 40, no. 8, pp. 1605–1617, 2005.

[64] V. Loddo, G. Marcì, C. Martín, L. Palmisano, V. Rives, and A. Sclafani, "Preparation and characterisation of $TiO_2$ (anatase) supported on $TiO_2$ (rutile) catalysts employed for 4-nitrophenol photodegradation in aqueous medium and comparison with $TiO_2$ (anatase) supported on $Al_2O_3$," *Applied Catalysis B: Environmental*, vol. 20, no. 1, pp. 29–45, 1999.

[65] S. Bakardjieva, J. Šubrt, V. Štengl, M. J. Dianez, and M. J. Sayagues, "Photoactivity of anatase-rutile $TiO_2$ nanocrystalline mixtures obtained by heat treatment of homogeneously precipitated anatase," *Applied Catalysis B: Environmental*, vol. 58, no. 3-4, pp. 193–202, 2005.

[66] S.-C. Jung, S.-J. Kim, N. Imaishi, and Y.-I. Cho, "Effect of $TiO_2$ thin film thickness and specific surface area by low-pressure metal–organic chemical vapor deposition on photocatalytic activities," *Applied Catalysis B: Environmental*, vol. 55, no. 4, pp. 253–257, 2005.

[67] X. Wang, J. C. Yu, P. Liu, X. Wang, W. Su, and X. Fu, "Probing of photocatalytic surface sites on $SO^{42-}/TiO_2$ solid acids by in situ FT-IR spectroscopy and pyridine adsorption," *Journal of Photochemistry and Photobiology A: Chemistry*, vol. 179, no. 3, pp. 339–347, 2006.

[68] M. Zheng, M. Gu, Y. Jin, and G. Jin, "Preparation, structure and properties of $TiO_2$–PVP hybrid films," *Materials Science and Engineering: B*, vol. 77, no. 1, pp. 55–59, 2000.

[69] K. Tennakone and K. G. U. Wijayantha, "Heavy-metal extraction from aqueous medium with an immobilized $TiO_2$ photocatalyst and a solid sacrificial agent," *Journal of Photochemistry and Photobiology A: Chemistry*, vol. 113, no. 1, pp. 89–92, 1998.

[70] M. N. Pons, A. Alinsafi, F. Evenou et al., "Treatment of textile industry wastewater by supported photocatalysis," *Dyes and Pigments*, vol. 74, no. 2, pp. 439–445, 2007.

[71] N. Keller, G. Rebmann, E. Barraud, O. Zahraa, and V. Keller, "Macroscopic carbon nanofibers for use as photocatalyst support," *Catalysis Today*, vol. 101, no. 3-4, pp. 323–329, 2005.

[72] H. Zollinger, Ed., *Color Chemistry: Synthesis, Properties and Applications of Organic Dyes and Pigments*, VCH, 2nd edition, 1991.

[73] E. J. Weber and V. C. Stickney, "Hydrolysis kinetics of reactive blue 19-vinyl sulfone," *Water Research*, vol. 27, no. 1, pp. 63–67, 1993.

[74] C. Ràfols and D. Barceló, "Determination of mono- and disulphonated azo dyes by liquid chromatography–atmospheric pressure ionization mass spectrometry," *Journal of Chromatography A*, vol. 777, no. 1, pp. 177–192, 1997.

[75] A. Houas, H. Lachheb, M. Ksibi, E. Elaloui, C. Guillard, and J.-M. Herrmann, "Photocatalytic degradation pathway of methylene blue in water," *Applied Catalysis B: Environmental*, vol. 31, no. 2, pp. 145–157, 2001.

[76] U. Pagga and D. Brown, "The degradation of dyestuffs: part II behaviour of dyestuffs in aerobic biodegradation tests," *Chemosphere*, vol. 15, no. 4, pp. 479–491, 1986.

[77] A. B. Prevot, C. Baiocchi, M. C. Brussino et al., "Photocatalytic degradation of acid blue 80 in aqueous solutions containing $TiO_2$ suspensions," *Environmental Science and Technology*, vol. 35, no. 5, pp. 971–976, 2001.

[78] M. Saquib and M. Muneer, "$TiO_2$-mediated photocatalytic degradation of a triphenylmethane dye (gentian violet), in aqueous suspensions," *Dyes and Pigments*, vol. 56, no. 1, pp. 37–49, 2003.

[79] W. Z. Tang and H. An, "UV/$TiO_2$ photocatalytic oxidation of commercial dyes in aqueous solutions," *Chemosphere*, vol. 31, no. 9, pp. 4157–4170, 1995.

[80] V. Meshko, L. Markovska, M. Mincheva, and A. E. Rodrigues, "Adsorption of basic dyes on granular acivated carbon and natural zeolite," *Water Research*, vol. 35, no. 14, pp. 3357–3366, 2001.

[81] W. S. Kuo and P. H. Ho, "Solar photocatalytic decolorization of methylene blue in water," *Chemosphere*, vol. 45, no. 1, pp. 77–83, 2001.

[82] C. Galindo, P. Jacques, and A. Kalt, "Photooxidation of the phenylazonaphthol AO20 on $TIO_2$: kinetic and mechanistic investigations," *Chemosphere*, vol. 45, no. 6-7, pp. 997–1005, 2001.

[83] S. S. Patll and V. M. Shinde, "Biodegradation studies of aniline and nitrobenzene in aniline plant wastewater by gas chromatography," *Environmental Science and Technology*, vol. 22, no. 10, pp. 1160–1165, 1988.

[84] A. T. Moore, A. Vira, and S. Fogel, "Biodegradation of trans-1,2-dichloroethylene by methane-utilizing bacteria in an aquifer simulator," *Environmental Science and Technology*, vol. 23, no. 4, pp. 403–406, 1989.

[85] V. M. Correia, T. Stephenson, and S. J. Judd, "Characterization of textile wastewaters—a review," *Environmental Technology*, vol. 15, no. 10, pp. 917–929, 1994.

[86] I. Arslan and I. A. Balcioglu, "Degradation of commercial reactive dyestuffs by heterogenous and homogenous advanced oxidation processes: a comparative study," *Dyes and Pigments*, vol. 43, no. 2, pp. 95–108, 1999.

[87] S. H. Lin and C. M. Lin, "Treatment of textile waste effluents by Ozonation and chemical coagulation," *Water Research*, vol. 27, no. 12, pp. 1743–1748, 1993.

[88] S. H. Lin and W. Y. Liu, "Continuous treatment of textile water by ozonation and coagulation," *Journal of Environmental Engineering*, vol. 120, no. 2, pp. 437–446, 1994.

[89] F. Strickland and S. Perkins, "Decolorization of continuous dyeing wastewater by ozonation," *Textile Chemist & Colorist*, vol. 27, no. 5, pp. 11–15, 1995.

[90] P. Aranyosi, Z. Csepregi, I. Rusznák, L. Töke, and A. Víg, "The light stability of azo dyes and azo dyeings. III. The effect of artificial perspiration on the light stability of reactive and non-reactive derivatives of two selected azo chromophores in aqueous solution," *Dyes and Pigments*, vol. 37, no. 1, pp. 33–45, 1998.

[91] W. G. Kuo, "Decolorizing dye wastewater with Fenton's reagent," *Water Research*, vol. 26, no. 7, pp. 881–886, 1992.

[92] E. Balanosky, J. Fernadez, J. Kiwi, and A. Lopez, "Degradation of membrane concentrates of the textile industry by fenton like reactions in iron-free solutions at biocompatible pH values (pH $\approx 7-8$)," *Water Science & Technology*, vol. 40, no. 4-5, pp. 417–424, 1999.

[93] W. Feng, D. Nansheng, and Z. Yuegang, "Discoloration of dye solutions induced by solar photolysis of ferrioxalate in aqueous solutions," *Chemosphere*, vol. 39, no. 12, pp. 2079–2085, 1999.

[94] C. Morrison, J. Bandara, and J. Kiwi, "Sun light induced decolouration/degradation of non-biodegradation Orange dye by advanced Oxidation technologies in homogenous and heterogeneous media," *Journal of Advanced Oxidation Technologies*, vol. 1, no. 2, pp. 160–169, 1996.

[95] S.-F. Kang, C.-H. Liao, and S.-T. Po, "Decolorization of textile wastewater by photo-fenton oxidation technology," *Chemosphere*, vol. 41, no. 8, pp. 1287–1294, 2000.

[96] I. Arslan, I. A. Balcioglu, and T. Tuhkanen, "Advanced oxidation of synthetic dyehouse effluent by $O_3$, $H_2O_2/O_3$ and $H_2O_2$/UV processes," *Environmental Technology*, vol. 20, no. 9, pp. 921–931, 1999.

[97] N. H. Ince and D. T. Gönenç, "Treatability of a textile azo dye by UV/$H_2O_2$," *Environmental Technology*, vol. 18, no. 2, pp. 179–185, 1997.

[98] K. Vinodgopal and P. V. Kamat, "Combine electrochemistry with photocatalysis," *CHEMTECH*, vol. 26, no. 4, pp. 18–22, 1996.

[99] C. Lizama, M. C. Yeber, J. Freer, J. Baeza, and H. D. Mansilla, "Reactive dyes decolouration by $TiO_2$ photo-assisted catalysis," *Water Science and Technology*, vol. 44, no. 5, pp. 197–203, 2001.

[100] F. Zhang, J. Zhao, T. Shen, H. Hidaka, E. Pelizzetti, and N. Serpone, "$TiO_2$-assisted photodegradation of dye pollutants II. Adsorption and degradation kinetics of eosin in $TiO_2$ dispersions under visible light irradiation," *Applied Catalysis B: Environmental*, vol. 15, no. 1-2, pp. 147–156, 1998.

[101] P. Qu, J. Zhao, T. Shen, and H. Hidaka, "$TiO_2$-assisted photodegradation of dyes: a study of two competitive primary processes in the degradation of RB in an aqueous $TiO_2$ colloidal solution," *Journal of Molecular Catalysis A: Chemical*, vol. 129, no. 2-3, pp. 257–268, 1998.

[102] G. Liu, T. Wu, J. Zhao, H. Hidaka, and N. Serpone, "Photoassisted degradation of dye pollutants. 8. Irreversible degradation of alizarin red under visible light radiation in air-equilibrated aqueous $TiO_2$ dispersions," *Environmental Science and Technology*, vol. 33, no. 12, pp. 2081–2087, 1999.

[103] J. Zhao, T. Wu, K. Wu, K. Oikawa, H. Hidaka, and N. Serpone, "Photoassisted degradation of dye pollutants. 3. Degradation of the cationic dye rhodamine B in aqueous anionic surfactant/$TiO_2$ dispersions under visible light irradiation: evidence for the need of substrate adsorption on $TiO_2$ particles," *Environmental Science and Technology*, vol. 32, no. 16, pp. 2394–2400, 1998.

[104] M. R. Hoffmann, S. T. Martin, W. Choi, and D. W. Bahnemann, "Environmental applications of semiconductor photocatalysis," *Chemical Reviews*, vol. 95, no. 1, pp. 69–96, 1995.

[105] I. K. Konstantinou and T. A. Albanis, "Photocatalytic transformation of pesticides in aqueous titanium dioxide suspensions using artificial and solar light: intermediates and degradation pathways," *Applied Catalysis B: Environmental*, vol. 42, no. 4, pp. 319–335, 2003.

[106] A. L. Linsebigler, G. Lu, and J. T. Yates, "Photocatalysis on $TiO_2$ surfaces: principles, mechanisms, and selected results," *Chemical Reviews*, vol. 95, no. 3, pp. 735–758, 1995.

[107] P. Reeves, R. Ohlhausen, D. Sloan et al., "Photocatalytic destruction of organic dyes in aqueous $TiO_2$ suspensions using concentrated simulated and natural solar energy," *Solar Energy*, vol. 48, no. 6, pp. 413–420, 1992.

[108] X. Z. Li and M. Zhang, "Decolorization and biodegradability of dyeing wastewater treated by a $TiO_2$-sensitized photo-oxidation process," *Water Science and Technology*, vol. 34, no. 9, pp. 49–55, 1996.

[109] Y. Wang, "Solar photocatalytic degradation of eight commercial dyes in $TiO_2$ suspension," *Water Research*, vol. 34, no. 3, pp. 990–994, 2000.

[110] S. Sakthivel, B. Neppolian, M. V. Shankar, B. Arabindoo, M. Palanichamy, and V. Murugesan, "Solar photocatalytic degradation of azo dye: comparison of photocatalytic efficiency of ZnO and $TiO_2$," *Solar Energy Materials and Solar Cells*, vol. 77, no. 1, pp. 65–82, 2003.

[111] V. Augugliaro, C. Baiocchi, A. B. Prevot et al., "Azo-dyes photocatalytic degradation in aqueous suspension of $TiO_2$ under solar irradiation," *Chemosphere*, vol. 49, no. 10, pp. 1223–1230, 2002.

[112] M. Stylidi, D. I. Kondarides, and X. E. Verykios, "Pathways of solar light-induced photocatalytic degradation of azo dyes in aqueous $TiO_2$ suspensions," *Applied Catalysis B: Environmental*, vol. 40, no. 4, pp. 271–286, 2003.

[113] S. Gomes de Moraes, R. Sanches Freire, and N. Durán, "Degradation and toxicity reduction of textile effluent by combined photocatalytic and ozonation processes," *Chemosphere*, vol. 40, no. 4, pp. 369–373, 2000.

[114] I. A. Balcioglu and I. Arslan, "Treatment of textile waste water by heterogenous photocatalytic oxidation processes," *Environmental Technology*, vol. 18, no. 10, pp. 1053–1059, 1997.

[115] X. Z. Li and F. B. Li, "Study of Au/$Au^{3+}$-$TiO_2$ photocatalysts toward visible photooxidation for water and wastewater treatment," *Environmental Science and Technology*, vol. 35, no. 11, pp. 2381–2387, 2001.

[116] H. Zollinger, *Properties of Organic Dyes and Pigments in Color Chemistry*, VCH Publishers, New York, NY, USA, 1978.

[117] B. Kraeutler and A. J. Bard, "Heterogeneous photocatalytic decomposition of saturated carboxylic acids on titanium dioxide powder. Decarboxylative route to alkanes," *Journal of the American Chemical Society*, vol. 100, no. 19, pp. 5985–5992, 1978.

[118] W. Dunn, K. Wilboun, F. Fan, and A. Brad, "Heterogeneous photocatalytic oxidation of hydrocarbons on platinized $TiO_2$ powders," *The Journal of Physical Chemistry A*, vol. 84, no. 32, pp. 7–10, 1980.

[119] Y. Gao, M. Hu, and B. Mi, "Membrane surface modification with $TiO_2$-graphene oxide for enhanced photocatalytic performance," *Journal of Membrane Science*, vol. 455, pp. 349–356, 2014.

[120] C. Xu, A. Cui, Y. Xu, and X. Fu, "Graphene oxide-$TiO_2$ composite filtration membranes and their potential application for water purification," *Carbon*, vol. 62, pp. 465–471, 2013.

[121] A. L. Ahmad, M. A. Majid, and B. S. Ooi, "Functionalized PSf/$SiO_2$ nanocomposite membrane for oil-in-water emulsion separation," *Desalination*, vol. 268, no. 1–3, pp. 266–269, 2011.

[122] Z. Wang, H. Yu, J. Xia et al., "Novel GO-blended PVDF ultrafiltration membranes," *Desalination*, vol. 299, pp. 50–54, 2012.

[123] C. P. Athanasekou, G. E. Romanos, F. K. Katsaros, K. Kordatos, V. Likodimos, and P. Falaras, "Very efficient composite titania membranes in hybrid ultrafiltration/photocatalysis water treatment processes," *Journal of Membrane Science*, vol. 392-393, pp. 192–203, 2012.

[124] Y. S. Lin, "Microporous and dense inorganic membranes: current status and prospective," *Separation and Purification Technology*, vol. 25, no. 1–3, pp. 39–55, 2001.

[125] L. G. A. van de Water and T. Maschmeyer, "Mesoporous membranes—a brief overview of recent developments," *Topics in Catalysis*, vol. 29, no. 1-2, pp. 67–77, 2004.

[126] W. S. Hummers and R. E. Offeman, "Preparation of graphitic oxide," *Journal of the Preparation of Graphitic Oxide, American Chemical Society*, vol. 80, no. 6, p. 1339, 1958.

[127] P. Gao, J. Liu, S. Lee, T. Zhang, and D. D. Sun, "High quality graphene oxide-CdS-Pt nanocomposites for efficient photocatalytic hydrogen evolution," *Journal of Materials Chemistry*, vol. 22, no. 5, pp. 2292–2298, 2012.

[128] J. Liu, H. Jeong, K. Lee et al., "Reduction of functionalized graphite oxides by trioctylphosphine in non-polar organic solvents," *Carbon*, vol. 48, no. 8, pp. 2282–2289, 2010.

[129] D. C. Marcano, D. V. Kosynkin, J. M. Berlin et al., "Improved synthesis of graphene oxide," *ACS Nano*, vol. 4, no. 8, pp. 4806–4814, 2010.

[130] Y. Gao, M. Hu, and B. Mi, "Membrane surface modification with $TiO_2$-graphene oxide for enhanced photocatalytic performance," *Journal of Membrane Science*, vol. 455, pp. 349–356, 2014.

[131] C. P. Athanasekou, N. G. Moustakas, S. Morales-Torres et al., "Ceramic photocatalytic membranes for water filtration under UV and visible light," *Applied Catalysis B: Environmental*, vol. 178, pp. 12–19, 2015.

[132] M.-L. Luo, J.-Q. Zhao, W. Tang, and C.-S. Pu, "Hydrophilic modification of poly (ether sulfone) ultrafiltration membrane surface by self-assembly of $TiO_2$ nanoparticles," *Applied Surface Science*, vol. 249, no. 1–4, pp. 76–84, 2005.

[133] H.-B. Yao, L.-H. Wu, C.-H. Cui, H.-Y. Fang, and S.-H. Yu, "Direct fabrication of photoconductive patterns on LBL assembled graphene oxide/PDDA/titania hybrid films by photothermal and photocatalytic reduction," *Journal of Materials Chemistry*, vol. 20, no. 25, pp. 5190–5195, 2010.

[134] G. Williams, B. Seger, and P. V. Kamt, "$TiO_2$-graphene nanocomposites. UV-assisted photocatalytic reduction of graphene oxide," *ACS Nano*, vol. 2, no. 7, pp. 1487–1491, 2008.

[135] S. Morales-Torres, L. M. Pastrana-Martínez, J. L. Figueiredo, J. L. Faria, and A. M. T. Silva, "Design of graphene-based $TiO_2$ photocatalysts—a review," *Environmental Science and Pollution Research*, vol. 19, no. 9, pp. 3676–3687, 2012.

[136] X. B. Ke, H. Y. Zhu, X. P. Gao, J. W. Liu, and Z. F. Zheng, "High-performance ceramic membranes with a separation layer of metal oxide nanofibers," *Advanced Materials*, vol. 19, no. 6, pp. 785–790, 2007.

[137] X. Chen and S. S. Mao, "Titanium dioxide nanomaterials: synthesis, properties, modifications, and applications," *Chemical Reviews*, vol. 107, no. 7, pp. 2891–2959, 2007.

[138] Y. Xu, H. Bai, G. Lu, C. Li, and G. Shi, "Flexible graphene films via the filtration of water-soluble noncovalent functionalized graphene sheets," *Journal of the American Chemical Society*, vol. 130, no. 18, pp. 5856–5857, 2008.

[139] R. Bissessur, P. K. Y. Liu, W. White, and S. F. Scully, "Encapsulation of polyanilines into graphite oxide," *Langmuir*, vol. 22, no. 4, pp. 1729–1734, 2006.

[140] C. Xu, Z. Chen, and X. Fu, "Graphene oxide-mediated formation of freestanding, thickness controllable metal oxide films," *Journal of Materials Chemistry*, vol. 21, no. 34, pp. 12889–12893, 2011.

[141] H.-K. Jeong, P. L. Yun, R. J. W. E. Lahaye et al., "Evidence of graphitic AB stacking order of graphite oxides," *Journal of the American Chemical Society*, vol. 130, no. 4, pp. 1362–1366, 2008.

[142] O. Akhavan, "Photocatalytic reduction of graphene oxides hybridized by ZnO nanoparticles in ethanol," *Carbon*, vol. 49, no. 1, pp. 11–18, 2011.

[143] O. Akhavan and E. Ghaderi, "Photocatalytic reduction of graphene oxide nanosheets on $TiO_2$ thin film for photoinactivation of bacteria in solar light irradiation," *Journal of Physical Chemistry C*, vol. 113, no. 47, pp. 20214–20220, 2009.

[144] J. Zhao, W. Ren, and H.-M. Cheng, "Graphene sponge for efficient and repeatable adsorption and desorption of water contaminations," *Journal of Materials Chemistry*, vol. 22, no. 38, pp. 20197–20202, 2012.

[145] C. Chen, W. Cai, M. Long et al., "Synthesis of visible-light responsive graphene oxide/$TiO_2$ composites with p/n heterojunction," *ACS Nano*, vol. 4, no. 11, pp. 6425–6432, 2010.

[146] J.-F. Li, Z.-L. Xu, H. Yang, L.-Y. Yu, and M. Liu, "Effect of $TiO_2$ nanoparticles on the surface morphology and performance of microporous PES membrane," *Applied Surface Science*, vol. 255, no. 9, pp. 4725–4732, 2009.

[147] N. A. A. Hamid, A. F. Ismail, T. Matsuura et al., "Morphological and separation performance study of polysulfone/titanium dioxide (PSF/$TiO_2$) ultrafiltration membranes for humic acid removal," *Desalination*, vol. 273, no. 1, pp. 85–92, 2011.

[148] E. Yuliwati, A. F. Ismail, T. Matsuura, M. A. Kassim, and M. S. Abdullah, "Characterization of surface-modified porous PVDF

hollow fibers for refinery wastewater treatment using microscopic observation," *Desalination*, vol. 283, pp. 206–213, 2011.

[149] H. Park and W. Choi, "Photocatalytic reactivities of nafion-coated $TiO_2$ for the degradation of charged organic compounds under UV or visible light," *Journal of Physical Chemistry B*, vol. 109, no. 23, pp. 11667–11674, 2005.

[150] J. S. Chen, C. Chen, J. Liu, R. Xu, S. Z. Qiao, and X. W. Lou, "Ellipsoidal hollow nanostructures assembled from anatase $TiO_2$ nanosheets as a magnetically separable photocatalyst," *Chemical Communications*, vol. 47, no. 9, pp. 2631–2633, 2011.

[151] H. B. Wu, H. H. Hng, and X. W. D. Lou, "Direct synthesis of anatase $TiO_2$ nanowires with enhanced photocatalytic activity," *Advanced Materials*, vol. 24, no. 19, pp. 2567–2571, 2012.

[152] P. Gao, J. Liu, T. Zhang, D. D. Sun, and W. Ng, "Hierarchical $TiO_2$/CdS 'spindle-like' composite with high photodegradation and antibacterial capability under visible light irradiation," *Journal of Hazardous Materials*, vol. 229-230, pp. 209–216, 2012.

[153] J.-Y. Liao, B.-X. Lei, D.-B. Kuang, and C.-Y. Su, "Tri-functional hierarchical $TiO_2$ spheres consisting of anatase nanorods and nanoparticles for high efficiency dye-sensitized solar cells," *Energy and Environmental Science*, vol. 4, no. 10, pp. 4079–4085, 2011.

[154] T. Zhang, J. Liu, and D. D. Sun, "A novel strategy to fabricate inorganic nanofibrous membranes for water treatment: use of functionalized graphene oxide as a cross linker," *RSC Advances*, vol. 2, no. 12, pp. 5134–5137, 2012.

[155] C. Ruan, L. Zhang, Y. Qin et al., "Synthesis of porphyrin sensitized $TiO_2$/graphene and its photocatalytic property under visible light," *Materials Letters*, vol. 141, pp. 362–365, 2015.

[156] C. P. Athanasekou, S. Morales-Torres, V. Likodimos et al., "Prototype composite membranes of partially reduced graphene oxide/$TiO_2$ for photocatalytic ultrafiltration water treatment under visible light," *Applied Catalysis B: Environmental*, vol. 158-159, pp. 361–372, 2014.

[157] C. Heitner-Wirguin, "Recent advances in perfluorinated ionomer membranes: structure, properties and applications," *Journal of Membrane Science*, vol. 120, no. 1, pp. 1–33, 1996.

[158] S. J. Sondheimer, N. J. Bunce, M. E. Lemke, and C. A. Fyfe, "Acidity and catalytic activity of Nafion-H," *Macromolecules*, vol. 19, no. 2, pp. 339–343, 1986.

[159] N. Zhang, Y. Zhang, and Y.-J. Xu, "Recent progress on graphene-based photocatalysts: current status and future perspectives," *Nanoscale*, vol. 4, no. 19, pp. 5792–5813, 2012.

[160] J. Wiszniowski, D. Robert, J. Surmacz-Gorska, K. Miksch, and J.-V. Weber, "Photocatalytic decomposition of humic acids on $TiO_2$ part I: discussion of adsorption and mechanism," *Journal of Photochemistry and Photobiology A: Chemistry*, vol. 152, no. 1–3, pp. 267–273, 2002.

[161] W. Yuan and A. L. Zydney, "Humic acid fouling during microfiltration," *Journal of Membrane Science*, vol. 157, no. 1, pp. 1–12, 1999.

[162] A. W. Zularisam, A. F. Ismail, and R. Salim, "Behaviours of natural organic matter in membrane filtration for surface water treatment-a review," *Desalination*, vol. 194, no. 1–3, pp. 211–231, 2006.

# 2

# Synthesis, Characterization, and Evaluation of Antimicrobial and Antifertility Efficacy of Heterobimetallic Complexes of Copper (II)

**Ashu Chaudhary and Anshul Singh**

*Department of Chemistry, Kurukshetra University, Kurukshetra 136 119, India*

Correspondence should be addressed to Ashu Chaudhary; ashuchaudhary21@gmail.com

Academic Editor: Hakan Arslan

Heterobimetallic complexes of the copper with remarkable antimicrobial and antifertility action have been synthesized by the conventional method by reacting $CuCl_2$ and o-phenylenediamine in the methanolic medium forming the mononuclear complex of the type $[Cu(C_6H_8N_2)_2]Cl_2$ which was further treated with group 4 or 14 organometallic dichlorides ($Ph_2M'Cl_2$, $Me_2M'Cl_2$, and $Cp_2M''Cl_2$) [$M'$ = Sn, Si, and $M''$ = Ti, Zr] leading to formation of heterobimetallic complexes. All the synthesized complexes were characterized in the light of various physicochemical techniques like elemental analyses, conductivity measurements, molecular weight determination, electronic and infrared spectra, electronic spin resonance, X-ray, and mass spectral studies. The *in vitro* antibacterial activity of synthesized complexes against *Escherichia coli* and *Staphylococcus aureus* was tested to evaluate their inhibiting activities. The antifungal activities of mono- and heterobimetallic complexes were studied by screening the compounds against *Fusarium oxysporum* and *Alternaria alternata*. These complexes were then investigated to reveal their effect on the reproductive organs of male rats. The treatment has shown substantial decline in the sperm motility and density signifying the imperious antifertility activity.

## 1. Introduction

Over the past few decades, studies on heterobimetallic complexes have aroused as a conspicuous field in the quest for new chemistry because of their plausible applications in fundamental, applied sciences, and coordination chemistry beneficial in industrial and synthetic processes such as catalysis, photochemistry, and biological systems. Heterobimetallic complexes have gained augmented research interest in recent years owing to the presence of two different metals which contributes to captivating structures and results in imperative applications.

In recent years, bimetallic complexes encompassing two different metal centers have acknowledged escalated consideration owing to the assumption that their reactivity should vary significantly from that of monometallic complexes or homobimetallic complexes [1]. The physical properties of these complexes are greatly affected by the presence of two metals in the same molecule along with an appreciable change in their reactivity. This is moreover due to the substantial modification in the individual properties of the metals or in the development of unique characteristics, which do not occur in monometallic compounds [2]. Varying applications of heterobimetallic complexes have been thoroughly explored which are found to be arising from the bifunctionality of the differing metal atoms [3].

Heterobimetallic complexes with transition metal ions have been subject of interest and entangled in biological processes of life [4]. Among the essential heavy metal ions in the human body, $Cu^{2+}$ is third in abundance after $Fe^{3+}$ and $Zn^{2+}$, and it plays very important role in several biological processes. Furthermore, many bimetallic complexes embroiling transition metals have been found to have potentials as antibacterial, anticancer, and antidiabetic agents [5].

The treatment of infectious diseases still remains an important and exigent problem because of an amalgamation of factors comprising emerging contagious ailments and the escalating number of multidrug resistant microbial pathogens. Despite the availability of enormous number of antibiotics and chemotherapeutics for medical use, the emergence of deep-rooted antibiotic resistance created in the last decades revealed an extensive medical need for new classes of antimicrobial agents. There is an essential need for the discovery of new compounds proficient at antimicrobial activity, possibly acting through mechanism of action and dissimilar from well-known classes of antimicrobial agents to which numerous clinically relevant pathogens are now resistant. Keeping all these fascinating facts under consideration, we have synthesized and characterized the monometallic and heterobimetallic complexes of copper along with group 4 and 14 metals which were further screened against a variety of pathogenic organisms.

Expeditiously growing population and inadequate sources are believed to be the most critical worldwide problems today. This upturn in the world population has burgeoned the benefits of economic and technological advancement. Fertility control is very essential for maintaining satisfactory standards in the developing countries. Hence, there is an abrupt requirement for a reasonable, safe, and effective contraceptive. Many inorganic compounds have been examined and applied for antifertility activity. All this work has emanated our interest in the development of a new series of heterobimetallic complexes with an aim to assess their effect on fertility and other reproductive organs of male albino rats. All the synthesized complexes were found possessing remarkable antifertility activity with discernable reduction in sperm motility and density.

## 2. Experimental

### 2.1. Synthesis

*2.1.1. Materials and Methods.* All the solvents used were first distilled and dried before use. Anhydrous $CuCl_2$, o-phenylenediamine, and organometallic dichlorides of group 4 and 14 ($Ph_2M'Cl_2$, $Me_2M'Cl_2$, and $Cp_2M''Cl_2$) were used as obtained. All the chemicals used for the synthesis of complexes were of AR grade.

*2.1.2. Synthesis of Monometallic Complex $[Cu(C_6H_8N_2)_2]Cl_2$.* Anhydrous $CuCl_2$ (1.34 g, 0.01 mol) was dissolved in hot methanol in a 100 mL round bottomed flask. A solution of o-phenylenediamine in methanol was added to the above solution in 1:2 stoichiometric proportions with a constant stirring of about 10–12 hrs. The complex thus obtained was washed and recrystallized from methanol and dried under vacuum.

*2.1.3. Synthesis of Heterobimetallic Complex $[Cu(C_6H_6N_2)_2Sn_2(Ph)_4Cl_2]$.* Monometallic complex (3.50 g, 0.01 mol) dissolved in hot methanol was then treated with solution of 6.88 g, 0.02 mol of $Ph_2SnCl_2$ in hot methanol in 1:2 stoichiometric proportions. In the reaction, a solid product was obtained after keeping the reaction mixture overnight at room temperature. The products formed were washed with hot methanol and then dried under vacuum.

On replacing $Ph_2SnCl_2$, the other complexes were obtained in the similar manner using $Me_2SnCl_2$, $Ph_2SiCl_2$, $Cp_2TiCl_2$, and $Cp_2ZrCl_2$ respectively.

*2.2. Analytical Methods and Physical Measurements.* Molar conductance of the compounds was measured in $10^{-3}$ M DMF solution on a Systronic model 305 conductivity bridge. The IR spectra (4000–200 $cm^{-1}$) were recorded on a model Nicolet Magna FTIR-550 spectrophotometer in the form of KBr pellets. The far infrared spectra of the complexes were recorded on the same spectrophotometer in Nujol Mulls using CsI cell. Molecular weights were determined by the Rast Camphor method. Metal contents were determined gravimetrically. Electronic spectra in dimethyl sulphoxide were recorded, using method as the solvent on a UV-160A Shimadzu spectrophotometer. X-ray powder diffraction spectra of the compound were obtained on a Philips (Model P.W. 1840) automatic diffractometer while EPR spectra of the complexes were recorded on a Varian E-4 EPR spectrometer. Nitrogen and chlorine were estimated by Kjeldahl's method and Volhard's method, respectively.

*2.3. Antimicrobial Assay.* Heterobimetallic complexes with macrocyclic framework have engendered copious research in the current years, due to their probable applications in catalysis, applied sciences, photochemistry, and biological systems. An escalating interest in developing better fungicides and bactericides has prompted us to assess the microbial efficacy of all the synthesized monometallic and heterobimetallic complexes.

*(i) Antifungal Activity Studies.* The antifungal activities were evaluated against *Alternaria alternata* and *Fusarium oxysporum* by Radial growth method. The potato dextrose agar medium was prepared by dissolving glucose (20 g), starch (20 g), agar-agar (20 g), and 1000 mL of distilled water at 25 ± 2°C. Bavistin was used as standard fungicide. The compounds at 100 and 200 ppm concentrations in methanol were then mixed in the medium using a small amount of DMF for initial solution (Figure 1). The linear growth of the fungus was determined by measuring the diameter of the colony after 96 hrs [6]. Growth inhibition was calculated using formula:

$$\% \text{ inhibition} = C - \frac{T}{C} \times 100, \quad (1)$$

where $C$ and $T$ are the diameters of the fungal colony in the control and test plates, respectively (Table 1).

*(ii) Antibacterial Activity Studies.* The synthesized complexes were tested against *Escherichia coli* and *Staphylococcus aureus* at 500 and 1000 ppm concentrations. In this case, streptomycin was used as standard (Figure 2). Inhibition zone technique was used to evaluate antibacterial activities [7]. In this technique, nutrient agar medium (peptone, beef extract, NaCl, and agar-agar) and 5 mm diameter paper discs (Whatman No. 1) filter paper were used. The compounds were dissolved in methanol in 500 and 1000 ppm concentrations. The filter paper discs were soaked in these solutions of the compounds, dried, and then placed in the Petri plates

TABLE 1: Fungicidal screening data of the Cu(II) complexes (percent growth inhibition after 96 hrs at 25 ± 2°C, Conc. in ppm).

| Complex | Alternaria alternata | | Fusarium oxysporum | |
|---|---|---|---|---|
| | 100 ppm | 200 ppm | 100 ppm | 200 ppm |
| $[Cu(C_6H_8N_2)_2]Cl_2$ | 23 | 30 | 26 | 34 |
| $[Cu(C_6H_6N_2)_2Sn_2(Ph)_4Cl_2]$ | 36 | 41 | 40 | 51 |
| $[Cu(C_6H_6N_2)_2Sn_2(Me)_4Cl_2]$ | 42 | 52 | 47 | 59 |
| $[Cu(C_6H_6N_2)_2Si_2(Ph)_4Cl_2]$ | 49 | 60 | 54 | 75 |
| $[Cu(C_6H_6N_2)_2Ti_2(Cp)_4Cl_2]$ | 26 | 34 | 30 | 36 |
| $[Cu(C_6H_6N_2)_2Zr_2(Cp)_4Cl_2]$ | 30 | 37 | 39 | 41 |
| Bavistin | 100 | 100 | 100 | 100 |

TABLE 2: Bacterial screening data of Cu(II) complexes (percent growth inhibition after 36 hrs at 30 ± 1°C).

| Complex | *Escherichia coli* | | *Staphylococcus aureus* | |
|---|---|---|---|---|
| | 500 ppm | 1000 ppm | 500 ppm | 1000 ppm |
| $[Cu(C_6H_8N_2)_2]Cl_2$ | 3 | 5 | 2 | 4 |
| $[Cu(C_6H_6N_2)_2Sn_2(Ph)_4Cl_2]$ | 5 | 7 | 3 | 5 |
| $[Cu(C_6H_6N_2)_2Sn_2(Me)_4Cl_2]$ | 6 | 9 | 5 | 6 |
| $[Cu(C_6H_6N_2)_2Si_2(Ph)_4Cl_2]$ | 7 | 11 | 8 | 5 |
| $[Cu(C_6H_6N_2)_2Ti_2(Cp)_4Cl_2]$ | 4 | 5 | 2 | 3 |
| $[Cu(C_6H_6N_2)_2Zr_2(Cp)_4Cl_2]$ | 4 | 7 | 3 | 4 |
| Streptomycin | 17 | 18 | 15 | 17 |

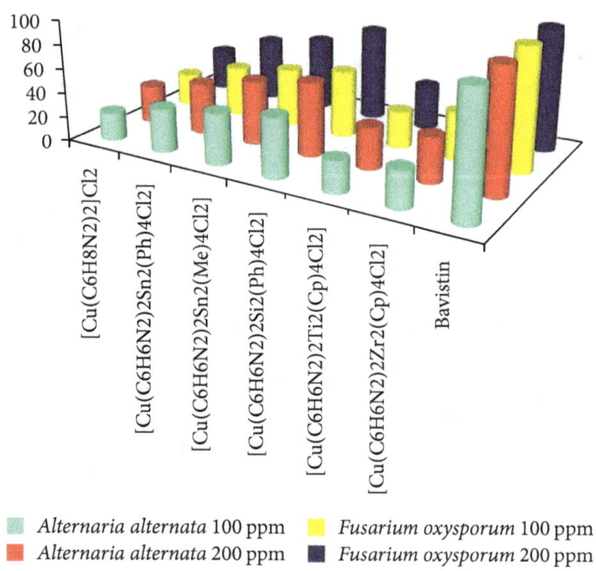


FIGURE 1: Antifungal activity of synthesized monometallic and heterobimetallic complexes.

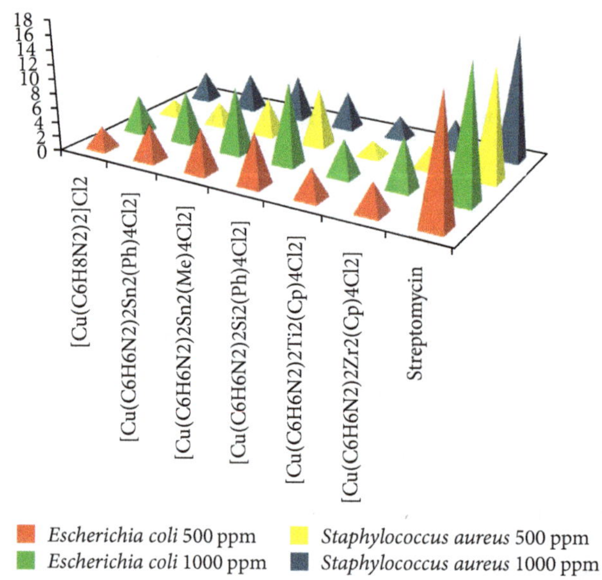


FIGURE 2: Antibacterial activity of synthesized monometallic and heterobimetallic complexes.

previously seeded with the test organism. The Petri dishes were stored in an incubator at 30 ± 1°C for 36 hrs. The zone of inhibition thus formed around each disc containing the test compound was measured accurately (Table 2).

*2.4. Antifertility Activity Studies.* The antifertility activity of monometallic and heterobimetallic complexes was evaluated by using male albino rats weighing 180–200 g which were housed in cages by maintaining the standard conditions (12 : 12 light and dark cycle; 25 ± 3°C and 35–60% relative humidity). The animals of groups were fed with food pellet acquired from Ashirwad Industries, Chandigarh, and the tap water was supplied ad libitum. These rats were equally divided into groups with six rats present in each group. Group A animals were kept as control and were given olive oil only. The animals of group B were treated with monometallic complex of copper (II) with a dose of about 30 mg/kg body weight for a period of about 60

TABLE 3: Sperm dynamics and fertility after the administration of synthesized complexes.

| Treatment | Number of males | Number of females | Number of pregnant females | Sperm motility (cauda epididymis) (%) | Sperm density (cauda epididymis) (million/ml) | Fertility Test (%) |
|---|---|---|---|---|---|---|
| Group A | 6 | 12 | 12/12 | 66.73 ± 1.33 | 54.5 ± 0.49 | 100 (+ve) |
| Group B | 6 | 12 | 8/12 | 48.59 ± 1.67 | $31.69^{**}$ ± 2.48 | 35 (−ve) |
| Group C | 6 | 12 | 5/12 | 38.84 ± 2.36 | 18.17 ± 3.39 | 60 (−ve) |
| Group D | 6 | 12 | 3/12 | $35.45^{**}$ ± 2.35 | $14.5^{**}$ ± 0.29 | 75 (−ve) |
| Group E | 6 | 12 | 4/12 | 30.65 ± 3.66 | $17.22^{**}$ ± 0.69 | 70 (−ve) |
| Group F | 6 | 12 | 1/12 | $26.59^{**}$ ± 1.67 | $10.01^{**}$ ± 0.22 | 90 (−ve) |

$^{**}P \leq 0.01$ = highly-significant.

TABLE 4: Effect of synthesized complexes on weight of reproductive organs of male rats.

| Treatment | Body weight (gm) | | Tissue weight (mg/100 gm body weight) | | | | |
|---|---|---|---|---|---|---|---|
| | Initial | Final | Testes | Epididymus | Seminal Vesicle | Ventral Prostrate | Vas Deference |
| Group A | 192.5 ± 5.45 | 212.5 ± 4.51 | 1032.09 ± 12.50 | 298.13 ± 4.86 | 237.21 ± 8.33 | 160.48 ± 4.32 | 72.05 ± 1.50 |
| Group B | 186.7 ± 18.00 | 206 ± 17.04 | $641.09^{*}$ ± 26.87 | $243.13^{*}$ ± 5.78 | $207.21^{*}$ ± 14.28 | $128.92^{*}$ ± 1.34 | $40.04^{**}$ ± 0.89 |
| Group C | 186.25 ± 3.35 | 208 ± 8.62 | $545.16^{**}$ ± 12.19 | $208.21^{**}$ ± 9.20 | $158.97^{**}$ ± 8.00 | $127.27^{**}$ ± 5.05 | $37.3^{**}$ ± 0.32 |
| Group D | 190.5 ± 8.84 | 212 ± 9.87 | $489.03^{**}$ ± 12.50 | $191.95^{**}$ ± 4.86 | $138.43^{**}$ ± 8.33 | $95.41^{**}$ ± 3.82 | $42.05^{**}$ ± 1.50 |
| Group E | 185 ± 6.96 | 202.75 ± 8.70 | $549.35^{**}$ ± 8.62 | 203.39 ± 2.88 | $149.67^{**}$ ± 4.27 | $109.49^{**}$ ± 0.93 | $33.41^{**}$ ± 2.73 |
| Group F | 187.5 ± 5.38 | 205 ± 17.38 | $451.08^{**}$ ± 6.23 | $162.75^{**}$ ± 1.86 | $123.73^{**}$ ± 4.82 | $97.45^{**}$ ± 3.29 | $41.24^{**}$ ± 0.80 |

$^{*}P \leq 0.01$ = significant; $^{**}P \leq 0.001$ = highly-significant.

days, while groups C, D, E, and F were treated with heterobimetallic complexes {$[Cu(C_6H_6N_2)_2Sn_2(Ph)_4Cl_2]$, $[Cu(C_6H_6N_2)_2Sn_2(Me)_4Cl_2]$ $[Cu(C_6H_6N_2)_2Si_2(Ph)_4Cl_2]$, and $[Cu(C_6H_6N_2)_2Ti_2(Cp)_4Cl_2]$} of about the same dose dissolved in olive oil and for relatively analogous time period. Near the end of the experiment, that is, on day 55, each male was housed individually with two estrous females for 3–5 days. The females were scheduled for laparotomy 2 weeks later to determine the number and viability of fetuses. After mating, males were removed for autopsy. The motility and density of cauda epididymal spermatozoa was assessed using a hemocytometer. All the recorded values of body/organs weight, sperm dynamics, hematological parameters, and tissue biochemistry were expressed in the terms of mean ± SEM. The data were analyzed statistically by using Student's "$t$-"test and the significance of difference was set at $P < 0.01$ and $P < 0.001$ as observed in Tables 3–5 and graphically represented in Figures 3–5.

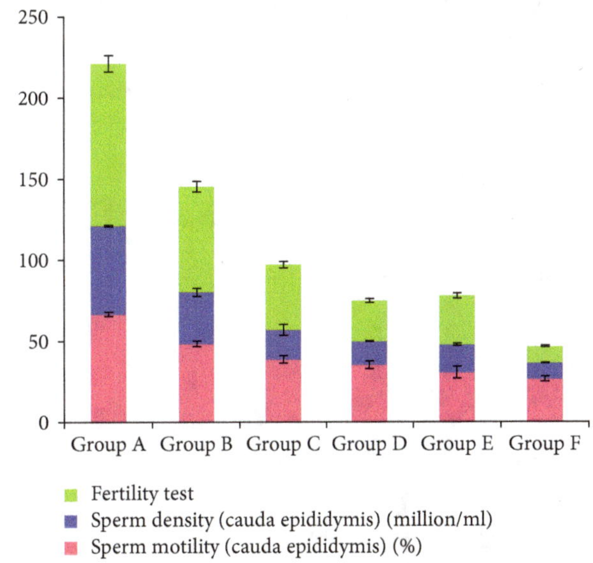


FIGURE 3: Altered sperm dynamics and fertility test after treatment with precursors and their complexes.

## 3. Results and Discussions

Cupric chloride reacts with o-phenylenediamine leading to formation of mononuclear complex which on further treatment with organometallic dichlorides of group 4 or 14 resulted in formation of novel series of heterobimetallic complexes of copper.

All the reactions proceeded slickly. The complexes were isolated as coloured solids, and their formation was ascertained on the basis of elemental analysis, infrared, esr, mass, and electronic spectral data. The molar conductance values of heterobimetallic complexes measured in DMF (18–33 $ohm^{-1}$ $mol^{-1}$ $cm^2$) showed them to be nonelectrolytes.

The analytical and physical data of the synthesized complexes have been summarized in Table 6.

*3.1. Infrared Spectra.* The IR spectra of the mononuclear as well as bimetallic complexes were recorded (Table 7) and a relative study confirmed the formation of heterobimetallic complexes with the anticipated coordination pattern. The primary amine exhibits a band at higher frequency than that of the corresponding secondary amine. In case of

Table 5: Biochemical changes in the male rats after treatment with the synthesized complexes.

| Treatment | Cholesterol (mg/gm) | | Glycogen (mg/gm) | |
|---|---|---|---|---|
| | Testes | Liver | Testes | Liver |
| Group A | $8.03 \pm 0.16$ | $16.98 \pm 0.04$ | $4.77 \pm 0.38$ | $22.83 \pm 0.71$ |
| Group B | $9.83 \pm 0.26$ | $15.84 \pm 0.07$ | $4.13 \pm 0.22$ | $21.99 \pm 0.44$ |
| Group C | $9.45^{*} \pm 0.33$ | $14.64^{*} \pm 0.03$ | $3.62^{**} \pm 0.19$ | $21.13^{*} \pm 0.30$ |
| Group D | $10.06^{**} \pm 0.28$ | $13.92^{**} \pm 0.07$ | $3.19^{**} \pm 0.32$ | $20.04^{**} \pm 0.85$ |
| Group E | $11.6^{**} \pm 0.14$ | $13.82^{**} \pm 0.05$ | $3.53^{**} \pm 0.26$ | $18.85^{**} \pm 0.67$ |
| Group F | $12.7^{**} \pm 0.34$ | $11.98^{**} \pm 0.04$ | $2.44^{*} \pm 0.46$ | $18.06^{**} \pm 0.47$ |

$^{*}P < 0.01$ = significant; $^{**}P < 0.001$ = highly-significant.

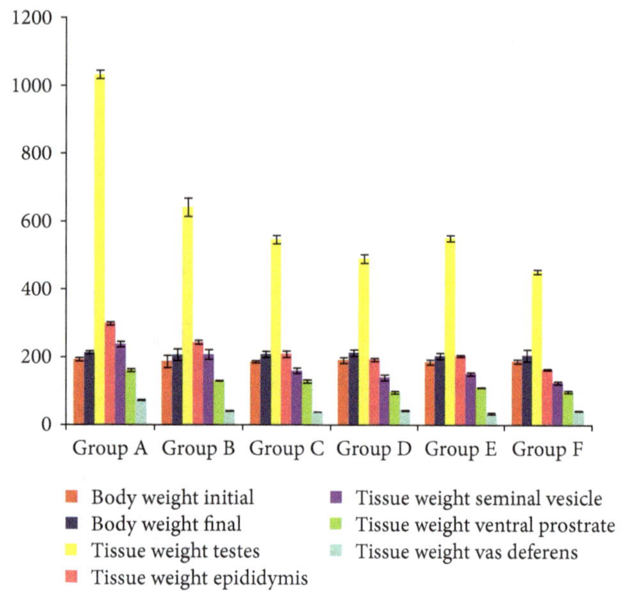


Figure 4: Alteration in the body weight and weight of reproductive organs after treatment with the synthesized monometallic and heterobimetallic complexes.

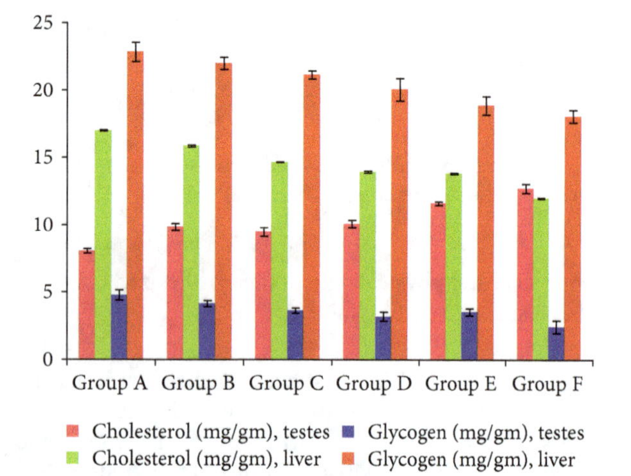


Figure 5: Effects of the synthesized complexes on the cholesterol and glycogen content of tissues.

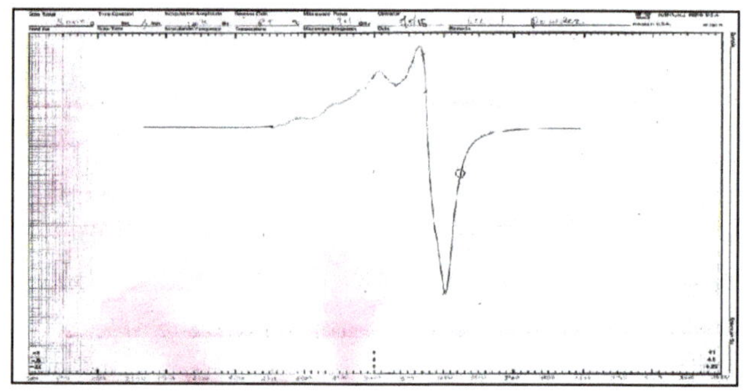

Figure 6: The X band ESR spectrum of mononuclear complex $[Cu(C_6H_8N_2)_2]Cl_2$.

mononuclear complex, a broad and strong band is observed for $\nu$(N-H) in the range of 3218–3255 cm$^{-1}$ which swings over to lower frequency region in case of bimetallic complexes [8] evidently specifying the formation of bond between group 14 metal and nitrogen, while bands due to $\delta$(N-H) appear in the region of 1529–1540 cm$^{-1}$ with no evident change after chelation [9]. Aromatic ring (C=C) stretching appears at 1645, 1529, and 1453 cm$^{-1}$, while (C-H) and (C-N) stretching were assigned at 3059 and 843 cm$^{-1}$, respectively. Bands of medium intensity in the range of 422–580 cm$^{-1}$ are attributed to $\nu$(M-N) vibrations [10]. In case of heterobimetallic complexes with cyclopentadienyl ring, IR bands appear at 3000 for $\nu$(C-H), 1435 for $\nu$(C-C), 1028 for $\delta$(C-H) in plane, and 812 for $\delta$(C-H) out of plane vibrations. In addition, bands due to (Ti-$C_5H_5$) and (Zr-$C_5H_5$) appear at 446 and 443 cm$^{-1}$, respectively. A medium intensity band observed in the far IR region of the metal complexes (460–470) cm$^{-1}$ was assigned to $\nu$(Cu-N). The far IR spectra show a distinct band at 315 cm$^{-1}$ attributed to (Cu-Cl) band [11].

*3.2. Electron Spin Resonance Spectra.* The ESR spectra of monometallic and heterobimetallic complexes were recorded at the room temperature. The observed data showed that in case of monometallic complex $g_{\parallel} = 2.09$ and $g_{\perp} = 2.06$, respectively, with $g_{av} = 2.07$. As $g_{\parallel} > g_{\perp} > 2.0023$, it is suggested that the unpaired electron is localized in $d_{x^2-y^2}$ orbital which indicates the square planar geometry around the Cu(II) ion [12] (Figure 6). The $g$ values witnessed for heterobimetallic complexes has $g_{\perp} > g_{\parallel}$ suggesting

TABLE 6: Analytical data and physical properties of the synthesized complexes.

| Complex | Color & M.P. (°C) | Analysis: found (calc.) (%) | | | | | | Mol. Wt. found (calc.) |
|---|---|---|---|---|---|---|---|---|
| | | C | H | N | Cl | Cu | M′ | |
| $[Cu(C_6H_8N_2)_2]Cl_2$ | Brown 240 | 41.12 (41.09) | 4.65 (4.60) | 15.98 (15.97) | 20.24 (20.22) | 18.14 (18.12) | — | 355.62 (350.73) |
| $[Cu(C_6H_6N_2)_2Sn_2(Ph)_4Cl_2]$ | Dark brown 232 | 48.25 (48.23) | 4.09 (4.05) | 6.27 (6.25) | 7.94 (7.91) | 7.15 (7.09) | 26.52 (26.48) | 898.50 (896.57) |
| $[Cu(C_6H_6N_2)_2Sn_2(Me)_4Cl_2]$ | Dark brown 209 | 29.86 (29.83) | 3.76 (3.75) | 8.74 (8.70) | 11.2 (11.0) | 9.88 (9.86) | 36.88 (36.85) | 650.00 (644.26) |
| $[Cu(C_6H_6N_2)_2Si_2(Ph)_4Cl_2]$ | Brown 220 | 60.48 (60.45) | 5.09 (5.07) | 7.88 (7.83) | 9.96 (9.91) | 8.92 (8.88) | 7.86 (7.85) | 723.42 (715.33) |
| $[Cu(C_6H_6N_2)_2Ti_2(Cp)_4Cl_2]$ | Blackish brown 231 | 54.75 (54.69) | 4.63 (4.59) | 7.98 (7.97) | 10.14 (10.09) | 9.06 (9.04) | 13.64 (13.62) | 708.41 (702.81) |
| $[Cu(C_6H_6N_2)_2Zr_2(Cp)_4Cl_2]$ | Shiny brown 228 | 48.72 (48.68) | 4.14 (4.08) | 7.13 (7.10) | 9.12 (8.98) | 8.07 (8.05) | 23.15 (23.11) | 792.00 (789.53) |

TABLE 7: IR spectral data (in $cm^{-1}$) of the synthesized monometallic and heterobimetallic complexes.

| Complex | $\nu$(N-H) | $\delta$(N-H) | $\nu$(M-N) | $\nu$(Cu-N) | $\nu$(Cu-Cl) |
|---|---|---|---|---|---|
| $[Cu(C_6H_8N_2)_2]Cl_2$ | 3250 | 1540 | — | 468 | — |
| $[Cu(C_6H_6N_2)_2Sn_2(Ph)_4Cl_2]$ | 3185 | 1536 | 470 | 465 | 315 |
| $[Cu(C_6H_6N_2)_2Sn_2(Me)_4Cl_2]$ | 3187 | 1534 | 468 | 465 | 315 |
| $[Cu(C_6H_6N_2)_2Si_2(Ph)_4Cl_2]$ | 3182 | 1536 | 581 | 463 | 315 |
| $[Cu(C_6H_6N_2)_2Ti_2(Cp)_4Cl_2]$ | 3185 | 1536 | 525 | 464 | 315 |
| $[Cu(C_6H_6N_2)_2Zr_2(Ph)_4Cl_2]$ | 3190 | 1536 | 530 | 465 | 315 |

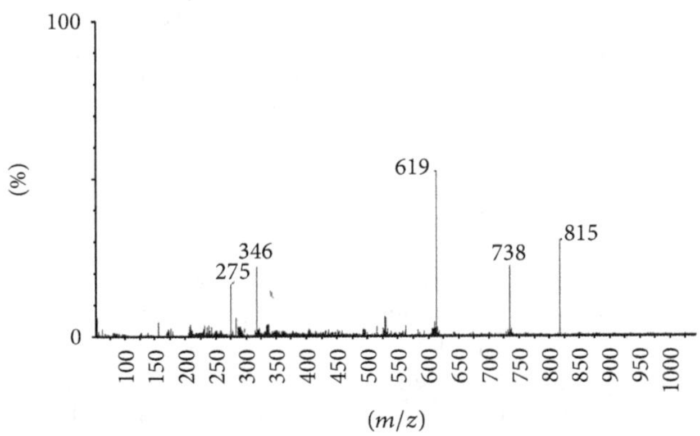


FIGURE 7: Mass spectrum of heterobimetallic complex $[Cu(C_6H_6N_2)_2Sn_2(C_6H_5)_4Cl_2]$.

a distorted octahedral geometry for these complexes with $d_{z^2}$ as the ground state of the system [13]. On this basis, it is concluded that chloride ions coordinate with Cu(II) on complexation with $R_2M'Cl_2$ achieving an octahedral environment for copper.

*3.3. Electronic Spectra.* The electronic spectrum of monometallic complex exhibits a strong *d-d* band at 594 nm which is characteristic of square planar complexes. A band is also obtained at 247 nm which is ascribed to L → M charge transfer. The electronic spectrum of complexes exhibits a strong band at 277 nm and shoulders assignable to intraligand transition and ligand to metal charge transfer [14]. In bimetallic octahedral complexes of copper, distortion may lead to dislodgment of *d-d* bands to lower energies and hence two bands are observed at 920 nm and 880 nm.

*3.4. Molar Conductance.* The molar conductance values of heterobimetallic complexes in DMF solution fall in the range of 18–33 $ohm^{-1} mol^{-1} cm^2$ suggesting that they are nonelectrolytes while higher conductance value of mononuclear complex of about 220 $ohm^{-1} mol^{-1} cm^2$ indicates the electrolytic character [15].

*3.5. Mass Spectroscopy.* The FAB mass spectra of the heterobimetallic complex $[Cu(C_6H_6N_2)_2Sn_2(C_6H_5)_4Cl_2]$ exhibited molecular ion peaks at $m/z$ 815, 738, 619, and 346 assigned to $[Cu(C_6H_6N_2)_2Sn_2(C_6H_5)_3Cl_2]^+$, $[Cu(C_6H_6N_2)_2Sn_2(C_6H_5)_2Cl_2]^+$, $[Cu(C_6H_6N_2)_2Sn(C_6H_5)_2Cl_2]^+$, and $[Cu(C_6H_6N_2)_2Cl_2]^+$, respectively. The two coordinated chlorides are removed with a mass loss of $m/z = 71$ with peak obtained at $m/z = 275$ (Figure 7).

*3.6. X-Ray Powder Diffraction Studies.* The possible geometry of the finely powdered sample of the compound $[Cu(C_6H_8N_2)_2]Cl_2$ has been inferred on the basis of X-ray powder diffraction studies (Figure 8). The observed interplanar spacing values ("$d$" in Å) have been measured from the diffractogram of this compound and the Miller indices $h$, $k$, and $l$ have been assigned to each $d$ value as reported in Table 8. The results show that compound belongs

Table 8: X-ray powder diffraction data for the compound $[Cu(C_6H_8N_2)_2]Cl_2$.

| Peak number | $2\theta$ (deg.) (obs.) | $h$ | $k$ | $l$ | $d$-spacing (obs.) |
|---|---|---|---|---|---|
| (1) | 11.30 | 1 | 2 | 0 | 9.765 |
| (2) | 11.50 | 2 | 1 | 0 | 9.669 |
| (3) | 15.80 | 0 | 5 | 0 | 7.048 |
| (4) | 19.80 | 2 | 5 | 0 | 5.634 |
| (5) | 21.92 | 4 | 0 | 0 | 5.095 |
| (6) | 21.98 | 4 | 0 | 0 | 5.125 |
| (7) | 25.80 | 3 | 3 | 1 | 4.339 |
| (8) | 26.28 | 1 | 6 | 1 | 4.261 |
| (9) | 26.27 | 1 | 5 | 1 | 4.41 |
| (10) | 27.20 | 3 | 4 | 1 | 4.120 |
| (11) | 30.40 | 2 | 7 | 1 | 3.695 |
| (12) | 30.48 | 7 | 2 | 1 | 3.799 |
| (13) | 33.40 | 6 | 0 | 0 | 3.371 |
| (14) | 34.42 | 0 | 6 | 0 | 3.426 |
| (15) | 38.10 | 3 | 9 | 1 | 2.968 |
| (16) | 43.30 | 5 | 10 | 0 | 2.626 |
| (17) | 58.90 | 4 | 3 | 3 | 1.970 |
| (18) | 58.10 | 2 | 0 | 1 | 1.706 |

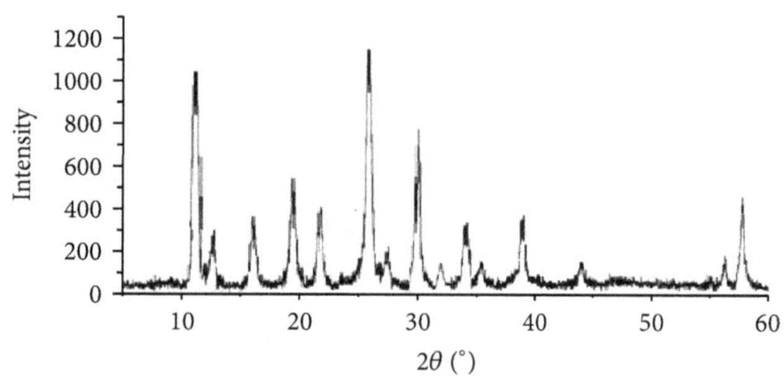


Figure 8: X-ray diffraction pattern of compound $[Cu(C_6H_8N_2)_2]Cl_2$.

to "orthorhombic" crystal system having unit call parameters as

$$a = 21.430,\quad b = 33.640,\quad c = 5.430;\quad \alpha = \beta = \gamma = 90^\circ. \tag{2}$$

Refined values of $a = 21.430$, $b = 33.640$, $c = 5.430$; $\alpha = \beta = \gamma = 90^\circ$ (orthorhombic system).

On the basis of above spectral studies, the structure can be suggested for mononuclear and heterobimetallic complex, respectively. See Figure 9.

### 3.7. Biological Evaluation

*(i) Antimicrobial Activities.* The results described in Tables 1 and 2 reveal that all the compounds are active against these organisms even at low concentrations, and the inhibition of the growth of microorganism was found to be dependent on the concentration of the compounds [6]. The results of the biological screening indicated that the heterobimetallic complexes exhibit effective antifungal activity against fungal strain *F. oxysporum*; however, antibacterial activity was found to be brilliant in case of Gram-negative bacteria as compared to Gram-positive bacteria. The difference arises due to variation in the thickness of peptidoglycan layer which is more in Gram-positive as compared to Gram-negative bacteria leading to difficult penetration across the cell wall.

*(ii) Antifertility Activity*

*Effect on Organ Weights.* The current study revealed that administration of monometallic complex and their various heterobimetallic complexes at the dose level of 30 mg/Kg body weight for 60 days ensued in a substantial reduction in weight of testes and other sex organs (Table 4). The reduction in testicular weight may be due to reduced tubule size, spermatogenic arrest, and inhibition of steroid biosynthesis of Leydig cells [16]. Diminutions in the weight of accessory reproductive organs directly support the curtailed availability of androgen [17].

*Effect on Sperm Density.* Sperm density of the testis and cauda epididymides was decreased significantly ($P \le 0.001$) in rats treated with the complexes at all the dose levels (Table 3). The low sperm density in cauda epididymis is caused by the alternation in androgen level.

*Effect on Fertility.* A 35–90% negative fertility was observed for heterobimetallic complexes (Table 3). The negative fertility test is credited to lack of forward progression and reduction in density of spermatozoa and altered biochemical milieu of cauda epididymis [18].

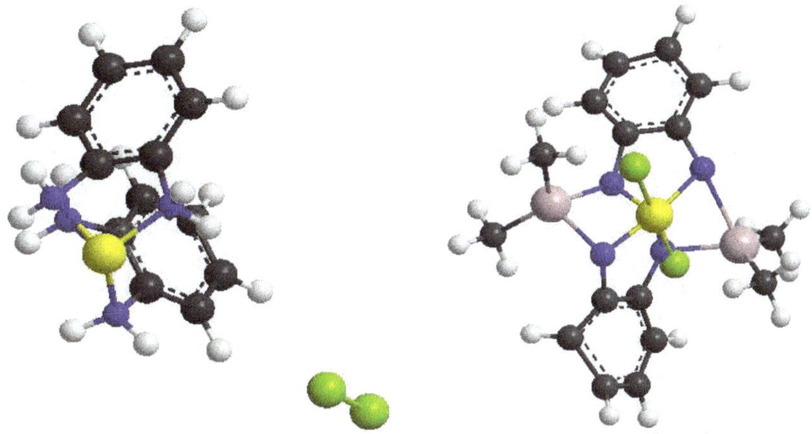

*Mononuclear complex*
$[Cu(C_6H_8N_2)_2]Cl_2$

*Heteronuclear complex*
$[Cu(C_6H_6N_2)_2Sn_2(Me)_4Cl_2]$

FIGURE 9

*Effect on Cholesterol Levels.* It has been observed that the concentration of cholesterol increases in testes (Table 5). This increased concentration of testicular cholesterol may be the result of nonutilization leading to the reduction of production of the hormone which is involved in the control of fertility of animals including rats [19]. It is therefore concluded that the heterobimetallic complexes of copper are more effective than the monometallic complexes in inhibiting the fertility due to the synergistic action.

## 4. Conclusion

In this paper, we describe the synthesis of monometallic and heterobimetallic complexes of copper which are further characterized by IR, ESR, and electronic and mass spectral studies. These studies evidently propose square planar and octahedral geometry for mono- and heterobimetallic complexes, respectively. These complexes were then evaluated for their antifungal, antibacterial, and antifertility activity. The antimicrobial results indicated that the heterobimetallic complexes exhibited promising activity against microbes as compared to mononuclear complexes. However, the results obtained revealed that these complexes exhibited utmost activity against fungal strain *F. oxysporum* and bacterial strain *E. coli*. The results drawn from antifertility activity suggested the efficiency of the synthesized heterobimetallic complexes in reducing fertility more effectively than the mononuclear complexes of copper from which they have been derived and thus play an imperative role in improving bioactivity.

## Conflicts of Interest

The authors declare that they have no conflicts of interest.

## Acknowledgments

The authors (Ashu Chaudhary and Anshul Singh) wish to express gratitude to the Council of Scientific and Industrial Research (CSIR), New Delhi, India, for financial assistance in the form of JRF vide Letter no. 09/105(0221)/2015-EMR-I.

## References

[1] C. M. Thomas, "Metal-metal multiple bonds in early/late heterobimetallic complexes: Applications toward small molecule activation and catalysis," *Comments on Inorganic Chemistry*, vol. 32, no. 1, pp. 14–38, 2011.

[2] M. Rakkiyasamy, K. Kalimuthu, and S. Rangasamy, "Heterobimetallic complexes of transition metals-synthesis and characterization," *Journal of Chemical Biological and Physical Sciences Section A*, vol. 3, no. 1, pp. 16–19, 2012.

[3] B. G. Cooper, J. W. Napoline, and C. M. Thomas, "Catalytic applications of early/late heterobimetallic complexes," *Catalysis Reviews - Science and Engineering*, vol. 54, no. 1, pp. 1–40, 2012.

[4] R. K. Al-Shemary, A. M. A. Al-Khazraji, and A. F. S. S. M. Lateef, "Synthesis, characterization and antimicrobial activity studies of mixed-1,10-phenanthroline-Mn(II),Co(II), Cu(II), Ni(II) and Hg(II) complexes with schiff base[2,2'-(1Z,1'Z)-(biphenyl-4,4'-diylbis(azan-1-yl-1-ylidene))bis(methan-1-yl-1-ylidene)diphenol]," *International Journal of Science and Research (IJSR)*, vol. 5, no. 5, pp. 1787–1793, 2016.

[5] A. A. Osowole, O. B. A. Agbaje, and B. O. Ojo, "Synthesis, characterization and antibacterial properties of some heteroleptic metal(II) complexes of paracetamol and vanillin," *Asian Journal of Pharmaceutical and Clinical Research*, vol. 7, no. 3, pp. 145–149, 2014.

[6] K. Sharma, R. V. Singh, and N. Fahmi, "Palladium(II) and platinum(II) derivatives of benzothiazoline ligands: synthesis, characterization, antimicrobial and antispermatogenic activity," *Spectrochimica Acta—Part A: Molecular and Biomolecular Spectroscopy*, vol. 78, no. 1, pp. 80–87, 2011.

[7] N. Fahmi, A. Kumari, and R. V. Singh, "Synthesis, characterization, antimicrobial and pesticidal activity of organoarsenic(III) and organoantimony(III) complexes with N∩S donor ligands," *International Journal of Pharmaceutical Sciences and Research*, vol. 5, no. 12, pp. 5260–5266, 2014.

[8] S. Yadav, A. Moheman, and K. S. Siddiqi, "Synthesis and characterization of heterobimetallic complexes of the type [Cu(pn)2][MCl4] where M = Co(II), Ni(II), Cu(II), Zn(II), Cd(II), and Hg(II)," *Arabian Journal of Chemistry*, vol. 9, no. 2, pp. S1747–S1754, 2016.

[9] M. Chauhan, K. Banerjee, and F. Arjmand, "DNA binding studies of novel copper(II) complexes containing L-tryptophan as chiral auxiliary: In vitro antitumor activity of Cu-Sn2 complex in human neuroblastoma cells," *Inorganic Chemistry*, vol. 46, no. 8, pp. 3072–3082, 2007.

[10] S. Tabassum, M. Zaki, F. Arjmand, and I. Ahmad, "Synthesis of heterobimetallic complexes: In vitro DNA binding, cleavage and antimicrobial studies," *Journal of Photochemistry and Photobiology B: Biology*, vol. 114, pp. 108–118, 2012.

[11] U. Kumar and S. Chandra, "Synthesis, spectral and antifungal studies of some coordination compounds of cobalt(II) and copper(II) of a novel 18-membered octaaza [N8] tetradentate macrocyclic ligand," *Journal of Saudi Chemical Society*, vol. 15, no. 2, pp. 187–193, 2011.

[12] F. Arjmand, B. Mohani, and S. Parveen, "New Dihydro OO'Bis (Salicylidene) 2,2' aminobenzothiazolyl borate complexes: kinetic and voltammetric studies of dimethyltin copper complex with guanine, adenine, and calf thymus DNA," *Bioinorganic Chemistry and Applications*, vol. 2006, Article ID 32896, 10 pages, 2006.

[13] K. S. Siddiqi, H. Afaq, S. A. A. Nami, and A. Umar, "Heterobimetallic complexes containing Cu and Si," *Synthesis and Reactivity in Inorganic and Metal-Organic Chemistry*, vol. 33, no. 8, pp. 1459–1468, 2003.

[14] K. Nagaraj, S. Sakthinathan, and S. Arunachalam, "Synthesis, CMC determination, antimicrobial activity and nucleic acid binding of a surfactant Copper(II) complex containing phenanthroline and alanine schiff-base," *Journal of Fluorescence*, vol. 24, no. 2, pp. 589–598, 2014.

[15] A. Chaudhary, N. Bansal, N. Fahmi, and R. V. Singh, "Toxicological aspects of newly designed macrocyclic complexes of iron(II)," *Indian Journal of Chemistry*, vol. 43A, pp. 320–322, 2004.

[16] S. C. Joshi, P. Tibrewal, A. Sharma, and P. Sharma, "Evaluation of toxic effect of 2,4-d (2,4-dichlorophenoxyacetic acid) on fertility and biochemical parameters of male reproductive system of albino rats," *International Journal of Pharmacy and Pharmaceutical Sciences*, vol. 4, no. 3, pp. 338–342, 2012.

[17] S. C. Joshi, R. Mathur, A. Gajraj, and T. Sharma, "Influence of methyl parathion on reproductive parameters in male rats," *Environmental Toxicology and Pharmacology*, vol. 14, no. 3, pp. 91–98, 2003.

[18] S. C. Joshi, B. Bansal, and N. D. Jasuja, "Evaluation of reproductive and developmental toxicity of cypermethrin in male albino rats," *Toxicological and Environmental Chemistry*, vol. 93, no. 3, pp. 593–602, 2011.

[19] P. Kapoor, S. C. Joshi, R. V. Singh, and N. Fahmi, "Antispermatogenic activity of later transition metal complexes of Benzothiazolines derived from substituted coumarins in male rats," *World Journal of Pharmaceutical Research*, vol. 5, no. 2, pp. 1056–1068, 2016.

# Online Solid-Phase Extraction LC-MS/MS: A Rapid and Valid Method for the Determination of Perfluorinated Compounds at Sub ng·L$^{-1}$ Level in Natural Water

**Salvatore Barreca, Maddalena Busetto, Matteo Vitelli, Luisa Colzani, Laura Clerici, and Pierluisa Dellavedova**

*Agenzia Regionale per la Protezione dell'Ambiente della Lombardia (ARPA Lombardia), U.O.C. Laboratorio di Milano, Sede laboratoristica di Monza e Brianza via Solferino N° 16, 20900 Monza, MB, Italy*

Correspondence should be addressed to Salvatore Barreca; salvatore.barreca@unipa.it

Academic Editor: Amit Bhatnagar

In this research paper, we report a method able to detect perfluorinated compound at pg·L$^{-1}$ levels in superficial and underground water samples. An online solid phase extraction HPLC-MS/MS method was developed for the analysis of 12 perfluoroalkyl acids (PFASs). The method is based on injection of 5000 $\mu$L of water sample in SPE online WAX column followed by chromatography separation and mass spectrometry determination. SPE online elution was performed by water + 0.05% $NH_4OH$ and methanol + 0.05% $NH_4OH$, while separation of target compounds was achieved within 10 min by Gemini chromatographic column operating from 1 to 12 pH range and using a mixture of water-methanol + 0.05% $NH_4OH$. Sub ng·L$^{-1}$ method detection limits (from 0.2 to 5 ng·L$^{-1}$), linearity (from 0.2 to 250 ng·L$^{-1}$), accuracy (from 80 to 120%), and precision (RSD less than 15% at LOQs levels) were achieved. The method is capable of measuring PFAS at trace levels, but above all it can reach the limit of 200 pg·L$^{-1}$ required by European regulation for PFOS determination in surface and underground waters. The method was validated for quantitative analysis of PFASs in real water samples.

## 1. Introduction

In the last decades, a wide variety of different pollutants have been found in several environment samples [1–3]. The research community has made significant efforts in order to develop analytical techniques capable of detecting these pollutants at trace levels [4–6].

The term "perfluorinated compounds" (PFCs) is referred to a class of organic substances in which all the hydrogens of the hydrocarbon backbones are substituted with fluorine atoms [7].

PFCs have been synthesized and commercialized since the 1960s and are used in numerous industrial and consumer products especially for intermediates and/or additives in the synthesis of certain fluorine compounds [8]. Such compounds are widely used in various domestic and industrial products including adhesives, polishes, paints, waxes, textiles stain/water/grease repellents in carpets and clothing, or in cooking utensils such as nonstick coatings [9–11].

PFC's extensive uses may depend on whether the fluorine-carbon bonds are extremely stable, and it should imply very high thermal and chemical stability [7]. Due to these properties, PFCs are classified as persistent pollutants, and some of these substances bioaccumulate in the environment. In addition, former research suggests that impact of exposure from consumer products is usually lower than impact of exposure in contaminated drinking water or contaminated food, such as fish [12, 13].

Although several PFAs such as perfluorooctane sulfonate (PFOS), perfluorooctanoic acid (PFOA), perfluorohexane sulfonate (PFHxS), and perfluorononanoic acid (PFNA) have been more widely studied, research has not yet determined the health effects and mechanisms resulting from human exposure to PFAs at levels typically found in water and food [14, 15].

Generally, PFAs accumulate in the human body, and their levels decrease slowly over time.

Some studies in humans have shown that certain PFAs may have an impact on the developing fetus and child and decrease fertility, interfere with the natural hormones of the body, increase cholesterol, affect the immune system, and increase cancer risk [16, 17].

Further studies are required to determine whether the same effects are caused by the same mechanism of action.

In the last ten years, the scientific and the health communities have focused interest in perfluorinated compounds by investigating in particular perfluorooctane sulfonate (PFOS) and perfluorooctanoic acid (PFOA). The most common PFCs have been identified at considerable levels in various environmental matrices, predominantly in aqueous environment [18, 19]. Indeed, these compounds may present a potential risk to humans via drinking water and to wildlife via receiving waters.

More recently, detection of PFAs relevant levels in the area of Veneto, Italy, have highlighted the necessity of undertaking studies evaluating the concentration of perfluorinated compounds in different water matrices.

In this respect, several Environmental Protection Agency of Italy (ARPA) are endeavoring to develop and validate methods able to assess PFAs concentrations in environmental water samples.

Generally, the analysis of PFAs at trace levels in different environmental matrices represents a challenge in analytical chemistry. Indeed, quantification of PFAs at concentration of $ng{\cdot}L^{-1}$ or $pg{\cdot}L^{-1}$ in drinking water requires extraction techniques and cleanup procedures to provide enrich the sample and reduce matrix interferences. In this respect, the most effective analytical technique for the determination of perfluorinated alkyl acids (PFAs) in drinking water consists of extraction procedures by offline solid-phase extraction (SPE) cartridge containing polystyrene divinylbenzene followed by elution in methanol, concentration to dryness with nitrogen, injection and analysis using liquid chromatography/tandem mass spectrometry (LC/MS/MS) [20–22].

Unfortunately, due to several challenges such as matrix effects and cross-contamination, classic analytical techniques cannot detect PFAs at levels below $10\,ng{\cdot}L^{-1}$, especially for PFOS and PFAs at long chain [23, 24]. For this reason, it is necessary to develop and validate an analytical procedure capable of improving detection limits for all PFAs.

Different research groups have endeavored to develop innovative online SPE techniques to detect PFASs in water samples.

Within this context, several methods employ formic acid or other substances to improve LODs, but at ultratrace level analysis, the risk of cross-contaminations caused by external additions such as acid or other substances represent a critical point.

Recently, Dasu et al. have reported an ultrasensitive method for the analysis of perfluorinated alkyl acids in drinking water using an SPE online column coupled with high-performance liquid chromatography-tandem mass spectrometry. This method uses water and methanol containing 0.1% of $NH_4OH$ [25] as mobile phases. Unfortunately, eluents at pH value above 8 are not totally compatible with classic C18 or polymeric columns employed in chromatographic separations. To the best of our knowledge, the best LOQ value for PFOS obtained by using SPE online technique values were reported at $1\,ng{\cdot}L^{-1}$ [25]. The research centre at ARPA Lombardia pioneers the monitoring of activities on superficial water samples on the Italian territory. It has recently developed an analytical method based on solid phase extraction online coupled with HPLC/MS-MS to detect PFCs concentration in natural water samples by using a C18 column with pH working range from 1 to 12.

## 2. Experimental

*2.1. Chemicals and Materials.* Native standard solutions of the individual analytes in methanol were purchased from Wellington Laboratories. $^{13}C$ mass-labeled methanolic solutions of each analyte were used as internal standards at $25\,ng{\cdot}L^{-1}$, while $^{13}C$ M3PFBA, $^{13}C$ MPFDA, $^{13}C$ MPFOA, and $^{13}C$ MPFOS were used as surrogate standards in order to check percentage recovery by final concentration of surrogate standard solution at $25\,ng{\cdot}L^{-1}$. All mass-labeled standards were purchased from Wellington Laboratories.

Stock solutions were prepared in methanol and were stored at −18°C in amber glassware. Methanol HPLC grade was purchased from Sigma-Aldrich. High purity water (Milly-Q water grade) was prepared using a Millipore Milli-Q purification system. All standard solutions were purchased at concentration of $2000\,mg{\cdot}L^{-1}$. Mass-labeled solution and calibration solutions were prepared by serial dilution of stock solutions in tap water (S. Anna). For the preconcentration and cleanup experiments, an Oasis WAX 150 mg/6 mL was used for SPE online, relying upon the good recoveries reported in existing literature by using Oasis WAX-SPE cartridges including short-chain (C4–C6) compounds.

Two LC Column Poroshell 120 EC-C18 $3 \times 50$ mm, 2.7 $\mu$m, were used to eluent cleanup. $NH_4OH$ concentrate solution was purchased from Sigma-Aldrich.

A Gemini C18 3 $\mu$m, $100 \times 3$ mm, made Phenomenex operating a pH value ranged 1–12 was used for chromatographic separation column. Samples were collected in a propylene bottle of 0.250 mL and refrigerated at a temperature of 4°C until analyses.

*2.2. Solvent Cleanup.* Methanol and Milli-Q water were precleaned by connecting separate online scrubber columns for each mobile phase. In this context, two columns were used. The first one was positioned before injection valve to clean up the eluent used for SPE online, and the second one was used to clean up the eluent employed in chromatographic separation.

*2.3. Liquid Chromatograph and Mass Spectrometric Instruments.* Analysis were performed using high-performance liquid chromatograph (HPLC) consisting of binary and quaternary pump Nexera LC pump, a DGU-20A 5R degassing unit, a SIL-30AC autosampler equipped with a 5000 μL loop, a CTO/20AC thermostated column compartment, and a CBM-20A module controller.

The Nexera LC system was coupled with a 6500 QTrap mass spectrometer (Sciex), equipped with a Turbo V® interface by an ESI probe. Elution conditions and mass spectrometry optimal parameters were reported in Tables 1–3, respectively. About SPE online column, load processes occur from 0.8 to 3.5 min, while elution in analytical column occurs from 3.5 to 10 min.

In order to minimize MS source contamination, the first 3.5 min and the last 3.5 min of the chromatographic run have been diverted to waste by means of a two-position six-port valve installed before the mass spectrometer.

Analysis was performed by multiple reaction monitoring (MRM) in negative ionization mode by using m/z, declustering potential, and collision energy as reported in Table 4.

Moreover, analytes were identified comparing both their retention times (RT) with the RT of the internal standards and by qualification ions.

## 3. Results and Discussion

Sample preparation and analytical operating conditions play an important role to ensure accuracy, precision, and repeatability recovery and yield the finest possible LOD.

Optimization of analytical separation was carried out by direct injection of standard solutions with a 100 μL loop, while the separation gradient elution for SPE online were investigated by injection of 2000 and 5000 μL, respectively. The best results, in terms of S/N and standard area value, were obtained by injection of 5000 μL. In detail, for standard at 0.2 ng·L$^{-1}$, the S/N for 2000 and 5000 μL was 2.0 and 7.5, respectively.

Moreover, laboratory has valuated different SPE online columns as reported in Section 2.1. Best results were obtained by using an Oasis WAX SPE online.

In the first step of the optimization process, the influence of eluent pH on elution procedures of PFAs from SPE online using Oasis WAX cartridge was jointly investigated. Best conditions were obtained by using $NH_4OH$ 0.05% in water-methanol eluent. Unfortunately, the most C 18 and polyvinyl column employed in HPLC operate a pH range 2–8 and mobile phase used to eluate total PFAS (C-4/C-12) from Oasis WAX cartridge has a pH value above 8. Within this context, to overcome such challenges, a Gemini C18 column able to operate at pH near 12 was used.

A representative chromatogram for PFASs by using developed analysis conditions is represented in Figure 1.

An acceptable chromatographic separation was achieved for 12 PFAs investigation, while best ionization responses were obtained for perfluoroalkyl acid compounds.

Calibration solutions were run at the beginning of each analytical session to ensure instrument response. The percent relative standard deviation between different replicate levels of calibration curves ranged from 0.5% to 10%.

Prior to analysis and every ten samples, blank solutions were analyzed, and for each analyte, the obtained signal was

Table 1: Elution conditions employed in SPE online column.

| Time (min) | Total flow (mL·min$^{-1}$) | % eluent A1 water 0.5% of $NH_4OH$ | % eluent B1 methanol 0.5% of $NH_4OH$ | Valve |
|---|---|---|---|---|
| 0.0 | 0.8 | 50 | 50 | A: exhaust |
| 3.5 | 0.8 | 50 | 50 | B: mass instrument |
| 6.5 | 0.8 | 0 | 100 | B: mass instrument |
| 8.0 | 0.8 | 0 | 100 | B: mass instrument |
| 8.5 | 0.8 | 0 | 100 | B: mass instrument |
| 8.6 | 0.8 | 50 | 50 | B: mass instrument |
| 9.5 | 0.8 | 50 | 50 | B: mass instrument |
| 10 | 0.8 | 50 | 50 | B: mass instrument |
| 12 | End | | | A: exhaust |

Table 2: Elution conditions employed in Gemini HPLC Column.

| Time (min) | Flow (mL·min$^{-1}$) | % eluent A Milli-Q water loading | % eluent B water/methanol 50 : 50 w/w water | Valve |
|---|---|---|---|---|
| 0.0 | 1.6 | 100 | 0 | A: exhaust |
| 3.5 | 1.6 | 100 | 0 | B: column |
| 6.5 | 1.6 | 100 | 0 | B: column |
| 8.0 | 1.6 | 100 | 0 | B: column |
| 8.5 | 1.6 | 0 | 100 | A: discharge |
| 8.6 | 1.6 | 0 | 100 | A: discharge |
| 9.5 | 1.6 | 0 | 100 | A: discharge |
| 10.0 | 1.6 | 100 | 0 | A: exhaust |
| 12.0 | End | | | |

Table 3: Parameter used in ESI-MS/MS.

| Source | Unit | Value |
|---|---|---|
| Curtain gas (CUR) | psi | 25 |
| Collision gas | — | High |
| Ion spray voltage (IS) | V | −4500 |
| Temperature TEM | °C | 450 |
| Ion source gas (GS1) | psi | 45 |
| Ion source gas (GS2) | psi | 55 |
| Entrance potential (EP) | V | −10 |
| Collision exit potential (CXP) | V | −15 |

TABLE 4: m/z, declustering potential and collision energy used.

| Analyte | Q1 precursion ion $[M\text{-}h]^-$ (m/z) | Q3 product ion (m/z) | Declustering potential (DP) | Q2 collision energy (CE) |
|---|---|---|---|---|
| PFBA-1 perfluorobutanoic acid | 213 | 169.1 | −10 | −14 |
| PFPeA-1 perfluoropentanoic acid | 263 | 219 | −10 | −12 |
| PFBS-1 perfluorobutane sulfonic acid | 299 | 80 | −40 | −70 |
| PFBS-2 perfluorobutane sulfonic acid | 299 | 98.8 | −40 | −40 |
| PFHxA-1 perfluorohexanoic acid | 313 | 269 | −10 | −13 |
| PFHxA-2 perfluorohexanoic acid | 313 | 119 | −10 | −30 |
| PFHpA-1 perfluoroheptanoic acid | 363 | 319 | −10 | −15 |
| PFHpA-2 perfluoroheptanoic acid | 363 | 169 | −10 | −20 |
| PFHxS-1 perfluorohexansulfonic acid | 399 | 79.9 | −40 | −100 |
| PFHxS-2 perfluorohexansulfonic acid | 399 | 98.9 | −40 | −76 |
| PFOA-1 perfluorooctanoic acid | 413 | 369 | −10 | −15 |
| PFOA-2 perfluorooctanoic acid | 41o3 | 169 | −10 | −20 |
| PFNA-1 perfluorononanoic acid | 463 | 419 | −10 | −15 |
| PFNA-2 perfluorononanoic acid | 463 | 168.9 | −10 | −25 |
| PFOS-1 perfluorooctanesulfonic acid | 499 | 80 | −60 | −120 |
| PFOS-2 perfluorooctanesulfonic acid | 499 | 99 | −60 | −100 |
| PFDA-1 perfluorodecanoic acid | 513 | 469.1 | −10 | −15 |
| PFDA-2 perfluorodecanoic acid | 513 | 218.9 | −10 | −22 |
| PFUnDA-1 perfluoroundecanoic acid | 563 | 519 | −10 | −15 |
| PFUnDA-2 perfluoroundecanoic acid | 563 | 269 | −60 | −140 |
| PFDoDA-1 perfluorododecanoic acid | 613 | 569 | −10 | −15 |
| PFDoDA-2 perfluorododecanoic acid | 613 | 269 | −10 | −20 |
| MPFBA-labeled perfluorobutanoic acid | 217 | 171.9 | −10 | −14 |
| $M_5$PFPeA-labeled perfluoropentanoic acid | 268 | 233 | −10 | −12 |
| $M_3$PFBS-labeled perfluorobutane sulfonic acid | 302 | 99 | −40 | −70 |
| $M_5$PFHxA-labeled perfluorohexanoic acid | 318 | 273 | −10 | −13 |
| $M_4$PFHpA-labeled perfluoroheptanoic acid | 367 | 322 | −10 | −20 |
| $M_3$PFHxS-labeled perfluorohexansulfonic acid | 402 | 99 | −40 | −100 |
| $M_8$PFOA-labeled perfluorooctanoic acid | 421 | 376 | −10 | −14 |
| $M_8$PFOS-labeled perfluorooctanesulfonic acid | 507 | 99 | −60 | −100 |
| $M_9$PFNA-labeled perfluorononanoic acid | 472 | 427 | −10 | −15 |
| $M_6$PFDA-labeled perfluorodecanoic acid | 519 | 474 | −10 | −15 |
| $M_7$PFUdA-labeled perfluoroundecanoic acid | 570 | 525 | −10 | −15 |
| MPFDoDA-labeled perfluorododecanoic acid | 615 | 570 | −10 | −15 |

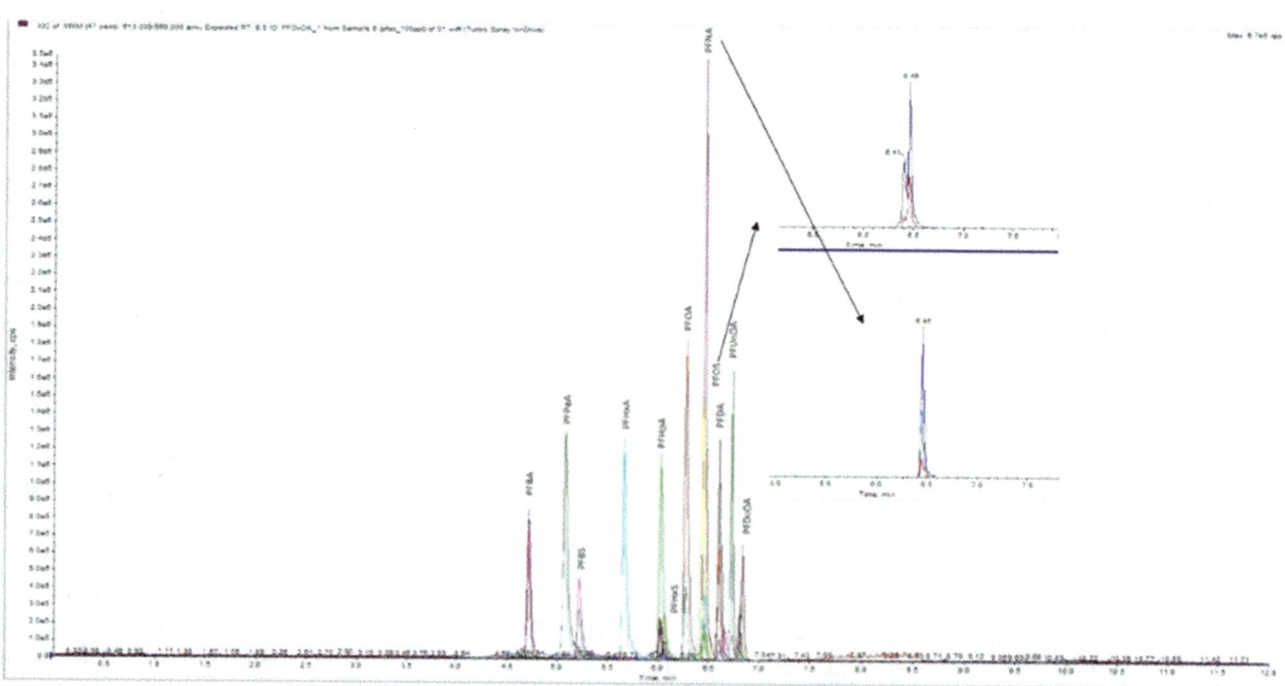

FIGURE 1: Chromatogram of PFAs solutions at 50 ng·L$^{-1}$.

<1/2 LOQ. According to ISO 17025 criteria, an independent solution was used to check standards used in calibration curves.

Moreover, to check possible instrumental drift, low and middle level calibration solutions were run after every ten sample analyses to verify the previously established calibration curve. Bias of each quality control ranged from ±20 to 30% for low and middle levels, respectively.

The method was validated in terms of detection limits, accuracy, and precision. Analyses were conducted by using real superficial water samples.

The limits of quantification (LOQs) were estimated by 10-time standard deviation calculated at first level of calibration curve prepared in real superficial water sample.

Table 5 summarizes the results obtained for these performance parameters. In all analyses, analyte recovery was checked by mass-labeled standard. Accuracy and repeatability were calculated on data set of six analyses conducted on water-spiked samples.

All data obtained were subject to different statistical tests. Normal distribution was evaluated by using the Shapiro–Wilk test, while outlier data were checked by the Dixon test.

Linearity ranges were obtained from 0.2–250 to 5–250 $ng{\cdot}L^{-1}$ depending on the different analytes considered.

PFBA shows a linear range from 5 to 250 $ng{\cdot}L^{-1}$. Due to its short alkyl chain, chemical-physical properties of PFBA are very different compared to other PFASs compounds. In this regard, PFBA is adsorbed less than other PFASs in SPE online, and in this context, the first level corresponding to 5 $ng{\cdot}L^{-1}$. In any cases, by comparing this method with other direct injection or SPE methods, a good detection limit for PFBA was achieved. For the most PFASs, good correlation can be obtained in calibration curve ranged from 1 to 250 $ng{\cdot}L^{-1}$.

For PFOS, one of the most important analytes monitored, linearity of method was evaluated at low levels over concentration range from 0.2 to 250 $ng{\cdot}L^{-1}$ finding a linear response (correlation coefficients in Table 5). Generally, in the literature, detection limit for PFOS, by direct injection or SPE online, ranged from 5 to 10 $ng{\cdot}L^{-1}$, and only few cases reported LOD at the level of 1 $ng{\cdot}L^{-1}$ [25, 26]. In this context, quantification limits of 0.2 $ng{\cdot}L^{-1}$ without sample preparations can be considered an interesting analytical result.

Calibration curves for PFOS in the range from 0.2 to 250 $ng{\cdot}L^{-1}$ for quantitative and qualitative ions (499 m/z and 80 m/z resp.) are reported in Figure 2.

Moreover, accuracy was evaluated by performing recovery tests in terms of percentage recovery by using spiked samples (tap and superficial water) fortified at various levels (from 10 to 50 $ng{\cdot}L^{-1}$), and the results are shown in Figures 3(a)–3(c) for low-, middle-, and high-level (10, 25, and 50 $ng{\cdot}L^{-1}$) concentrations, respectively.

For all analyzed samples, analytical recovery, determined by surrogate standard, ranged from 80 to 120%.

For all PFASs, recovery ranged from 80 to 120% in good agreement with accuracy required by the regulations on the determination of PFASs in water which is a satisfactory result.

To determine repeatability of method, different experiments were conducted in replicated analysis at low level, middle levels, and high levels using spiked-water samples with native standard solution at different concentrations. At LOQ levels, RSD were less than 15% (Table 5), while at other levels, RSD were less than 10%.

Moreover, in order to verify the robustness of the analytical method, the matrix effect was investigated by comparing calibration curves for each compound by using tap and superficial water samples. Three replicates for each level were used to obtain calibration curves, and robustness was studied by ME (matrix effect) index evaluated according to the following equation:

$$\mathrm{ME}\% = \frac{m(\text{sample})}{m(\text{standard})} * 100 - 100, \quad (1)$$

where $m(\text{sample})$ is the slope of the calibration curve obtained in matrix and $m(\text{standard})$ is the slope of the calibration curve in solvent; accordingly, ME values <0 indicate ion suppression, whereas ME values >0 indicate ion enhancement.

The ME index calculated for each analyte in tap water and real superficial water is reported in Table 6.

As reported in the literature [27], ME index value in the range of ±20 are considered to have a negligible influence on the performance of an analytical method. For the proposal method, for all analytes, ME index is less than ±20. In this regard, the method developed can be considered robust.

## 4. Conclusions

This paper reports the development, validation, and application of a new SPE online HPLC-MS/MS method able to detect perfluorinated compound in water samples.

The proposed method provides an easy, rapid, and without samples preconcentration application to analyze PFASs. The proposed method is accurate and precise, shows lower limits of quantification and presents a good linearity at the range of $pg{\cdot}L^{-1}$-$ng{\cdot}L^{-1}$ levels.

The method can measure PFASs at the trace levels in water samples and attain most of the adopted quality standards except for European EQS for PFOS (0.65 ng/L) which is considered a challenging limit for every LC-MS analytical method. Furthermore, compared with other SPE online analytical methods, the new analytical method achieves a new real limit for PFOS (200 $pg{\cdot}L^{-1}$). Moreover, working under pH conditions above 10, it is better to use a column operating at pH range between 1 and 12.

Finally, a matrix effect depending on superficial water was not detected.

The optimized method was successfully applied to the quantitation of PFASs in natural water samples, and the value of method was successfully tested by using spiked samples.

Table 5: Limits of quantification (LOQs), linearity range, R2, accuracy, and repeatability parameters evaluated at LOQ.

| Compound | LOQ (ng·L$^{-1}$) | Linear range (ng·L$^{-1}$) | $R^2$ | Accuracy (%) | Repeatability CV% |
|---|---|---|---|---|---|
| PFBA | 5 | 5–200 | 0.999 | 101.1 | 10.7 |
| PFPeA | 1 | 1–250 | 0.999 | 91.7 | 10.6 |
| PFBS | 1 | 1–250 | 0.999 | 95.6 | 11.5 |
| PFHxA | 1 | 1–250 | 0.999 | 97.4 | 8.28 |
| PFHpA | 1 | 1–250 | 0.999 | 97.2 | 10.4 |
| PFHxS | 1 | 1–250 | 0.999 | 88.94 | 9.7 |
| PFOA | 1 | 1–250 | 0.999 | 99.9 | 11.5 |
| PFNA | 1 | 1–250 | 0.999 | 93.0 | 13.8 |
| PFOS | 0.2 | 0.2–250 | 0.999 | 85.0 | 12.4 |
| PFDA | 5 | 5–250 | 0.999 | 81.0 | 12.1 |
| PFUnDA | 5 | 5–250 | 0.999 | 107.2 | 11.3 |
| PFDoDA | 5 | 5–250 | 0.999 | 107.4 | 9.5 |

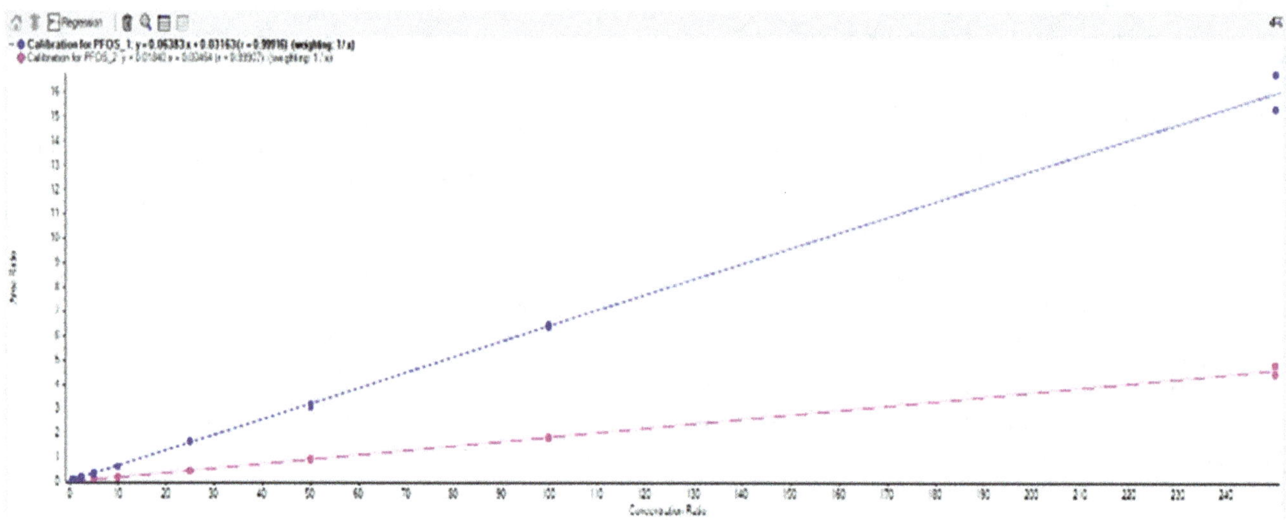

Figure 2: PFOS calibration curve in the range from 0.2 to 250 ng·L$^{-1}$. Dotted blue line represents m/z transition 499 ⟶ 80; dotted red line represents m/z transition 499 ⟶ 99.

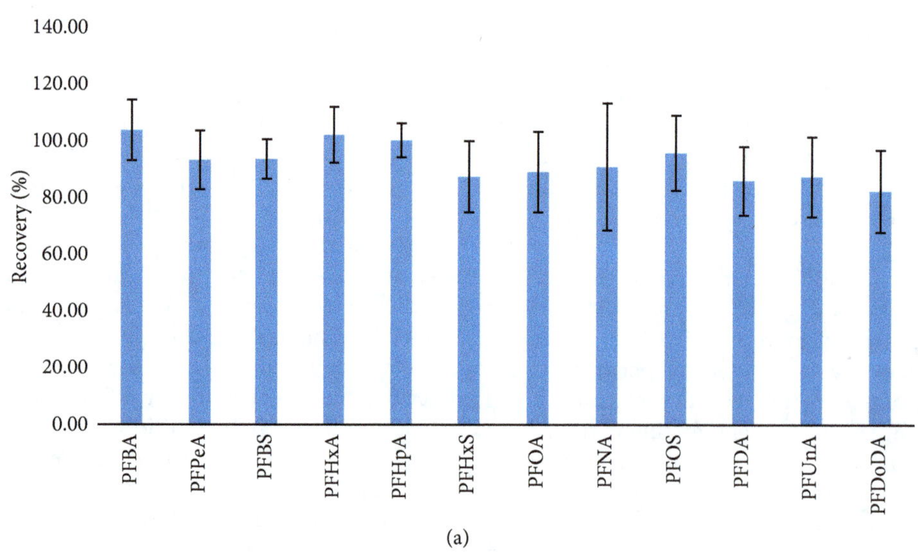


(a)

Figure 3: Continued.

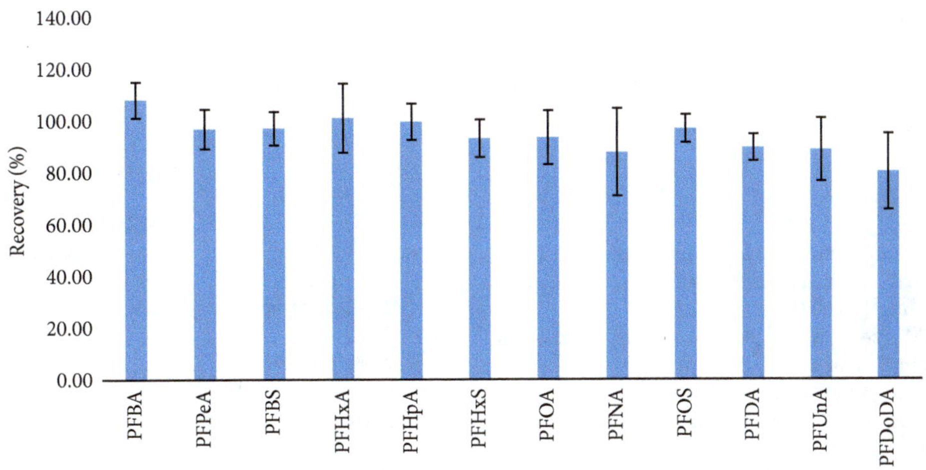


(b)

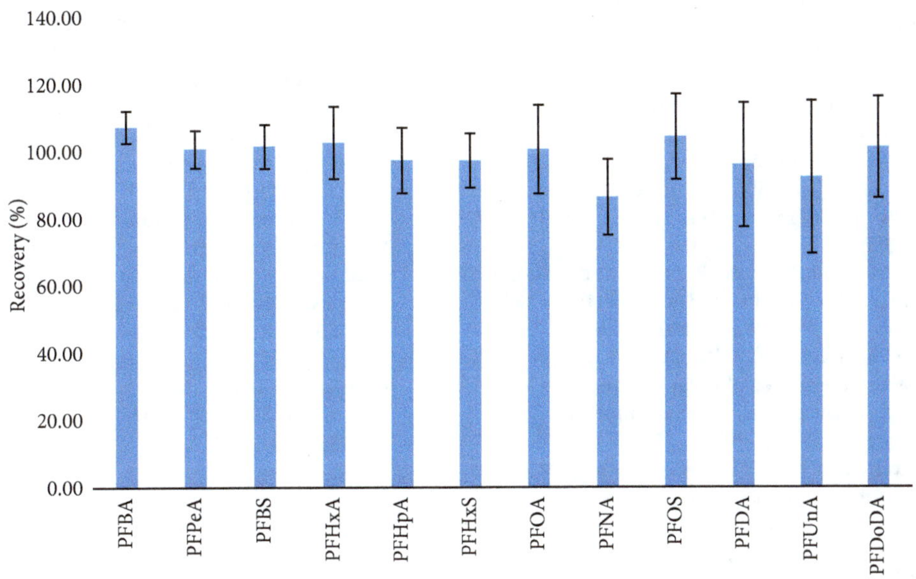


(c)

FIGURE 3: (a) Percentage recovery for all analytes at 10 ng·L$^{-1}$. (b). Percentage recovery for all analytes at 25 ng·L$^{-1}$. (c). Percentage recovery for all analytes at 50 ng·L$^{-1}$.

TABLE 6: Matrix effect index for all analytes.

| Analyte | ME index |
|---|---|
| PFBA | −1.4 |
| PFPeA | 4.5 |
| PFBS | 1.7 |
| PFHxA | −2.2 |
| PFHpA | −0.67 |
| PFHxS | 5.1 |
| PFOA | −0,20 |
| PFNA | −1.7 |
| PFOS | 0.08 |
| PFDA | −0.64 |
| PFUnDA | −2.4 |
| PFDoDA | −6.8 |

## Disclosure

The findings and conclusions of this article are solely the responsibility of the authors and do not represent the official views of ARPA Lombardia.

## Conflicts of Interest

The authors declare that they have no conflicts of interest.

## Acknowledgments

The study was supported by Regione Lombardia financial source.

## References

[1] R. Kallenborn, *Perfluorinated Alkylated Substances (PFAS) in the Nordic Environment*, Nordic Council of Ministers, Copenhagen, Denmark, 2004.

[2] S. Barreca, A. Mazzola, S. Orecchio, and N. Tuzzolino, "Polychlorinated biphenyls in sediments from Sicilian coastal area (Scoglitti) using automated soxhlet, GC-MS, and principal component analysis," *Polycyclic Aromatic Compounds*, vol. 34, no. 3, pp. 237–262, 2014.

[3] E. Magi, M. Di Carro, C. Mirasole, and B. Benedetti, "Combining passive sampling and tandem mass spectrometry for the determination of pharmaceuticals and other emerging pollutants in drinking water," *Microchemical Journal*, vol. 136, pp. 56–60, 2018.

[4] V. Paiano, E. Fattore, A. Carra, C. Generoso, R. Fanelli, and R. Bagnati, "Liquid chromatography-tandem mass spectrometry analysis of perfluorooctane sulfonate and perfluorooctanoic acid in fish fillet samples," *Journal of Analytical Methods in Chemistry*, vol. 2012, Article ID 719010, 5 pages, 2012.

[5] S. Barreca, S. Orecchio, and A. Pace, "Photochemical sample treatment for extracts clean up in PCB analysis from sediments," *Talanta*, vol. 103, pp. 349–354, 2013.

[6] R. Loos, B. M. Gawlik, G. Locoro, E. Rimaviciute, S. Contini, and G. Bidoglio, "EU-wide survey of polar organic persistent pollutants in European river waters," *Environmental Pollution*, vol. 157, no. 2, pp. 561–568, 2009.

[7] T. Stahl, D. Mattern, and H. Brunn, "Toxicology of perfluorinated compounds," *Environmental Sciences Europe*, vol. 23, p. 38, 2011.

[8] M. Fricke, "U. Lahl Risikobewertung von perfluortensiden als beitrag zur aktuellen diskussion zum REACH-Dossier der EU-Kommission," *Umweltwissenschaften und Schadstoff-Forschung*, vol. 17, no. 1, pp. 36–49, 2005.

[9] H. Fromme, M. Schlumer, J. Ungewiss, and E. Roscher, "Materialien zur Umweltmedizin-Umweltmedizinische Bedeutung perfluorierter Kohlenwasserstoffe (PFC)," in *Bayerisches Landesamt für Gesundheit und Lebensmittelsicherheit (Hrsg.), Band 16 der Schriftenreihe*, Bavarian State Office for Health and Food Safety, Bavaria, Germany, 2006.

[10] J. M. O'Brien, A. C. Carew, S. Chu, R. J. Letcher, and S. W. Kennedy, "Perfluorooctane sulfonate (PFOS) toxicity in domestic chicken (Gallus gallus domesticus) embryos in the absence of effects on peroxisome proliferator activated receptor alpha (PPARα)-regulated genes," *Comparative Biochemistry and Physiology Part C: Toxicology & Pharmacology*, vol. 149, no. 4, pp. 524–530, 2009.

[11] F. Xiao, T. R. Halbach, M. F. Simcik, and J. S. Gulliver, "Input characterization of perfluoroalkyl substances in wastewater treatment plants: source discrimination by exploratory data analysis," *Water Research*, vol. 46, no. 9, pp. 3101–3109, 2012.

[12] J. Holzer, O. Midasch, K. Rauchfuss et al., "Biomonitoring of perfluorinated compounds in children and adults exposed to perfluorooctanoate-contaminated drinking water," *Environmental Health Perspectives*, vol. 116, no. 5, pp. 651–657, 2008.

[13] S. P. J. Van Leeuwen, C. P. Swart, I. Van der Veen, and J. De Boer, "Significant improvements in the analysis of perfluorinated compounds in water and fish: results from an interlaboratory method evaluation study," *Journal of Chromatography A*, vol. 1216, no. 3, pp. 401–409, 2009.

[14] S. T. Wolf and W. K. Reagen, "Method and validation for the analysis of perfluorinated compounds in water by pre-sampling isotope dilution-direct injection-LC/MS/MS," *Analytical Methods*, vol. 5, no. 10, pp. 2444–2454, 2013.

[15] A. K. Rosenmai, C. Taxvig, T. Svingen et al., "Fluorinated alkyl substances and technical mixtures used in food paper-packaging exhibit endocrine-related activity in vitro," *Andrology*, vol. 4, no. 4, pp. 662–672, 2016.

[16] A. P. Starling, J. L. Adgate, R. F. Hamman et al., "Perfluoroalkyl substances during pregnancy and offspring weight and adiposity at birth: examining mediation by maternal fasting glucose in the healthy start study," *Environmental Health Perspectives*, vol. 125, no. 6, 2017.

[17] G. W. Olsen, D. C. Mair, C. C. Lange et al., "Per-and polyfluoroalkyl substances (PFAS) in American Red Cross adult blood donors, 2000–2015," *Environmental Research*, vol. 157, pp. 87–95, 2017.

[18] S. P. Van Leeuwen, A. Kärrman, B. Van Bavel, J. De Boer, and G. Lindström, "Struggle for quality in determination of perfluorinated contaminants in environmental and human samples," *Environmental Science & Technology*, vol. 40, no. 24, pp. 7854–7860, 2006.

[19] J. Thompson, G. Eaglesham, and J. Mueller, "Concentrations of PFOS, PFOA and other perfluorinated alkyl acids in Australian drinking water," *Chemosphere*, vol. 83, no. 10, pp. 1320–1325, 2011.

[20] X. Zhao, Y. Cai, F. Wu, Y. Pan, H. Liao, and B. Xu, "Determination of perfluorinated compounds in environmental water samples by high-performance liquid chromatography-electrospray tandem mass spectrometry using surfactant-coated $Fe_3O_4$ magnetic nanoparticles as adsorbents," *Microchemical Journal*, vol. 98, no. 2, pp. 207–214, 2011.

[21] S. Valsecchi, S. Polesello, M. Mazzoni, M. Rusconi, and M. M. Petrovic, "On-line sample extraction and purification for the LC–MS determination of emerging contaminants in environmental samples," *Trends in Environmental Analytical Chemistry*, vol. 8, pp. 27–37, 2015.

[22] K. Bielicka-Daszkiewicz, "Extraction techniques based on solid state and connected with liquid chromatography," *Journal of Liquid Chromatography & Related Technologies*, vol. 39, no. 10, pp. 477–487, 2016.

[23] J. L. Wilkinson, P. S. Hooda, J. Swinden, J. Barker, and S. Barton, "Spatial distribution of organic contaminants in three rivers of Southern England bound to suspended particulate material and dissolved in water," *Science of the Total Environment*, vol. 593-594, pp. 487–497, 2017.

[24] P. Zhao, X. Xia, J. Dong et al., "Short-and long-chain perfluoroalkyl substances in the water, suspended particulate matter, and surface sediment of a turbid river," *Science of the Total Environment*, vol. 568, pp. 57–65, 2016.

[25] K. Dasu, S. F. Nakayama, M. Yoshikane, M. A. Mills, J. M. Wright, and S. Ehrlich, "An ultra-sensitive method for the analysis of perfluorinated alkyl acids in drinking water

using a column switching high performance liquid chromatography tandem mass spectrometry," *Journal of Chromatography A*, vol. 1494, pp. 46–54, 2017.

[26] N. P. H. Lien, S. Fujii, S. Tanaka et al., "Perfluorinated substances in tap water of Japan and several countries and their relationship to surface water contamination," *Environmental Engineering Research*, vol. 43, pp. 611–618, 2006.

[27] L. Ciofi, L. Renai, D. Rossini et al., "Applicability of the direct injection liquid chromatographic tandem mass spectrometric analytical approach to the sub-ng $L^{-1}$ determination of perfluoro-alkyl acids in waste, surface, ground and drinking water samples," *Talanta*, vol. 176, pp. 412–421, 2018.

4

# Adsorptive Removal of Copper by Using Surfactant Modified Laterite Soil

**Tien Duc Pham,[1] Hoang Hiep Nguyen,[1] Ngoc Viet Nguyen,[1] Thanh Tu Vu,[1] Thi Ngoc Mai Pham,[1] Thi Hai Yen Doan,[1] Manh Ha Nguyen,[1] and Thi Mai Viet Ngo[2]**

[1] *Faculty of Chemistry, VNU-University of Science, Vietnam National University-Hanoi, 19 Le Thanh Tong, Hoan Kiem, Hanoi 10000, Vietnam*
[2] *Thai Nguyen University of Education, Thai Nguyen University, 20 Luong Ngoc Quyen, Quang Trung, Thai Nguyen, Vietnam*

Correspondence should be addressed to Tien Duc Pham; tienduchphn@gmail.com

Academic Editor: Khalid Z. Elwakeel

Removal of copper ion ($Cu^{2+}$) by using surfactant modified laterite (SML) was investigated in the present study. Characterizations of laterite were examined by X-ray diffraction (XRD), Fourier transform infrared spectroscopy (FT-IR), inductively coupled plasma mass spectrometry (ICP-MS), and total carbon analysis. The optimum conditions for removal of $Cu^{2+}$ by adsorption using SML were systematically studied and found as pH 6, contact time 90 min, adsorbent dosage 5 mg/mL, and ionic strength 10 mM NaCl. The equilibrium concentration of copper ions was measured by flame atomic absorption spectrometry (F-AAS). Surface modification of laterite by anionic surfactant sodium dodecyl sulfate (SDS) induced a significant increase of the removal efficiency of $Cu^{2+}$. The surface modifications of laterite by preadsorption of SDS and sequential adsorption of $Cu^{2+}$ were also evaluated by XRD and FT-IR. The adsorption of $Cu^{2+}$ onto SML increases with increasing NaCl concentration from 1 to 10 mM, but at high salt concentration this trend is reversed because desorption of SDS from laterite surface was enhanced by increasing salt concentration. Experimental results of $Cu^{2+}$/SML adsorption isotherms at different ionic strengths can be represented well by a two-step adsorption model. Based on adsorption isotherms, surface charge effects, and surface modification, we suggest that the adsorption mechanism of $Cu^{2+}$ onto SML was induced by electrostatic attraction between $Cu^{2+}$ and the negatively charged SML surface and nonelectrostatic interactions between $Cu^{2+}$ and organic substances in the laterite.

## 1. Introduction

Removal of heavy metal ion from aqueous solution is important in environmental concern because heavy metal can induce serious problems to human health through the water resources. Therefore, so many projects focus on the removal of heavy metal ion to protect water resources [1–7]. Numerous treatment techniques have been used for heavy metal ion removal from the aquatic environment such as adsorption, ion exchange, coagulation/flocculation, chemical precipitation, photodegradation, and electrochemical oxidation [8–11]. Among these techniques, adsorption is one of the most common technologies for removing heavy metal ions [8, 10, 12–15]. Although activated carbon and activated metal oxides are effective adsorbent for the removal of heavy metal ions [2, 4, 14–17], a big challenge is costly. Thus, cheap adsorbents or/and natural adsorbents or modified cheap adsorbents are preferable in developing countries [5, 8, 12, 18]. To increase the removal efficiency of heavy metal ion by natural minerals, a modification of adsorbent surfaces is needed because the surface of such materials is nonactive. The modified solid adsorbents to remove metal ion was successfully investigated by many researches and,in these cases, not only adsorption capacity but also the adsorptive selectivity was significantly enhanced [19–21].

Copper, which is well-known heavy metal, is toxic and nonbiodegradable to aquatic ecosystems and living organisms [22]. Therefore, copper can accumulate in sediments and tissues of living organisms, and separation of copper ions is essential to discharge wastewater into environment. Recently, many studies investigated the removal of copper ion, Cu(II), by adsorption techniques using novel adsorbent

[3, 16, 22–25]. Adsorptive removal of Cu(II) by chemically modified adsorbent was investigated in previous studies [3, 26].

Laterite is a common soil in tropical countries. Laterite is also easily collected in Vietnam, meaning that laterite is very low cost. Basically, laterite has positive charge in the acid and neutral media so it has been widely used for the removal of many toxic anionic ions including anionic heavy metal [5, 27, 28]. Nevertheless, the removal of cations is rather hard due to the strong electrostatic repulsive force. Thus, in order to increase the removal efficiency of the heavy metal cations by adsorption technique, surface modification of laterite is necessary. Anionic surfactant such as sodium dodecyl sulfate (SDS) is an ecofriendly chemical that can be used in environmental remediation to remove both inorganic and organic pollutants [29]. However, SDS modified laterite to remove $Cu^{2+}$ has not been studied [5].

Adsorption is normally conducted under isothermal condition so that adsorption isotherms fitted by theoretical models are useful to better understand adsorption mechanisms and to explain the interactions between the laterite and copper ion. As for describing adsorption characteristics of heavy metal ions, Langmuir and Freundlich isotherm models are often discussed [30, 31]. However, Langmuir and Freundlich models cannot be applied for adsorption isotherms of surfactants. Thus, adsorption of inorganic pollutants onto surfactant modified laterite could not be fitted by Langmuir and Freundlich models. Fortunately, a two-step model presented by Zhu and Gu [32] with a general adsorption isotherm equation was successfully applied to various types of surfactants, polymers, and dyes adsorption isotherms for numerous systems [32–37]. Adsorptive removal of both organic and inorganic pollutants by using surfactant modified alumina was thoroughly studied by Pal and coworkers [2, 16, 38–41], indicating that surfactant modified solid adsorbent is a novel adsorbent. Nevertheless, the authors have not studied adsorption of Cu(II) on surfactant modified laterite.

In this study, we investigate adsorptive removal copper ion using surfactant modified laterite (SML). To the best of our knowledge, this is the first systematic study on the adsorption of copper ion from aqueous solution on SML. Characterizations of laterite and SML were determined by X-ray diffraction (XRD), inductively coupled plasma mass spectrometry (ICP-MS), Fourier transform infrared spectroscopy (FT-IR), and total carbon analysis. The optimum parameters for adsorptive removal copper ion by using SML are systematically studied. The surface modification before and after adsorption is also investigated by XRD and FT-IR. The adsorption mechanism is also proposed on the basis of adsorption isotherms and surface modification.

## 2. Experimental

*2.1. Materials.* Raw laterite was collected from a local place of Thach That, Hanoi, Vietnam. The laterite was treated before measurements as follows: the laterite was washed various times by ultrapure water to reach neutral pH and then dried at 110°C. The treated laterite was cooled in a desiccator at room temperature and stored in a polyethylene container. The dried laterite was sieved in order to collect the particles with the size lower than 0.1 mm. The laterite which was modified with 0.01 M SDS (solid/liquid ratio 200 mg/mL) in 0.01 M NaCl at pH 4 by shaking for 3 h and then washed with ultrapure water was called surfactant modified laterite (SML).

Standard solution of copper (1000 ppm in 0.5 M $HNO_3$) was supplied by Merck (Germany). All chemicals were of analytical reagent grade and were used without further purification. Copper nitrate salt, $Cu(NO_3)_2{\cdot}3H_2O$, was purchased from Merck (Germany). Anionic surfactant, sodium dodecyl sulfate (SDS, with purity > 95%), from Scharlau (Spain, EU), was used to modify the surface of laterite. The effect of ionic strength was studied by the addition of NaCl (Merck). In order to adjust solution pH, HCl, and NaOH (Merck) were used. Ultrapure water system (Labconco, USA) with resistivity 18.2 MΩ was used to produce ultrapure water in preparing all aqueous solutions.

*2.2. Adsorption Studies.* All adsorption experiments were conducted by batch technique. Initially, the copper stock solution of 2000 ppm was prepared by dissolving precisely calculated amount of copper nitrate salt. Then, the stock solution was appropriately diluted based on experimental requirement.

A known amount of adsorbents solution was thoroughly mixed with 50 mL aqueous copper solution ($Cu^{2+}$) of 10 ppm in 250 mL Erlenmeyer flasks at 25 ± 2°C controlled by an air conditioner. The effect of operating condition (pH, adsorbent dosages, contact time, ionic strength, and initial absorbate concentration) on removal of $Cu^{2+}$ was studied. The concentration of $Cu^{2+}$ was determined by flame atomic absorption spectrometry (F-AAS). The removal ($R$, %) of $Cu^{2+}$ was calculated by

$$\text{Removal}\,(R, \%) = \frac{C_i - C_f}{C_i} \times 100\% \qquad (1)$$

where $C_i$ and $C_f$ are initial concentration and final concentration of $Cu^{2+}$, respectively.

To study adsorption isotherms, the concentration of $Cu^{2+}$ was varied from 10 ppm to 2000 ppm and pH was adjusted to the desired value. The adsorption capacity of $Cu^{2+}$ ($\Gamma\ Cu^{2+}$) onto SDS modified laterite was determined by the different concentrations of $Cu^{2+}$ solutions before adsorption and after equilibrium process by F-AAS.

*2.3. Instrumental Analytical Methods.* The concentration of $Cu^{2+}$ was determined by using an atomic absorption spectrometer (AA-6800, Shimadzu, Japan). Hollow cathode lamp (HCL) was used to emit a narrow wavelength of 324.8 nm. The slit width was kept as constant of 0.5 nm for all AAS measurements. The linear relationship between the absorbance and concentrations of $Cu^{2+}$ had a correlation coefficient of at least 0.999.

Concentration of anionic surfactant, SDS used for surface modification of laterite, was determined by spectrophotometry with Ultraviolet Visible spectrophotometer (UV-1650 PC, Shimadzu, Japan) followed our previous paper [36].

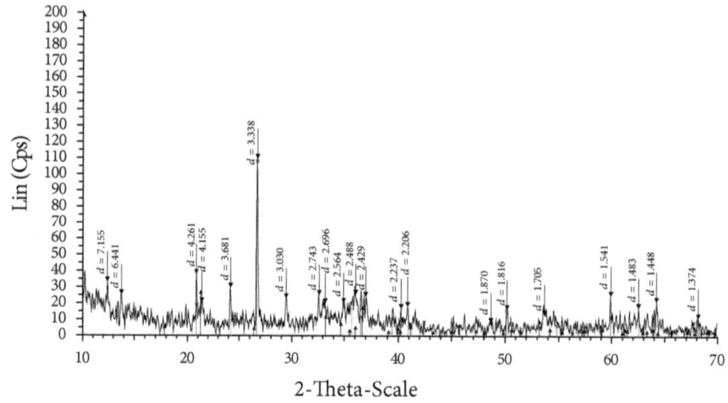


FIGURE 1: The XRD pattern of raw laterite.

Potentiometric method was used to determine pH of all solutions. The method was carried out using a HI 2215 Hanna Instruments pH meter by a glass combination electrode. We use three standard buffers (Hanna) to calibrate the electrode before measuring pH of solutions. All measurements were carried out at 25 ± 2°C.

*2.4. Characterization Methods.* The chemical compositions of laterite were examined by inductively coupled plasma mass spectrometer (ICP-MS Elan 9000, Perkin Elmer, USA) and total carbon analyzer using a TOC $V_{CPH}$ (Shimadzu, Japan).

X-ray diffraction (XRD) was collected on a Bruker D8 Advance X-ray diffractometer, CuK$\alpha$ radiation ($\lambda$ = 0.1549 nm). Intensity for the diffraction peaks was recorded in the 10°–70° ($2\theta$) range with a step size of 0.03°.

To evaluate functional groups of laterite and to confirm surface modification of laterite after $Cu^{2+}$ adsorption, Fourier transform infrared spectroscopy was performed with an Affinity-1S (Shimadzu, Japan). The FT-IR spectra were obtained under at the same conditions: 25°C, atmospheric pressure, and resolution of 4 $cm^{-1}$.

## 3. Modeling by General Isotherm Equation

The obtained isotherms were fitted by a general isotherm equation. The equation was derived by assuming that two adsorption steps can occur at the solid-liquid interface [32].

The general isotherm equation is

$$\Gamma = \frac{\Gamma_{\infty} k_1 C \left(1/n + k_2 C^{n-1}\right)}{1 + k_1 C \left(1 + k_2 C^{n-1}\right)}, \quad (2)$$

where $\Gamma$ is amount of $Cu^{2+}$ adsorbed; $\Gamma_{\infty}$ is the maximum adsorption amount; $k_1$ and $k_2$ are equilibrium constants for the first layer adsorption and clusters of $n$ molecules or multilayer adsorption. $C$ denotes the equilibrium concentration of $Cu^{2+}$ in solution.

The selected fitting parameters were described in our previously published papers [35–37].

TABLE 1: The chemical composition of raw laterite.

| Content (%) | Value |
|---|---|
| $Al_2O_3$ | 15.0 |
| $Fe_2O_3$ | 40.1 |
| $SiO_2$ | 33.2 |
| $MnO_2$ | 0.7 |
| $P_2O_5$ | 0.5 |
| $Na_2O$, | |
| $K_2O$, | ~0.8 |
| CaO | |
| C | 0.7 |

## 4. Results and Discussion

### *4.1. Characterization of Laterite*

*4.1.1. Characterization of Raw Laterite.* Table 1 shows the chemical compositions of raw laterite material. The inorganic components were determined by ICP-MS while carbon content was measured by TC. The results in Table 1 indicate that the main chemical compositions of laterite are metal oxides in which $Fe_2O_3$, $SiO_2$, and $Al_2O_3$ are dominated. The composition of laterite used in the present study is similar to the published papers [45, 46].

The XRD pattern of laterite is represented in Figure 1. The predominated morphologies were found as quartz ($SiO_2$), hematite ($Fe_2O_3$), and goethite (FeO(OH)). The peaks at $2\theta$ = 45, 66 indicated $Al_2O_3$ with very low intensity. The mineralogical phases of laterite used in this study is close to laterite used in other researches [5, 45].

Figure 2 indicates the FT-IR spectra of laterite without any modification. As can be seen in Figure 2, the band at 3622.32 and 3406.29 was assigned to –OH group of Si and Al. Another band at 1645.28 was assigned to inner layer water molecules [5, 45]. The bands at 1031.92, 1036.84, 912.33, and 796.60 appeared in the spectra because of the presence of Si–O–Fe, Al–OH, Fe–OH vibrations. The Fe–O bonds stretching at 534.28 and 464.84 were also obtained due to the presence of hematite in structure of laterite. The results of FT-IR are in good agreement with ICP-MS and XRD given above.

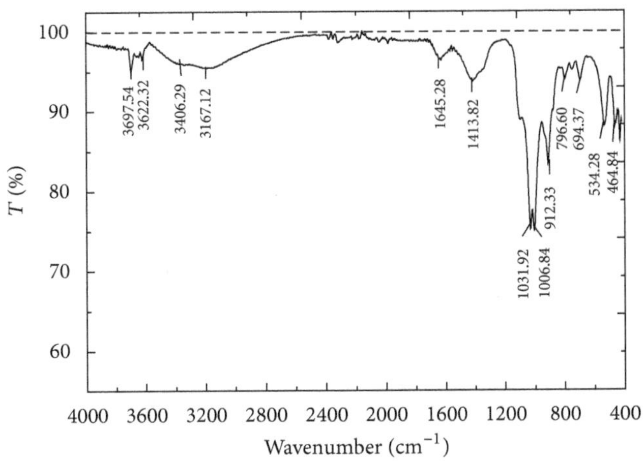


FIGURE 2: The FT-IR spectra of raw laterite.

*4.1.2. Characterization of Laterite after SDS Preadsorption and Copper Adsorption.* Anionic surfactant sodium dodecyl sulfate (SDS) with a concentration of 0.01 M (higher than critical micelle concentration, CMC) was used to modify the surface of laterite in 0.01 M NaCl (pH 5). In order to evaluate the change in mineral phase and surface modification and of laterite after preadsorption of SDS (SML) and sequential adsorption of $Cu^{2+}$ with initial concentration of 100 ppm in 10 mM NaCl (pH 5), the XRD and FT-IR techniques are also used.

Figures 3(a) and 3(b) show that the morphologies still contain quartz ($SiO_2$), hematite ($Fe_2O_3$), and goethite (FeO(OH)). These are similar to raw laterite, demonstrating that the structure of laterite does not change after adsorption SDS and adsorption of $Cu^{2+}$. However, the intensity of special peaks assigned for each mineral phase was changed. The intensity of specific peak for $SiO_2$ is enhanced while the others of $Fe_2O_3$ and FeO(OH) are reduced after preadsorption of SDS (Figure 3(a)) compared with original laterite. SDS is an anionic surfactant that can easily attract positive mineral ($Fe_2O_3$ and FeO(OH)) than negative one of $SiO_2$. For the case of laterite after consequential adsorption of cation $Cu^{2+}$, the peak indicated for $SiO_2$ decreased dramatically while the others for $Fe_2O_3$ and FeO(OH) slightly increased [5, 27].

Figure 4 shows that FT-IR spectra of modified laterite by preadsorption of SDS are similar to the raw one (Figure 4(a)). In additive, the relative intensity of asymmetrical and symmetrical stretching of $-CH_2-$ presented at 2926.01 and 2854.65 $cm^{-1}$ decreases dramatically in the spectra of SML while these peaks appear with very high intensity in spectra of SDS powder (data not shown). This confirms that the hydrophobic interaction can work on the surface of laterite. In addition, the characteristic peaks of $SO_4^{2-}$ at about 1247 $cm^{-1}$ and 1218 $cm^{-1}$ appear very strong in spectra of SDS while all bands disappear in the spectra of SML. It is demonstrated that SDS has sulfate head groups in contact with the surface of laterite via the electrostatic attraction at the current salt concentration (10 mM NaCl). In other words, the modification of laterite was successful due to the presence of bilayer and/or admicelles on the surface of laterite [36].

As can be seen in Figure 4(b), the band at 3442.94 assigned to –OH is very big due to the hydroxo complex of Cu(II). The bands at 1031.92, 1036.84, and 912.33 that appeared in the spectra of raw laterite are shifted to lower wavenumber at 1029.99, 1004.91, and 910.40 because the surface of laterite is changed. In addition, the peaks of $-CH_2-$ present at 2926.01 and 2854.65 $cm^{-1}$ of SML could not be seen in the spectra of SML after $Cu^{2+}$ adsorption. These results suggest the adsorption of copper molecules onto SML via electrostatic attraction and nonelectrostatic interactions between $Cu^{2+}$ and organic compounds in laterite.

*4.2. Surface Modification of Laterite by SDS.* Anionic surfactant SDS with a concentration of 0.01 M was used to modify the surface of laterite in 0.01 M NaCl (pH 5). The obtained adsorption capacity of SDS on laterite (with solid-liquid ratio 200 mg/mL) is 0.01 mmol/g. Although the adsorption SDS on laterite is small, the loading implies the presence of bilayer and/or admicelles of SDS [36]. As a result, the surface charge of laterite is negative which can enhance adsorption of cationic heavy metal $Cu^{2+}$. Figure 5 indicates that the removal efficiency of $Cu^{2+}$ in 1 mM NaCl (pH 5) with initial concentration of 10 ppm increases from 61.6% to 89.5% after the surface modification of laterite by SDS.

Typical concentration of Cu(II) in wastewater and surface water is quite low. The limited concentration according to national technical regulation on water quality of Vietnam is equal to 1 mg/L for class B (low quality water). It should be noted that when the concentration of Cu(II) in aqueous solution is less than 2.5 mg/L the removal efficiency of copper through adsorption using SML is approximately 99%. Furthermore, the maximum contaminant level (MCL) standard according to USEPA for Cu is 0.25 mg/L. It implies that the SML is a promising material to remove Cu(II) from aqueous solutions.

Effective parameters on adsorptive removal of copper ion by using SML are systematically studied, as given below.

### *4.3. Adsorptive Removal of Copper Ion by Using Surfactant Modified Laterite*

*4.3.1. Effect of pH.* Solution pH plays an important role in the adsorption of copper ion onto surfactant modified laterite (SML) because it can affect the surface charge of SML and charging behavior of chemical speciation of copper. At $pH > 7$, the high precipitation of $Cu^{2+}$ can take place which influences the concentration of $Cu^{2+}$ in solution. The effect of initial pH on the adsorption of $Cu^{2+}$ by SML was investigated in the pH range of 3–7 in 1 mM NaCl (Figure 6). This implies that the negatively charged SML surface can easily attract $Cu^{2+}$ by electrostatic attraction.

As can be seen in Figure 6, the removal of $Cu^{2+}$ using SML increases with increasing solution pH from pH 4 to pH 6 because of the competition between $H^+$ (at low pH) and cation $Cu^{2+}$ on the surface of SML. However, the desorption of SDS is increased at $pH > 7$ so that the removal efficiency of $Cu^{2+}$ decreases. Thus, optimum pH for removal of $Cu^{2+}$ by SML is pH 6 which agrees well with optimum pH in the

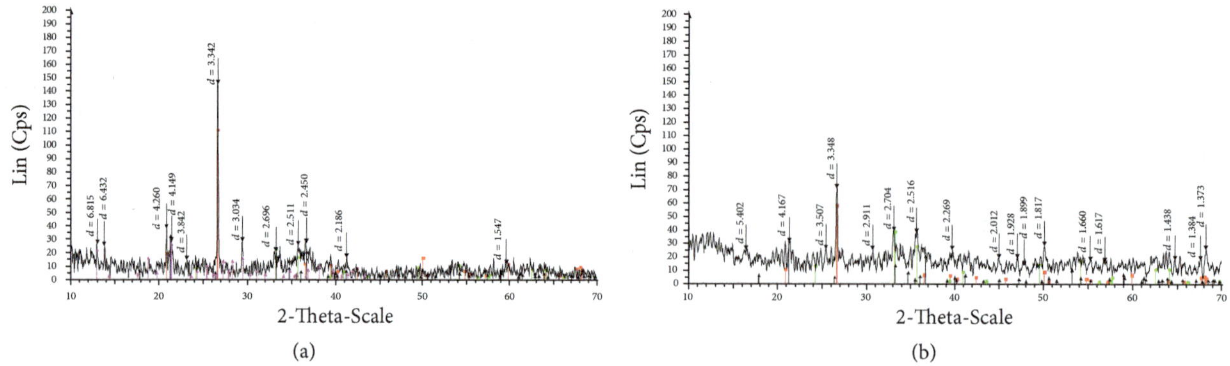


FIGURE 3: The XRD patterns of laterite after preadsorption of SDS (a) and after sequential adsorption of $Cu^{2+}$ (b).

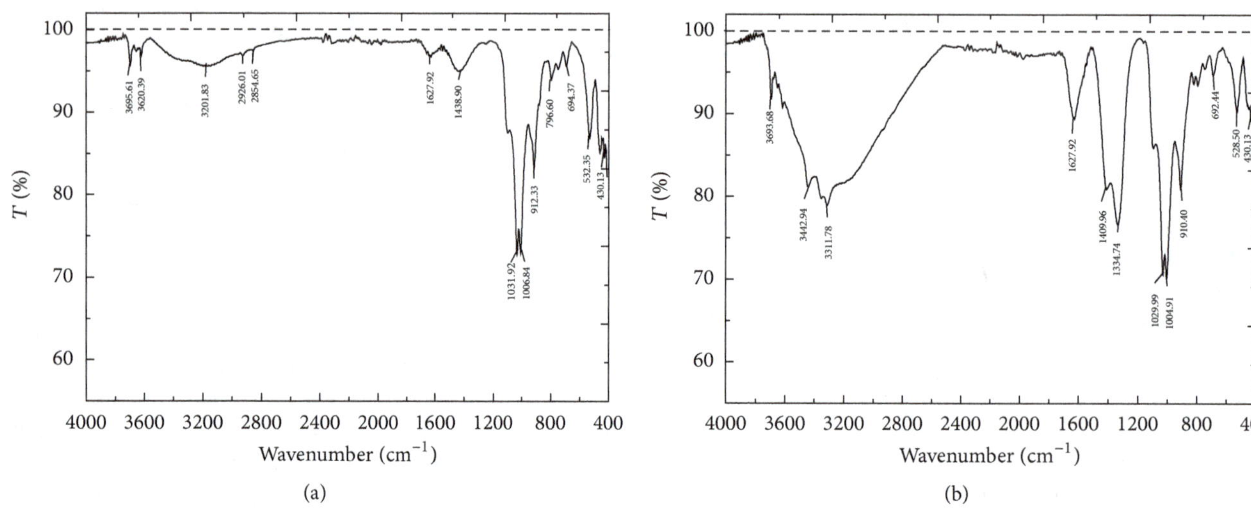


FIGURE 4: The FT-IR spectra of laterite after preadsorption of SDS (a) and after sequential adsorption of $Cu^{2+}$ (b).

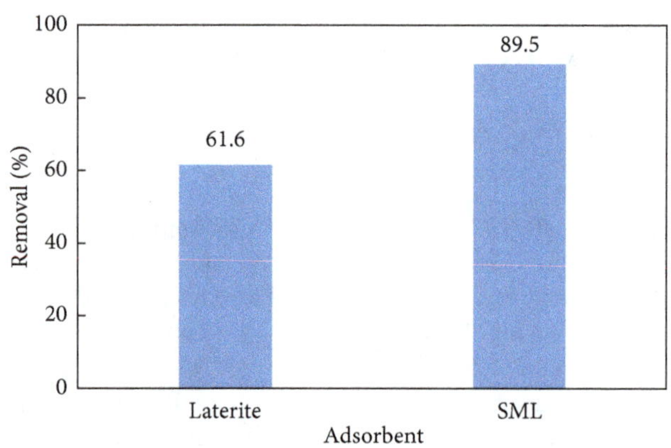


FIGURE 5: The removal of $Cu^{2+}$ at the initial concentration of 10 ppm in 1 mM NaCl (pH 5) using laterite and surfactant modified laterite (SML).

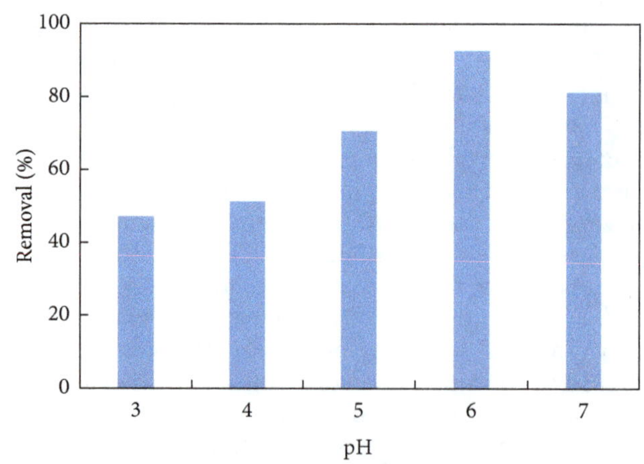


FIGURE 6: The removal of $Cu^{2+}$ by surfactant modified laterite (SML) as a function of pH (Ci ($Cu^{2+}$) = 10 ppm; contact time 90 min; adsorbent dosage 5 mg/mL; 1 mM NaCl).

case of adsorption heavy metal ions on surfactant modified alumina (SMA) [2, 3]. However, alumina is much more expensive compared with laterite so SML is more suitable than SMA for removing heavy metal ions in developing countries.

The point of zero charge of laterite is about 7.4 [45, 46], meaning that at pH < 7.4 laterite has positive charge, since surface modification of laterite with SDS was conducted at pH

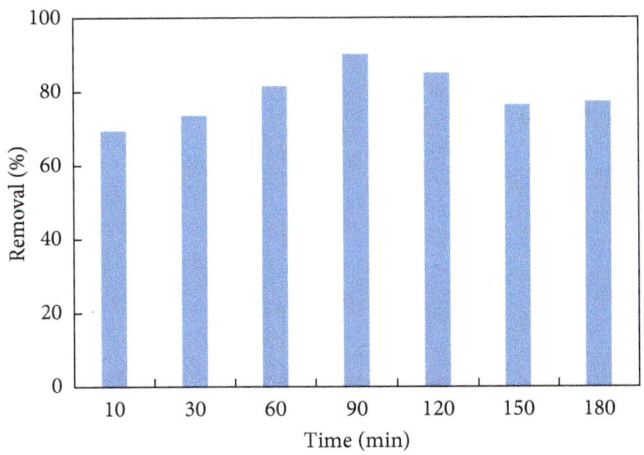


FIGURE 7: The removal of $Cu^{2+}$ by SML at different contact time (Ci ($Cu^{2+}$) = 10 ppm; pH 6; adsorbent dosage 5 mg/mL; 1 mM NaCl).

4.0 to promote preadsorption of SDS on laterite surface. After preadsorption of SDS, the concomitant of proton occurred so pH of solution increased around 0.6 units (from pH 5.0 to 5.6). This trend is similar to the case of SDS adsorption on $Al_2O_3$ [36]. However, net surface charge of SML is negative which enhances adsorption of $Cu^{2+}$ significantly. After copper adsorption, pH of solutions decreased (from 5.0 to 4.7), demonstrating that adsorption of cation $Cu^{2+}$ induces the desorption of SDS with proton into solution.

*4.3.2. Effect of Contact Time.* Contact time affects the completeness of adsorption equilibration. The effect of contact time on the adsorptive removal of $Cu^{2+}$ by using SML is presented in Figure 7. Figure 7 shows the removal efficiency of $Cu^{2+}$ from aqueous solution by using SML growing with time from 10 min to 180 min. It is suggested that adsorption reaches equilibrium at 90 min. After 180 min, the removal decreases because high concentration of $Na^+$ in NaCl background can displace $Cu^{2+}$. The equilibrium adsorption in this study is longer than the case of adsorptive removal of $Cu^{2+}$ by natural kaolin clay (only 30 min) and the removal of $Cu^{2+}$ by goethite mineral (60 min) [47, 48]. Nevertheless, 90 min is acceptable and is selected as the optimum contact time for removal of $Cu^{2+}$ by using SML in 1 mM NaCl (pH 6).

*4.3.3. Effect of Adsorbent Dosage.* The adsorbent dosage has a significant effect on the adsorption process because it can influence the total surface area of adsorbent and number of binding site. The amount of SML was varied from 0.05 to 6.0 g of adsorbent that corresponded from 1.0 to 12 mg/mL (Figure 8).

Figure 8 reveals that the removal of $Cu^{2+}$ by SML increases with increasing adsorbent dosage from 1 to 12 mg/mL. It may be explained by the increased large number of available binding sites for adsorption or increased net specific surface area with an increase of dosage [49]. However, an increase in adsorbent causes the increase of the adsorption capacity; then the adsorption capacity decreases due to the aggregation of colloidal particles at high adsorbent dosage

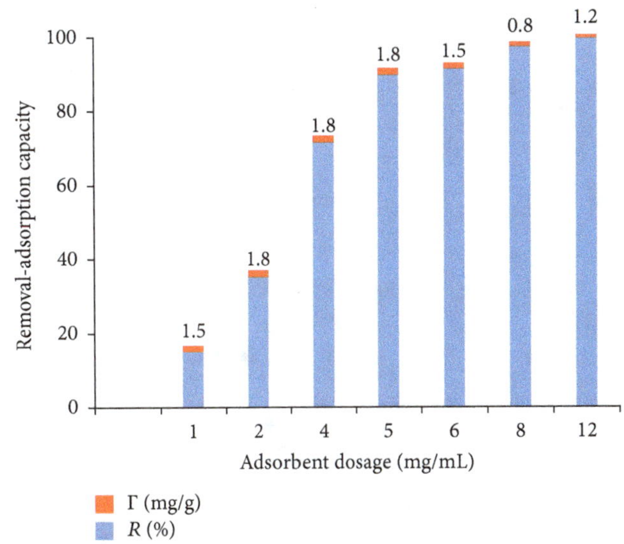


FIGURE 8: Removal ($R$) and adsorption capacity ($\Gamma$) of $Cu^{2+}$ by SML as a function of adsorbent dosage (Ci ($Cu^{2+}$) = 10 ppm; pH 6; contact time 90 min; 1 mM NaCl).

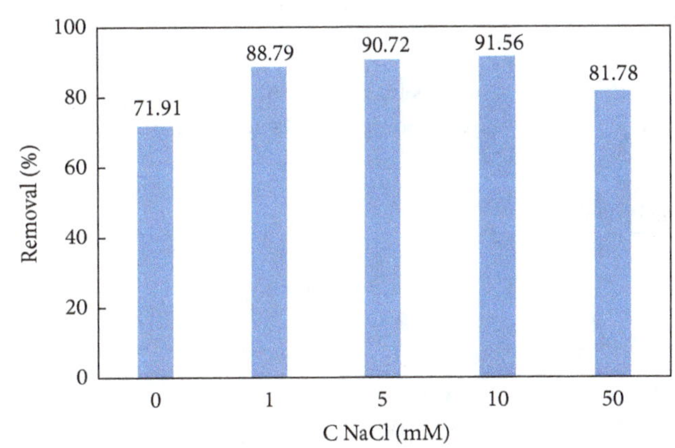


FIGURE 9: The removal of $Cu^{2+}$ by SML at different NaCl concentrations (Ci ($Cu^{2+}$) = 10 ppm; pH 6; contact time 90 min; adsorbent dosage 5 mg/mL).

[50]. Optimum adsorbent dosage is found to be 5 mg/mL and is fixed for the remaining studies.

*4.3.4. Effect of Ionic Strength.* Ionic strength affects electrostatic attraction between ionic adsorbates and charged surface adsorbent. For adsorption of $Cu^{2+}$ onto SML, ionic strength also induces a change of existence of SDS molecules on laterite surface. As can be seen in Figure 9, the removal efficiency is lowest in the absence of salt, demonstrating that the electrostatic interaction causes adsorption. Furthermore, adsorption at 50 mM NaCl concentration is lower than that at 10 mM. This result can be explained by desorption of SDS at high salt concentration. Desorption of SDS is enhanced with increasing NaCl concentration from 10 to 50 mM. When increasing salt, the concentration of $Cl^-$ is high so the removal is decreased. However, the removal increases with an increase of NaCl concentration from 1 to 10 mM, suggesting that not

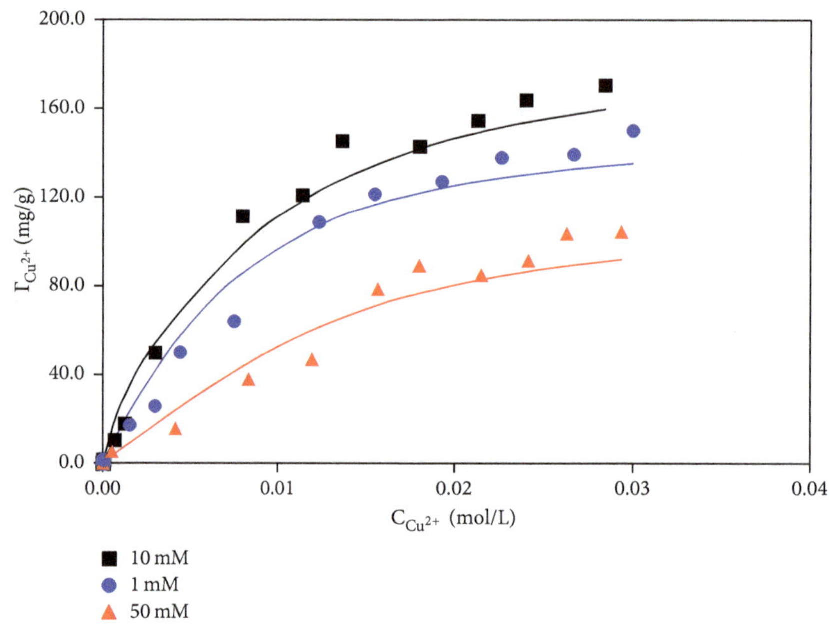


FIGURE 10: Adsorption isotherms of $Cu^{2+}$ onto SDS modified laterite (SML) as a function of equilibrium concentration of $Cu^{2+}$ at different NaCl concentrations. Points are experimental data; solid lines are results of the 2-step adsorption model.

TABLE 2: The fit parameters for $Cu^{2+}$ adsorption onto SML which are maximum adsorbed amount $\Gamma_{\infty,Cu^{2+}}$, the equilibrium constants $k_{1,Cu^{2+}}$ and $k_{2,Cu^{2+}}$ for first layer adsorption and multilayer adsorption, respectively, and the number of cluster of $Cu^{2+}$ ion $n_{Cu^{2+}}$.

| C salt (mM NaCl) | $\Gamma_{\infty,Cu^{2+}}$ (mg/g) | $\mathbf{k}_{1,Cu^{2+}}$ (g/mg) | $k_{2,Cu^{2+}}$ $(g/mg)^{n-1}$ | $\mathbf{n}_{Cu^{2+}}$ |
|---|---|---|---|---|
| 1 | 150 | $3.0 \times 10^2$ | $8.0 \times 10^2$ | 2.4 |
| 10 | 185 | $5.0 \times 10^2$ | $8.0 \times 10^2$ | 2.5 |
| 50 | 110 | $1.5 \times 10^2$ | $7.5 \times 10^2$ | 2.5 |

only electrostatic but also nonelectrostatic between Cu(II) and/or Cu(II) speciation and organic substances in laterite induces adsorption. The effect of salt concentration will be discussed in the adsorption isotherms section.

### 4.4. Characteristics and Mechanisms of Copper Ion Adsorption Isotherms on SDS Modified Laterite

*4.4.1. Adsorption Isotherms of Copper Ion on SDS Modified Laterite by a Two-Step Adsorption Model.* The effect of ionic strength on adsorption of copper ion on SDS modified laterite (SML) is clearly demonstrated in the isotherms (Figure 10). At pH 6, the $Cu^{2+}$ adsorption capacity increases with increasing NaCl from 1 to 10 mM. At higher salt concentration (from 10 to 50 mM), the adsorption capacity reduces with an increase of salt. The increase in salt concentration increases the number of cations $Na^+$ (counterions) on the negatively charged layer of SML, reducing the electrostatic effect of SML $Cu^{2+}$ ion. It is quite different from the case of adsorption of copper ion on strawberry leaf powder [51] in which the effect of $Na^+$, $K^+$, $Mg^{2+}$, and $Ca^{2+}$ on the copper adsorption is not significant. In our case, the electrostatic attraction between the positive charge of $Cu^{2+}$ ion and negative charge of SML is effectively screened by increasing NaCl concentrations from 10 to 50 mM. Nonelectrostatic interactions are more important in adsorption at low ionic strength (1 to 10 mM NaCl). These results agree well with the effect of ionic strength on the removal of copper ion using SML shown in previous section.

Figure 10 indicates that, at different salt concentrations, the experimental results can be represented well by the general isotherm equation (2) with the fitting parameters in Table 2.

As shown in Table 2, increasing ionic strength from 1 to 10 mM causes an increase in $\mathbf{k}_{1,Cu^{2+}}$ but a decrease in $\mathbf{k}_{1,Cu^{2+}}$ when increasing salt from 10 to 50 mM. It is noted that the changes in $k_{2,Cu^{2+}}$ and the number of $Cu^{2+}$ clusters are not significant ($k_{2,NH_4^+} \approx 8.0 \times 10^2$ and $\mathbf{n}_{Cu^{2+}} \approx 2.5$). The value of $\mathbf{k}_{1,Cu^{2+}}$ is also related to the slope of isotherm. As a result, at high salt concentration, the slope is lower than that at low salt. It suggests that desorption of SDS from the bilayer of admicelles is enhanced by increasing salt concentration from 10 to 50 mM while at low salt concentration the link between sulfate groups of SDS an $Cu^{2+}$ is not strong as other interactions.

*4.4.2. Adsorption Mechanisms and Advantages of Copper Ion onto SDS Modified Laterite.* Adsorptive removal of copper

TABLE 3: Adsorption capacity of $Cu^{2+}$ on natural materials or modified natural materials and SML.

| Adsorbent | Adsorption capacity (mg/g) | References |
|---|---|---|
| Zeolite, clinoptilolite | 1.64 | [12] |
| HCl-treated clay | 83.3 | [42] |
| Marine alga, *Ecklonia maxima* | 90 | [43] |
| Red paddy soil | 0.21 | [44] |
| SDS modified laterite | 185 | This study |

ion ($Cu^{2+}$) is enhanced by using anionic surfactant, SDS modified laterite (SML). The effective conditions on adsorption of $Cu^{2+}$ ion by SML were systematically studied in Section 4.3. The two-step model was established to describe the $Cu^{2+}$ adsorption onto SML, suggesting that adsorption of $Cu^{2+}$ could replace admicellar bilayers of SDS molecules on the laterite surface at high salt concentration. Adsorption of $Cu^{2+}$ decreases with an increase of ionic strength at high salt concentration because desorption of SDS is enhanced by increasing salt concentration [35] which is in good agreement with discussion in Section 4.3.1. It is also represented by two-step adsorption model with decreasing $\mathbf{k}_{1,Cu^{2+}}$ when increasing NaCl concentration from 10 to 50 mM. At low salt concentration, the main interactions inducing adsorption may be lateral, hydrogen bonding, surface complexation, and Van der Waals interactions between $Cu^{2+}$ and organic substances in the laterite.

Adsorption mechanisms of $Cu^{2+}$ ion on SML are also supported by the results of FT-IR spectra (see Section 4.1.2). The bands of $CH_2-$ of SML did not occur after adsorption of $Cu^{2+}$ in 10 mM NaCl, suggesting that the adsorption of $Cu^{2+}$ onto SML is mainly controlled by the electrostatic attraction between the negatively charged sulfate groups of bilayer admicelles and positive charge of $Cu^{2+}$. It seems to be similar to the case of $Cu^{2+}$ on natural kaolinite clays [48]. Nevertheless, at low ionic strength, the attraction between positive $Cu^{2+}$ and negative SML is not so strong compared with other interactions. In this case, nonelectrostatic interactions between $Cu^{2+}$ and/or speciation of $Cu^{2+}$ with organic substances in laterite mainly contribute to adsorption.

Some published papers focused on sorption of Cu(II) onto different kinds of sorbents. Nevertheless, sorption of Cu(II) onto SML has not been reported. Furthermore, SML used in the present study has very high adsorption capacity (185 mg/g) compared with other natural sorbents or modified natural materials (Table 3). It demonstrates that SML is novel material for removal of Cu(II) from aqueous solution.

## 5. Conclusions

For the first time, we investigated adsorption characteristics of copper ion ($Cu^{2+}$) on sodium dodecyl sulfate (SDS) surfactant modified laterite (SML) in aqueous solution. The adsorption experiments were quantified by flame atomic absorption spectrometry (F-AAS). The raw laterite, laterite after preadsorption of SDS, and sequential adsorption of $Cu^{2+}$ were characterized by X-ray diffraction (XRD) and Fourier transform infrared spectroscopy (FT-IR). The optimum conditions for adsorptive removal of $Cu^{2+}$ using SML were found as pH 6, contact time 90 min, adsorbent dosage 5 mg/mL, and ionic strength 10 mM NaCl. Adsorption isotherms of $Cu^{2+}$ on SML at different NaCl concentrations were fitted well by a two-step adsorption model. The adsorption of $Cu^{2+}$ on SML decreased with an increase of NaCl concentration due to the enhancement of SDS desorption with increasing salt concentration but at low NaCl concentrations (1 to 10 mM), adsorption $Cu^{2+}$ on SML increased with an increase of salt. Adsorption mechanisms of $Cu^{2+}$ onto SML were mainly controlled by electrostatic attraction between the positive charge of $Cu^{2+}$ and the negatively charged layer SML at high ionic strength while at low ionic strength nonelectrostatic interactions between $Cu^{2+}$ and organic compounds in laterite induced adsorption. The SML was demonstrated as a novel adsorbent to remove $Cu^{2+}$ from aqueous solutions.

## Competing Interests

The authors declare that they have no competing interests.

## Acknowledgments

This research is funded by the Vietnam National University, Hanoi (VNU), under Project no. QG.16.12.

## References

[1] V. K. Gupta, I. Ali, T. A. Saleh, A. Nayak, and S. Agarwal, "Chemical treatment technologies for waste-water recycling—an overview," *RSC Advances*, vol. 2, no. 16, pp. 6380–6388, 2012.

[2] M. U. Khobragade and A. Pal, "Adsorptive removal of Mn(II) from water and wastewater by surfactant-modified alumina," *Desalination and Water Treatment*, vol. 57, no. 6, pp. 2775–2786, 2016.

[3] M. U. Khobragade and A. Pal, "Fixed-bed column study on removal of Mn(II), Ni(II) and Cu(II) from aqueous solution by surfactant bilayer supported alumina," *Separation Science and Technology (Philadelphia)*, vol. 51, no. 8, pp. 1287–1298, 2016.

[4] M. MacHida, B. Fotoohi, Y. Amamo, and L. Mercier, "Cadmium(II) and lead(II) adsorption onto hetero-atom functional mesoporous silica and activated carbon," *Applied Surface Science*, vol. 258, no. 19, pp. 7389–7394, 2012.

[5] S. Mitra, L. S. Thakur, V. K. Rathore, and P. Mondal, "Removal of Pb(II) and Cr(VI) by laterite soil from synthetic waste water: single and bi-component adsorption approach," *Desalination and Water Treatment*, vol. 57, no. 39, pp. 18406–18416, 2016.

[6] B. Prélot, W. Janusz, F. Thomas et al., "Adsorption of cadmium ions at the electrolyte/silica interface I. Experimental study of surface properties," *Applied Surface Science*, vol. 196, no. 1—4, pp. 322–330, 2002.

[7] M. K. Sahu, S. Mandal, S. S. Dash, P. Badhai, and R. K. Patel, "Removal of Pb(II) from aqueous solution by acid activated red mud," *Journal of Environmental Chemical Engineering*, vol. 1, no. 4, pp. 1315–1324, 2013.

[8] S. E. Bailey, T. J. Olin, R. M. Bricka, and D. D. Adrian, "A review of potentially low-cost sorbents for heavy metals," *Water Research*, vol. 33, no. 11, pp. 2469–2479, 1999.

[9] H. B. Bradl, "Adsorption of heavy metal ions on soils and soils constituents," *Journal of Colloid and Interface Science*, vol. 277, no. 1, pp. 1–18, 2004.

[10] F. Fu and Q. Wang, "Removal of heavy metal ions from wastewaters: a review," *Journal of Environmental Management*, vol. 92, no. 3, pp. 407–418, 2011.

[11] A. Radian, D. Michaeli, C. Serban, R. Nechushtai, and Y. G. Mishael, "Bioactive apo-ferredoxin-polycation-clay composites for iron binding," *Journal of Materials Chemistry*, vol. 20, no. 21, pp. 4361–4365, 2010.

[12] S. Babel and T. A. Kurniawan, "Low-cost adsorbents for heavy metals uptake from contaminated water: a review," *Journal of Hazardous Materials*, vol. 97, no. 1–3, pp. 219–243, 2003.

[13] H.-P. Chao, C.-K. Lee, L.-C. Juang, and Y.-L. Han, "Sorption of organic compounds, oxyanions, and heavy metal ions on surfactant modified titanate nanotubes," *Industrial and Engineering Chemistry Research*, vol. 52, no. 29, pp. 9843–9850, 2013.

[14] C. Faur-Brasquet, Z. Reddad, K. Kadirvelu, and P. Le Cloirec, "Modeling the adsorption of metal ions ($Cu^{2+}$, $Ni^{2+}$, $Pb^{2+}$) onto ACCs using surface complexation models," *Applied Surface Science*, vol. 196, no. 1–4, pp. 356–365, 2002.

[15] M. Visa, C. Bogatu, and A. Duta, "Simultaneous adsorption of dyes and heavy metals from multicomponent solutions using fly ash," *Applied Surface Science*, vol. 256, no. 17, pp. 5486–5491, 2010.

[16] M. U. Khobragade and A. Pal, "Fixed-bed column study on removal of Mn(II), Ni(II) and Cu(II) from aqueous solution by surfactant bilayer supported alumina," *Separation Science and Technology*, vol. 51, no. 8, pp. 1287–1298, 2016.

[17] T. A. Saleh and V. K. Gupta, "Processing methods, characteristics and adsorption behavior of tire derived carbons: a review," *Advances in Colloid and Interface Science*, vol. 211, pp. 93–101, 2014.

[18] M. Ahmaruzzaman, "Industrial wastes as low-cost potential adsorbents for the treatment of wastewater laden with heavy metals," *Advances in Colloid and Interface Science*, vol. 166, no. 1-2, pp. 36–59, 2011.

[19] K. Z. Elwakeel, "Removal of Cr(VI) from alkaline aqueous solutions using chemically modified magnetic chitosan resins," *Desalination*, vol. 250, no. 1, pp. 105–112, 2010.

[20] K. Z. Elwakeel and E. Guibal, "Potential use of magnetic glycidyl methacrylate resin as a mercury sorbent: from basic study to the application to wastewater treatment," *Journal of Environmental Chemical Engineering*, vol. 4, no. 3, pp. 3632–3645, 2016.

[21] M. M. H. Khalil, K. Z. Al-Wakeel, S. S. A. E. Rehim, and H. A. E. Monem, "Efficient removal of ferric ions from aqueous medium by amine modified chitosan resins," *Journal of Environmental Chemical Engineering*, vol. 1, no. 3, pp. 566–573, 2013.

[22] M. Ozmen, K. Can, G. Arslan, A. Tor, Y. Cengeloglu, and M. Ersoz, "Adsorption of Cu(II) from aqueous solution by using modified $Fe_3O_4$ magnetic nanoparticles," *Desalination*, vol. 254, no. 1–3, pp. 162–169, 2010.

[23] W. Sun, K. Yin, and X. Yu, "Effect of natural aquatic colloids on Cu(II) and Pb(II) adsorption by $Al_2O_3$ nanoparticles," *Chemical Engineering Journal*, vol. 225, pp. 464–473, 2013.

[24] X. Sun, X. Huang, X.-P. Liao, and B. Shi, "Adsorptive removal of Cu(II) from aqueous solutions using collagen-tannin resin," *Journal of Hazardous Materials*, vol. 186, no. 2-3, pp. 1058–1063, 2011.

[25] M.-W. Wan, C.-C. Kan, B. D. Rogel, and M. L. P. Dalida, "Adsorption of copper (II) and lead (II) ions from aqueous solution on chitosan-coated sand," *Carbohydrate Polymers*, vol. 80, no. 3, pp. 891–899, 2010.

[26] M. Li, M.-Y. Li, C.-G. Feng, and Q.-X. Zeng, "Preparation and characterization of multi-carboxyl-functionalized silica gel for removal of Cu (II), Cd (II), Ni (II) and Zn (II) from aqueous solution," *Applied Surface Science*, vol. 314, pp. 1063–1069, 2014.

[27] A. Maiti, J. K. Basu, and S. De, "Development of a treated laterite for arsenic adsorption: effects of treatment parameters," *Industrial and Engineering Chemistry Research*, vol. 49, no. 10, pp. 4873–4886, 2010.

[28] M. Sarkar, A. Banerjee, P. P. Pramanick, and A. R. Sarkar, "Use of laterite for the removal of fluoride from contaminated drinking water," *Journal of Colloid and Interface Science*, vol. 302, no. 2, pp. 432–441, 2006.

[29] M. J. Rosen and J. T. Kunjappu, *Surfactants and Interfacial Phenomena*, John Wiley & Sons, 2012.

[30] H. M. F. Freundlich, "Über die adsorption in lösungen," *Zeitschrift für Physikalische Chemie*, vol. 57, pp. 385–470, 1906.

[31] I. Langmuir, "The adsorption of gases on plane surfaces of glass, mica and platinum," *The Journal of the American Chemical Society*, vol. 40, no. 9, pp. 1361–1403, 1918.

[32] B.-Y. Zhu and T. Gu, "Surfactant adsorption at solid-liquid interfaces," *Advances in Colloid and Interface Science*, vol. 37, no. 1-2, pp. 1–32, 1991.

[33] I. Hoffmann, C. Oppel, U. Gernert, P. Barreleiro, W. von Rybinski, and M. Gradzielski, "Adsorption isotherms of cellulose-based polymers onto cotton fibers determined by means of a direct method of fluorescence spectroscopy," *Langmuir*, vol. 28, no. 20, pp. 7695–7703, 2012.

[34] R. S. Ndong and W. B. Russel, "Linear viscoelasticity of $ZrO_2$ nanoparticle dispersions with associative polymers," *Rheologica Acta*, vol. 51, no. 9, pp. 771–782, 2012.

[35] T. D. Pham, M. Kobayashi, and Y. Adachi, "Adsorption of polyanion onto large alpha alumina beads with variably charged surface," *Advances in Physical Chemistry*, vol. 2014, Article ID 460942, 9 pages, 2014.

[36] T. D. Pham, M. Kobayashi, and Y. Adachi, "Adsorption of anionic surfactant sodium dodecyl sulfate onto alpha alumina with small surface area," *Colloid and Polymer Science*, vol. 293, no. 1, pp. 217–227, 2015.

[37] T. D. Pham, M. Kobayashi, and Y. Adachi, "Adsorption characteristics of anionic azo dye onto large $\alpha$-alumina beads," *Colloid and Polymer Science*, vol. 293, no. 7, pp. 1877–1886, 2015.

[38] A. Adak, M. Bandyopadhyay, and A. Pal, "Removal of crystal violet dye from wastewater by surfactant-modified alumina," *Separation and Purification Technology*, vol. 44, no. 2, pp. 139–144, 2005.

[39] A. Adak, M. Bandyopadyay, and A. Pal, "Fixed bed column study for the removal of crystal violet (C. I. Basic Violet 3) dye from aquatic environment by surfactant-modified alumina," *Dyes and Pigments*, vol. 69, no. 3, pp. 245–251, 2006.

[40] A. Adak, A. Pal, and M. Bandyopadhyay, "Removal of phenol from water environment by surfactant-modified alumina through adsolubilization," *Colloids and Surfaces A: Physicochemical and Engineering Aspects*, vol. 277, no. 1–3, pp. 63–68, 2006.

[41] A. K. Das, S. Saha, A. Pal, and S. K. Maji, "Surfactant-modified alumina: an efficient adsorbent for malachite green removal from water environment," *Journal of Environmental Science and Health. Part A*, vol. 44, no. 9, pp. 896–905, 2009.

[42] T. Vengris, R. Binkiene, and A. Sveikauskaite, "Nickel, copper and zinc removal from waste water by a modified clay sorbent," *Applied Clay Science*, vol. 18, no. 3-4, pp. 183–190, 2001.

[43] D. Feng and C. Aldrich, "Adsorption of heavy metals by biomaterials derived from the marine alga Ecklonia maxima," *Hydrometallurgy*, vol. 73, no. 1-2, pp. 1–10, 2004.

[44] B. Huang, Z. Li, J. Huang et al., "Adsorption characteristics of Cu and Zn onto various size fractions of aggregates from red paddy soil," *Journal of Hazardous Materials*, vol. 264, pp. 176–183, 2014.

[45] A. Maiti, J. K. Basu, and S. De, "Experimental and kinetic modeling of As(V) and As(III) adsorption on treated laterite using synthetic and contaminated groundwater: effects of phosphate, silicate and carbonate ions," *Chemical Engineering Journal*, vol. 191, pp. 1–12, 2012.

[46] A. Maiti, H. Sharma, J. K. Basu, and S. De, "Modeling of arsenic adsorption kinetics of synthetic and contaminated groundwater on natural laterite," *Journal of Hazardous Materials*, vol. 172, no. 2-3, pp. 928–934, 2009.

[47] A. Jaiswal, S. Banerjee, R. Mani, and M. C. Chattopadhyaya, "Synthesis, characterization and application of goethite mineral as an adsorbent," *Journal of Environmental Chemical Engineering*, vol. 1, no. 3, pp. 281–289, 2013.

[48] M.-Q. Jiang, X.-Y. Jin, X.-Q. Lu, and Z.-L. Chen, "Adsorption of Pb(II), Cd(II), Ni(II) and Cu(II) onto natural kaolinite clay," *Desalination*, vol. 252, no. 1–3, pp. 33–39, 2010.

[49] F. Mazloomi and M. Jalali, "Ammonium removal from aqueous solutions by natural Iranian zeolite in the presence of organic acids, cations and anions," *Journal of Environmental Chemical Engineering*, vol. 4, no. 2, pp. 1664–1673, 2016.

[50] J. K. Wolterink, L. K. Koopal, M. A. C. Stuart, and W. H. Van Riemsdijk, "Surface charge regulation upon polyelectrolyte adsorption, hematite, polystyrene sulfonate, surface charge regulation: theoretical calculations and hematite-poly(styrene sulfonate) system," *Colloids and Surfaces A: Physicochemical and Engineering Aspects*, vol. 291, no. 1-3, pp. 13–23, 2006.

[51] H. Liu, Y. Dong, H. Wang, and Y. Liu, "Ammonium adsorption from aqueous solutions by strawberry leaf powder: equilibrium, kinetics and effects of coexisting ions," *Desalination*, vol. 263, no. 1-3, pp. 70–75, 2010.

# 5

# Optimization of the Extraction of the Volatile Fraction from Honey Samples by SPME-GC-MS, Experimental Design, and Multivariate Target Functions

**Elisa Robotti,[1] Federica Campo,[1] Marta Riviello,[1] Marco Bobba,[1,2] Marcello Manfredi,[1] Eleonora Mazzucco,[1] Fabio Gosetti,[1] Giorgio Calabrese,[3] Emanuele Sangiorgi,[2] and Emilio Marengo[1]**

[1]*Department of Sciences and Technological Innovation, University of Piemonte Orientale, Viale Michel 11, 15121 Alessandria, Italy*
[2]*Istituto Zooprofilattico Sperimentale della Lombardia e dell'Emilia Romagna, Via Bianchi 9, 25124 Brescia, Italy*
[3]*Department of Pharmacy, Università degli Studi di Napoli Federico II, Via Montesano 49, 80131 Napoli, Italy*

Correspondence should be addressed to Elisa Robotti; elisa.robotti@uniupo.it

Academic Editor: Shayessteh Dadfarnia

Head space (HS) solid phase microextraction (SPME) followed by gas chromatography with mass spectrometry detection (GC-MS) is the most widespread technique to study the volatile profile of honey samples. In this paper, the experimental SPME conditions were optimized by a multivariate strategy. Both sensitivity and repeatability were optimized by experimental design techniques considering three factors: extraction temperature (from 50°C to 70°C), time of exposition of the fiber (from 20 min to 60 min), and amount of salt added (from 0 to 27.50%). Each experiment was evaluated by Principal Component Analysis (PCA) that allows to take into consideration all the analytes at the same time, preserving the information about their different characteristics. Optimal extraction conditions were identified independently for signal intensity (extraction temperature: 70°C; extraction time: 60 min; salt percentage: 27.50% w/w) and repeatability (extraction temperature: 50°C; extraction time: 60 min; salt percentage: 27.50% w/w) and a final global compromise (extraction temperature: 70°C; extraction time: 60 min; salt percentage: 27.50% w/w) was also reached. Considerations about the choice of the best internal standards were also drawn. The whole optimized procedure was than applied to the analysis of a multiflower honey sample and more than 100 compounds were identified.

## 1. Introduction

Honey is gaining increasing commercial importance due to not only its nutritional properties but also its therapeutic and healing effects [1–3]. In the last years, increasing attention has been devoted to the characterization of the profile of volatile molecules in honey samples above all in authenticity and adulteration studies [4–22], since the commercial value of honey is usually strictly correlated to its botanical and/or geographical origin. Unifloral honeys are certainly more valuable than multifloral; products show a different commercial value based on their botanical origin, particularly when the origin is strictly defined, for example, for PDO (Protected Designation of Origin) products.

The volatile profile is usually determined by Head-Space Solid Phase Micro-Extraction (HS-SPME), followed by a gas chromatographic (GC) determination with Mass Spectrometry (MS) detection [4–19]. HS-SPME GC-MS is quite widespread both as a fingerprinting tool [23–26] and for identification and quantification of specific classes of compounds or single analytes [27]: it is applied to honey samples in food safety and adulteration studies (for the determination of residues of chemicals used to combat honeybee diseases or parasite infestations, as fungicides, insecticides, acaricides, and pesticides [28]) or in authenticity studies for the determination of botanical biomarkers [29–31].

In HS-SPME particular attention has to be paid to the optimization of the extraction procedure that is subjected to the effect of several experimental parameters as conditioning time, extraction time and temperature, ionic strength, amount of sample, water/sample ratio, sample solution/headspace volume ratio, and type of fiber used for SPME. The optimization step allows the selection of the best experimental conditions to be applied during the extraction to provide the best extraction yield avoiding in the meantime fiber saturation.

In SPME, optimization is usually accomplished by a monovariate approach in which one parameter is optimized at a time [32]. In this way, no evaluation of the interactions existing between the parameters can be accomplished. Experimental design techniques have also been applied [33–43] but few applications are present regarding honey [44–46].

One of the drawbacks of SPME is its low repeatability, if compared to other extraction techniques: to improve repeatability, the achievement of the equilibrium during the adsorption of the analytes on the fiber is mandatory. Notwithstanding the use of experimental design to improve repeatability is reported in literature [47] for other applications, optimization in SPME is usually carried out to maximize signal intensity alone, that is, gas-chromatographic peak areas.

Moreover, most of applications of SPME to honey report optimized conditions identified using two main strategies:

(1) when the monovariate approach is adopted, optimization is usually accomplished on a set of selected analytes, representative of the analytes present [30, 33, 35–38];

(2) both in monovariate optimization and when experimental design tools are exploited, global signal is defined for each experiment as the average or the sum of the areas of all the peaks identified in the chromatogram [34, 35, 39–46]. This approach takes into account the presence of several analytes but no information is retained about the different physical-chemical characteristics of each species.

A true multivariate target function could be defined by the use of desirability or utility functions [48, 49] or by using Principal Component Analysis (PCA) [49–51]. A multivariate procedure has been previously proposed by our research group for the optimization of the instrumental parameters in MS determinations and for the elemental profiling in Inductively Coupled Plasma Mass Spectrometry (ICP-MS) [51]. A similar procedure has been recently applied by Ribeiro et al. to the extraction of the volatiles from coffee and compared to desirability functions [49].

To overcome the above mentioned limitations, optimization of both signal intensity (expressed as chromatographic peak areas) and repeatability (expressed as the coefficient of variation, CV%, of each peak area across replications of the same experiment) was carried out here by experimental design. Three main factors underwent optimization: temperature and time of extraction and the ionic strength, expressed as amount of salt added to the sample. PCA was used as a dimensionality reduction tool, to provide a multivariate target function for the optimization phase so that all the analytes are taken into account, each one described by its own area (for signal intensity optimization) or coefficient of variation (for repeatability optimization). Optimal extraction conditions were identified for both signal intensity and repeatability separately, and a final global optimal compromise was also reached. Some considerations on the choice of the most suitable internal standard were also drawn. The main novelty here proposed is coupling experimental design with the establishment of a multivariate target function for the optimization of both signal intensity and repeatability; to the authors' knowledge the optimization of both these parameters is reported here for the first time to the SPME procedure for honey samples.

## 2. Theory

*2.1. Experimental Design.* Experimental design techniques allow the evaluation of the effect played by a set of experimental factors on a selected experimental response by the use of the most informative and most exiguous set of experiments [52]. Here, experimental design was exploited for the optimization of the SPME procedure [39–42] according to both the maximum signal intensity and repeatability.

A two-level full factorial design was chosen for studying the effect played by three factors: extraction temperature ($T$, in °C), extraction time ($t$, in min), and amount of salt (salt, expressed as % w/w). The experimental design requires a total of $2^p$ ($p$ = number of factors) experiments, that is, $2^3 = 8$ experiments. The center of the experimental domain was added to evaluate the potential curvature of the system behavior. The experimental domain explored for each factor is defined in Table 1, together with the experiments carried out.

The adopted design allows the evaluation of the effect of all principal factors and their two- and three-order interactions, that is, of the general Ordinary Least Squares (OLS) model:

$$y = b_0 + b_1 \cdot T + b_2 \cdot t + b_3 \cdot \text{salt} + b_4 \cdot T \cdot t + b_5 \cdot T \cdot \text{salt} + b_6 \cdot t \cdot \text{salt} + b_7 \cdot T \cdot t \cdot \text{salt}, \quad (1)$$

where the coefficients $b_i$ estimate the effect of the corresponding variable or interaction on the response $y$. The complete experimental design, including the center of the domain, was replicated three times (a total of 27 experiments) to provide information about the experimental repeatability.

Here the experimental response $y$ was defined by a multivariate target function, based on PCA.

Each experiment was described by the chromatographic profile obtained, consisting of a total of 143 peak areas. The optimization of signal intensity (peak area) is usually accomplished in literature by submitting to optimization the average [44] or the sum [34, 35, 39, 45, 46] of the areas calculated for all the identified peaks. This choice does not take into account the correlation structure of the data. The analytes belong to different chemical classes characterized by different physical-chemical properties and, in particular, different volatility: so the average or the sum does not allow considering the existence of groups of analytes with similar or opposite behavior. The same considerations can be driven about repeatability that is usually described for each $i$th

TABLE 1: Experimental domain investigated: low (−), intermediate (0), and high (+) levels adopted for each experimental factor.

| | Extr. temperature ($T$) | Extr. time ($t$) | Amount of salt (salt) |
|---|---|---|---|
| Exp 1 | − | − | − |
| Exp 2 | + | − | − |
| Exp 3 | − | + | − |
| Exp 4 | + | + | − |
| Exp 5 | − | − | + |
| Exp 6 | + | − | + |
| Exp 7 | − | + | + |
| Exp 8 | + | + | + |
| Exp 9 | 0 | 0 | 0 |
| | Experimental domain | | |
| | $T$ (°C) | $t$ (min) | Salt (% w/w honey) |
| − | 50 | 20 | 0 |
| + | 70 | 60 | 27.50 |
| **0** | 60 | 40 | 13.75 |

experiment and each $j$th peak as the coefficient of variation (CV) of the area of the $j$th peak calculated across three or more replications of the same $i$th experiment:

$$\text{CV}_{i,j}\% = \frac{s_{i,j}}{\overline{x}_{i,j}} * 100, \quad (2)$$

where $s_{i,j}$ is the standard deviation of the $j$th peak of the $i$th experiment and $\overline{x}_{i,j}$ is the average area of the $j$th peak of the $i$th experiment. Again, the description of each experiment by the use of the average CV% or the sum of all CV% is not an effective choice.

*2.2. Principal Component Analysis (PCA).* A true multivariate target function was defined here by means of PCA [53]. As commonly acknowledged, PCA represents the objects in a new reference system characterized by new variables called Principal Components (PCs), orthogonal one to each other and computed hierarchically, that is, in decreasing order of explained variance. The PCs therefore account for independent sources of information, relegating experimental noise and random variations in the last PCs. PCA can be exploited to provide a multivariate target function for signal intensity and repeatability optimization.

*(i) Signal Intensity.* PCA was carried out on the dataset consisting of all the experiments performed (9 × 3 = 27 experiments), each described by the areas of the 143 peaks. In this case, a target function can be identified if a meaningful direction of optimization can be defined in the space given by the relevant PCs. For this purpose two simulated experiments were defined, representing the "best" and the "worst" experiments for what regards signal intensity. They were calculated from the results of the experimental design as follows:

$$\begin{aligned} \text{best}_j &= \max(x)_j + 20 * \frac{\max(x)_j}{100}, \\ \text{worst}_j &= \min(x)_j - 20 * \frac{\min(x)_j}{100}, \end{aligned} \quad (3)$$

where $j$ identifies the $j$th peak; the minimum and maximum values are calculated over the areas of the $j$th peak among all the experiments of the design. In this way, the best experiment is defined as the one characterized by the largest areas recorded along the design experiments plus 20% (best signal intensity) and the worst is defined as the one characterized by the smallest areas minus 20% (worst signal intensity). These two experiments can be reprojected in the space given by the relevant PCs to identify the direction of optimization as the direction connecting them. All experiments can be therefore projected onto the optimization direction and described in terms of their distance from the best experiment along this direction; the distance thus defined can be used as a multivariate target function: the smallest the distance from the best experiment, the largest the signal intensity.

*(ii) Repeatability.* PCA was carried out on the dataset consisting of the 9 experiments of the design, each described by the CV% of each peak area. In this case, the CV% was calculated for each peak and each experiment from the three replications of the design. Also in this case a target function can be identified if a direction of optimization can be defined in the space of the relevant PCs. Again, the "best" and the "worst" experiments for what regards repeatability can be calculated and reprojected in the space of the relevant PCs:

$$\begin{aligned} \text{best}_j &= \min(\text{CV})_j - 20 * \frac{\min(\text{CV})_j}{100}, \\ \text{worst}_j &= \max(\text{CV})_j + 20 * \frac{\max(\text{CV})_j}{100}, \end{aligned} \quad (4)$$

where $j$ indicates the $j$th peak; the minimum and maximum values are calculated over the CV% of the $j$th peak calculated for all the experiments of the design. The best experiment is therefore defined as the one characterized by the smallest CV% recorded over the design experiments minus 20% (best repeatability) and the worst as the one characterized by the largest CV% plus 20% (worst repeatability). Again, the distance of each experiment from the best along the optimization direction thus defined can be used as a multivariate target function: the smallest this distance, the best the repeatability.

*2.3. Calculation of Regression Models.* Regression models were then built relating the distance of each experiment from "best" along the optimization functions established for signal intensity and repeatability, to the investigated factors and their interactions. Only statistically relevant ($p$-level $< 0.05$) coefficients of experimental factors and/or factors interactions were included in the final models and were identified by ANOVA (ANalysis Of VAriance) [52]. Thanks to the availability of replicated experiments, the model for signal intensity could be tested for curvature effects by ANOVA [52].

## 3. Materials and Methods

*3.1. SPME Conditions.* SPME was carried out in 20 mL Glass Headspace Vials (22.5 × 75 mm; Supelco, Bellefonte, PA, USA) with aluminum seals equipped with PTFE/silicone septa (Supelco, Bellefonte, PA, USA). SPME was carried out with a 50/30 $\mu$m DVB/CAR/PDMS fiber (Stableflex, Supelco, Bellefonte, PA, USA). Extraction was carried out on 4 g of honey, added with 5.5 mL milliQ water and internal standards (42.5 $\mu$L o-tolualdehyde 10.0 mg/L, 42.5 $\mu$L 2,4,6-trimethylphenol 10.0 mg/L, 85 $\mu$L benzophenone 10.0 mg/L). The volume ratio honey/water adopted was chosen to avoid saturation effects on the fiber and achieve a ratio solution : headspace of about 1 : 1, to guarantee the establishment of the equilibrium [54].

When prescribed by the experimental design, salt was added to the samples as aqueous solution at 100 g/L and 200 g/L. The addition to the sample of 5.5 mL of the first or second solution provides a final concentration of salt of 13.75 or 27.50% expressed as weight of salt over weight of honey (Table 1). Saturated salt solutions were avoided to improve repeatability: the use of saturated solutions causes in general a low repeatability probably due to adsorption effects at the surface of crystals that can be present in the solution.

All experiments were performed on the same commercial organic multiflower honey and preliminary results demonstrated that no degradation took place at the extraction temperature adopted in the study: comparing the experiments carried out while increasing the extraction temperature, all peaks areas increased and no new peak appeared at high temperature.

*3.2. GC-MS Analyses.* GC-MS analyses were performed by a GC system Saturn CP-3800 (Varian, Walnut Creek, CA, USA) equipped with an ion trap mass analyzer Saturn 2200 (Varian, Walnut Creek, CA, USA); ionization was accomplished by EI at 70 eV and MS acquisition was carried out in full scan mode in the range 35–400 $m/z$. GC separations were performed on a capillary column VF-5MS (30 m × 0.25 mm × 0.25 $\mu$m, Varian, Walnut Creek, CA, USA) and helium was used as carrier gas with a constant flow rate of 1.00 mL/min. The GC temperature program was as follows: from 35 to 70°C at 5°C/min, 70°C for 15 min, from 70 to 100°C at 5°C/min, 100°C for 15 min, from 100 to 250°C at 7°C/min, 250°C for 2 min. The injector temperature was set at 230°C in splitless mode for the first 5 min and then with a split ratio 1 : 100. Three internal standards (IS) were used: o-tolualdehyde, 2,4,6-trimethylphenol, and benzophenone (Sigma Aldrich, Milwaukee, WI, USA).

*3.3. Data Analysis.* PCA, regression models, calculations, and graphical representations were performed by The Unscrambler X 10.2 (Camo, Oslo, Norway), Statistica v.7.1 (StatSoft Inc, Tulsa, OK, USA), Matlab R2014b (The Mathworks, Natick, MA, USA), and Excel 2007 (Microsoft Corporation, Redmond, WA, USA).

## 4. Results and Discussion

*4.1. Evaluation of the Amount of Sample.* As commonly acknowledged, in SPME, the amount of sample used during extraction is a key point to provide the highest signal intensity while avoiding saturation effects occurring on the fiber. Starting from the most of applications present in literature, reporting amounts of sample ranging from about 2 to over 16 g of honey, we decided to investigate different amounts by keeping the sample solution : headspace volume ratio almost constant at about 1 [54]. Three experiments were therefore carried out in the same experimental conditions for what concerns both SPME and GC-MS, using 2.00, 4.00, and 8.00 g honey added with 7.0, 5.5, and 3.0 mL water, respectively. Each experiment was therefore described by the area of the 143 peaks recorded in the chromatogram. Figure 1 reports the results obtained: the peaks are reported on $x$-axis in order of elution, while the corresponding areas are reported on the $y$-axis (the darker the marker, the highest the sample amount). For most of the peaks the areas increase in the order 2–4–8 g of honey, showing that for these peaks no saturation effect is present. For other peaks two different behaviors occur: (1) an increase between 2 and 4 g but not between 4 and 8 g (saturation effects are present with 8 g); (2) an increase between 4 and 8 g but not between 2 and 4 g (low signal intensity in the range 2–4 g). From these considerations we decided to use 4 g of honey in the successive experiments.

The experiments of the replicated experimental design were performed in a randomized order.

*4.2. Optimization of Signal Intensity.* PCA was carried out on the (27 × 143) data matrix, 27 being the number of replicated experiments of the experimental design and 143 being the areas of the identified peaks. The data matrix was autoscaled before PCA. The first 2 PCs were considered significant (accounting for about 66% of the overall variance: $PC_1$ = 46.99%, $PC_2$ = 18.93%).

Figures 2(a) and 2(b) report the loading plots of the first two PCs: loadings are reported separately for each PC; the $x$-axis reports the peaks in order of elution from left to right while the $y$-axis reports the corresponding loadings. As it can be observed from the figures, the first PC is mainly related to the behavior of the peaks eluting in the second part of the chromatogram (long retention times), almost all characterized by large positive loadings; an opposite behavior is observed for $PC_2$, which shows large negative loadings of the peaks belonging to the first part of the chromatogram (short retention times). The first two PCs are therefore able to sum up the information about peak areas depending on retention time, that is, about signal intensity: experiments with large signal intensity (large peak areas) are characterized by large positive scores on $PC_1$ and large negative ones on $PC_2$. The information about signal intensity is therefore separated in the first two components, each one related to different retention times of the analytes.

Figure 2(c) reports the corresponding score plot but labels the scores according to temperature. The score plot reports also the two simulated experiments described in the theory section, labeled, respectively, "best" and "worst," indicating

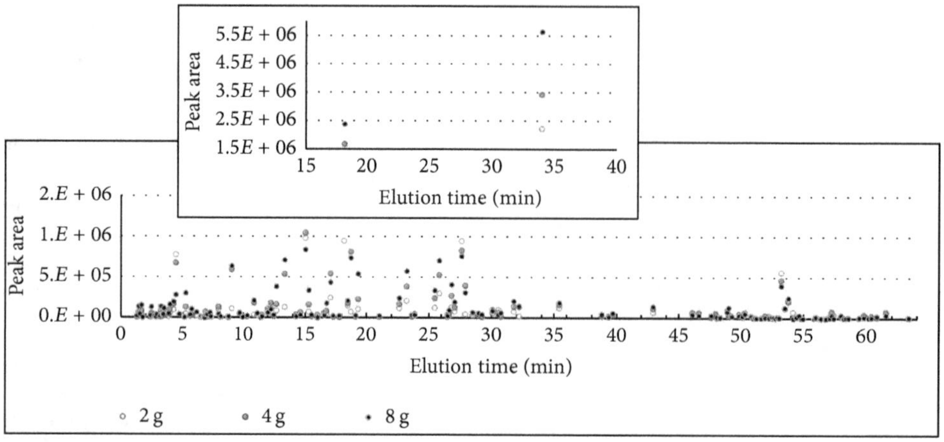


FIGURE 1: Evaluation of the optimal amount of sample in SPME GC-MS analysis: areas ($y$-axis) of each peak identified by its elution time ($x$-axis) using the experimental conditions reported in Section 4.1. Circles are represented on a color scale: the darker the circle, the larger the amount of honey. An enlargement is provided for the region of elution time between 15 and 40 minutes due to scale effects.

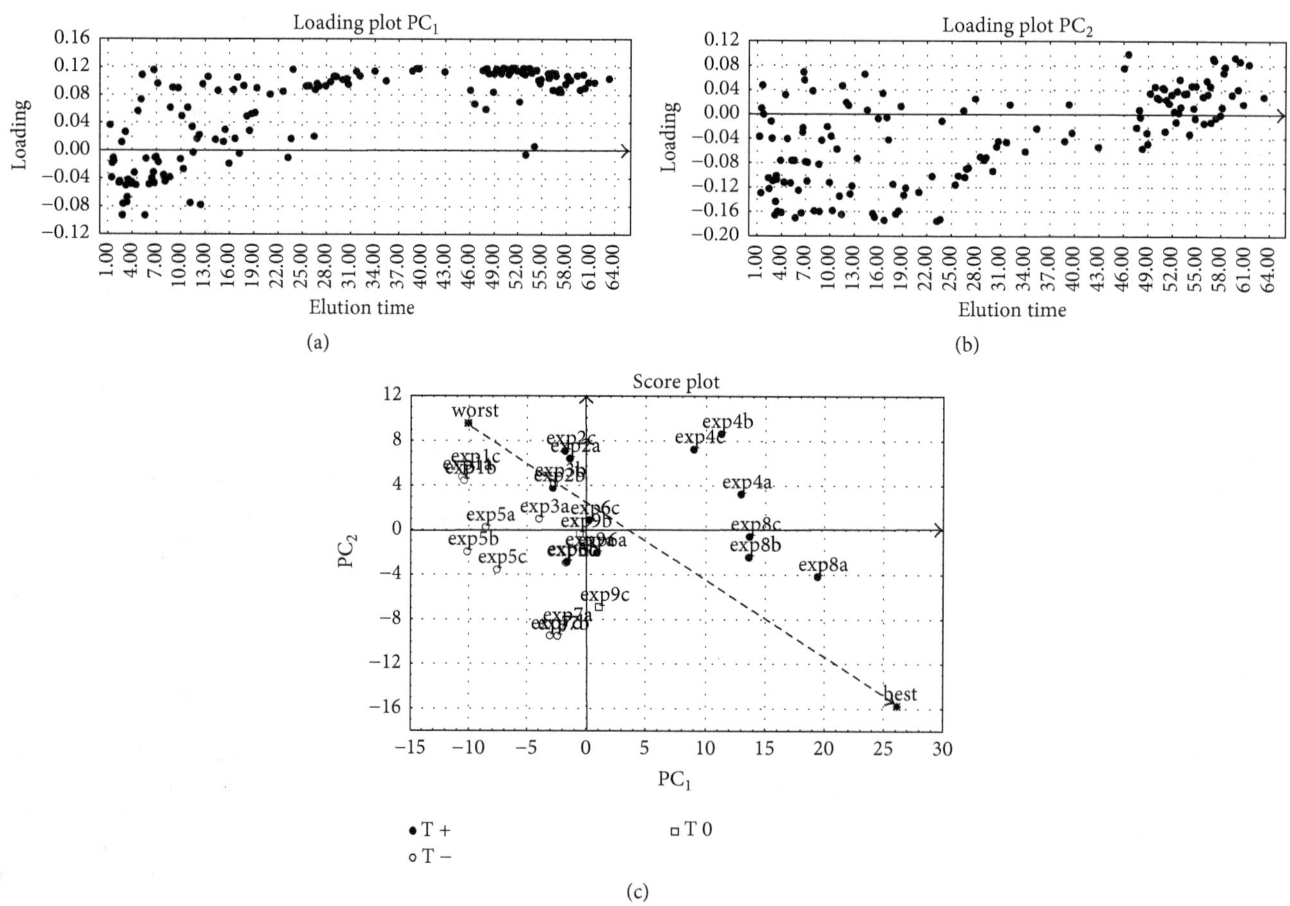


FIGURE 2: PCA results for signal intensity: loading plots for $PC_1$ (a) and $PC_2$ (b) and score plot of the first two PCs (c) with scores labeled according to extraction temperature. Experiments are numbered from 1 to 8 according to their order in Table 1; Exp 9 represents the center of the experimental domain. The last letter in the labels refers to the replication of the experimental design: "a," "b," or "c."

TABLE 2: ANOVA table for signal intensity (a) and repeatability (b); SS = sum of squares; df = degrees of freedom; $F$ = calculated $F$-Fisher value; $p$-level = calculated significance level.

(a) ANOVA: model for signal intensity

| | SS | df | MS | $F$ | $p$-level |
|---|---|---|---|---|---|
| $T$ | 388.88 | 1 | 388.88 | 80.20 | 0.0000 |
| $t$ | 656.79 | 1 | 656.79 | 135.46 | 0.0000 |
| salt | 240.95 | 1 | 240.95 | 49.70 | 0.0000 |
| Error | 106.67 | 22 | 4.85 | | |
| Total SS | 1404.79 | 26 | | | |

(b) ANOVA: model for repeatability

| | SS | df | MS | $F$ | $p$-level |
|---|---|---|---|---|---|
| $t$ * salt | 218.23 | 1 | 218.23 | 15.05 | 0.0081 |
| $T$ * $t$ * salt | 70.91 | 1 | 70.91 | 4.89 | 0.0689 |
| Error | 86.99 | 6 | 14.49 | | |
| Total SS | 376.13 | 8 | | | |

TABLE 3: Regression results for the signal intensity (a) and repeatability (b) models: effects estimates and their standard error, $t$-Student calculated, significance $p$-level calculated, regression coefficients, and their standard error.

(a) Regression results: model of the signal intensity

| | Effect | Std err (effect) | $t$-Student | $p$-level | Coeff | Std err (coeff) |
|---|---|---|---|---|---|---|
| Mean/interc. | 30.40 | 0.44 | 69.8 | 0.000 | 30.40 | 0.44 |
| $T$ | −8.00 | 0.92 | −8.70 | 0.000 | −4.00 | 0.46 |
| $t$ | −10.40 | 0.92 | −11.3 | 0.000 | −5.20 | 0.46 |
| salt | −6.30 | 0.92 | −6.85 | 0.000 | −3.20 | 0.46 |

(b) Regression results: model of the repeatability

| | Effect | Std err (effect) | $t$-Student | $p$-level | Coeff | Std err (coeff) |
|---|---|---|---|---|---|---|
| Mean/interc. | 11.0 | 1.3 | 8.75 | 0.000 | 11.0 | 1.3 |
| $t$ * salt | −10.0 | 2.7 | −3.88 | 0.008 | −5.0 | 1.3 |
| $T$ * $t$ * salt | 6.0 | 2.7 | 2.21 | 0.069 | 3.0 | 1.3 |

the direction of optimization (dotted arrow). Exp 1 and Exp 2 are the closest to the worst: Exp 1 is characterized by small areas of the peaks located both in the first (negative scores on $PC_2$) and in the second (negative scores on $PC_1$) part of the chromatogram. Exp 8 and Exp 4 are the best ones for what regards $PC_1$ (larger areas for the peaks of the second part of the chromatogram), while Exp 7 is the best one for what regards the peaks eluting in the first half of the chromatogram. The first PC is mainly related to the effect played by extraction temperature and time (experiments with high extraction temperature and time are at positive scores along $PC_1$): increasing temperature and time mainly plays a role on the peaks eluting in the second part of the chromatogram, as expected, since these compounds have higher boiling temperature and lower volatility. The second PC is instead mainly related to the effect played by salt (experiments with salt were at negative scores): salt plays a relevant effect mainly on the peaks eluting in the first half of the chromatogram.

The direction identified in the score plot between the "worst" and "best" experiments (dotted arrow in Figure 2(c)) represented the direction of the target function: the distance between each experiment and the "best" along the target function was used as a multivariate index of signal intensity and exploited to build a regression model relating signal intensity to the effect played by the examined factors.

Table 2(a) reports the ANOVA results obtained: only relevant effects are reported ($p$-level < 0.05). The final model contains only the principal effects: no second or third order interaction is significant and the model does not show a relevant curvature ($p$-level = 0.1379). Table 3(a) reports the coefficients of the final model together with their standard error and the corresponding $p$-level. The coefficients of the three factors show negative values, indicating that an increase of all the principal factors improves the experimental response. The final model shows a good fitting ability ($R^2$ = 0.9158). The plot of the predicted versus the observed values (Figure 3(a)) confirms the good fitting ability, while the residuals (Figure 3(b)) show no significant trend. Figure 3(c) reports the response surface plot of signal intensity ($z$-axis) as a function of temperature and time ($x$- and $y$-axes, resp.) when salt is at the highest level. The best overall conditions are obtained with all three factors at the highest level ($T$ = 70°C, $t$ = 60 min, and salt = 27.50%); either with or without the use of salt, a significant improvement in signal intensity can be reached when temperature and time are set at the highest level. The second best experiment within the experimental design is Exp 7 (temperature at low level, time at high level, and salt at high level).

Figure 4 reports the chromatograms obtained for Exp 8 and Exp 1, representing, respectively, the optimal and the worst experimental settings found in the experimental design: comparing the areas of the 143 identified peaks, more than 80% of the peaks show an increase in Exp 8 towards Exp 1; for about 54% the increase is two-fold and for 23%, three-fold, and the average value is about three-fold (with a maximum of 19 fold); Exp 1 shows a slight increase of some very early eluting peaks. Considering the signal intensity, the best conditions are those of Exp 8 (extraction temperature: 70°C; extraction time: 60 min; and salt percentage: 27.50% w/w).

*4.3. Optimization of Repeatability.* The three replications of each experiment of the design were used to estimate the experimental error (repeatability) in each condition of the experimental design. PCA was carried out on the (9 × 143) data matrix, 9 being the experiments of the experimental design and 143 the number of different CV%s calculated, one for each chromatographic peak. The data were autoscaled before PCA. The loading and score plot of the first two PCs (explaining, resp., 32.87 and 21.06% of the overall variance) are reported in Figures 5(a)–5(c). As already described for signal intensity, the score plot reports also the "best" and the "worst" simulated experiments.

From the score plot it is possible to observe that the optimization direction, consisting, as before, in the direction

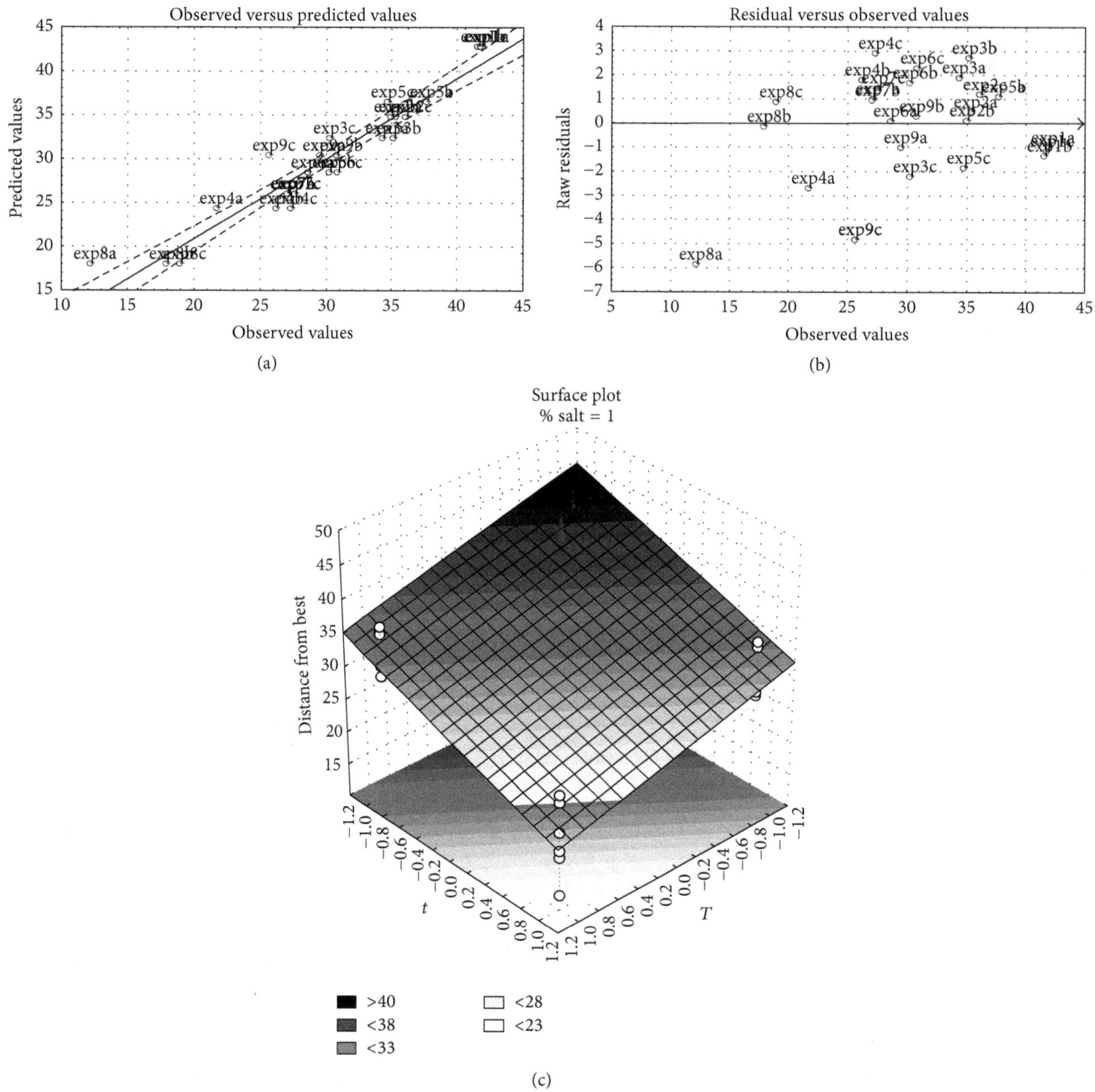


Figure 3: Model for signal intensity: predicted versus observed values (a), residuals versus observed values (b), and surface response for the interaction between temperature and time when salt is fixed at the highest level (c).

connecting the worst and best experiments, is represented by the first PC. The loading plot of the first PC shows most of the peaks with a large negative loading: experiments are therefore aligned along $PC_1$ from negative to positive scores according to a decreasing CV% value (i.e., increasing repeatability) for most of the peaks. The second PC accounts for the different behavior of the early eluting peaks (large positive loadings) and the last eluting ones (large negative loadings). Exp 7 and Exp 1 are therefore the best ones for what regards repeatability, showing the lowest overall CV% value (large positive scores on $PC_1$), while Exp 5 is the worst (largest negative scores on $PC_1$). Exp 2 and 5 are the most different according to $PC_2$: Exp 2 shows a repeatability in general above the average but its experimental conditions are particularly effective in improving repeatability of the last eluting peaks; Exp 5 is the overall worst experiment of the design but, according to $PC_2$, shows a behavior opposite to Exp 2. According to the second PC, among the experiments with an overall repeatability above the average, Exp 1, Exp 7, and Exp 8 improve in particular the repeatability of the early eluting peaks, while Exp 2 and Exp 6 show an opposite behavior. Anyway, the first PC is sufficient as the target function, $PC_2$ showing a negligible effect: the distance of the experiments from the best along $PC_1$ was therefore used for building a model for repeatability. Table 2(b) shows the results obtained for ANOVA, while Table 3(b) reports the

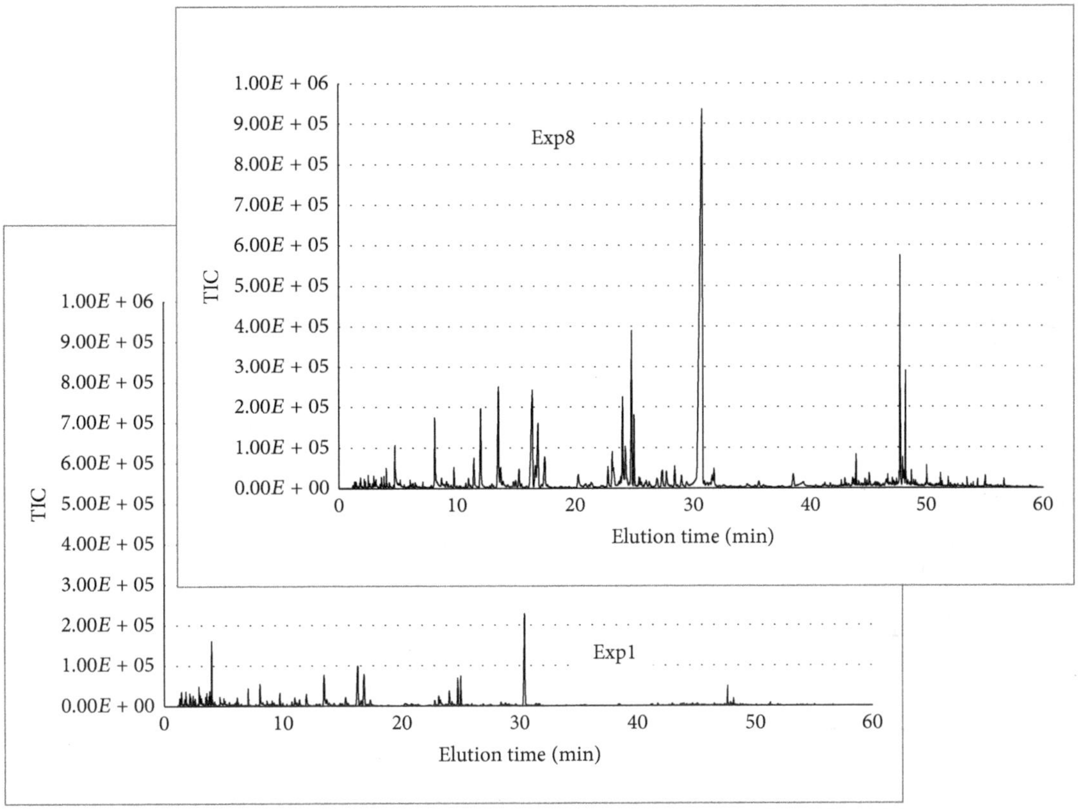


FIGURE 4: TIC chromatogram for Exp 1 (extraction temperature: 50°C, extraction time: 20 min, and no salt) and Exp 8 (extraction temperature: 70°C, extraction time: 60 min, 27.50% w/w of salt). Other details for the analytical method are reported in the experimental sections.

coefficients of the relevant effects together with their $p$-level and their standard deviation.

The final model shows an acceptable fitting ability ($R^2 = 0.7687$) allowing a discussion of the effects played by the factors on repeatability, devoted to the identification of the best operative conditions.

The final model contains the interaction between time and salt and the three-way interaction: this last one was retained as significant since it shows a $p$-level quite close to 0.05. The coefficient of the interaction between time and salt shows a negative coefficient, while the three-way interaction shows a positive coefficient. Figure 5(d) represents the three-way interaction: the experimental results obtained in terms of distance of each experiment from the best along the optimization direction are reported on a cube representing the experimental domain where each dimension corresponds to one of the investigated factors. When salt is not added, the best repeatability is obtained with a short extraction time, regardless of the level of temperature adopted. When salt is added instead and when temperature is low, increasing the extraction time strongly increases repeatability, which is slightly reduced with time at high temperature. On the other hand, with a short extraction time an increase in temperature strongly improves repeatability, while an opposite effect is recorded with a long extraction time. The overall best conditions correspond to Exp 7 ($T = 50°C$, $t = 60$ min, and salt = 27.5%). Table 4 resumes the repeatability results obtained for each experiment: each column reports the % of peaks in the chromatogram of the corresponding experiment showing a CV% smaller than four selected thresholds. As expected, Exp 7 (extraction temperature: 50°C; extraction time: 60 min; salt percentage: 27.50% w/w) provides the best results for the optimization of repeatability, showing more than 40% of peaks with a CV% smaller than 5%.

TABLE 4: Results obtained for repeatability for the experiments of the experimental design: % of peaks in the chromatogram with CV% smaller than 5%, 10%, 20%, and 30%. Results in bold represent the maxima recorded for each threshold level.

| | % CV < 5% | % CV < 10% | % CV < 20% | % CV < 30% |
|---|---|---|---|---|
| Exp 1 | 16 | 45 | **73** | **86** |
| Exp 2 | 21 | 42 | 63 | 76 |
| Exp 3 | 3.4 | 12 | 39 | 66 |
| Exp 4 | 2.1 | 20 | 51 | 72 |
| Exp 5 | 2.1 | 14 | 46 | 72 |
| Exp 6 | 16 | 45 | 66 | 79 |
| Exp 7 | **43** | **66** | **77** | **84** |
| Exp 8 | 5.5 | 29 | **76** | **87** |
| Exp 9 | 2.7 | 18 | 51 | **83** |

*4.4. Best Compromise.* It is useful to identify the experimental conditions allowing achievement of the best compromise between high signal intensity and high repeatability. To this purpose, the experiments of the experimental design can be

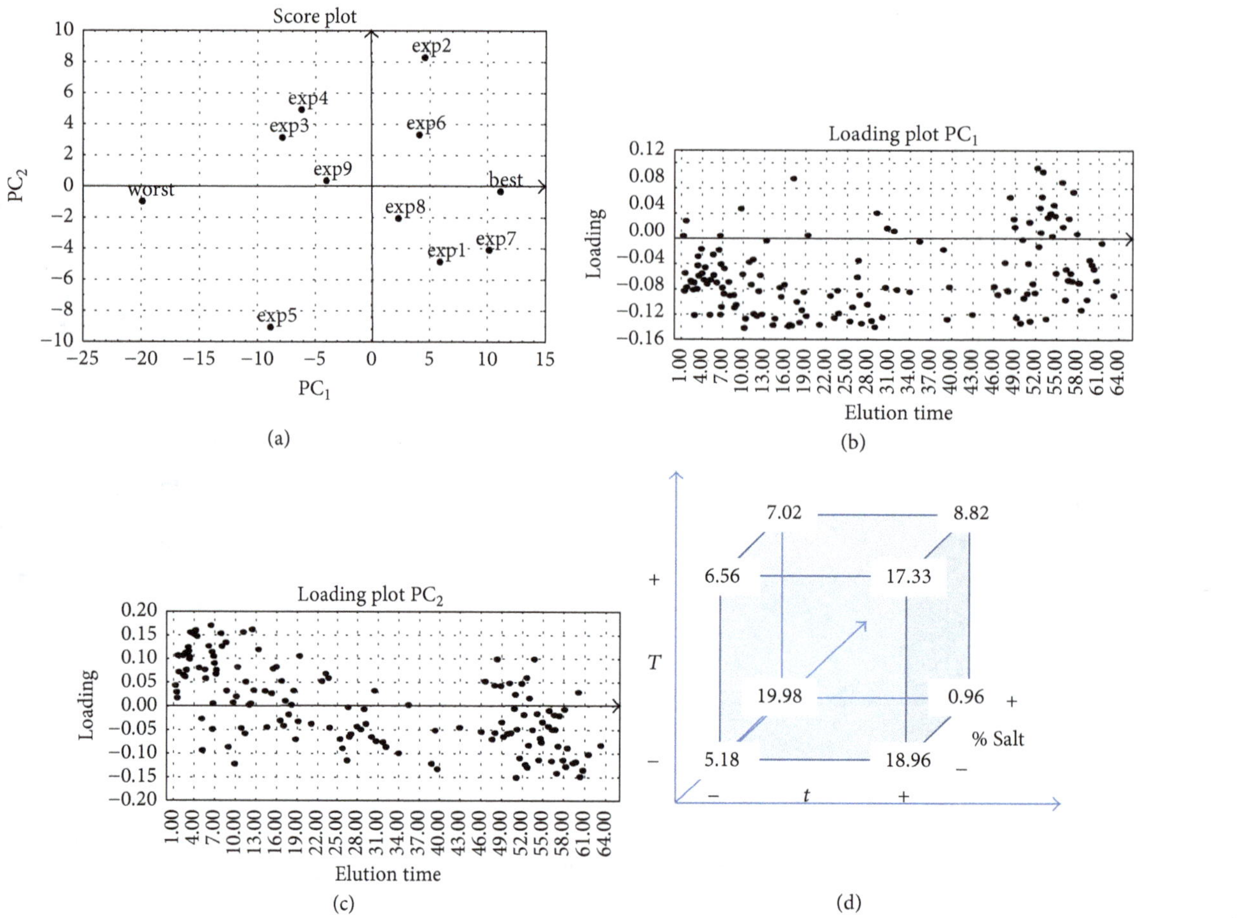


FIGURE 5: Results for repeatability: score plot of the first two PCs (a) and loading plots for $PC_1$ (b) and $PC_2$ (c); representation of the three-way interaction (d). Experiments are numbered from 1 to 8 according to their order in Table 1; Exp 9 represents the center of the experimental domain.

plotted as a function of their signal intensity and repeatability (Figure 6(a)). The scatter plot reports the repeatability of each experiment of the experimental design on the $y$-axis, expressed as distance from "best" along the first PC. Signal intensity is instead reported as the average distance from "best" in the space given by the first two PCs calculated over the three replications of the same experimental conditions. The plot shows also the best and the worst simulated experiments. Both repeatability and signal intensity are reported as range-scaled values computed with respect to "worst" and "best," respectively. The experimental conditions corresponding to the best compromise between optimization of signal intensity and of repeatability correspond to the experiment closest to "best."

Exp 8 and 7 appear as the best results: Exp 8 is the closest to "best" (distance = 0.96), while the extraction conditions adopted in Exp 7 provide a larger improvement of repeatability rather than of signal intensity (distance from the best = 1.23). The final global best compromise is therefore reached with the experimental settings of Exp 8 (extraction temperature: 70°C; extraction time: 60 min; salt percentage: 27.50% w/w).

*4.5. Exploration of the Experimental Domain around the Optimal Conditions.* Since the optimal experimental conditions identified, corresponding to Exp 8, are located at an extreme of the experimental domain, it is necessary to explore the experimental domain close to Exp 8 to check the model stability even outside the domain initially investigated. Therefore, three experiments were carried out: Exp 8, Exp 8−, and Exp 8+ (each experiment was replicated three times). Exp 8− ($T$ = 65°C, $t$ = 55 min, and salt = 24%) is located within the initial experimental domain, while Exp 8+ ($T$ = 75°C, $t$ = 65 min, and salt = 31%) is located outside the experimental domain; both experiments lay along the direction of optimization identified by the model. Figures 6(b) and 6(c) report the results obtained for signal intensity (b) and repeatability (c): the experiments are reported on the $x$-axis, while their distance from "best" is on the $y$-axis. As for signal intensity, the results show an improvement passing from Exp 8− to Exp 8 and Exp 8+, while there is no statistically significant difference between the three experiments regarding repeatability. Exp 8 does not represent a global optimum for signal intensity but its experimental conditions can be effectively applied since they represent a reasonable

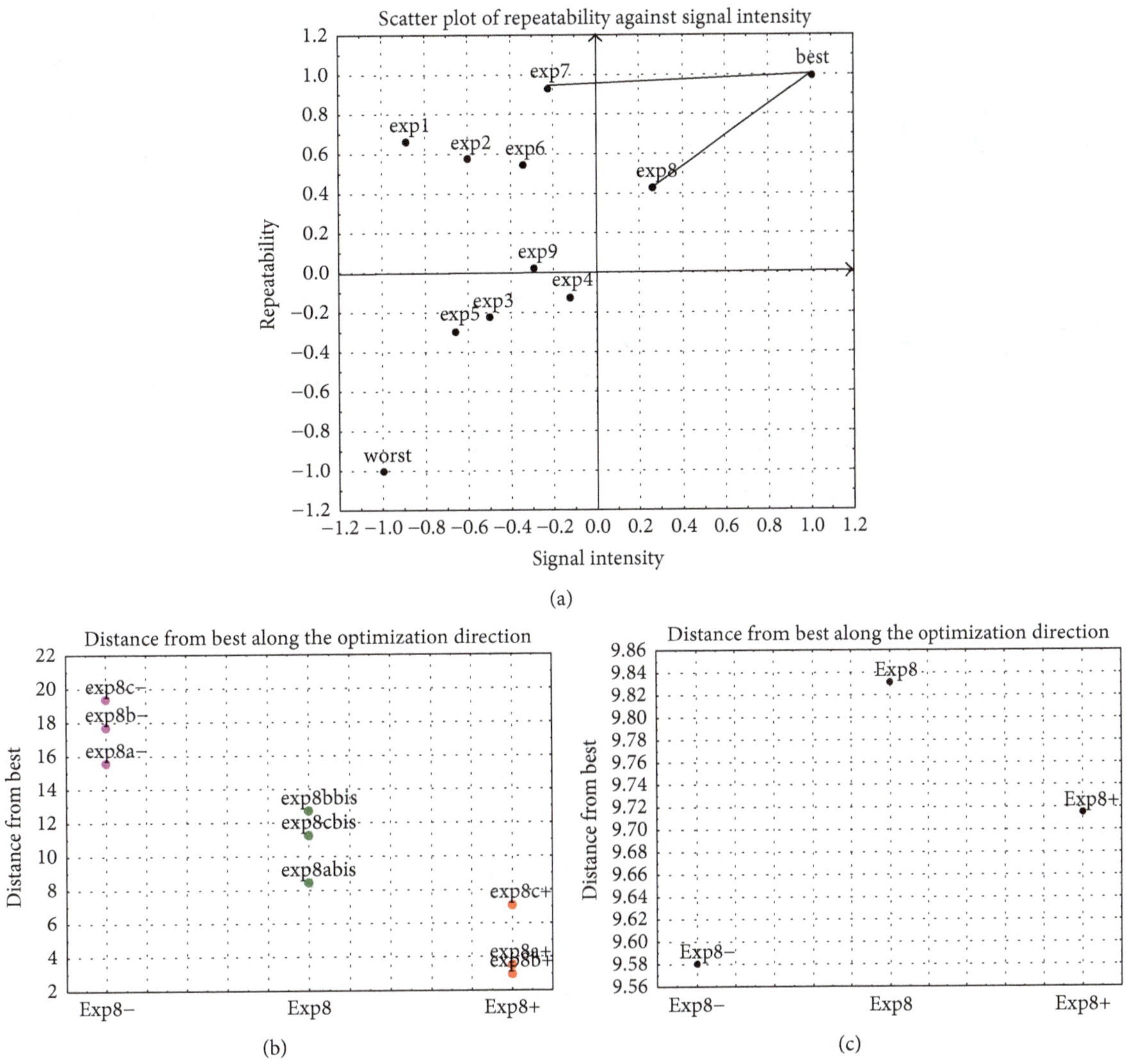


FIGURE 6: Repeatability versus signal intensity: identification of the best compromise between signal intensity and repeatability optimization (a). Exploration of the experimental domain around Exp 8: signal intensity (expressed as distance from best) (b) and repeatability (expressed as distance from best) (c) for Exp 8−, Exp 8, and Exp 8+. Experimental conditions were reported in Section 4.5.

compromise with time of analysis; moreover, repeatability is stable in the neighborhoods of Exp 8 and the linear model built for signal intensity is still valid in the region explored outside the initial experimental domain.

*4.6. Choice of the Suitable Internal Standard.* Some considerations can also be drawn about the choice of the suitable internal standard (IS). Here, three compounds were tested: o-tolualdehyde ($t_r$ ~ 15 min), 2,4,6-trimethylphenol ($t_r$ ~ 27 min), and benzophenone ($t_r$ ~ 54 min). The effectiveness of the three IS selected was tested by calculating the CV% of each peak from the peak areas normalized in turn for each of the available IS. The results are reported in Figure 7 for Exp 7 and Exp 8, considering three different time ranges in the chromatogram. Figure 7 reports for each case the % of peaks with CV% smaller than selected threshold values, namely, 2, 5, 10, and 20%.

For the early eluting peaks (0–20 min; Figure 7(a)), 2,4,6-trimethylphenol (indicated as "trime") seems the best choice, providing the largest % of peaks with small CV%; in particular, when it is used as IS in this time range, Exp 7 shows better results than Exp 8. For the intermediate part of the chromatogram (20–40 min; Figure 7(b)), 2,4,6-trimethylphenol is again the best choice but, in this case, its use as IS provides similar results for Exp 7 and Exp 8. Finally, for the last eluting peaks (40 min-end; Figure 7(c)), benzophenone (indicated as "benzo") is the best alternative and again Exp 7 and Exp 8 show similar results. O-Tolualdehyde (indicated as "o-tolu") never represents the best choice. So, two different IS can be used for peak area normalization: 2,4,6-Trimethylphenol for peaks eluting before 40 min and benzophenone for the last eluting ones. When IS are used for peak area normalization, Exp 7 and Exp 8 show similar behavior concerning repeatability: this consideration further confirms

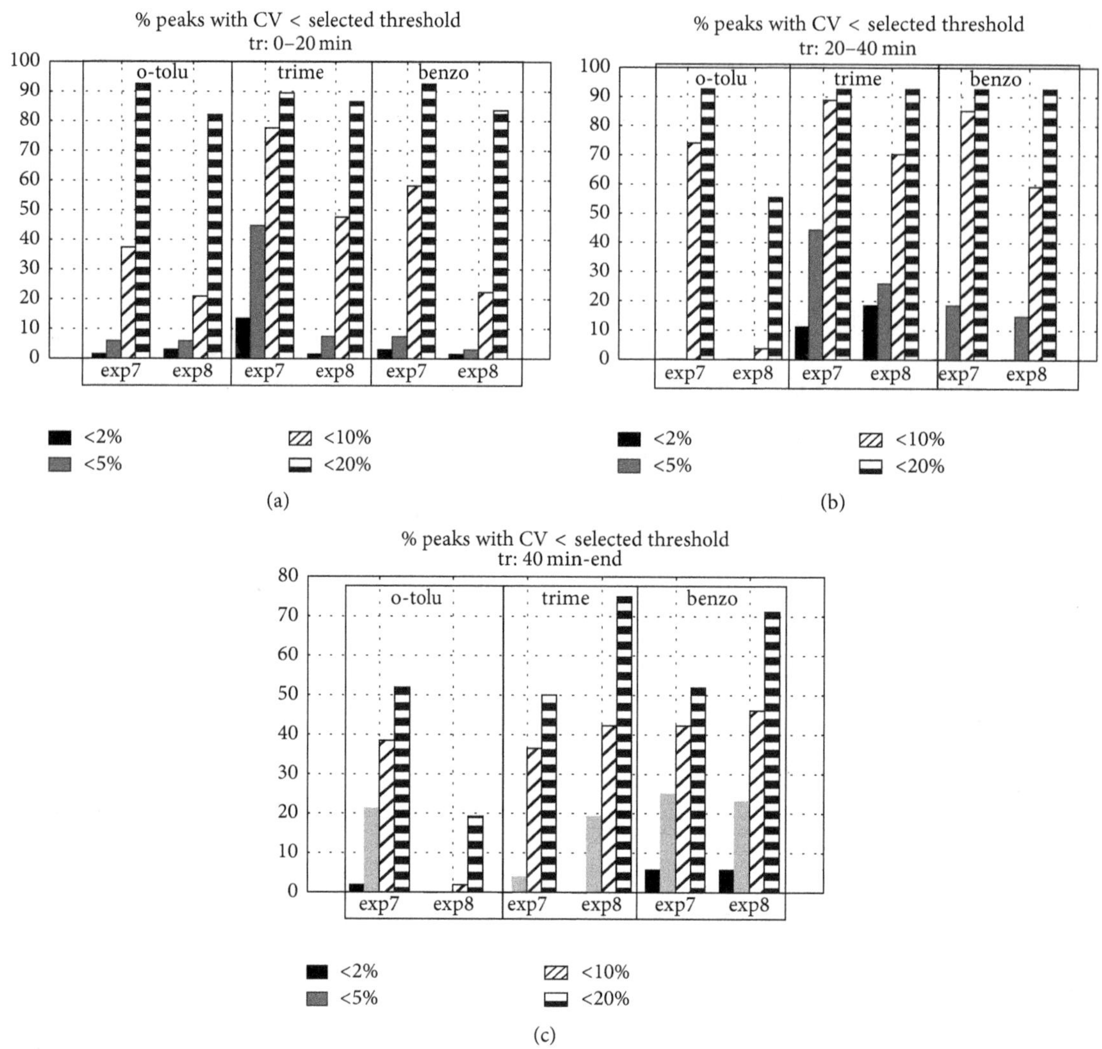


FIGURE 7: Comparison of three I.S.: % of peaks with CV% less than selected threshold levels (2, 5, 10, and 20) when o-tolualdehyde (o-tolu), 2,4,6-trimethylphenol (trime), and benzophenone (benzo) are used as I.S. for area correction: peaks eluting in the range 0–20 min (a), 20–40 min (b), and 40 min-end (c). The results are represented for Exp 7 ($T = 60^{\circ}$C, $t = 60$ min, and salt = 27.5%) and Exp 8 ($T = 70^{\circ}$C, $t = 60$ min, and salt = 27.5%).

that the identification of Exp 8 conditions as the global optimal ones is a good choice.

*4.7. GC-MS Analysis.* The optimized conditions of SPME and GC-MS were applied to the analysis of an organic commercial multiflower honey sample. From the results, a total of 143 peaks corresponding to 143 molecules were identified and reported in Table S1 in Supplementary Material available online at https://doi.org/10.1155/2017/6437857. Each species was identified by the comparison of its experimental mass spectrum with those present in the NIST 2.0 library and compared by the use of standards, if present. The identified molecules belonged to different classes as carboxylic acids (that give a "buttery" flavor), alcohols, aldehydes (associated to a "caramel" flavor), ketones, terpenes, esters, norisoprenoid, and furan derivatives. The presence of branched aldehydes as 2-methyl-butanal and 2-methyl-propanal can derive from the metabolism of aminoacids and from the fermentation by the action of yeasts, generally accepted in honey. The presence of short chain and branched alcohols (as 1-butanol and 3-buten-1-ol-3-methyl) can derive from the Maillard's reaction that takes place during honey storage and processing, while furfural and 2(3H)-furanone-5-ethenyldihydro-5-methyl derived from nonenzymatic browning reactions. The identification of terpenes as $\alpha$-terpinene and $\tau$-terpinene is associated to the floral nectar of honeydew gathered by honeybees [55, 56].

## 5. Conclusions

Optimization of SPME of honey samples was carried out here considering both signal intensity (expressed as chromatographic peak areas) and repeatability (expressed as the variation of each peak area across replications of the same experiment) by means of experimental design, studying the effect of three factors: temperature of extraction, time of extraction, and the amount of salt added to the sample. Each experiment was evaluated by a multivariate target function based on PCA, taking into account all the analytes at the same time. For both signal intensity and repeatability, a multivariate target function was defined in the space given by the significant PCs. The position of each experiment along the target function was therefore used as response to build regression models relating signal intensity and repeatability to the effects played by the relevant factors and/or their interactions. A model with good fitting ability was achieved for signal intensity, while the model for repeatability proved to be suitable for a general discussion of the effects played by the relevant factors. Optimal experimental conditions were finally identified as the best compromise between repeatability and signal intensity optimization: temperature of extraction at 70°C, time of extraction at 60 min, and 27.50% w/w of salt. It was also possible to identify the most suitable internal standards, namely, 2,4,6-trimethylphenol, which proved to be the best choice for peaks eluting before 40 min and benzophenone, that was optimal for the last eluting ones.

The optimized method was used for the SPME GC-MS analysis of a multiflower honey sample and more than a hundred compounds were identified.

The proposed multivariate approach represents an effective approach that can be adopted to accomplish multivariate optimization when solid phase microextraction has to be applied. The main advantage of this procedure is the achievement of global optimal experimental conditions providing the optimization of both signal intensity and repeatability; this is particularly important since low repeatability is one of the most serious drawbacks in SPME. The results obtained for what regards the experimental conditions identified as the best ones are in agreement with similar procedures reported in literature. Here we also provide indication about the best internal standard to be adopted, while in literature one single I.S. or no I.S. at all is used for volatiles extraction from honey samples by SPME.

## Competing Interests

The authors declare that there is no conflict of interests regarding the publication of this paper.

## Acknowledgments

The authors gratefully acknowledge support from Ministero della Sanità, Italy (Ricerca Finalizzata, Project PRF2009201, "untargeted approach in detecting food adulteration and contamination using spectroscopic techniques database and chemometrics").

## References

[1] E. Ranzato, S. Martinotti, and B. Burlando, "Epithelial mesenchymal transition traits in honey-driven keratinocyte wound healing: comparison among different honeys," *Wound Repair and Regeneration*, vol. 20, no. 5, pp. 778–785, 2012.

[2] N. J. Basson and S. R. Grobler, "Antimicrobial activity of two South African honeys produced from indigenous Leucospermum cordifolium and Erica species on selected micro-organisms," *BMC Complementary and Alternative Medicine*, vol. 8, article no. 41, 2008.

[3] A. B. Jull, N. Walker, and S. Deshpande, "Honey as a topical treatment for wounds," *Cochrane Database of Systematic Reviews*, vol. 2, Article ID CD005083, 2013.

[4] L. Castro-Vázquez, V. Leon-Ruiz, M. E. Alañon, M. S. Pérez-Coello, and A. V. González-Porto, "Floral origin markers for authenticating Lavandin honey (*Lavandula angustifolia x latifolia*). Discrimination from Lavender honey (*Lavandula latifolia*)," *Food Control*, vol. 37, no. 1, pp. 362–370, 2014.

[5] A. Agila and S. Barringer, "Effect of adulteration versus storage on volatiles in Unifloral Honeys from different floral sources and locations," *Journal of Food Science*, vol. 78, no. 2, pp. C184–C191, 2013.

[6] I. Špánik, A. Janáčová, Z. Šusterová et al., "Characterisation of VOC composition of Slovak monofloral honeys by GC×GC-TOF-MS," *Chemical Papers*, vol. 67, no. 2, pp. 127–134, 2013.

[7] Y. Yang, M.-J. Battesti, J. Paolini, A. Muselli, P. Tomi, and J. Costa, "Melissopalynological origin determination and volatile composition analysis of Corsican "*Erica arborea* spring maquis" honeys," *Food Chemistry*, vol. 134, no. 1, pp. 37–47, 2012.

[8] V. Kaškoniene and P. R. Venskutonis, "Floral markers in honey of various botanical and geographic origins: a review," *Comprehensive Reviews in Food Science and Food Safety*, vol. 9, no. 6, pp. 620–634, 2010.

[9] E. Alissandrakis, E. Mantziaras, P. A. Tarantilis, P. C. Harizanis, and M. Polissiou, "Generation of linalool derivatives in an artificial honey produced from bees fed with linalool-enriched sugar syrup," *European Food Research and Technology*, vol. 231, no. 1, pp. 21–25, 2010.

[10] D. Mannaş and T. Altuğ, "SPME/GC/MS and sensory flavour profile analysis for estimation of authenticity of thyme honey," *International Journal of Food Science and Technology*, vol. 42, no. 2, pp. 133–138, 2007.

[11] C. Guyot-Declerck, S. Renson, A. Bouseta, and S. Collin, "Floral quality and discrimination of *Lavandula stoechas*, *Lavandula angustifolia*, and *Lavandula angustifolia x latifolia* honeys," *Food Chemistry*, vol. 79, no. 4, pp. 453–459, 2002.

[12] T. Dymerski, T. Chmiel, A. Mostafa et al., "Botanical and geographical origin characterization of Polish honeys by headspace SPME-GC× GC-TOFMS," *Current Organic Chemistry*, vol. 17, no. 8, pp. 853–870, 2013.

[13] G. Papotti, D. Bertelli, and M. Plessi, "Use of HS-SPME-GC-MS for the classification of Italian lemon, orange and citrus spp. honeys," *International Journal of Food Science and Technology*, vol. 47, no. 11, pp. 2352–2358, 2012.

[14] L. Cuevas-Glory, E. Ortiz-Vázquez, J. A. Pino, and E. Sauri-Duch, "Floral classification of Yucatan Peninsula honeys by PCA & HS-SPME/GC-MS of volatile compounds," *International Journal of Food Science and Technology*, vol. 47, no. 7, pp. 1378–1383, 2012.

[15] I. Escriche, M. Kadar, M. Juan-Borrás, and E. Domenech, "Using flavonoids, phenolic compounds and headspace volatile profile for botanical authentication of lemon and orange honeys," *Food Research International*, vol. 44, no. 5, pp. 1504–1513, 2011.

[16] I. Stanimirova, B. Üstün, T. Cajka et al., "Tracing the geographical origin of honeys based on volatile compounds profiles

assessment using pattern recognition techniques," *Food Chemistry*, vol. 118, no. 1, pp. 171–176, 2010.

[17] E. de la Fuente, R. M. Valencia-Barrera, I. Martínez-Castro, and J. Sanz, "Occurrence of 2-hydroxy-5-methyl-3-hexanone and 3-hydroxy-5-methyl-2-hexanone as indicators of botanic origin in eucalyptus honeys," *Food Chemistry*, vol. 103, no. 4, pp. 1176–1180, 2007.

[18] E. de la Fuente, I. Martínez-Castro, and J. Sanz, "Characterization of Spanish unifloral honeys by solid phase microextraction and gas chromatography-mass spectrometry," *Journal of Separation Science*, vol. 28, no. 9-10, pp. 1093–1100, 2005.

[19] L. Piasenzotto, L. Gracco, and L. Conte, "Solid phase microextraction (SPME) applied to honey quality control," *Journal of the Science of Food and Agriculture*, vol. 83, no. 10, pp. 1037–1044, 2003.

[20] N. Jánošková, O. Vyviurska, and I. Špánik, "Identification of volatile organic compounds in honeydew honeys using comprehensive gas chromatography," *Journal of Food and Nutrition Research*, vol. 53, no. 4, pp. 353–362, 2014.

[21] A. Romano, M. Cuenca, S. Makhoul et al., "Comparison of e-Noses: the case study of honey," *Italian Journal of Food Science*, vol. 28, no. 2, pp. 326–337, 2016.

[22] M. Moniruzzaman, I. Rodríguez, M. Ramil, R. Cela, S. A. Sulaiman, and S. H. Gan, "Assessment of gas chromatography time-of-flight accurate mass spectrometry for identification of volatile and semi-volatile compounds in honey," *Talanta*, vol. 129, pp. 505–515, 2014.

[23] M. Pontes, J. C. Marques, and J. S. Câmara, "Screening of volatile composition from Portuguese multifloral honeys using headspace solid-phase microextraction-gas chromatography-quadrupole mass spectrometry," *Talanta*, vol. 74, no. 1, pp. 91–103, 2007.

[24] F. Pasini, S. Gardini, G. L. Marcazzan, and M. F. Caboni, "Buckwheat honeys: screening of composition and properties," *Food Chemistry*, vol. 141, no. 3, pp. 2802–2811, 2013.

[25] I. Jerković, A. Kasum, Z. Marijanović, and C. I. G. Tuberoso, "Contribution to the characterisation of honey-based Sardinian product abbamele: volatile aroma composition, honey marker compounds and antioxidant activity," *Food Chemistry*, vol. 124, no. 1, pp. 401–410, 2011.

[26] A. C. Soria, J. Sanz, and I. Martínez-Castro, "SPME followed by GC-MS: a powerful technique for qualitative analysis of honey volatiles," *European Food Research and Technology*, vol. 228, no. 4, pp. 579–590, 2009.

[27] A. C. Soria, I. Martínez-Castro, C. de Lorenzo, and J. Sanz, "Occurrence of nitriles in *Taraxacum* labelled honeys," *Food Chemistry*, vol. 107, no. 1, pp. 439–443, 2008.

[28] N. Campillo, P. Viñas, R. Peñalver, J. I. Cacho, and M. Hernández-Córdoba, "Solid-phase microextraction followed by gas chromatography for the speciation of organotin compounds in honey and wine samples: a comparison of atomic emission and mass spectrometry detectors," *Journal of Food Composition and Analysis*, vol. 25, no. 1, pp. 66–73, 2012.

[29] G. Papotti, D. Bertelli, M. Lolli, A. G. Sabatini, and M. Plessi, "Methyl anthranilate content in Italian citrus honeys determined by HS-SPME-GC," *International Journal of Food Science and Technology*, vol. 44, no. 10, pp. 1933–1938, 2009.

[30] D. Bertelli, G. Papotti, M. Lolli, A. G. Sabatini, and M. Plessi, "Development of an HS-SPME-GC method to determine the methyl anthranilate in Citrus honeys," *Food Chemistry*, vol. 108, no. 1, pp. 297–303, 2008.

[31] A. C. Soria, M. González, C. De Lorenzo, I. Martínez-Castro, and J. Sanz, "Estimation of the honeydew ratio in honey samples from their physicochemical data and from their volatile composition obtained by SPME and GC-MS," *Journal of the Science of Food and Agriculture*, vol. 85, no. 5, pp. 817–824, 2005.

[32] B. Plutowska, T. Chmiel, T. Dymerski, and W. Wardencki, "A headspace solid-phase microextraction method development and its application in the determination of volatiles in honeys by gas chromatography," *Food Chemistry*, vol. 126, no. 3, pp. 1288–1298, 2011.

[33] Q. L. Ma, N. Hamid, A. E. D. Bekhit, J. Robertson, and T. F. Law, "Optimization of headspace solid phase microextraction (HS-SPME) for gas chromatography mass spectrometry (GC-MS) analysis of aroma compounds in cooked beef using response surface methodology," *Microchemical Journal*, vol. 111, pp. 16–24, 2013.

[34] E. Paula Barros, N. Moreira, G. Elias Pereira, S. G. F. Leite, C. Moraes Rezende, and P. Guedes De Pinho, "Development and validation of automatic HS-SPME with a gas chromatography-ion trap/mass spectrometry method for analysis of volatiles in wines," *Talanta*, vol. 101, pp. 177–186, 2012.

[35] G. Basaglia and M. C. Pietrogrande, "Optimization of a SPME/GC/MS method for the simultaneous determination of pharmaceuticals and personal care products in waters," *Chromatographia*, vol. 75, no. 7-8, pp. 361–370, 2012.

[36] P. C. F. De Lima Gomes, J. Y. Barletta, C. E. D. Nazario et al., "Optimization of in situ derivatization SPME by experimental design for GC-MS multi-residue analysis of pharmaceutical drugs in wastewater," *Journal of Separation Science*, vol. 34, no. 4, pp. 436–445, 2011.

[37] C. Pizarro, N. Pérez-del-Notario, and J. M. González-Sáiz, "Optimisation of a simple and reliable method based on headspace solid-phase microextraction for the determination of volatile phenols in beer," *Journal of Chromatography A*, vol. 1217, no. 39, pp. 6013–6021, 2010.

[38] J. Januszkiewicz, H. Sabik, S. Azarnia, and B. Lee, "Optimization of headspace solid-phase microextraction for the analysis of specific flavors in enzyme modified and natural Cheddar cheese using factorial design and response surface methodology," *Journal of Chromatography A*, vol. 1195, no. 1-2, pp. 16–24, 2008.

[39] J. Salafranca, C. Domeño, C. Fernández, and C. Nerín, "Experimental design applied to the determination of several contaminants in Duero River by solid-phase microextraction," *Analytica Chimica Acta*, vol. 477, no. 2, pp. 257–267, 2003.

[40] A. Asfaram, M. Ghaedi, A. Goudarzi, M. Soylak, and S. Mehdizadeh Langroodi, "Magnetic nanoparticle based dispersive micro-solid-phase extraction for the determination of malachite green in water samples: optimized experimental design," *New Journal of Chemistry*, vol. 39, no. 12, pp. 9813–9823, 2015.

[41] A. Asfaram, M. Ghaedi, A. Goudarzi, and M. Soylak, "Comparison between dispersive liquid-liquid microextraction and ultrasound-assisted nanoparticles-dispersive solid-phase microextraction combined with microvolume spectrophotometry method for the determination of Auramine-O in water samples," *RSC Advances*, vol. 5, no. 49, pp. 39084–39096, 2015.

[42] S. Bahrani, M. Ghaedi, M. J. Mansoorkhani, A. Asfaram, A. A. Bazrafshan, and M. K. Purkait, "Ultrasonic assisted dispersive solid-phase microextraction of eriochrome cyanine R from water sample on ultrasonically synthesized lead (II) dioxide nanoparticles loaded on activated carbon: experimental design methodology," *Ultrasonics Sonochemistry*, vol. 34, pp. 317–324, 2017.

[43] A. Asfaram, M. Ghaedi, and A. Goudarzi, "Optimization of ultrasound-assisted dispersive solid-phase microextraction based on nanoparticles followed by spectrophotometry for the simultaneous determination of dyes using experimental design," *Ultrasonics Sonochemistry*, vol. 32, pp. 407–417, 2016.

[44] J. N. Bianchin, G. Nardini, J. Merib, A. N. Dias, E. Martendal, and E. Carasek, "Screening of volatile compounds in honey using a new sampling strategy combining multiple extraction temperatures in a single assay by HS-SPME-GC-MS," *Food Chemistry*, vol. 145, pp. 1061–1065, 2014.

[45] L. Ceballos, J. A. Pino, C. E. Quijano-Celis, and A. Dago, "Optimization of a HS-SPME/GC-MS method for determination of volatile compounds in some cuban unifloral honeys," *Journal of Food Quality*, vol. 33, no. 4, pp. 507–528, 2010.

[46] F. Pellati, S. Benvenuti, F. Yoshizaki, D. Bertelli, and M. C. Rossi, "Headspace solid-phase microextraction-gas chromatography-mass spectrometry analysis of the volatile compounds of *Evodia* species fruits," *Journal of Chromatography A*, vol. 1087, no. 1-2, pp. 265–273, 2005.

[47] S. Furlanetto, S. Orlandini, P. Mura, M. Sergent, and S. Pinzauti, "How experimental design can improve the validation process. Studies in pharmaceutical analysis," *Analytical and Bioanalytical Chemistry*, vol. 377, no. 5, pp. 937–944, 2003.

[48] M. M. Bravo, A. S. Valenzuela, E. P. Fuentes, and W. V. Quiroz, "Critical evaluation of fiber coatings for organotin determination by using solid phase microextraction in headspace mode," *Journal of Chromatography A*, vol. 1223, pp. 9–14, 2012.

[49] J. S. Ribeiro, R. F. Teófilo, F. Augusto, and M. M. C. Ferreira, "Simultaneous optimization of the microextraction of coffee volatiles using response surface methodology and principal component analysis," *Chemometrics and Intelligent Laboratory Systems*, vol. 102, no. 1, pp. 45–52, 2010.

[50] E. Robotti, F. Gosetti, E. Mazzucco, D. Zampieri, and E. Marengo, "Iterative optimization of an ESI IT mass spectrometer using regular simplex and a multivariate target function representing the S/N ratio," *Journal of the American Society for Mass Spectrometry*, vol. 22, no. 1, pp. 118–129, 2011.

[51] E. Marengo, M. Aceto, E. Robotti, M. Oddone, and M. Bobba, "Optimisation of sensitivity in the multi-elemental determination of 83 isotopes by ICP-MS as a function of 21 instrumental operative conditions by modified simplex, principal component analysis and partial least squares," *Talanta*, vol. 76, no. 5, pp. 1224–1232, 2008.

[52] G. Box, S. Hunter, and W. G. Hunter, *Statistics for Experimenters*, John Wiley & Sons, Hoboken, NJ, USA, 2005.

[53] D. L. Massart, B. G. M. Vanderginste, S. M. Deming, Y. Michotte, and L. Kaufman, *Chemometrics: A Textbook*, Elsevier, Amsterdam, The Netherlands, 1988.

[54] J. Pawliszyn, *Solid Phase Microextraction: Theory and Practice*, Wiley-VCH, New York, NY, USA, 1997.

[55] A. C. Soria, I. Martínez-Castro, and J. Sanz, "Some aspects of dynamic headspace analysis of volatile components in honey," *Food Research International*, vol. 41, no. 8, pp. 838–848, 2008.

[56] H. E. Tahir, Z. Xiaobo, H. Xiaowei, S. Jiyong, and A. A. Mariod, "Discrimination of honeys using colorimetric sensor arrays, sensory analysis and gas chromatography techniques," *Food Chemistry*, vol. 206, pp. 37–43, 2016.

6

# Development of a Methodology for the Determination of Pesticide Residues in Cajá-Manga Pulp (*Spondias dulcis* L.) Using Solid-Liquid Extraction with Low-Temperature Partitioning

**Escarlet T. D. Santana,[1] Dayane F. Soares,[2] and Anizio M. Faria[1]**

[1]*Instituto de Ciências Exatas e Naturais do Pontal, Universidade Federal de Uberlândia, Ituiutaba, MG 38304-402, Brazil*
[2]*Instituto Federal de Educação, Ciência e Tecnologia do Triângulo Mineiro, Ituiutaba, MG 38305-200, Brazil*

Correspondence should be addressed to Anizio M. Faria; anizio@ufu.br

Academic Editor: Maurice Millet

Emergent fruits with functional properties and sui generis flavours are increasingly gaining international market, mainly due to the pleasant taste and its nutritional properties. However, methodologies for determination of pesticide residues in these fruits are still incipient or nonexistent. In this work, a simple, rapid, and accessible methodology to small laboratories was developed, optimized, and validated for the determination of four pesticide residues in the cajá-manga pulp: chlorpyrifos, thiabendazole, carboxin, and difenoconazole. The extraction of pesticides from the cajá-manga pulp was performed by the liquid-solid extraction with low-temperature partitioning, achieving efficiency in the range of 75.3–129.3% with limits of quantification between 170 and 430 $ng \cdot g^{-1}$ for all compounds. The methodology was validated according to the SANTE/11813/2017 document, demonstrating good detectability, selectivity, precision, and accuracy with limits of quantification within the range of the maximum residual limits preconized for the compounds. The quantification was performed by gas chromatography with flame ionization detection which, although it does not present the best detectability for the compounds, is a lower cost instrumentation and is available in several analysis laboratories in Brazil, making the method more accessible to evaluate the pesticide residues in fruit pulp. Thus, in this work, a methodology for the determination of pesticides in the cajá-manga pulp is available for the monitoring of pesticide residues in fruit pulp, in an efficient and accessible way.

## 1. Introduction

The healthy foods consumption has been growing every year at rates higher than those of the world economy. Among these foods, fruits stand out as the preferred and most consumed by the population [1–3]. Sources of vitamins, fibers, minerals, and various other substances of nutritional and/or therapeutic value, fruits, have pleasant sensorial characteristics that intensify their consumption. A fruit that has had a growing commercialization in several countries around the world, whether in natural or processed form as pulps, juices, popsicles, and ice creams, is the cajá-manga [4, 5]. The cajá-manga (*Spondias dulcis*) belongs to the Anacardiaceae family, is native of the Society Islands in Polynesia, has a *sui generis* flavour and a high nutritional value, and is rich in vitamins A and B, fiber, calcium, phosphorus, and iron, essential nutrients for a healthy lifetime [5–7].

The exploitation of the cajá-manga is mainly carried out in an extractivistic way in Brazil [8], and the cajazeira is frequently affected by diseases and plagues, such as anthracnose (*Colletotrichum gloeosporioides*), verrugose (*Sphaceloma fawcetti*), fruit flies, and sauva ants, which affect the production and the quality of fruits [9–11]. To supply an increasingly growing market for cajá-manga consumption, the use of pesticides is a faster and more effective solution to guarantee productivity and meet the growing demand. However, when applied incorrectly, pesticides leave residues in foods that cause harmful effect to human health, being one of the main problems of food intoxication today [12–17].

In Brazil, the monitoring of pesticide residues in food is carried out by the National Agency for Sanitary Vigilance (ANVISA) through the Programa de Análise de Resíduos de Agrotóxicos em Alimentos (PARA) [18]. Annually, the PARA publishes reports with the indexes of pesticide residues in the main foods of vegetal origin. Nevertheless, emerging fruits such as cajá-manga do not present methodologies for the determination of pesticide residues and, therefore, are not included in the reports of the PARA, despite the increasing popularization of its consumption.

In this work, a simple, fast, and easy-to-implement method was developed for the determination of four pesticide residues in the cajá-manga pulp by solid-liquid extraction with low-temperature partitioning (SLE-LTP) and quantification by gas chromatography with flame ionization detection (GC-FID). SLE-LTP is a simple extraction method involving a sequence of steps: addition of a small volume of water to the solid sample, addition of a water-miscible extractor solvent, vortex homogenization, and freezing of the system for phase separation [19–23]. The organic extract, which will remain liquid after the freezing step, is removed by using a syringe and introduced into a scientific instrumentation suitable for quantification of the analyte(s). SLE-LTP has already been used for the extraction of pesticides from different matrices, such as yacare eggs [19], soil [20], fresh food [21, 23, 24], and processed foods [22]. Due to the simplicity of use and efficiency in the extraction process, the optimal conditions for extraction of pesticide residues from the cajá-manga pulp by SLE-LTP will be determined in this work. The gas chromatography with flame ionization detection will be the analytical technique used to quantify the pesticide residues due to the possibility of applying the method in various laboratories, from those with a simpler structure to the larger ones. So, one of the objectives of this work is to provide a simple, efficient, and, mainly, accessible method to laboratories that do not have sophisticated instrumentation for pesticide residue analysis. The pesticides studied were difenoconazole, carboxin, chlorpyrifos, and thiabendazole, as they have an efficient action against the most common plagues and diseases that affect the cajá tree and are recommended by the Ministry of Agriculture in Brazil. The methodology was optimized, validated, and applied to cajá-manga samples acquired from local greengrocers in Triângulo Mineiro, Brazil.

## 2. Materials and Methods

*2.1. Reagents and Solvents.* HPLC-grade methanol, acetonitrile, hexane, and pesticide-grade ethyl acetate were purchased from Tedia (Rio de Janeiro, Brazil), while HPLC-grade dichloromethane was from Merck (Darmstadt, Germany). Ultrapure water (Millipore, São Paulo, Brazil) was used throughout. The solvents used in this work were filtered using a 0.22 $\mu$m Teflon membrane from Millipore. The pesticide standards chlorpyrifos (99%), carboxin (99%), difenoconazole (98%), and thiabendazole (98%) were obtained from Pestanal (Riedel-de-Häen, PA, USA). Standard stock solutions (1000 mg·L$^{-1}$) of these pesticides were prepared in methanol. Cajá-manga samples for application testing were collected from the greengrocers in Ituiutaba, MG, Brazil.

*2.2. Solid-Liquid Extraction with Low-Temperature Partitioning.* Cajá-manga fruits were washed, peeled, and pitted, and their pulps were cut into little pieces. Cajá-manga pulp pieces were processed in a conventional multiprocessor (Mallory, Oggy Black model; Madrid, Spain). 2.000 g of the cajá-manga processed pulp was spiked with stock solutions of the pesticide mixture. 0.500 g of the fortified pulp was submitted to the solid-phase extraction with low-temperature partition (SLE-LTP). Spiked samples were disposed off in Falcon tubes adding 4 mL of distilled water and different volumes (4, 6, 8, and 10 mL) of the solvent (methanol, acetonitrile, acetone, or ethyl acetate). The mixture was stirred in a vortex mixer for 1 min and allowed to stand by at room temperature for 5 min. After that, Falcon tubes were cooled at −4°C for different periods (0.5, 1.5, 3.0, and 6.0 h). All measures were performed in triplicates.

After the congelation time, the aqueous phase containing the cajá-manga pulp was frozen and the organic counterparts were maintained in the liquid form. 1 mL of the organic phase was collected and filtered in a paper filtration containing a clean-up sorbent (Si(PDAS), $C_{18}$, PSA, or silica). Si (PDAS) is a sorbent obtained by thermal immobilization of poly(alkylmethylsiloxane-co-dimethylsiloxane), a copolymer with $C_{16}/C_{18}$ groups intercalated by $C_1$ groups, onto silica particles [25, 26]. The filtered organic extract was injected into a gas chromatographic system.

*2.3. Method Validation.* The methodology for the determination of the pesticide residues: chlorpyrifos, carboxin, thiabendazole, and difenoconazole in the cajá-manga pulp, using solid-liquid extraction with low-temperature partitioning, was validated according to the Guidance Document on Analytical Quality Control and Method Validation Procedures for Pesticides Residues Analysis in Food and Feed (SANTE/11813/2017 document) [27]. The analytical parameters of the methodology evaluated in this validation study were the selectivity, linearity, range, limits of detection and quantification, matrix effect, precision (repeatability and within-laboratory reproducibility), and accuracy.

*2.4. Chromatographic Analyses.* The extraction of the samples by the SLE-LTP method was performed in a gas chromatography (GC) model Focus GC, equipped with a flame ionization detector (FID) (Thermo, Austin, TX, USA) using a dimethylpolysiloxane OV-5-fused silica capillary column (30 m × 0.25 mm i.d. × 0.25 $\mu$m film thickness) using the following conditions: $N^2$ was used as the carrier gas at a flow rate of 1.5 mL·min$^{-1}$, column inlet split ratio of 1 : 5, and temperature program at 100°C for 1 min and then heated at a rate of 25°C min$^{-1}$ to 200°C and heated again at a rate of 40°C min$^{-1}$ to 300°C and held isothermal for 4.5 min, injector temperature of 280°C, and a detector with a temperature of 300°C.

The presence of each pesticide residue in the sample was determined by comparing its retention time with that obtained for the pesticide reference standard, under the same GC analysis conditions formerly described. The GC retention times for chlorpyrifos, thiabendazole, carboxin, and difenoconazole were 4.87, 5.37, 6.40, and 8.87 min, respectively.

## 3. Results and Discussion

The pesticides evaluated in this study, carboxin, chlorpyrifos, difenoconazole, and thiabendazole, are those recommended for the diseases mitigation that usually affects the cajá-manga fruit. The main physicochemical properties of compounds are presented in Table 1, as well as their chemical structures in Figure 1.

According to the physicochemical properties of compounds in Table 1, the four pesticides have strong retention in the GC system with the Kováts retention index above 1900, which require a higher column temperature for elutions of them. The order of the compound elutions in the 5% phenylmethylsiloxane-95% dimethylsiloxane capillary column is observed by the values of the Kováts retention index (RI): chlorpyrifos, thiabendazole, carboxin, and difenoconazole, requiring temperature programming to separate substances with retention indexes close to each other and accelerate the elution of difenoconazole, which presents higher RI, justifying the GC temperature programming performed in this work. It is important to mention that these substances are normally analyzed by both HPLC and GC systems [29–33], and therefore, gas chromatography was chosen because it is a cleaner, less-expensive, and more accessible technique for the pesticide quantifications compared to HPLC. Typically, GC detection is performed by mass spectrometry, but flame ionization detection is also used for detection and quantification of these species and was chosen in this work due to the greater availability of this system in smaller laboratories, making the methodology more accessible.

*3.1. Optimization of Solid-Liquid Extraction with Low-Temperature Partitioning.* The solid-liquid extraction with low-temperature partitioning (SLE-LTP) was used in this study because it is an efficient method in the extraction of pesticide residues from different samples [19–23, 34–37] and, mainly, because it is an easy application method involving few steps and low consumption of solvents and sorbents compared to other extraction methods. The SLE-LTP was optimized for the extraction of pesticide residues from the cajá-manga pulp to obtain maximum recovery of the analytes. The studied parameters were the extractor solvent, the extractor solvent volume, the freezing time of the sample, and the addition of the clean-up sorbent.

SLE-LTP is a solvent extraction method in which the extractor solvents are usually miscible in water and their choice is essential to obtain an efficient extraction because the selectivity of the extraction method is associated with the polarity characteristics of them. In addition to the efficiency, two other characteristics are required in the solvent extractor: the first one is that its melting temperature should be below the melting temperature of the water so that in the freezing step, the extract can be separated from the sample matrix and remain in the liquid state during the freezing step, and the second one is that its density should be lower than that of the water, as this fact facilitates the removal of the extract in the extraction process. The ethyl acetate, acetone, acetonitrile, and methanol solvents were evaluated in this step, and the results are shown in Figure 2.

Acetonitrile was the solvent which resulted in the higher recoveries of pesticide residues from the cajá-manga pulp. Acetonitrile was also the solvent that extracted the least amount of lipophilic interferences from the matrices when compared with acetone and ethyl acetate, besides being completely miscible in water, being its density lower than that of water, and remaining liquid at temperatures lower than 0°C. The polarity indices of evaluated solvents are ethyl acetate 4.4, methanol 5.1, acetone 5.1, acetonitrile 5.8, and water 10.2. As can be observed, the polarity index of acetonitrile is the closest to that of water (more polar among the studied solvents), whereas methanol and acetone have similar polarity indices. These three solvents solubilize better in water than ethyl acetate, which may have led to similar behaviors in the extraction of pesticide residues compared to this latter solvent (Figure 2). Thus, considering all these characteristics, acetonitrile was chosen as the extractor solvent of the SLE-LTP method of pesticide residues from the cajá-manga pulp.

In the SLE-LTP, it is necessary to add water so that the sample matrix remains frozen in the partitioning step. For the sample used in the SLE-LTP process, a volume of 4.0 mL was required to ensure freezing of the sample and facilitate the process of compound partitioning into the organic phase. According to the pKa values of pesticides shown in Table 1, all compounds are in their nonionized molecular forms at pH 5-6 (pH from the aqueous matrix of cajá-manga), and therefore, the extraction does not require pH adjustment for a higher partition into the organic phase. In the literature, the ratio of water volume and extractor solvent volume suggested for the maximum extraction efficiency should be 1 : 2 [19, 34]. Thus, the volume of acetonitrile used in the SLE-LTP for a higher recovery of pesticide residues from the cajá-manga pulp was evaluated. The results obtained are shown in Figure 3.

Extraction using a smaller volume of acetonitrile did not allow adequate separation of the liquid and frozen phases in the freezing step. Despite the addition of 4 mL of acetonitrile, the nonfrozen liquid phase had a lower volume than 1 mL, and an amount of mass of the sample remained dispersed in this small acetonitrile volume, making it difficult to obtain a clean extract (free of fractions of the cajá-manga pulp). Thus, matrix components were concentrated in this small volume, causing strong interference in the elution of the compounds and leading to high extraction percentages, as seen in Figure 3. On the other hand, using volumes greater than 6 mL of acetonitrile, recovery of the residues becomes practically constant, with a satisfactory efficiency between 70 and 110%, for the four evaluated compounds. As the

Table 1: Physicochemical properties of the studied pesticides.

| Compounds | log $P$ | Molar mass ($g{\cdot}mol^{-1}$) | Water solubility ($mg{\cdot}L^{-1}$)* | Density ($g{\cdot}cm^{-3}$)* | Kováts retention index | pKa |
|---|---|---|---|---|---|---|
| Chlorpyrifos | 4.96 | 350.58 | 2 | 1.4 | 1947 | — |
| Thiabendazole | 2.47 | 201.25 | 50 | 1.1 | 2010 | 4.08 |
| Carboxin | 2.14 | 235.30 | 147 | 1.7 | 2138 | 13.24 |
| Difenoconazole | 4.67 | 406.26 | 15 | 1.5 | 3136 | 1.95 |

*Measured at 25°C. All data are extracted from Reference [28].

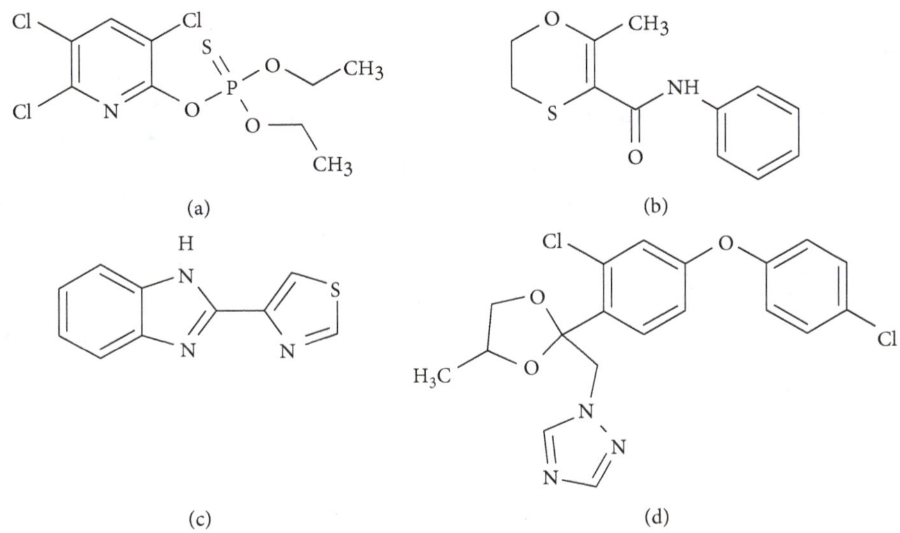


Figure 1: Chemical structures of pesticides: (a) chlorpyrifos, (b) carboxin, (c) thiabendazole, and (d) difenoconazole.

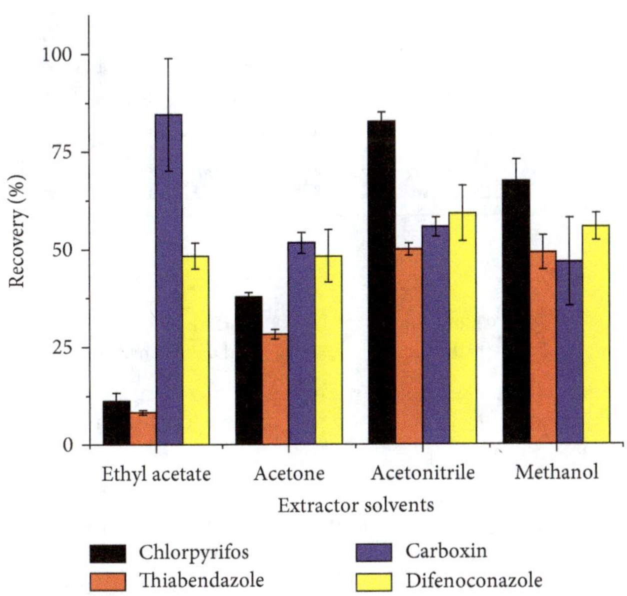


Figure 2: Pesticide residue recoveries in cajá-manga by the SLE-LTP method using different extractor solvents.

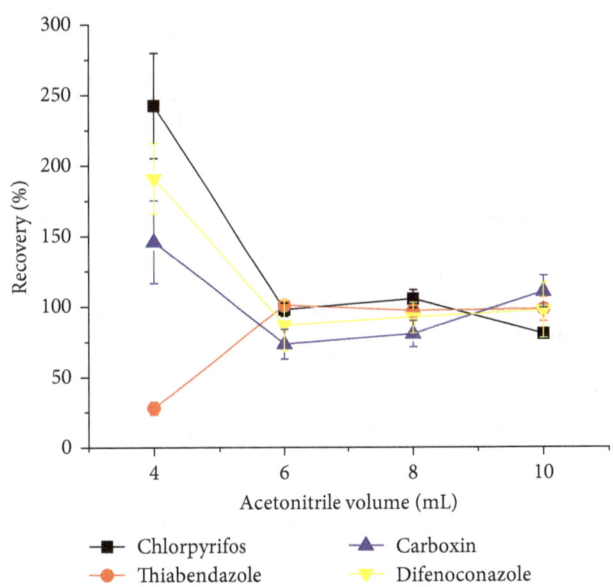


Figure 3: Pesticide residue recoveries in cajá-manga by the SLE-LTP method using different volumes of acetonitrile.

pesticide recoveries did not improve significantly with increasing acetonitrile volume from 6 mL, the lowest volume was defined for the SLE-LTP method. Therefore, the relation between the volumes of water and extractor solvent, 1 : 1.5, obtained for the extraction of pesticide residues from the cajá-manga pulp is lower than that suggested for some determinations in the literature using this method, minimizing costs and organic waste generation [19–22, 34]. It is also important to mention that according to the pesticide solubilities (Table 1), the four compounds are fully solubilized in the aqueous fruit matrix at the concentrations evaluated in this study. However, their partitioning for less-polar solvents is favored since they have partition coefficients (log $P$) with a tendency to distribute to nonpolar phases. For this reason, the recovery rates obtained by the ESL-PBT method for the four pesticides were close to 100%.

The freezing in the solid-liquid extraction for partitioning the compounds at low temperature is a very important step for an efficient and concomitant-free extraction from the sample matrix. In this step, the separation of the phases occurs due to lowering of the temperature, condition in which there is the decrease of the solubility of acetonitrile in an aqueous phase. For the phase separation to take place efficiently, the freezing time of the sample should be long enough for the cajá-manga pulp solidification to occur preventing their transfer to the organic phase, and the distribution of the pesticides to the organic phase be as complete as possible. Thus, the minimum freezing time required for maximum recovery of pesticide residues from the cajá-manga pulp was evaluated, and the results are found in Figure 4.

The freezing time of 30 min was not enough to separate the phases by lowering the temperature because the aqueous phase and the pulp of the sample remained liquefied at this time. From 1 h and 30 min, the whole aqueous phase was frozen, during which time the highest recoveries of pesticide residues were recorded. From 3 h, there is a continuous drop in the pesticide residue recoveries. After 3 h of freezing time, the compounds probably decant within the organic liquid phase, since pesticides have higher densities than acetonitrile (Table 1). This decantation appears to increase gradually with increasing freezing time, leading to the pesticide adsorptions on the frozen phase surface and reducing their recoveries. In the literature, freezing times greater than 3 h are recorded for efficient extraction of pesticides from other matrices [19, 22, 23, 35]. As the composition of the cajá-manga pulp is approximately 70% water and low content of liposoluble components, the extraction of residues of the compounds studied occurred more quickly [5]. This lower freezing time leads to a reduction of the total sample preparation time, increasing the analytical frequency for this determination.

The cajá-manga pulp has an intense yellow coloration and constituents potentially interfering with the quantification of pesticides in the chromatographic system. Removal of these interferents in the sample preparation step is important to reduce the matrix effect on the response of compounds. Therefore, after the partition of pesticide residues into the organic phase of acetonitrile, different sorbents were evaluated for the removal of the constituents from the sample through a filtration step of the organic extract. The pesticides were recovered in the filtrate, and the results are presented in Figure 5.

According to Figure 5, the use of clean-up sorbents significantly reduces the amount extracted from the four compounds analyzed, independently of the polarity of evaluated sorbents. This result, however, does not represent a negative factor in the analyses, since the extracts obtained in this work presented clear and colorless aspects and, therefore, do not require a previous purification step for GC analysis. The cajá-manga has a pulp rich in carbohydrates, sugars, proteins, fibers, volatiles (terpenes), and minerals, besides an intense yellow coloration. Thus, a major advantage of SLE-LTP, in the case of a complex matrix such as cajá-manga pulp, is that the freezing of the sample presents itself as a clean-up step for the extracts.

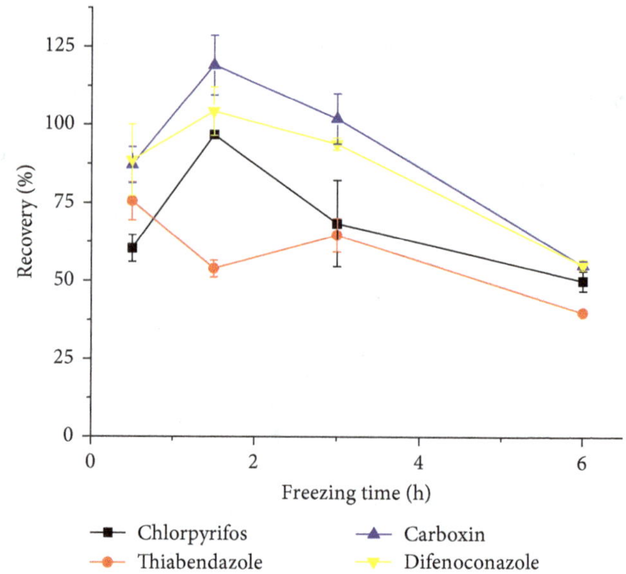


FIGURE 4: Pesticide residue recoveries in cajá-manga by the SLE-LTP method after different freezing times of samples.

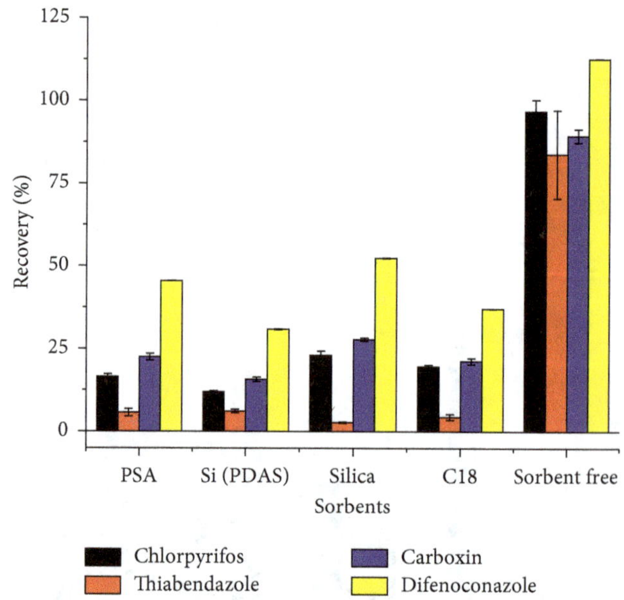


FIGURE 5: Pesticide residue recoveries in cajá-manga by the SLE-LTP method after the clean-up step of the extract, using different sorbents.

*3.2. SLE-LTP Method Validation.* After the optimization of SLE-LTP of pesticide residues from the cajá-manga pulp, the analytical parameters of the method were validated according to the Guidance Document on Analytical Quality Control and Method Validation Procedures for Pesticides Residues Analysis in Food and Feed (SANTE/11813/2017 document) in relation to selectivity, linearity, range, limits of detection and quantification, accuracy, and precision [27].

The selectivity of the method was evaluated by the separation of the analytes by gas chromatography with a flame ionization detector. The chromatogram of the separation of the four pesticides extracted from the cajá-manga pulp is presented in Figure 6, in which the discrimination of the peaks for each pesticide from other eluted substances can be observed. So, the method can be considered selective for the pesticides studied.

Some analytical parameters of the method were determined from the calibration curve plots for each compound by the standard addition method, in which portions of the cajá-manga pulp were spiked with pesticide standards in different concentrations, 0.17 to 11.75 $\mu g \cdot g^{-1}$, submitted to the optimized SLE-LTP, and the extracts were analyzed by GC-FID. The analytical parameters obtained for the method are presented in Table 2. The limit of detection (LOD) and limit of quantification (LOQ) for each pesticide were also estimated from the analytical curves for each pesticide. The LOD was calculated by using the following equation: LOD = $3.3 \times s_y/b$, where $b$ is the slope of the analytical curve and $s$ is the residual standard deviation of the analytical curve. The LOQ was calculated as LOQ = 3 × LOD.

The LOQ for the four pesticide residues was in the range of 0.17 to 0.43 $\mu g \cdot g^{-1}$ in the cajá-manga pulp using the SLE-LTP method with analysis by GC-FID. As there is no reference for determination of residues of pesticides in cajá-manga, the values obtained were compared with the maximum residue limits (MRLs) of these pesticides in different cultures preconized by Brazilian legislation. For different food types, the MRLs are in the range of 0.05–0.20 $mg \cdot kg^{-1}$ for carboxin, 0.01–2.00 $mg \cdot kg^{-1}$ for chlorpyrifos, 0.01–1.00 $mg \cdot kg^{-1}$ for difenoconazole, and 0.01–10 $mg \cdot kg^{-1}$ for thiabendazole [38]. Therefore, it is observed that the LOQ values found for the pesticides in the proposed method are within the range of MRLs for different types of food monitored by ANVISA, although GC-FID does not have a high detectability for the compounds studied. The analytical curves presented excellent adjustments, with determination coefficients higher than 0.996. The adjustments were linear for the chlorpyrifos, carboxin, and difenoconazole compounds, whereas for thiabendazole, the curve showed a better fit in the quadratic model. Figure 7 shows the analytical curves obtained by the standard addition method and by the external standard method for each pesticide to evaluate a possible matrix effect from the cajá-manga pulp.

According to the slopes of the analytical curves (Figure 7), comparing each compound by the two calibration methods constructed under the same concentrations, all compounds showed a positive matrix effect [36]; that is, the sensitivity of the proposed method increases when the analytical curve is prepared in the cajá-manga pulp itself and submitted to the SLE-LTP process. The positive matrix effect indicates that components of the sample matrix have a synergistic effect on the compound response in the sample, enhancing the analytical signal. This matrix effect is suppressed by the calibration using the standard addition method.

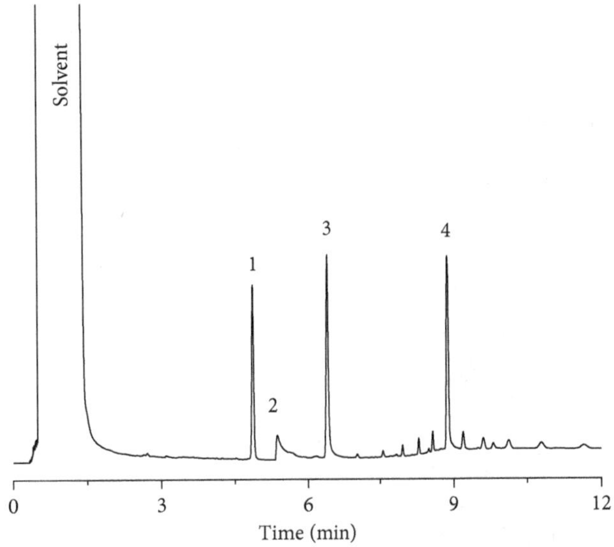


FIGURE 6: Separation of pesticide residues extracted from the cajá-manga pulp by the SLE-LTP method, using gas chromatography with flame ionization detection. Conditions: capillary column 30 m × 0.32 mm i.d. × 0.25 $\mu m$ film thickness, 5% phenylmethylsiloxane-95% dimethylsiloxane, split injection 1:5, injected volume 1 $\mu L$, injector temperature 280°C, FID temperature 300°C, $N_2$ flow rate 1.5 $mL \cdot min^{-1}$, and oven temperature 100°C (1 min), 25°C $min^{-1}$, 200°C, 40°C $min^{-1}$, 300°C (4.5 min). (1) Chlorpyrifos; (2) thiabendazole; (3) carboxin; (4) difenoconazole.

The precision and accuracy of the method for the determination of pesticide residues in the cajá-manga pulp were evaluated according to the recommendations of the SANTE guidelines [27]. The precision of the method was evaluated based on its repeatability and within-laboratory reproducibility, which were obtained by performing nine sample extractions on the same day in three different levels and by nine sample extractions in the same level on three different days, respectively. The relative standard deviation for repeatability ($RSD_r$) and relative standard deviation for within-laboratory reproducibility ($RSD_{wR}$) are presented in Table 3. The accuracy of the method was investigated by the recovery of pesticide residues from the cajá-manga pulp fortified at three different levels. Accuracy was obtained by percent recovery of each pesticide, and the results are presented in Table 3.

The SLE-LTP method presented adequate precision for the extraction of residues from the four pesticides selected for the study. As shown in Table 3, the method repeatability ($RSD_r$) was generally less than 20% at the different concentration levels evaluated, except for carboxin and chlorpyrifos at the lowest concentration level where the dispersion of the results was higher than 20%. The laboratory reproducibility ($RSD_{wR}$) evaluated at the 4 × LOQ level showed the low dispersion of the results with RSD values lower than 8% for all pesticides, indicating that the

TABLE 2: Analytical parameters of the SLE-LTP method of pesticide residues from the cajá-manga pulp.

| Pesticides | Linearity ($R^2$) | Range ($\mu g \cdot g^{-1}$) | LOD ($\mu g \cdot g^{-1}$) | LOQ ($\mu g \cdot g^{-1}$) |
|---|---|---|---|---|
| Chlorpyrifos | 0.996 | 0.18/9.05 | 0.06 | 0.18 |
| Thiabendazole | 0.999 | 0.43/10.80 | 0.13 | 0.43 |
| Carboxin | 0.996 | 0.17/8.35 | 0.06 | 0.17 |
| Difenoconazole | 0.998 | 0.24/11.75 | 0.14 | 0.24 |

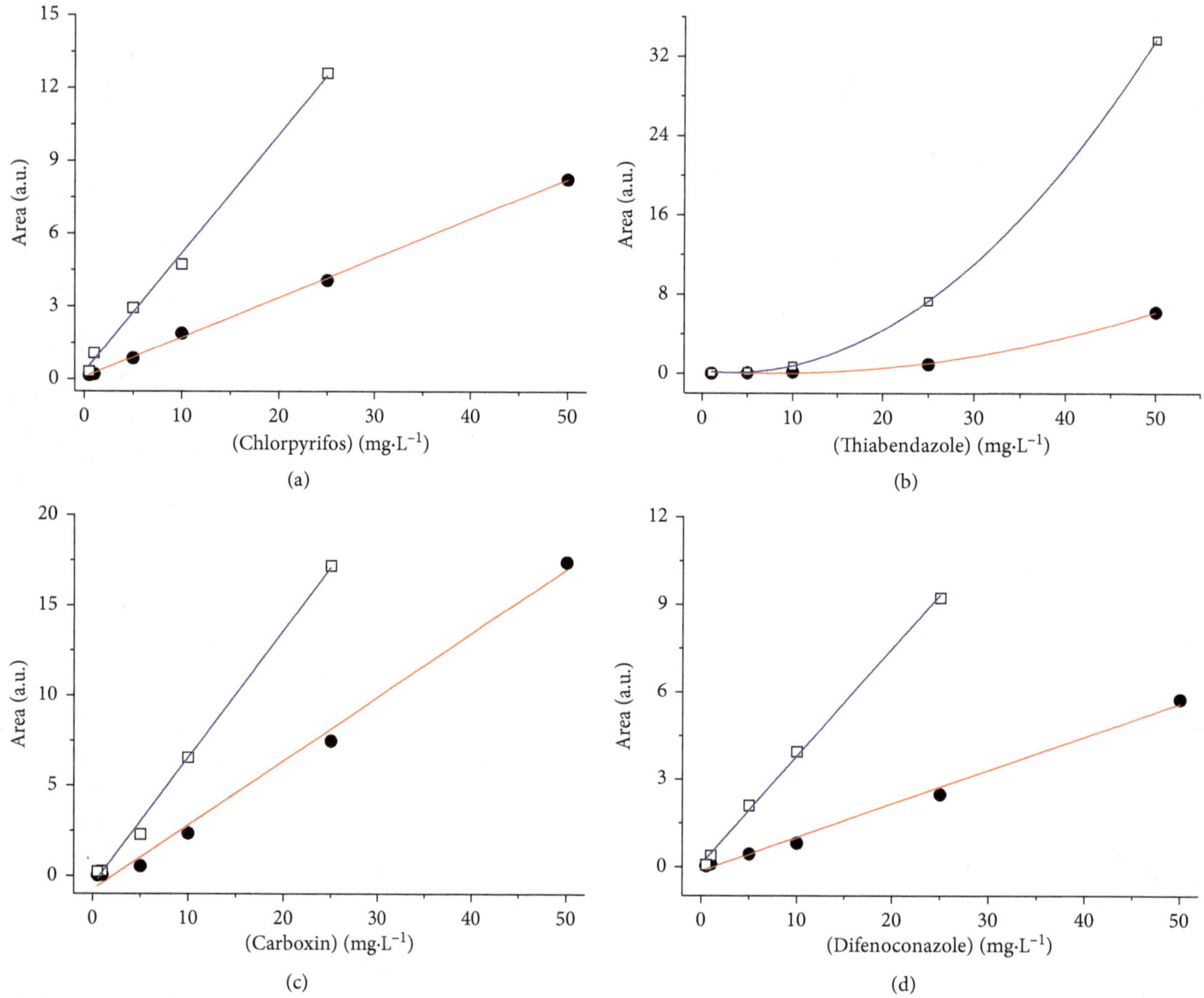


FIGURE 7: Analytical curves prepared by the external standard method and by the standard addition method for the compounds: (a) chlorpyrifos, (b) thiabendazole, (c) carboxin, and (d) difenoconazole.

TABLE 3: Accuracy (% recoveries) and precision (repeatability and within-laboratory reproducibility) of the SLE-LTP method developed ($n = 5$).

| Pesticides | Repeatability 2 × LOQ | | Repeatability 3 × LOQ | | Repeatability 4 × LOQ | | Within-lab reproducibility 4 × LOQ | |
|---|---|---|---|---|---|---|---|---|
| | Recovery (%) | $RSD_r$ (%) | Recovery (%) | $RSD_r$ (%) | Recovery (%) | $RSD_r$ (%) | Recovery (%) | $RSD_{wR}$ (%) |
| Chlorpyrifos | 96 | 22 | 95 | 13 | 89 | 6 | 88 | 2 |
| Thiabendazole | 188* | 14 | 129 | 7 | 117 | 6 | 111 | 5 |
| Carboxin | 69 | 26 | 78 | 8 | 82 | 11 | 76 | 7 |
| Difenoconazole | 80 | 13 | 90 | 4 | 109 | 9 | 102 | 6 |

* = Outlier value. A value above the acceptable range of recoveries for compounds in complex matrices.

application of the method on different days does not affect the recovery of the residues. The method also showed good accuracy, except for thiabendazole that showed recoveries always above 100%, probably due to the long tailing peak for its peak in the chromatogram, which makes more difficult the peak integration due to the possibility of coelution with constituents from the matrix or even due to the baseline instability. The lower the concentration of thiabendazole in the cajá-manga pulp, the more pronounced the interference becomes in the thiabenzadole quantification, reaching approximately 190% of recovery (Table 3). For the other compounds, the residue recoveries were in the range of 69 to 109%, being in accordance with the criteria recommended by the SANTE guidelines for this parameter, 70–120% [27].

*3.3. Analysis of Real Samples.* Two samples of different cajá-manga varieties, obtained from the greengrocer in Triângulo Mineiro, Brazil, were submitted to SLE-LTP with analysis by GC-FID to evaluate the accuracy and precision of the method. Both samples were spiked with a concentration equivalent to 4 × LOQ for each pesticide, and the recovery results are shown in Table 4.

The pesticide residue recoveries from the cajá-manga pulp of two commercially available varieties presented acceptable recovery ranges of 75.3 to 129.3%, considering the complexity of the matrices. Only, thiabendazole was not recovered in any of the three replicates of sample A. In this way, the method has a good accuracy of the measurements performed for the recovery of pesticides from the cajá-manga pulp, in which all the relative standard deviation (RSD) values were below 20%, as recommended by the SANTE document [27].

TABLE 4: Pesticide residue recoveries from two samples of the cajá-manga pulp fortified with pesticides, purchased from greengrocers in Triângulo Mineiro, Brazil ($n = 3$).

| Pesticides | Sample A | | Sample B | |
|---|---|---|---|---|
| | Recovery (%) | RSD | Recovery (%) | RSD |
| Chlorpyrifos | 106.8 | 11.1 | 129.3 | 4.2 |
| Thiabendazole | — | — | 75.3 | 3.8 |
| Carboxin | 76.7 | 7.2 | 104.3 | 8.0 |
| Difenoconazole | 86.8 | 12.7 | 104.4 | 1.8 |

## 4. Conclusions

In this work, an SLE-LTP method was proposed to extract carboxin, chlorpyrifos, difenoconazole, and thiabendazole from the cajá-manga pulp. SLE-LTP has been successfully used to extract pesticide residues from different matrices, such as maize grains, potato, carrot, soil, tomato, lettuce, and yacare eggs. However, the amount of the sample, the organic extractor volume, the freezing time, and the need for an additional cleaning step, after partitioning at low temperature, aroused the interest in simplifying the method by reducing the amounts normally used and applying it in a complex matrix such as the cajá-manga pulp. The cajá-manga has a pulp rich in carbohydrates, sugars, proteins, fibers, volatiles (terpenes), and minerals, besides an intense yellow coloration. Allied to this fact, no methodology for the pesticide residue determination is available to monitor the potential cajá-manga contamination, despite the increasing consumption of the fruit. After optimization and validation, it was possible to obtain an efficient methodology, with good precision and accuracy, based on solid-liquid extraction with low-temperature partitioning, for the determination of pesticide residues in the cajá-manga pulp. The SLE-LTP method proposed reduced the extractor organic volume by at least 25%, the freezing time by half, and the sample mass by at least 75% when compared with previous works. The extraction method did not require an additional step for the cleaning of the extract, even though the cajá-manga was a matrix with a high content of soluble substances and intense color. The quantification performed by GC-FID makes the methodology more accessible, cleaner, and less expensive, making it possible to determine the pesticides below the MRLs established by Brazilian legislation.

## Conflicts of Interest

The authors declare that there are no conflicts of interest regarding the publication of this paper.

## Acknowledgments

The authors are grateful to FINEP (Financiadora de Estudos e Projetos) for providing funding for the laboratorial infrastructure (MCT/Finep/CT-INFRA-Campi Regionais-01/2010) and for the acquisition of equipments (MCTI/Finep/CT-INFRA 01/2013) used in the development of this work. The authors also would like to acknowledge the FAPEMIG (Fundação de Amparo à Pesquisa do Estado de Minas Gerais, APQ-01425-14) for financial support and fellowships.

## References

[1] M. S. L. Ferreira, M. S. P. Santos, T. M. A. Moro, G. J. Basto, R. M. S. Andrade, and E. C. B. A. Gonçalves, "Formulation and characterization of functional foods based on fruit and vegetable residue flour," *Journal of Food and Science Technology*, vol. 52, no. 2, pp. 822–830, 2015.

[2] K. E. Bradburry, P. N. Appleby, and T. J. Key, "Fruit, vegetable, and fiber intake in relation to cancer risk: findings from the European prospective investigation into cancer and nutrition (EPIC)," *American Journal of Clinical Nutrition*, vol. 100, no. 1, pp. 394S–398S, 2014.

[3] S. L. Fulton, M. C. McKinley, I. S. Young, C. R. Cardwell, and J. V. Woodside, "The effect of increasing fruit and vegetable consumption on overall diet: a systematic review and meta-analysis," *Critical Reviews in Food Science and Nutrition*, vol. 56, no. 5, pp. 802–816, 2016.

[4] T. L. L. Silva, E. P. Silva, E. R. Asquieri et al., "Physicochemical characterization and behavior of biocompounds of *caja-*

*manga* fruit (*Spondias mombin* L.)," *Food Science Technology*, vol. 38, no. 3, pp. 399–406, 2018.

[5] E. S. L- Vanzela, P. Ramin, M. A. U- Guez, G. V. Santos, E. Gomes, and R. Silva, "Chemical and sensory characteristics of pulp and peel '*cajá-manga*' (Sonn.) jelly. *Spondias cytherea*," *Ciência e Tecnologia de Alimentos*, vol. 31, no. 2, pp. 398–405, 2011.

[6] N. Benkeblia and M. G. Lopez, "Saccharides and fructooligosaccharides composition of green and ripe *Averrhoa carambola*, *Blighia sapida* and *Spondias dulcis* fruits," *Food Chemistry*, vol. 176, pp. 314–318, 2015.

[7] M. Mohammed, P. Bridgemohan, M. S. Mohamed, R. S. H. Bridgemohan, and Z. Mohammed, "Postharvest physiology and storage of golden apple (*Spondias cythera sonnerat* or *Spondias dulcis forst*): a review," *Journal of Food Processing and Technology*, vol. 8, no. 12, pp. 1–8, 2017.

[8] E. F. Silva, L. S. S. Martins, and V. R. Oliveira, "Diversity and genetic structure in cajá tree (*Spondias mombin* L.) populations in northeastern Brazil," *Revista Brasileira de Fruticultura*, vol. 31, no. 1, pp. 171–181, 2009.

[9] O. S. Graham, M. Mohammed, and L. D. Wickham, "Effects of heat treatments on the quality of miniature golden apples (*Spondias cytherea Sonn)* during low temperature storage," *Food, Agriculture and Environment*, vol. 2, no. 2, pp. 48–53, 2004.

[10] S. B- Nascimento, "Emergence and pupal mortality factors of *Anastrepha obliqua* (macq.) (Diptera: Tephritidae) along the fruiting season of the host *Spondias dulcis* L.," *Neotropical Entomology*, vol. 30, no. 2, pp. 207–215, 2001.

[11] A. Gordon, "Exporting traditional fruits and vegetables to the united states: trade, food science, and sanitary and phytosanitary/technical barriers to trade considerations," in *Food Safety and Quality Systems in Developing Countries, Volume One: Export Challenges and Implementation Strategies*, Academic Press, Cambridge, MA, USA, 2015.

[12] A. Masiá, M. M. Suarez Varela, A. Llopis Gonzalez, and Y. Picó, "Determination of pesticides and veterinary drug residues in food by liquid chromatography-mass spectrometry: a review," *Analytica Chimica Acta*, vol. 936, pp. 40–61, 2016.

[13] S. Grimalt and P. Dehouck, "Review of analytical methods for the determination of pesticide residues in grapes," *Journal of Chromatography A*, vol. 1433, pp. 1–23, 2016.

[14] M. Ledoux, "Analytical methods applied to the determination of pesticide residues in foods of animal origin. A review of the past two decades," *Journal of Chromatography A*, vol. 1218, no. 8, pp. 1021–1036, 2011.

[15] B. M. Keikotlhaile, P. Spanoghe, and W. Steurbaut, "Effects of food processing on pesticide residues in fruits and vegetables: a meta-analysis approach," *Food and Chemical Toxicology*, vol. 48, no. 1, pp. 1–6, 2010.

[16] J. H. Syed, A. Alamdar, A. Mohammad et al., "Pesticide residues in fruits and vegetables from Pakistan: a review of the occurrence and associated human health risks," *Environmental Science and Pollution Research*, vol. 21, no. 23, pp. 13367–13393, 2014.

[17] M. Kazemi, A. M. Tahmasbi, R. Valisadeh, A. A. Naserian, and A. Soni, "Organophosphate pesticides: a general review," *Agricultural Science Research Journals*, vol. 2, no. 9, pp. 512–522, 2012.

[18] R. M. Rigotto, D. P. Vasconcelos, and M. M. Rocha, "Pesticide use in Brazil and problems for public health," *Caderno de Saúde Pública*, vol. 30, no. 7, pp. 1360–1362, 2014.

[19] D. A. Sousa, R. M. Gonçalves, F. F. Heleno, M. E. L. R. Queiroz, and M. R. R. Marchi, "Chemometric optimization of solid–liquid extraction with low-temperature partition (SLE-LTP) for determination of persistent organic pollutants in *Caiman yacare* eggs," *Microchemical Journal*, vol. 114, pp. 266–272, 2014.

[20] L. M. B. Rigueira, K. L. Ribeiro, M. E. L. R. Queiroz et al., "Determination of chlorpyrifos and thiamethoxam in potato tuber (*Solanum tuberosum* L.) and soil of Brazil using solid-liquid extraction with low temperature partitioning (SLE/LTP)," *Journal of the Brazilian Chemical Society*, vol. 24, no. 12, pp. 2042–2049, 2013.

[21] G. P. Pinho, A. A. Neves, M. E. L. R. Queiroz, and F. O. Silvério, "Pesticide determination in tomatoes by solid–liquid extraction with purification at low temperature and gas chromatography," *Food Chemistry*, vol. 121, no. 1, pp. 251–256, 2010.

[22] R. S. Freitas, M. E. L. R. Queiroz, L. R. D. Faroni, F. F. Heleno, and V. V. Moura, "Development of a solid-liquid extraction method with low-temperature partitioning for the determination of insecticides in ozonized maize grain," *Química Nova*, vol. 37, no. 2, pp. 238–243, 2014.

[23] A. I. G. Costa, M. E. L. R. Queiroz, A. A. Neves, F. A. Sousa, and L. Zambolim, "Determination of pesticides in lettuce using solid–liquid extraction with low temperature partitioning," *Food Chemistry*, vol. 181, pp. 64–71, 2015.

[24] E. H. C. Morais, A. A. Z. Rodrigues, M. E. L. R. Queiroz, A. A. Neves, and P. H. D. Morais, "Determination of thiamethoxam, triadimenol and deltamethrin in pineapple using SLE-LTP extraction and gas chromatography," *Food Control*, vol. 42, pp. 9–17, 2014.

[25] A. S. Novais, J. F. Ribeiro Filho, E. M. F. Amaral, and A. M. Faria, "Novo sorvente de hidrofobicidade reduzida para extração em fase sólida: preparação e caracterização," *Química Nova*, vol. 38, no. 2, pp. 274–279, 2015.

[26] G. O. Carvalho, C. G. A. Silva, and A. M. Faria, "A new stationary phase for analysis of hydrophobic compounds by RP-LC," *Chromatographia*, vol. 79, no. 1-2, pp. 19–27, 2016.

[27] SANTE/11813/2017, *Guidance Document on Analytical Quality Control and Method Validation Procedures for Pesticides Residues Analysis in Food and Feed*, European Commission Directorate-General for Health and Food Safety, 2017, https://ec.europa.eu/food/sites/food/files/plant/docs/pesticides_mrl_guidelines_wrkdoc_2017-11813.pdf, May 2018.

[28] http://www.chemicalize.org, September 2018.

[29] M. A. Farajzadeh, D. Djozan, M. R. A. Mogaddam, and M. Bamorowat, "Extraction and preconcentration technique for triazole pesticides from cow milk using dispersive liquid–liquid microextraction followed by GC-FID and GC-MS determinations," *Journal of Separation Science*, vol. 34, no. 11, pp. 1309–1316, 2011.

[30] M. A. Farajzadeh, D. Djozan, N. Nouri, M. Bamorowat, and M. S. Shalamzari, "Coupling stir bar sorptive extraction-dispersive liquid–liquid microextraction for preconcentration of triazole pesticides from aqueous samples followed by GC-FID and GC-MS determinations," *Journal of Separation Science*, vol. 33, no. 12, pp. 1816–1828, 2010.

[31] Y. Yamazaki and T. Ninomiya, "Determination of benomyl, diphenyl, *o*-phenylphenol, thiabendazole, chlorpyrifos, methidathion, and methyl parathion in oranges by solid-phase extraction, liquid chromatography, and gas chromatography," *Journal of AOAC International*, vol. 82, no. 6, pp. 1474–1478, 1999.

[32] M. A. Farajzadeh, M. R. A. Mogaddam, S. R. Aghdam, N. Nouri, and M. Bamorowat, "Application of elevated temperature-dispersive liquid-liquid microextraction for determination of organophosphorus pesticides residues in aqueous samples followed by gas chromatography-flame ionization detection," *Food Chemistry*, vol. 212, pp. 198–204, 2016.

[33] W. M. Draper, "Optimizing nitrogen-phosphorus detector gas chromatography for pesticide analysis," *Journal of Agricultural and Food Chemistry*, vol. 43, no. 8, pp. 2077–2082, 1995.

[34] R. P. Rodrigues, W. R. Farias, S. M. Goulart et al., "Optimization of solid-liquid extraction with low temperature partition for determination of carbofuran in *Cucurbita pepo* L. ("zucchini") by high performance liquid chromatography," *Química Nova*, vol. 41, no. 2, pp. 213–218, 2018.

[35] E. A. Araújo, M. C. R. Lara, M. C. Reis et al., "Determination of haloxyfop-methyl, linuron, and procymidone pesticides in carrot using SLE-LTP extraction and GC-MS," *Food Analytical Methods*, vol. 9, no. 5, pp. 1344–1352, 2016.

[36] G. P. Pinho, A. A. Neves, M. E. L. R. Queiroz, and F. O. Silvério, "Efeito de matriz na quantificação de agrotóxicos por cromatografia gasosa," *Química Nova*, vol. 32, no. 4, pp. 987–995, 2009.

[37] A. I. G. Costa, M. E. L. R. Queiroz, A. A. Neves et al., "Mobility and persistence of the herbicide fomesafen in soils cultivated with bean plants using SLE/LTP and HPLC/DAD," *Environmental Science and Pollution Research*, vol. 22, no. 5, pp. 3457–3466, 2015.

[38] Agência Nacional de Vigilância Sanitária (ANVISA), *Limites Máximos de Resíduos de Agrotóxicos em Alimentos*, Agência Nacional de Vigilância Sanitária (ANVISA), Brazil, http://portal.anvisa.gov.br/programa-de-analise-de-registro-de-agrotoxicos-para, 2018.

# 7

# Three-Dimensional Excitation and Emission Fluorescence-Based Method for Evaluation of Maillard Reaction Products in Food Waste Treatment

Jiaze Liu,[1,2] Jun Yin,[1,2] Xiaozheng He,[1,2] Ting Chen,[1,2] and Dongsheng Shen[1,2]

[1]*School of Environmental Science and Engineering, Zhejiang Gongshang University, Hangzhou 310012, China*
[2]*Zhejiang Provincial Key Laboratory of Solid Waste Treatment and Recycling, Hangzhou 310012, China*

Correspondence should be addressed to Jun Yin; jun.yin77@gmail.com

Guest Editor: Gassan Hodaifa

Hydrothermal treatment (HT) of food waste (FW) can form Maillard reaction products (MRPs), the biorefractory organic matter due to the occurrence of Maillard reaction. However, the integrating qualitative and quantitative approach to assess MRPs is scarce. The goal of this study was to develop a method to characterize and quantify MRPs created by HT of FW. MRPs were identified by molecular weight fractionation, indirect spectrometric indicators, and three-dimensional excitation-emission fluorescence (3DEEM) analysis. The 3DEEM method combined with fluorescence regional integration (FRI) and parallel factor (PARAFAC) analyses was able to differentiate clearly between MRPs and other dissolved organic compounds compared to other approaches. The volume of fluorescence $\Phi$ from FRI and maximum fluorescence intensity $F_{max}$ from PARAFAC were found to be suitable quantitative parameters for determination of MRPs in the hydrothermal FW system. These two parameters were validated with samples from hydrothermal FW under various operating temperatures and pH.

## 1. Introduction

In China, the stacking of FW has become a major issue to cause environmental problems. Recently, anaerobic digestion (AD) as an attractive waste treatment practice has been used to decrease the amount of biowaste and recover energy [1]. Due to the high biodegradability and water content of FW, it becomes a good candidate for AD [2]. During the process of the AD, the hydrolysis step is generally considered as the rate-limiting step for complex organic substrates degradation. Therefore, hydrothermal treatment (HT) was ordinarily used as a pretreatment to promote the solubilization of complicated macromolecular solid organic matters, thus improving the AD process [3].

Despite the acceleration of dissolved properties of FW, it has been documented that HT is responsible for the formation of Maillard reaction products (MRPs) [4]. On the one hand, the formation of MRPs can lead to the substrate loss during the HT of FW. On the other hand, the influence of MRPs themselves on AD merits further investigation [5]. Therefore, to optimize the HT process and enhance the efficiency of AD, it is essential to provide an integrating quantitative and qualitative approach to assess the MRPs production.

Numerous methods have been developed to characterize the occurrence of MRPs by using the precision analysis instrument [6, 7]. However, the quantitative determination could not be achieved because there is no pure standard for the measurement of MRPs [6]. In addition, these devices are time-consuming and labor-intensive, limiting their application and spreading. Thus, some easy-to-use and convenient characterization techniques became most commonly used methods, including the $UVA_{254}$ and color intensity [8, 9].

Nowadays, three-dimensional excitation and emission fluorescence (3DEEM) is regarded as a promising tool to offer characteristic information for signature chemical structures in a complex mixture of chromophores [10]. And, the qualitative characterization of MRPs has been achieved by the traditional 3DEEM method [11]. However, a quantitative determination could not be realized because only one excitation/emission intensity value can be used for analysis. Recent studies have demonstrated that fluorescence regional integration (FRI) method and parallel factor analysis (PARAFAC) method were developed to integrate the area beneath EEM spectra and semiquantitatively assess the specific components in a complex system [12–14]. However, the application of 3DEEM to the semiquantitative characterization of MRPs in the complicated hydrothermal FW system is scarce. Therefore, the utilization of 3DEEM to distinguish between MRPs and other dissolved organic matter under various hydrothermal conditions is supposed to be further explored.

This study aimed at developing a method to characterize and quantify MRPs created by HT of FW. Firstly, MRPs were characterized and evaluated with different methods. Then, MRPs production was further assessed by the applicability of 3DEEM combined with FRI and PARAFAC, hence exploring the suitable fluorescence parameters for semi-quantifying the MRPs in the hydrothermal FW system.

## 2. Materials and Methods

*2.1. Food Waste Sample Preparations.* The FW, containing rice (44%), noodles (16%), vegetables (23%), meat (6%), and tofu (11%), was compounded based on the characteristics similar to FW collected from a canteen of Zhejiang Gongshang University (Hangzhou, China) in our previous study [15]. The five components came from the same vendor at Cui Yuan farmers' market (Hangzhou, China). The FW was cut into small pieces first by hand-breaking and then crushed using a mangler. The untreated FW sample was stored at −18°C before the HT. The main characteristics of the FW are listed in Table S1.

*2.2. Hydrothermal Treatment.* Hydrothermal treatment of FW was performed in an 80 mL airtight pressure digestion vessel as described by our previous study [16] at separate batch operations at each temperature. During HT, about 30 g crushed FW was placed in the vessel. Each batch was processed for 30 min. In the first experiment, the temperature manipulations were made at 110, 120, 130, 140, 150, and 160°C in an oil bath to explore the effect of temperature on MRPs production. In the second experiment, the different initial pH values (3.0, 4.0, 5.0, 6.0, 7.0, 8.0, 9.0, and 10.0) were tested at 130°C for 30 min. The time was measured from when the oil bath reached the set temperature. The vessels were cooled to ambient temperature after HT. Each treatment was performed in triple vessels.

*2.3. Extraction of WEOM.* WEOM was obtained with deionized water (solid-to-water ratio of 1 : 10 w/v), and the mixture was shaken for 1 h in a horizontal shaker at 35 ± 2°C. The extracts were separated from the mixture by centrifugation at 10,000 rpm for 5 min and filtered using the microfiltration membrane (0.45 $\mu$m).

*2.4. Synthetic MRPs Solution.* The synthetic MRPs solution was made by a concentrated solution of melanoidins, which are defined as brown substances formed during the final stage of the Maillard reaction. The formula of the concentrate was made with a 1 : 1 molar ratio of glucose and glycine with a buffer of 0.5 M $Na_2CO_3$ according to the previous research [17]. The solution was heated at 120°C for 3 hours as the record [17]. This synthetic MRPs solution has been used as model MRPs to analyze the properties of MRPs [9], and it was employed to examine the availability of MRPs characterization method in the present study.

### 2.5. Analytical Methods of MRPs

*2.5.1. Spectrometric Indicators. (1) $UVA_{254}$.* UVA was a measure of absorbance at 254 nm, measured in a 1 cm path length quartz cell. It can measure unsaturated bonds or aromaticity within dissolved organic matters [18]. Therefore, this spectrometric index was useful for this study as MRPs were linked to the presence of unsaturated double bonds and aromatic compounds and expressed as $cm^{-1}$·mL/g dry weight.

*(2) Color Intensity.* A spectrophotometer at a wavelength of 475 nm was used to determine color intensity in a 1 cm path length cell. The absorbance at this wavelength was characteristic of brown color. Characteristic color intensity was recorded in a platinum-cobalt (PtCo) unit as previously described [9].

*(3) Browning Index.* The Browning index of the FW solid was measured by an enzymatic digestion method which releases the brown pigments. Samples were dried for 24 h and grounded to a smaller size before use. The proposed method was modified based on pronase proteolysis created by Palombo et al. [19]. The procedure was as follows: 0.3 g of the dried sample was added into a test tube which contains 5 mL deionized distilled water at 45°C and mixed thoroughly. Then, another 0.4 mL of pronase solution was added into the mixture. After that, the test tubes were placed in a water bath, incubated for 120 min at 45°C, and then cooled in ice water, and 1 mL trichloroacetic acid (80% TCA) was added to each tube. Finally, centrifugation (20 min at 7000 rpm) and filtration were used before the spectrometric determination. The optical density of the filtrates was determined on a spectrophotometer. Samples were measured in a 1 mL cuvette with 1 cm pass length. The OD of the brown index was calculated as $OD = OD_{420nm} - OD_{550nm}$ and expressed as OD/g dry weight.

*2.5.2. Molecular Weight Fractionation.* Molecular weight fractionation was applied for a better separation and characterization of dissolved organic matters. Fractionation of samples was performed using an ultrafiltration centrifuge tube with different molecular weight cutoffs: 3 kDa, 10 kDa, and 30 kDa. The samples were filtered in series from 30 kDa to 3 kDa.

*2.5.3. 3DEEM Analyses.* The 3DEEM of WEOM was measured in a 1 cm cuvette using a Hitachi F-4600 fluorescence spectrometer at room temperature (25 ± 2°C). The scanning ranges were 200–500 nm for excitation and 250–500 nm for emission. Scanning was recorded at 5 nm intervals for excitation and 1 nm steps for emission, respectively, using a scanning speed of 2400 nm/min. The Milli-Q water blanks were subtracted in order to eliminate the effect of Raman scattering. In addition, exported EEMs were normalized by the Raman area and eliminated the primary and second Rayleigh scattering.

*(1) Peak-Picking Method.* A peak-picking method is used for the detection of the fluorescence intensity of easily-identifiable peaks and their locations within the EEMs. And then, the observed peaks were analyzed by comparing their fluorescence properties with the change of operating condition.

Besides, as an important humification index in EEM, HIX can reflect the degree of humification of the sample. HIX was calculated by dividing the emission intensity into the 435–480 nm regions by intensity in 300–345 nm when excitation intensity is at a wavelength of 255 nm and is shown as follows:

$$\mathrm{HIX} = \frac{\sum I_{(435-480)}}{\sum I_{(300-345)}}. \quad (1)$$

*(2) FRI Analysis Method.* The FRI approach was employed in this work to characterize the five excitation-emission regions of EEM spectra [12]. According to previous studies, EEM spectra are usually divided into five areas (Table S2): aromatic protein-like fluorophores (regions I and II), fulvic acid-like fluorophores (region III), soluble microbial product-like fluorophores (region IV), and humic acid-like fluorophores (region V). By normalizing the cumulative excitation-emission area volumes to relative regional areas ($nm^2$), the volume of fluorescence ($\Phi_{(i)}$) was calculated according to Chen et al. [12] within each region ($i$), applying the following equation:

$$\Phi_{(i)} = \mathrm{MF}_{(i)} \sum \sum I(\lambda_{ex}\lambda_{em})\Delta\lambda_{ex}\Delta\lambda_{em}, \quad (2)$$

where $\mathrm{MF}_{(i)}$ is a multiplication factor calculated by Equation (3); $\Delta\lambda_{ex}$ is the excitation wavelength interval (taken as 5 nm); $\Delta\lambda_{em}$ is the emission wavelength interval (taken as 1 nm); and $I(\lambda_{ex}\lambda_{em})$ is the fluorescence intensity at each excitation-emission pair (Raman units):

$$\mathrm{MF}_{(i)} = \frac{\text{total spectra area}}{\text{specific region area}_{(i)}}. \quad (3)$$

And the normalized excitation-emission area volumes referred to the value of region $i$. The entire region was calculated, and the percent fluorescence response was then determined using the following equation:

$$P_{i,n} = \frac{100 \times \Phi_{i,n}}{\Phi_{T,n}}. \quad (4)$$

*(3) PARAFAC Component Analysis Method.* PARAFAC is a method that decomposes EEMs of complex WEOM into available components. PARAFAC modeling was performed in MATLAB followed by the procedure recommended in the DOMFluor toolbox [20]. The PARAFAC models with two to seven components were computed, and core consistency was often applied to select the optimal number of components [21]. The concentration scores of each component were represented by maximum fluorescence intensity ($F_{max}$, R.U.).

*2.6. Other Analytical Methods.* The TS, volatile solids (VSs), dissolved organic carbon (DOC), soluble carbohydrate, and soluble protein were all determined. The diluent samples were operated as the same as mentioned in Section 2.3. SCOD was analyzed using standard methods [22]. The DOC concentrations of the WEOM solutions were determined with a total organic carbon analyzer (TOC-L CPH, Shimadzu, Japan). Soluble protein was quantified by the Lowry–Folin method using bovine serum albumin as the standard, and carbohydrate was determined using the phenol-sulfuric acid method with glucose as the standard [23, 24].

## 3. Results and Discussion

*3.1. Variation of Food Waste after Hydrothermal Treatment.* FW was treated under high pressure for 30 min at 110, 120,130, 140, 150, and 160°C. The SCOD concentrations of the hydrothermal-treated FW are shown in Figure S1. Increasing the temperature from 110 to 160°C caused an increase in SCOD from 85.55 g/kg to 119.91 g/kg. Although the increment of SCOD was obtained, color generation by Maillard reaction was also observed in the HT process due to the high temperature and the presence of proteins and carbohydrates [25]. Figure S2 shows that the treated solid gradually produced a darker brown color than the untreated FW with increasing treatment temperature. The significant color variation of hydrothermal FW was highly reliant on the formation of MRPs which depends on the reaction temperature. Nevertheless, owing to the complex and heterogeneous nature of the MRPs, it has not been possible to isolate or purify MRPs [26]. Therefore, there is a lack of methods for rapid, effective, and quantitative estimation of MRPs in the hydrothermal FW system.

*3.2. Characterization and Evaluation of MRPs in Food Waste after Hydrothermal Treatment*

*3.2.1. Distribution of Soluble Organic Compounds in WEOM.* Figure 1 presents the molecular weight distribution of WEOM before and after HT. DOC of the original sample was evenly in the MW analysis range and remained unchanged in each fraction after HT at 120°C. However, molecular weight fractionation distribution was modified at 140°C compared with 120°C. The total DOC content had a large promotion, but significant increase of DOC from 12 to 27 g/L only occurred in the MW > 30 kDa fraction. The

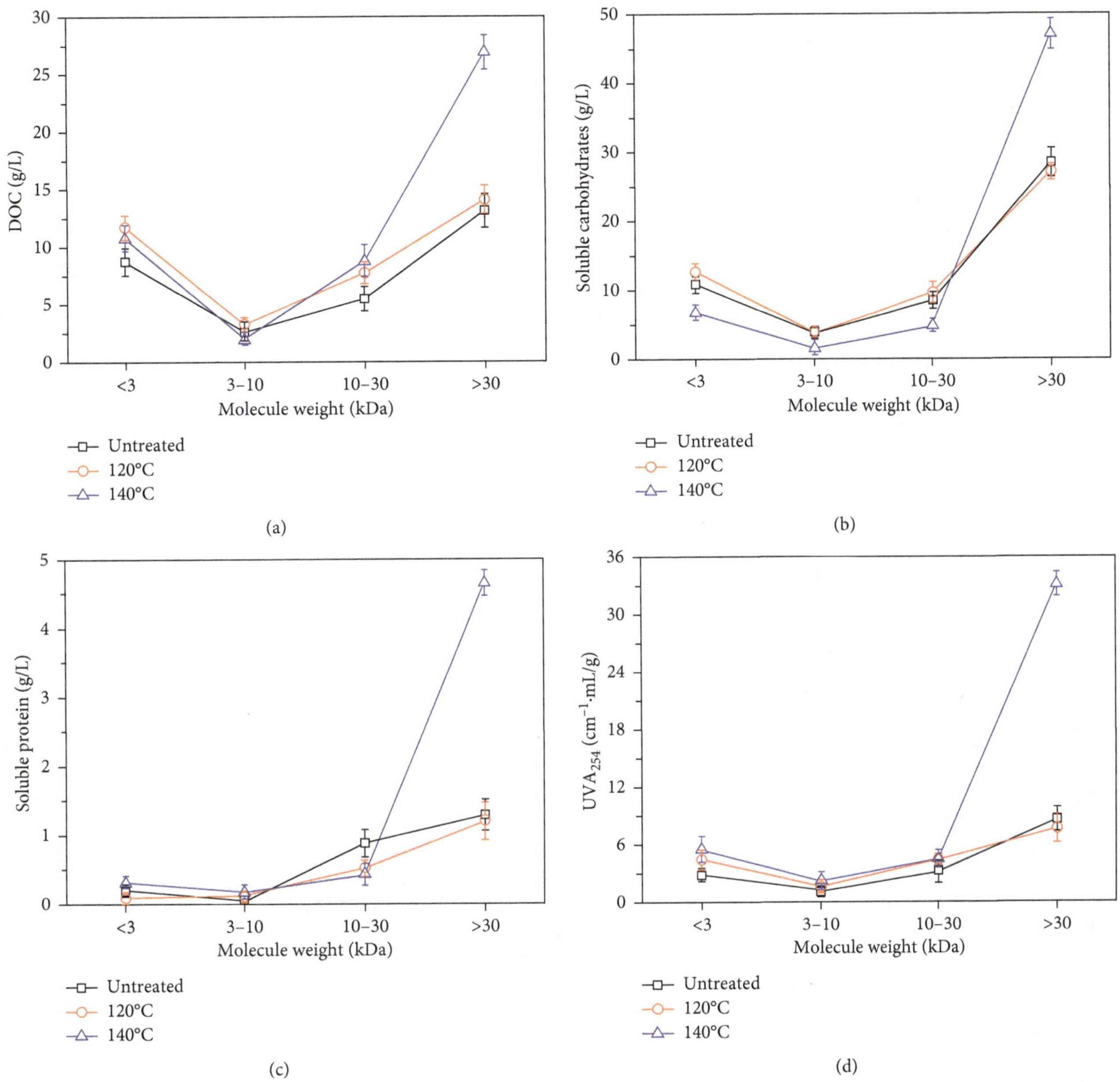


FIGURE 1: Different molecular weight distributions of the soluble organic compounds. (a) DOC. (b) Soluble carbohydrates. (c) Soluble protein. (d) $UVA_{254}$.

significant increment highlighted that dissolved organic compounds, residing in the MW > 30 kDa fraction, were expected as MRPs which have a molecular weight between 40 and 70 kDa [27]. Moreover, it is well known that the formation of MRPs has a positive relationship with the carbohydrates and proteins [28]. In 140°C group, the most prevalent fraction of soluble carbohydrates and soluble proteins also resided in the MW > 30 kDa fraction, which accounts for 78% and 83%, respectively. In addition, the characteristic absorbance $UVA_{254}$ of WEOM in the MW > 30 kDa fraction was 8, 7.5, and 34 $cm^{-1}$ mL/g for untreated FW, 120°C, and 140°C groups, respectively. The rise of this characteristic absorbance indicator showed that HT could result in the generation of MRPs from organic matter mixtures at 140°C. Similar conclusions were obtained by Barber [29] that MRP production occurred in the typical reaction range of thermal hydrolysis of 140–165°C.

*3.2.2. Spectrometric Characterization of the MRPs from Food Waste.* Nowadays, various easy-to-use quantitative spectrometric characterization methods have been applied to determine the MRP production in complicated systems [19, 30]. Three representative indicators ($\Delta UVA_{254}$, $\Delta$browning index, and $\Delta$color intensity) were employed to demonstrate the modification of MRPs content of hydrothermal FW (Figure 2), and raw material was taken as the control. At first, $\Delta UVA_{254}$ did not vary significantly ($p > 0.05$) after HT at 110°C and 120°C. The $\Delta UVA_{254}$ increased significantly ($p < 0.05$) at 130°C, which enhanced by

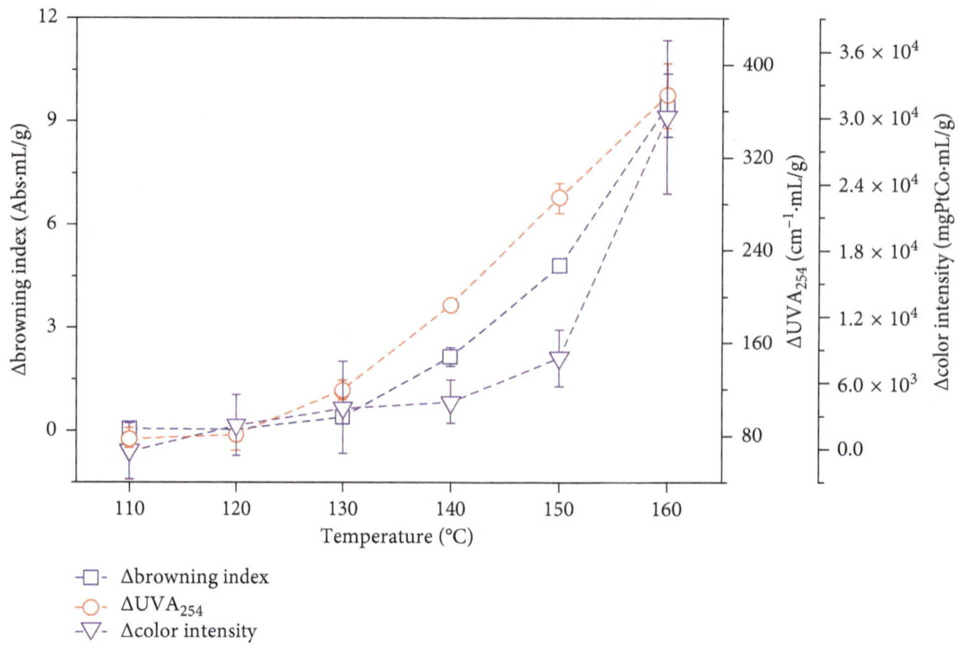


FIGURE 2: Three spectrometric indicators of MRPs production from FW at different temperatures.

1.5-fold compared with 120°C. Then, this surrogate index continued ascending with the increase of treatment temperature, indicating a significant number of compounds accumulated when the hydrothermal temperature above 120°C. In addition, a similar tendency was found in the result of the pronase method, and the Δbrowning index was utilized for determination of nonenzymatic browning. The changing of the Δbrowning index showed that the brown pigment production was slim to zero in 110°C and 120°C treatments and exhibited a regular increase beyond the critical temperature (120°C). The tendencies of these two parameters were consistent with the transformation of the surface color of hydrothermal-treated FW. However, the accuracy of these two indicators was still in doubt because of the complexity of hydrothermal FW systems. Moreover, the Δcolor intensity of WEOM was also recorded. Nevertheless, the relative standard deviation of Δcolor intensity was higher than other methods, and this color indicator did not demonstrate the significant difference ($p > 0.05$) between the 130°C and 140°C groups, which was different from the appearance changes of hydrothermal-treated FW. Although spectrometric characterization could quickly verify the accumulation of MRPs in an indirect way, the influence of other organic matters was still not solved.

*3.2.3. Characterization of MRPs by 3DEEM Fluorescence Analysis.* The objective of using 3DEEM fluorescence was to monitor the change of fluorescence properties in WEOM and to analyze the effect of temperature on MRPs generation in the hydrothermal FW system. Typical 3DEEM spectra of treated WEOM samples at different temperatures are shown in Figure 3, and detailed fluorescence properties are summarized in Table 1 by the "peak-picking" method. As expected, there were obvious differences in the 3DEEM spectra of the WEOM with the rise of temperature, and the maximal transformation was found between 130°C and 140°C groups. Moreover, Table 1 shows the change of peak intensity at the soluble microbial region ($\lambda_{ex}/\lambda_{em}$ = 285/357–361), which decreased from 3.58 R.U. (raw material) to 0.10 R.U. (160°C), and the sharp decrease trend started at the 140°C group. To the best of our knowledge, soluble microbial products (SMPs) contain various complex organic materials, such as proteinaceous material, carbohydrates, humic and fulvic substances, and organic acids [31, 32]. Thus, the organic matters resided in the region of soluble microbial by-products were considered as soluble carbohydrates and proteins because other organic matters would not appear in this region. Conversely, the peak intensity at the humic acid-like region ($\lambda_{ex}/\lambda_{em}$ = 320–360/419–438) increased from 0.43 to 3.34 R.U. as the temperature increased, and obvious transformation also occurred at 140°C group. The peak ratio (V/IV) ascended with the increase of temperature, which implied that the generation and accumulation of humic acid-like organic fluorescent molecules were attributed to the polymerization of the soluble carbohydrates and proteins during the HT. In addition, the movement of humic-like peaks location within the 3DEEM ($\lambda_{ex}/\lambda_{em}$) to longer wavelength (red-shifting), from $\lambda_{ex}/\lambda_{em}$ = 320/428 to 360/438, also indicated the humic acid-like fluorophores were concentrating and becoming more refractory [33]. Besides, the HIX value has been applied to evaluate the humification extent of dissolved organic matter [34]. From Table 1, the increase of HIX values after 140, 150 and 160°C treatments were indicative of more humic WEOM. It is generally known that MRPs are highly aromatic and resemble humic substances; it suggested that 3DEEM might serve as a good descriptor to prove the MRPs production during the HT.

Overall, compared with the 3DEEM analysis method, spectrometric approaches cannot provide a direct

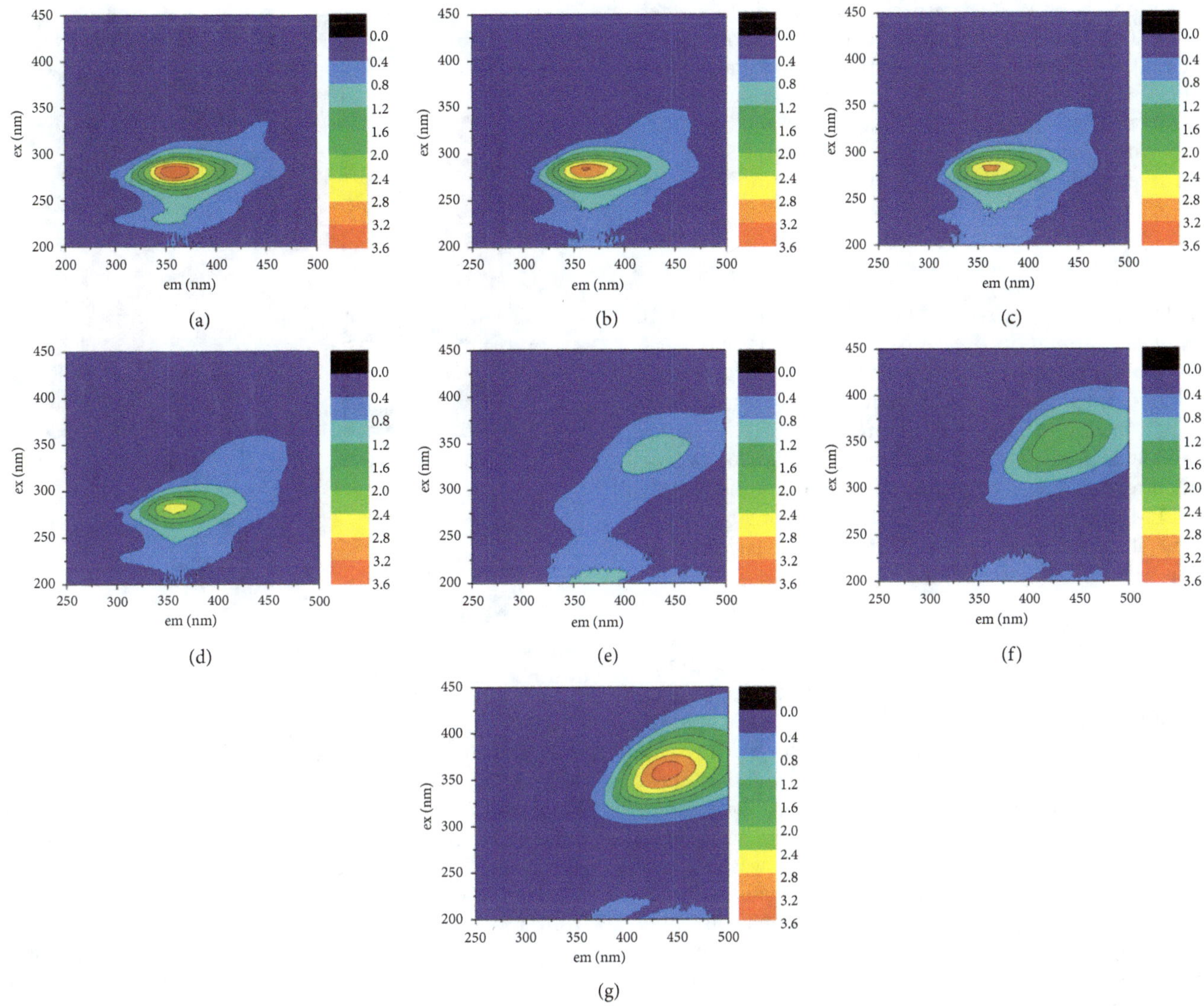

FIGURE 3: Typical EEM spectra of the WEOM extracted from FW treated at different temperatures. (a) Untreated, (b) 110°C, (c) 120°C, (d) 130°C, (e) 140°C, (f) 150°C, and (g) 160°C.

TABLE 1: Fluorescent properties of treated groups at each temperature.

| Sample name | Region IV peak location | | Region V peak location | | Peak ratio (IV/V) | HIX |
|---|---|---|---|---|---|---|
| | Ex/em | Fluorescence intensity (R.U.) | Ex/em | Fluorescence intensity (R.U.) | | |
| Untreated sample | 280/357 | 3.58 | 320/438 | 0.43 | 0.12 | 0.49 |
| 110°C treated sample | 285/360 | 3.24 | 320/434 | 0.58 | 0.18 | 0.62 |
| 120°C treated sample | 285/360 | 3.67 | 320/436 | 0.52 | 0.14 | 0.62 |
| 130°C treated sample | 285/359 | 2.56 | 325/419 | 0.60 | 0.23 | 0.45 |
| 140°C treated sample | 285/359 | 0.75 | 335/421 | 0.96 | 1.28 | 0.88 |
| 150°C treated sample | 285/361 | 0.42 | 345/430 | 1.96 | 4.67 | 1.68 |
| 160°C treated sample | 285/360 | 0.10 | 360/438 | 3.43 | 34.30 | 6.70 |

assessment of MRPs and potentially influence the determination of MRPs due to the fluctuation from other organic or colored compounds. The molecular weight fractionation technique was used to eliminate the interference of smaller molecular weight constituents, which caused the assessment of MRPs costlier and more time-consuming. Besides, Figures 1(b) and 1(c) show that even though the molecular weight fractionation was used, the existence of organic macromolecular compounds (carbohydrates and proteins) might also affect the accuracy of $UVA_{254}$ measurement. Accordingly, 3DEEM fluorescence analysis was simple, had good selectivity, and provided a wealth of information about WEOM, especially helpful in characterizing WEOM and interpreting the complex information in EEM [35]. However, fluorescence information was still restricted because only one peak location and fluorescence intensity value was used for analysis, and it should be noticed that the direct quantification of MRPs using 3DEEM can be of major interest for investigation of WEOM from FW.

### 3.3. Quantitative Characterization of MRPs Production

*3.3.1. Fluorescence EEMs-Based FRI Analysis and PARAFAC Component Estimation.* Recently, FRI analysis and PARAFAC component estimation were developed to provide a more complete data analysis than the traditional "peak-picking" technique [36]. Therefore, the FRI analytical method was used to reveal the transformation of organic compounds in WEOM by integration of the volume beneath each EEM region. According to the five regions of the FRI analysis, it can be found that the obvious characteristic peaks were presented in region IV and region V. The variation of percent fluorescence response $P_{(i,n)}$ of the WEOM is shown in Figure 4. The $P_{(i,n)}$ values of region IV and region V remained almost constant in 110, 120, and 130°C groups. When temperature increased from 130 to 160°C, the $P_{(i,n)}$ value of region IV decreased from 49.8% to 16.9%, and the $P_{(i,n)}$ of region V increased from 12.1% to 66.1%. Moreover, the $P_{(i,n)}$ value of region I also had a little reduction above 130°C. These results implied that humic acid-like material regarded as MRPs in region V were efficiently accumulated when HT temperature was beyond 130°C. It was assumed that the disappearance of soluble carbohydrates and proteinaceous products (region IV) and tryptophan-like substances (region I) were the source material for the polymerization of MRPs (region V). Therefore, region IV (carbohydrates and proteinaceous products) matters and region V (MRPs) matters can be effectively differentiated based on the FRI method.

Furthermore, PARAFAC can capture the heterogeneity of WEOM samples, thereby decomposing EEM spectra into various individual fluorescent components. Two fluorescent components evaluated by PARAFAC using the CORCONDIA procedure were the soluble microbial by-product-like component ($C_1$) and humic acid-like component ($C_2$). Individual components are shown in Figure 5 as contour plots. The excitation and emission loadings are also given in Figure 5, and excitation and emission characteristics of the components identified in this study and probable source of components are depicted in Table S3. In Table S3, two PARAFAC components are pointed out in detail: $C_1$ with peaks of ex/em = 280/360 nm was associated with soluble microbial products, defined as carbohydrates and proteins which could be easily biodegradable [37, 38], and $C_2$ with specific peaks of ex/em = 360/441 nm originated from humic acid-like substances [9, 28]. In the current study, the humic acid-like component identified by PARAFAC was MRPs, which had higher ex/em than the other synthetic or isolated humic compounds [11, 27, 39]. Furthermore, the observed humic-like compounds were forcefully identified as MRPs

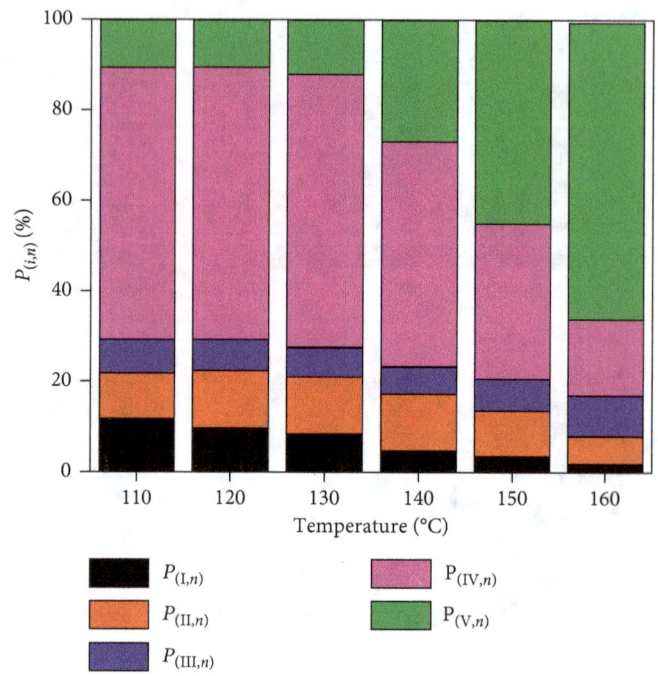


FIGURE 4: The distribution of FRI in WEOM from food waste after hydrothermal treatment.

by comparing its fluorescence property (ex/em) with synthetic MRPs solution created in our study (Figure S3). The maximum fluorescence intensities ($F_{max}$) of component $C_2$ remained constant at a low content when the temperature was in the range of 110°C to 130°C and then increased from 22.37 to 155.79 R.U. by further increasing the temperature from 130 to 160°C (Figure S4), which indicated little accumulation of nonbiodegradable MRPs at temperatures below 130°C and mass production of MRPs when temperature beyond 130°C. Meanwhile, the results in Figure 6 also show that the $F_{max}$ of $C_1$ had a slight reduction at 110°C, 120°C, and 130°C groups but substantially dropped with the further increase of temperature. Results from PARAFAC were consistent with the FRI analysis, which suggested biodegradable organic compounds and MRPs in WEOM can be separated by PARAFAC.

As mentioned above, the distribution of $P_{(i,n)}$ and the 2-component PARAFAC estimation provided complementary information, showing that humic acid-like matter (MRPs) and SMP materials (carbohydrates and proteins) which dominated the WEOM underwent an increase and decrease, respectively, as the operating temperature increased. These results proved FRI and PARAFAC can differentiate fluorescence characteristic disparities between MRPs and other organic compounds. And, it also can be seen that Maillard reaction ($C_2$) was accelerated at elevated temperature by consuming organics in $C_1$. Therefore, 3DEEM combined with FRI/PARAFAC could be employed to analysis of MRPs generated from hydrothermal-treated FW.

*3.3.2. Semiquantitative Characterization of Fluorescent Parameters for MRPs Production.* The volume of fluorescence

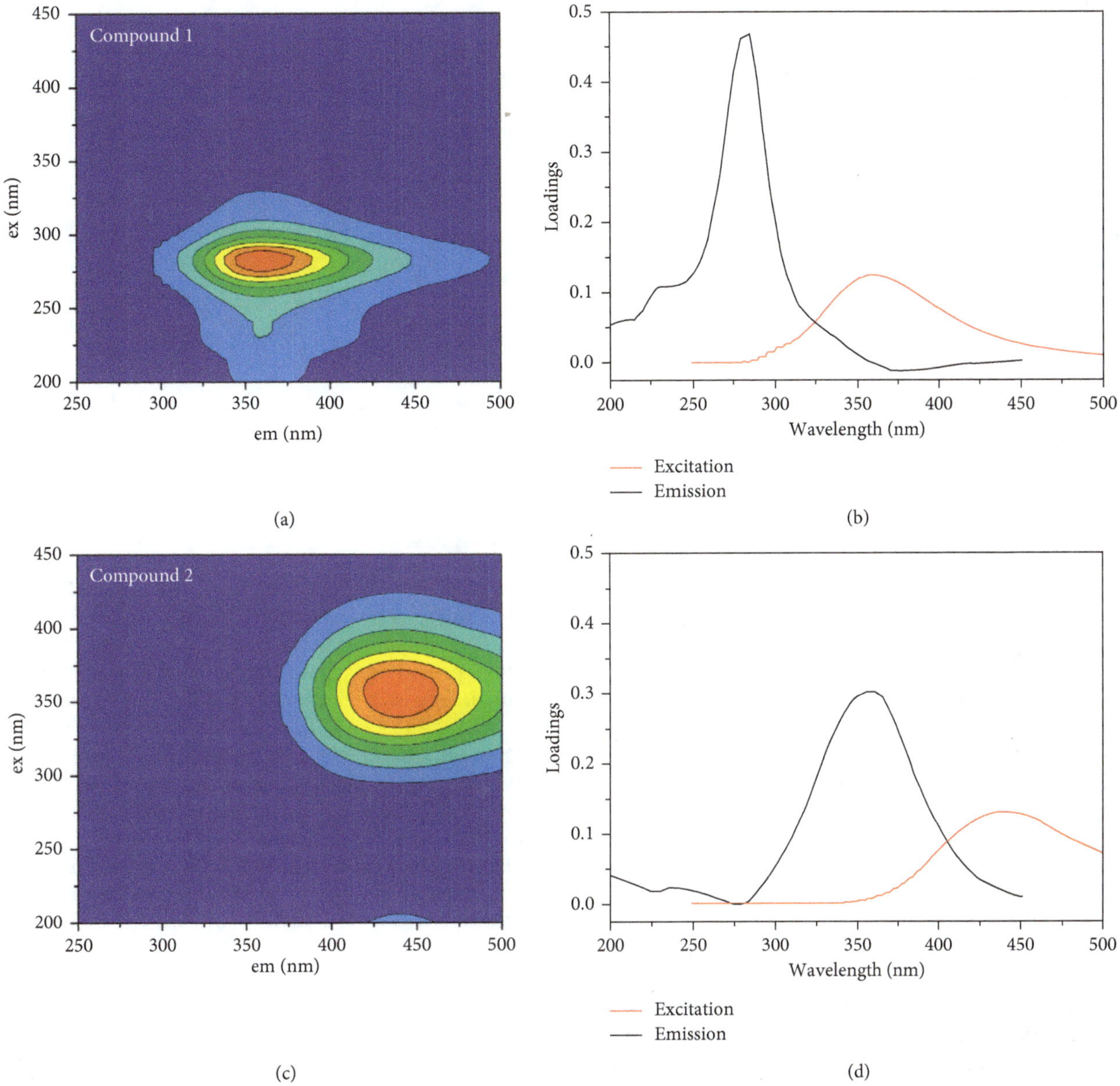


Figure 5: Two-component PARAFAC model generated from samples of WEOM: (a and b) contour plots of the spectral shapes; (c and d) line plots of the loadings.

$\Phi_{(V)}$ parameter from FRI and maximum fluorescence intensity in the $C_2$ component $F_{max}$ ($C_2$) parameter from PARAFAC were used as the characterization parameters for assessing the generation of MRPs. All the data were based on dry weight, and raw material was taken as control. The corresponding results of fluorescent indictors are presented in Figure 6. It can be seen from Figure 5 that the impact of temperature on the $\Delta\Phi_{(V)}$ was minimal in 110, 120, and 130°C groups ($p > 0.05$), staying at a low level from $5.1 \times 10^4$ to $9.0 \times 10^4$ R.U.·nm$^2$·mL/g. But then $\Delta\Phi_{(V)}$ increased significantly from $2.9 \times 10^5$ (140°C) to $1.2 \times 10^6$ R.U.·nm$^2$·mL/g (160°C). A similar tendency was found in $\Delta F_{max}$ ($C_2$) (Figure 5). The $\Delta F_{max}$ ($C_2$) remained unchanged as temperature increased from 110 to 130°C and then went through a successive and relatively large increase from 33.5 to 141.3 R.U.·mL/g as temperature further increased. These results were consistent with the fact that MRPs are temperature dependent [29].

Besides, the correlations between 3DEEM fluorescence parameters and spectrometric parameters were analyzed to check whether fluorescence parameters could effectively characterize the MRPs in FW after HT. Pearson correlations for 3DEEM fluorescence parameters and spectrometric parameters are summarized in Table S4. Generally, spectrometric parameters were strongly correlated with 3DEEM fluorescence parameters ($r > 0.880$, $p < 0.01$), which indicated that these two fluorescent analytical techniques could be employed to monitor the generation of MRPs and semiquantitatively determine the content of MRPs in the hydrothermal FW system. In addition, these two indicators were expected to be tested at various situations, which help their generalization and application.

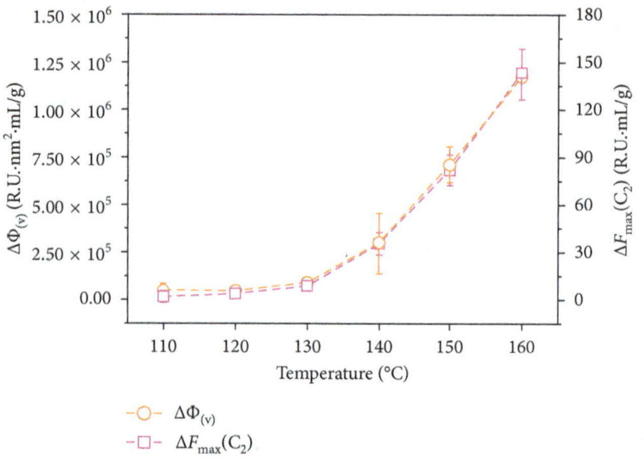


FIGURE 6: Semiquantitative characterization of 3DEEM fluorescent parameters for MRPs production from FW at different temperatures.

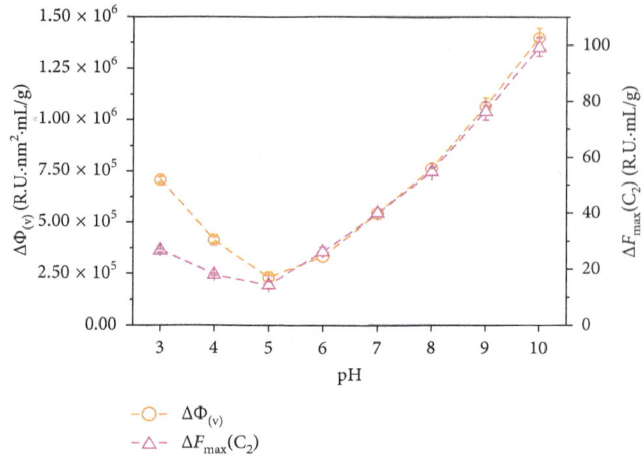


FIGURE 7: Fluorescence indicators of MRPs from FW at different hydrothermal-treated pH value.

*3.3.3. Examples of Application for MRPs Determination by 3DEEM Semiquantitative Fluorescence Parameters.* Apart from temperature, pH also has an impact on MRPs formation. From previous studies, pH and alkalinity were reported to increase the degree of polymerization, thereby causing the accumulation of MRPs [40]. Therefore, it is meaningful to evaluate the effect of pH on MRPs production during HT in FW. In order to semiquantitatively determine their production, 3DEEM-FRI/PARAFAC methods were applied. FW with initial pH from 3.0 to 10 was treated at 130°C for 30 min (Figure 7). As pH increased from 3.0 to 5.0, the mean $\Delta\Phi_{(V)}$ decreased from $7.05 \times 10^5$ to $2.29 \times 10^5$ R.U.·nm$^2$·mL/g and reached the minimum at pH 5.0. Nevertheless, an obvious increase of MRPs was observed from a minimum of $2.29 \times 10^5$ to $1.39 \times 10^6$ R.U.·nm$^2$·mL/g as the pH increased from 5.0 to 10.0. The $\Delta F_{max}$ ($C_2$) from PARAFAC analysis, which was associated with the contribution of MRPs, had a similar tendency to $\Delta\Phi_{(V)}$ and increased from the minimum of 14.24 R.U. at pH 5.0 to the maximum of 90.04 R.U. at pH 10. Previous studies have stated that high initial pH (pH > 5) could accelerate the Maillard reaction rate because Schiff-base matter formed easily [41]. Moreover, low initial pH could also cause the formation of MRPs as it favors 1,2-enolisation reaction pathway and results in the ascent of compounds like furfural or 5-hydroxymethyl-2-furaldehyde (HMF) [42]. Therefore, the results above further verified that $\Delta\Phi_{(V)}$ and $\Delta F_{max}$ ($C_2$) allow an effective evaluation for MRPs production in the hydrothermal FW system.

Besides, hydrothermal time and composition of FW also affect the occurrence of Maillard reaction. Thus, the effect of these hydrothermal conditions on MRPs production can be evaluated according to these two parameters. In this way, the formation of MRPs can be effectively controlled under the optimized hydrothermal condition, thus preventing the substrate loss and promoting limited hydrolysis efficiency simultaneously. Therefore, the increase of bio-converted energy can be achieved by reducing substrate consuming from MRPs formation.

## 4. Conclusion

Compared to traditional methods, the 3DEEM analysis is proved to be a more sensitive method to estimate the occurrence of Maillard reaction in the hydrothermal FW system, providing valuable information when the MRPs are formed. However, its utility is limited to the quantifying of the MRPs. The FRI and PARAFAC analytic methods were established to further distinguish MRPs from the other dissolved organic compounds by integration of the volume beneath each EEM region and capturing the heterogeneity of samples. And $\Delta\Phi_{(V)}$ from FRI and $\Delta F_{max}$ ($C_2$) from PARAFAC can be considered liable parameters for semi-quantifying MRPs under various temperature and pH. The minimum MRPs were validated at temperature below 130°C and pH 5.0.

## Conflicts of Interest

The authors declare that they have no conflicts of interest.

## Acknowledgments

The authors appreciate the National Natural Science Foundation of China (Nos. 51778580 and 51608480), Natural Science Foundation of Zhejiang Province of China (No. LQ16E080001), Open Foundation of Key Laboratory for Solid Waste Management and Environment Safety (Tsinghua University), Ministry of Education of China (SWMES 2015–07), and China Scholarship Council (iCET 2017) for providing the funding support for this project.

## Supplementary Materials

Table S1: characteristics of FW. Table S2: excitation and emission (ex/em) wavelengths of fluorescence region and their associated names. Table S3: comparison of the fluorescence component peak location in the current study with those reported in the previous literature and probable source description. Table S4: Pearson correlations among spectrometric parameters and EEM fluorescence parameters. Figure S1: SCOD of hydrothermal-treated food waste at different temperatures. The data were based on wet weight. Figure S2: the appearance of hydrothermal-treated food waste (untreated; 110°C, 120°C, 130°C, 140°C, 150°C, and 160°C treated). Figure S3: EEM spectra of the synthetic MRP solution produced by glycine and glucose after hydrothermal treatment. The peak location of synthetic MRP solution was $\lambda_{ex}/\lambda_{em}$ = 355/435; the $\lambda_{ex}/\lambda_{em}$ value is higher than the other humic isolates tested, which explains why this product was MRPs. Figure S4: $F_{max}$ values of the two components ($C_1$ and $C_2$) in WEOM from food waste after hydrothermal treatment. (*Supplementary Materials*)

## References


[1] W. Wang, H. Hou, S. Hu, and X. Gao, "Performance and stability improvements in anaerobic digestion of thermally hydrolyzed municipal biowaste by a biofilm system," *Bioresource Technology*, vol. 101, no. 6, pp. 1715–1721, 2010.

[2] Y. Li, Y. Jin, J. Li, H. Li, and Z. Yu, "Effects of thermal pretreatment on the biomethane yield and hydrolysis rate of kitchen waste," *Applied Energy*, vol. 172, pp. 47–58, 2016.

[3] B. Grycová, I. Koutník, and A. Pryszcz, "Pyrolysis process for the treatment of food waste," *Bioresource Technology*, vol. 218, pp. 1203–1207, 2016.

[4] X. Liu, W. Wang, X. Gao, Y. Zhou, and R. Shen, "Effect of thermal pretreatment on the physical and chemical properties of municipal biomass waste," *Waste Management*, vol. 32, no. 2, pp. 249–255, 2012.

[5] L. Ding, J. Cheng, D. Qiao et al., "Investigating hydrothermal pretreatment of food waste for two-stage fermentative hydrogen and methane co-production," *Bioresource Technology*, vol. 241, pp. 491–499, 2017.

[6] R. Chandra, R. N. Bharagava, and V. Rai, "Melanoidins as major colourant in sugarcane molasses based distillery effluent and its degradation," *Bioresource Technology*, vol. 99, no. 11, pp. 4648–4660, 2008.

[7] GE. Leiva, G. B. Naranjo, and L. S. Malec, "A study of different indicators of Maillard reaction with whey proteins and different carbohydrates under adverse storage conditions," *Food Chemistry*, vol. 215, p. 410, 2017.

[8] S. Pavlovic, R. C. Santos, and M. B. A. Glória, "Maillard reaction during the processing of 'Doce de leite'," *Journal of the Science of Food and Agriculture*, vol. 66, no. 2, pp. 129–132, 2010.

[9] J. Dwyer, D. Starrenburg, S. Tait, K. Barr, D. J. Batstone, and P. Lant, "Decreasing activated sludge thermal hydrolysis temperature reduces product colour, without decreasing degradability," *Water Research*, vol. 42, no. 18, pp. 4699–4709, 2008.

[10] M. Uchimiya, J. E. Knoll, W. F. Anderson, and K. R. Harris-Shultz, "Chemical analysis of fermentable sugars and secondary products in 23 sweet sorghum cultivars," *Journal of Agricultural and Food Chemistry*, vol. 65, no. 35, pp. 7629–7637, 2017.

[11] P. G. Coble, "Characterization of marine and terrestrial DOM in seawater using excitation-emission matrix spectroscopy," *Marine Chemistry*, vol. 51, no. 4, pp. 325–346, 1996.

[12] W. Chen, P. Westerhoff, J. A. Leenheer, and K. Booksh, "Fluorescence excitation-emission matrix regional integration to quantify spectra for dissolved organic matter," *Environmental Science and Technology*, vol. 37, no. 24, p. 5701, 2003.

[13] L. J. Zhu, Y. Zhao, Y. N. Chen et al., "Characterization of atrazine binding to dissolved organic matter of soil under different types of land use," *Ecotoxicology and Environmental Safety*, vol. 147, no. 2, pp. 1065–1072, 2018.

[14] L. J. Zhu, H. X. Zhou, X. Y. Xie et al., "Effects of floodgates operation on nitrogen transformation in a lake based on structural equation modeling analysis," *Science of the Total Environment*, vol. 631-632, pp. 1311–1320, 2018.

[15] K. Wang, J. Yin, D. Shen, and N. Li, "Anaerobic digestion of food waste for volatile fatty acids (VFAs) production with different types of inoculum: effect of pH," *Bioresource Technology*, vol. 161, pp. 395–401, 2014.

[16] J. Yin, K. Wang, Y. Yang, D. Shen, M. Wang, and H. Mo, "Improving production of volatile fatty acids from food waste fermentation by hydrothermal pretreatment," *Bioresource Technology*, vol. 171, pp. 323–329, 2014.

[17] E. C. Bernardo, R. Egashira, and J. Kawasaki, "Decolorization of molasses' wastewater using activated carbon prepared from cane bagasse," *Carbon*, vol. 35, no. 9, pp. 1217–1221, 1996.

[18] J. Dwyer, L. Kavanagh, and P. Lant, "The degradation of dissolved organic nitrogen associated with melanoidin using a UV/$H_2O_2$ AOP," *Chemosphere*, vol. 71, no. 9, pp. 1745–1753, 2008.

[19] R. Palombo, A. Gertler, and I. Saguy, "A simplified method for determination of browning in dairy powders," *Journal of Food Science*, vol. 49, no. 6, p. 1609, 2010.

[20] C. A. Stedmon and R. Bro, "Characterizing dissolved organic matter fluorescence with parallel factor analysis: a tutorial," *Limnology and Oceanography Methods*, vol. 6, no. 11, pp. 572–579, 2008.

[21] D. Zhu, B. Ji, H. L. Eum, and M. Zude, "Evaluation of the non-enzymatic browning in thermally processed apple juice by front-face fluorescence spectroscopy," *Food Chemistry*, vol. 113, no. 1, pp. 272–279, 2009.

[22] APHA, *Standard Methods for the Examination of Water and Wastewater*, American Public Health Association/American Water Works Association/Water Environment Federation, Washington, DC, USA, 20th edition, 1998.

[23] O. H. Lowry, N. J. Rosebrough, A. L. Farr et al., "Protein measurement with the Folin phenol reagent," *Journal of Biological Chemistry*, vol. 193, no. 1, pp. 265–275, 1951.

[24] D. Herbert, P. J. Philipps, and R. E. Strange, "Carbohydrate analysis," in *Methods in Enzymology*, vol. 5B, pp. 265–277, Elsevier, Amsterdam, Netherlands, 1971.

[25] C. Bougrier, J. P. Delgenès, and H. Carrère, "Effects of thermal treatments on five different waste activated sludge samples solubilisation, physical properties and anaerobic digestion," *Chemical Engineering Journal*, vol. 139, no. 2, pp. 236–244, 2008.

[26] A. Adams, K. A. Tehrani, M. Kersiene, R. Venskutonis, and N. De Kimpe, "Characterization of model melanoidins by the thermal degradation profile," *Journal of Agricultural and Food Chemistry*, vol. 51, no. 15, pp. 4338–4343, 2003.

[27] J. Dwyer, P. Griffiths, and P. Lant, "Simultaneous colour and DON removal from sewage treatment plant effluent: alum coagulation of melanoidin," *Water Research*, vol. 43, no. 2, pp. 553–561, 2009.

[28] S. I. F. S. Martins and M. A. J. S. van Boekel, "Melanoidins extinction coefficient in the glucose/glycine Maillard reaction," *Food Chemistry*, vol. 83, no. 1, pp. 135–142, 2003.

[29] W. P. Barber, "Thermal hydrolysis for sewage treatment: a critical review," *Water Research*, vol. 104, pp. 53–71, 2016.
[30] J. Dwyer and P. Lant, "Biodegradability of DOC and DON for UV/$H_2O_2$ pre-treated melanoidin based wastewater," *Biochemical Engineering Journal*, vol. 42, no. 1, pp. 47–54, 2008.
[31] J. Li, J. Wei, H. H. Ngo et al., "Characterization of soluble microbial products in a partial nitrification sequencing batch biofilm reactor treating high ammonia nitrogen wastewater," *Bioresource Technology*, vol. 249, pp. 241–246, 2017.
[32] C. Le and D. C. Stuckey, "Impact of feed carbohydrates and nitrogen source on the production of soluble microbial products (SMPs) in anaerobic digestion," *Water Research*, vol. 122, p. 10, 2017.
[33] M. Esparza-Soto, S. Núñez-Hernández, and C. Fall, "Spectrometric characterization of effluent organic matter of a sequencing batch reactor operated at three sludge retention times," *Water Research*, vol. 45, no. 19, pp. 6555–6563, 2011.
[34] J. Hur and G. Kim, "Comparison of the heterogeneity within bulk sediment humic substances from a stream and reservoir via selected operational descriptors," *Chemosphere*, vol. 75, no. 4, pp. 483–490, 2009.
[35] W. T. Li, S. Y. Chen, Z. X. Xu, Y. Li, C. D. Shuang, and A. M. Li, "Characterization of dissolved organic matter in municipal wastewater using fluorescence PARAFAC analysis and chromatography multi-excitation/emission scan: a comparative study," *Environmental Science and Technology*, vol. 48, no. 5, pp. 2603–2609, 2014.
[36] H. Wu, Z. Zhou, Y. Zhang, T. Chen, H. Wang, and W. Lu, "Fluorescence-based rapid assessment of the biological stability of landfilled municipal solid waste," *Bioresource Technology*, vol. 110, no. 2, pp. 174–183, 2012.
[37] X. Jia, B. Xi, M. Li et al., "Evaluation of biogasification and energy consumption from food waste using short-term hydrothermal pretreatment coupled with different anaerobic digestion processes," *Journal of Cleaner Production*, vol. 152, pp. 364–368, 2017.
[38] J. Sun, L. Guo, Q. Li et al., "Structural and functional properties of organic matters in extracellular polymeric substances (EPS) and dissolved organic matters (DOM) after heat pretreatment with waste sludge," *Bioresource Technology*, vol. 219, pp. 614–623, 2016.
[39] M. Zhang, Z. Wang, P. Li, H. Zhang, and L. Xie, "Biorefractory dissolved organic matter and colorants in cassava distillery wastewater: characterization, coagulation treatment and mechanisms," *Chemosphere*, vol. 178, pp. 272–279, 2017.
[40] M. Coca, M. T. Garcia, G. Gonzalez et al., "Study of coloured components formed in sugar beet processing," *Food Chemistry*, vol. 86, no. 3, pp. 421–433, 2004.
[41] W. Jiang, Y. Chen, X. He, S. Hu, S. Li, and Y. Liu, "A study of the tyramine/glucose Maillard reaction: variables, characterization, cytotoxicity and preliminary application," *Food Chemistry*, vol. 239, pp. 377–384, 2017.
[42] C. Rannou, D. Laroque, E. Renault, C. Prost, and T. Sérot, "Mitigation strategies of acrylamide, furans, heterocyclic amines and browning during the Maillard reaction in foods," *Food Research International*, vol. 90, pp. 154–176, 2016.

# 8

# A Stability-Indicating RP-HPLC-UV Method for Determination and Chemical Hydrolysis Study of a Novel Naproxen Prodrug

**Mohamed H. M. Hamid**[1,2] **and Tilal Elsaman**[1]

[1] *Department of Pharmaceutical Chemistry, College of Pharmacy, Omdurman Islamic University, P.O. Box 2587, 11452 Khartoum, Sudan*
[2] *Department of Pharmaceutical Chemistry, College of Pharmacy, University of Medical Sciences and Technology, P.O. Box 12810, Khartoum, Sudan*

Correspondence should be addressed to Mohamed H. M. Hamid; drholo2@gmail.com

Academic Editor: Orazio Nicolotti

A new naproxen amide prodrug was synthesized and spectrally characterized and a simple, precise, and accurate stability-indicating RP-HPLC method was developed and validated for determination and chemical hydrolysis study of the prodrug. Forced degradation studies were conducted as per the International Conference on Harmonization (ICH) guidelines to establish the stability-indicating power of the method. Separations were performed on a $C_{18}$ column (150 × 4.6 mm i.d., 5 $\mu$m p.s.). The mobile phase consisted of acetonitrile and phosphate buffer pH 4.0 in the ratio 60 : 40. The flow rate and injection volume were 1.0 mL/min and 15 $\mu$L, respectively. The peaks were monitored at 272 nm. The average retention time is 5.136 min. The linearity of the method was investigated in the range of 10–50 $\mu$g/mL and $r^2$ was found to be larger than 0.9987. The LOD and LOQ were found to be 1.853 and 5.615 $\mu$g/mL, respectively. Results indicated that the degradants are well resolved and separated from the prodrug. Hydrolysis kinetics studies were carried out in buffer solutions (pH 1.2, 5.5 and 7.4) to establish the fate of the prodrug. The half-lives in the respective buffers were 23.5, 262, and 334 hours *indicating sufficient stability to attain the goal of oral delivery*.

## 1. Introduction

A prodrug is a modified biologically active compound that will release the active compound *in vivo* following enzymatic or chemical cleavage. Many drugs have unwanted properties that hinder their clinical use, such as susceptibility to acid hydrolysis in the stomach, causing damage to the stomach itself. There are many approaches to overcome those problems such as coating, different formulations, and different routes of administration. However, chemical derivatization of the drug into a prodrug remains the most common approach [1, 2]. Nonsteroidal anti-inflammatory drugs (NSAIDs) are the most commonly used agents for the symptomatic relief of fever, minor inflammations, headaches, swelling, sports related pains, and arthritis pain. They are also very helpful in postoperative, menstrual, and dental pains. Individually, they have different characteristics, but all of them share one pharmacological activity in-common, which is their ability to inhibit the cyclooxygenase (COX) enzyme family. COX enzymes have many isoforms, but only two are well studied. They are COX-1 and COX-2. COX-1 is responsible for prostaglandins generation, which have cytoprotective effects on the stomach. COX-2 is mainly activated by inflammation. Therefore, most of the anti-inflammatory, antipyretic, and analgesic effects of the NSAIDs are attributed to their inhibition of COX-2, while most of their side effects can be related to inhibition of COX-1. Because NSAIDs are almost always used chronically especially by the elderly, their dangerous gastrointestinal (GI), cardiac, and renal side effects constitute a problem hindering their use. The prodrug approach is an effective way to mask the carboxylic acid group of NSAIDs, which is the main culprit in the GI adverse effects [3, 4].

Naproxen is a NSAID of the phenylpropanoic acid class just like ibuprofen. Its GI toxic side effects are intermediate between ibuprofen (low) and indomethacin (high). Its structure is enantiomeric, which is (+)-2-(6-methoxy-2-naphthyl)-propanoic acid (Figure 1) [5]. Naproxen (**NP**) still has all the side effects of NSAIDs; therefore, naproxen

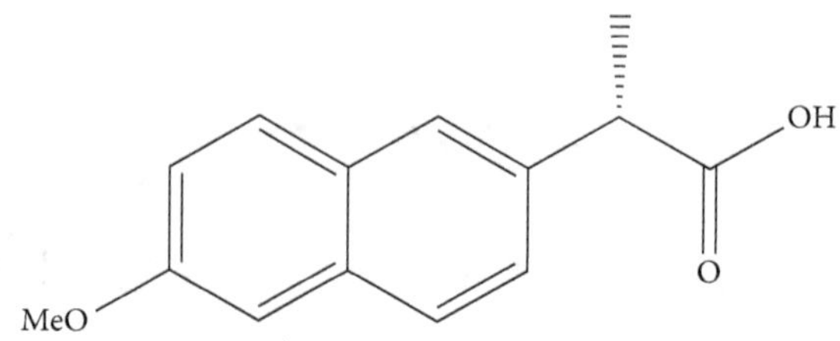

FIGURE 1: Chemical structure of S-naproxen.

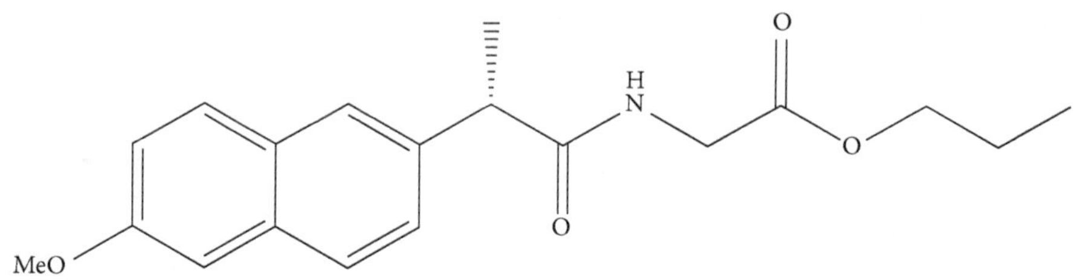

FIGURE 2: Chemical structure of pro-NP.

prodrugs were studied. Ester and amide prodrugs are the most studied prodrugs of naproxen. Studies show that they do reduce the GI adverse reactions of naproxen significantly [2, 5–7].

Quite a lot of different naproxen prodrugs were synthesized and reported in the literature [5, 6, 8–15]. However, the naproxen prodrug understudy here (referred to here after as **pro-NP**, Figure 2) contains and will release glycine *in vivo* which is not only safe but useful as a nutrient for the body. Furthermore, the intermediate of the synthesis of pro-NP is reported in the literature [16] to possess pharmacological activities. Microwave-assisted transesterification was used here for the first time to synthesize a naproxen prodrug. This reduces the transesterification time significantly and enhances the lipophilicity of the resulting prodrug which is crucial for oral activity.

Stability-indicating methods were described in the literature to study the chemical hydrolysis of naproxen prodrugs [12, 13, 15, 17–19]. Most of them employed HPLC with UV detection and $C_{18}$ columns for its simplicity the same as in the method below. However, Rautio et al. [13] used a fluorescence detector and Sheha et al. [15] used a $C_8$ column.

In light of the above and in continuation of our interest to synthesize a GI sparing NSAIDs prodrugs [2, 20], a new amide prodrug of naproxen (pro-NP) was synthesized and spectrally characterized and a validated stability-indicating RP-HPLC method was developed for studying the chemical stability of pro-NP.

## 2. Materials and Methods

*2.1. General.* S-Naproxen and ethyl glycinate were purchased from Sigma-Aldrich. Microwave reactor (CEM, USA) was used in final synthesis step. The reaction was monitored by TLC on precoated silica G plates using UV lamps. Melting point was determined on a Gallenkamp melting point apparatus and is uncorrected. Infrared (IR) Spectrum was recorded as KBr disc using the Perkin Elmer FT-IR Spectrum BX apparatus located at the Research Center, College of Pharmacy, King Saud University (Riyadh, Saudi Arabia). NMR spectrum was scanned in DMSO-d6 on a Bruker NMR spectrometer operating at 500 MHz for $^{1}$H and 125.76 MHz for $^{13}$C at the Research Center, College of Pharmacy, King Saud University (Riyadh, Saudi Arabia). Chemical shifts are expressed in $\delta$-values (ppm) relative to TMS as an internal standard. Coupling constants ($J$) are expressed in Hz. $D_2O$ was added to confirm the exchangeable protons. Mass spectra were measured on Agilent Triple Quadrupole 6410 QQQ LC/MS with ESI (Electrospray ionization) source. The acetonitrile *RS* used was HPLC grade, was purchased from Carlo Erba Reagents S.A.S. (F-27106 Val-de-Reuil Cedex, France), and was used without further purification. All other reagents used were of analytical grade.

*2.2. Synthesis.* Pro-NP was synthesized by coupling of S-naproxen with glycine ethyl ester hydrochloride using DCC as coupling agent according to the literature method [2] (Scheme 1). The resulting intermediate was further subjected to microwave-assisted transesterification reaction in propanol using potassium cyanide as catalyst for five minutes. The final product was purified by column chromatography using hexane : ethyl acetate as mobile phase.

*2.3. Chromatographic Conditions.* The HPLC method was developed on a Shimadzu prominence UFLC system (Shimadzu corporation, Kyoto, Japan) that consisted of a UV/Vis detector (SPD-20AV), a degasser (DGU-20A3R), column

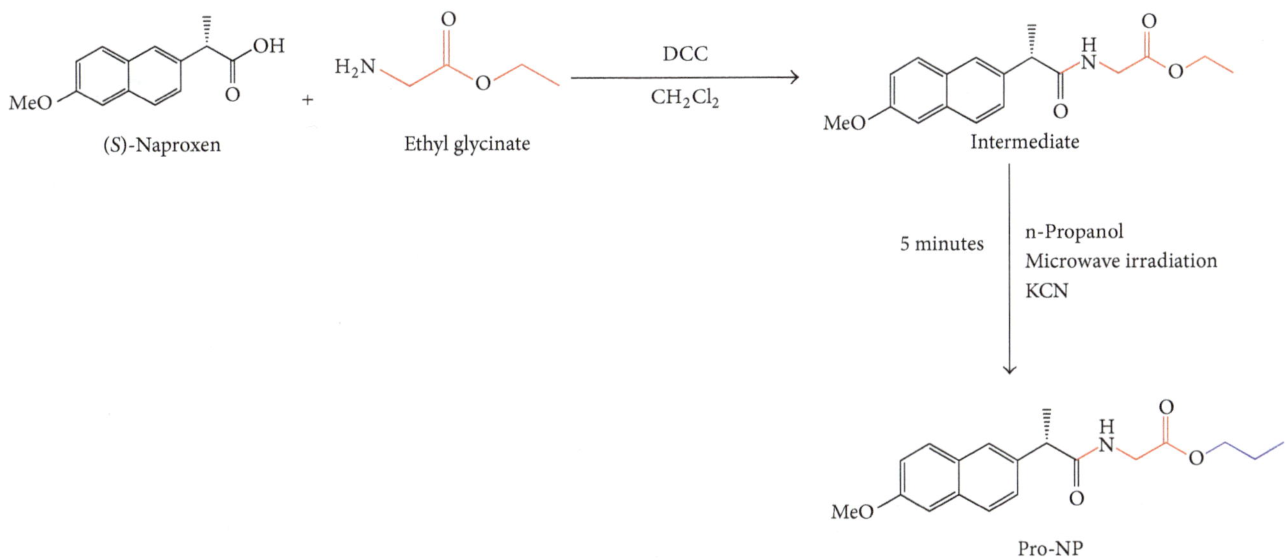


SCHEME 1: Synthesis of the targeted pro-NP.

oven (CTO-20A), and a pump (LC-20AB). The column used was $C_{18}$ (150 × 4.6 mm internal diameter, 5 $\mu$m particle size) at ambient temperature. Isocratic elution was used with the mobile phase consisting of acetonitrile and phosphate buffer pH 4.0 in the ratio 60 : 40. The flow rate and injection volume were 1.0 mL/min and 15 $\mu$L, respectively. The peaks were monitored at the wavelength of 272 nm. The average retention time is 5.136 min.

*2.4. Preparation of Standard Stock Solution.* A standard stock solution of pro-NP (100 $\mu$g/mL) was prepared by dissolving accurately weighed 10 mg of the drug in 100 mL methanol.

*2.5. Preparation of Phosphate Buffer pH 4.0.* 6.80 g of $KH_2PO_4$ was dissolved in 700 mL of water. Then the pH was adjusted when necessary with 10% v/v $H_3PO_4$. Water was added to complete to 1000 mL (BP 2013).

*2.6. Forced Degradation Stability Studies/Specificity.* Forced degradation stability study was conducted as per ICH guidelines Q1A (R2) [21] and Q1B [22], regarding hydrolytic degradation, thermal degradation, photolytic degradation, and oxidative degradation. Results for forced degradation study are shown in Table 2.

*2.6.1. Acidic and Alkaline Hydrolytic Degradation Study.* For the acid hydrolysis study, 2 mL of the pro-NP stock solution was transferred to a 20 mL volumetric flask and was completed up to the mark with 0.1 N HCl. A blank solution was also prepared. A determination of both solutions was made at zero time and after 24 h. For the base hydrolysis study, 2 mL separately of the pro-NP stock solution was transferred to two 20 mL volumetric flasks and completed up to the mark with 0.1 N and 0.01 N NaOH. Two blank solutions were made. A determination of all four solutions was made at only zero time.

*2.6.2. Oxidative Degradation Study.* 2 mL of the pro-NP stock solution was transferred to a 20 mL volumetric flask and made up to the mark with 3% $H_2O_2$ and an aliquot was injected into the HPLC at zero time and after 24 h. In another flask, 2 mL of the pro-NP stock solution was made to the mark with 6% $H_2O_2$ and put in an oven set at 50°C for 2 h and then an aliquot was injected into the HPLC. Blanks were made and analyzed.

*2.6.3. Thermal Degradation Study.* The solid pro-NP drug was put in an oven set at 70°C and an appropriate amount was taken, dissolved in methanol, and injected into the HPLC after 24 h and another after 48 h.

*2.6.4. Photolytic Degradation Study.* The solid pro-NP drug was subjected to different lamps that emit visible light, short UV, and long UV. After two days, an appropriate amount was taken, dissolved in methanol, and injected into the HPLC.

*2.7. Method Validation.* The proposed method was validated as per ICH guidelines [23] with respect to linearity, specificity, accuracy, precision, LOD, LOQ, and robustness.

*2.7.1. Specificity (Selectivity).* In addition to forced degradation studies specificity was also confirmed by spiking pro-NP with naproxen (NP), which is a known degradant of pro-NP. 10 mg of naproxen *RS* (NP) was dissolved in 100 mL of methanol to give 100 $\mu$g/mL stock solution of NP. 2 mL of the NP stock solution was mixed with 4 mL of pro-NP stock solution and completed to 20 mL with methanol and an aliquot of the mixture was injected into the HPLC. The resolution ($R_s$) between the two peaks was calculated.

*2.7.2. Linearity.* The standard solutions for the estimation of linearity were prepared as per Table 1. Standard solutions used in the estimation of linearity (calibration curve) were

TABLE 1: Preparation of standard solutions for the evaluation of linearity.

| Concentration ($\mu$g/mL) | Volume pipetted from the stock (mL) | Volume made up to (mL) |
|---|---|---|
| 10 | 2 | 20 |
| 15 | 3 | 20 |
| 20 | 4 | 20 |
| 25 | 5 | 20 |
| 30 | 6 | 20 |
| 50 | 10 | 20 |

TABLE 2: Results of forced degradation stability studies.

| Stress condition | Nature of pro-NP | Number of determinations | Time (hours) | Temperature (°C) | Degradation (%) |
|---|---|---|---|---|---|
| Hydrolysis, acidic | Solution | In 0.1 N HCl | 24 | 25 | 21.5 |
| Hydrolysis, alkali | Solution | In 0.1 N NaOH | 0 | 25 | 100 |
| | | In 0.01 N NaOH | 0 | 25 | 100 |
| Oxidation | Solution | In 3% $H_2O_2$ | 24 | 25 | 0 |
| | | In 3% $H_2O_2$ | 24 | 50 | 0 |
| | | In 6% $H_2O_2$ | 2 | 50 | 0 |
| Thermal | Solid | Once | 48 | 70 | 0 |
| Photolytic | Solid | Once | 48 | 25 | 0 |

TABLE 3: Data from the regression analysis.

| Parameter | Pro-NP |
|---|---|
| Linearity range | 10–50 $\mu$g/mL |
| Slope | 19173.985 |
| Intercept | −1695.79167 |
| Regression coefficient $R^2$ | 0.998740 |
| Regression equation | $y = 19174x - 1695.8$ |
| LOQ | 5.61457116 $\mu$g/mL |
| LOD | 1.852808483 $\mu$g/mL |

prepared by suitable dilutions from the pro-NP stock solution to attain solutions of different concentrations. A plot of the peak area versus the concentration of the standard solutions was made (Figure 7). Regression analysis was carried out on the plot. From the regression data, the correlation coefficient ($r^2$), the slope of the line, the $Y$-intercept, and line equation were calculated in Table 3.

*2.7.3. Precision.* Precision has been established in three steps: assessment of repeatability, interday precision assessment, and intraday precision assessment.

*2.7.4. Robustness.* Small but deliberate changes were made to the mobile phase pH, which is 4.0, and the detection wavelength, which is 272 nm. Twelve determinations were made as follows: 3 at 270 nm, 3 at 274 nm, 3 at pH 4.3, and 3 at pH 3.7. The % RSD was calculated for the determinations at the changed parameters with the determinations at the normal parameter.

*2.7.5. LOD and LOQ.* The LOD and LOQ were calculated from the equations

$$\text{LOD} = \frac{3.3\sigma}{s},$$
$$\text{LOQ} = \frac{10\sigma}{s}, \quad (1)$$

where $\sigma$ is the standard deviation of the responses and $s$ is the slope of the calibration curve.

*2.8. Determination of the Hydrolysis Reaction Kinetics In Vitro.* Chemical hydrolysis was studied at 37°C ± 1°C in pH 1.2 (SGF), pH 5.5 phosphate solution (duodenal pH), and pH 7.4. Each reaction was initiated by taking 3 mL of the stock methanolic solution of pro-NP, to 20 mL volumetric flask, one drop of Tween-80 was added, and the volume completed to the mark with the buffer solution of the intended pH. The solutions were transferred to screw-capped test tubes. Samples were taken at appropriate time intervals spread over eight hours and chromatographed and the residual concentrations were calculated.

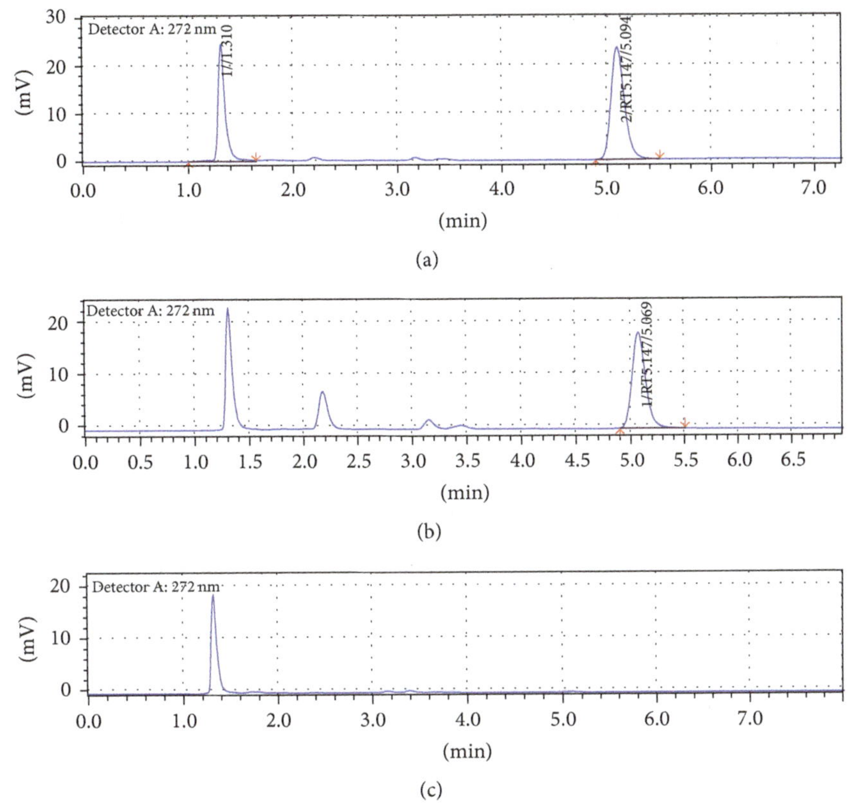


FIGURE 3: Chromatograms indicating acid hydrolysis degradation of pro-NP. (a) is the chromatogram of pro-NP in 0.1 N HCl at zero time. (b) is the chromatogram of pro-NP in 0.1 N HCl after 24 h. (c) is the chromatogram of a blank in the same study.

## 3. Results and Discussion

*3.1. Spectral Data for the Intermediate.* White powder; yield (76%), m.p. (84–86°C). IR (KBr): $\nu_{max}/cm^{-1}$ 3360 (NH), 1713 (C=O, ester), 1669 (C=O, amide). $^1$H NMR: ($CDCl_3$) $\delta$ 1.23 (t, 3H, $J$ = 7.0 Hz, $-OCH_2\mathbf{CH_3}$), $\delta$ 1.64 (d, 3H, $J$ = 7 Hz, $-CH\mathbf{CH_3}$), $\delta$ 3.76–3.79 (q, $\mathbf{CH}CH_3$), 3.94–4.05 (m, 5H, $-\mathbf{OCH_3}$ and $NH\mathbf{CH_2}$), 4.15–4.19 (q, 2H, $J$ = 7.0 Hz, $OCH_2$), 5.9 (s, 1H, **NH**, $D_2O$-exchangeable) 7.14–7.19 (m, 2H, Ar**H**), 7.43 (d, 1H, $J$ = 8 Hz, Ar**H**), 7.72–7.76 (m, 3H, Ar**H**). $^{13}$C NMR: $\delta$ 14.08 ($OCH_2\mathbf{CH_3}$), 18.4 ($CH\mathbf{CH_3}$), 41.5 ($NH\mathbf{CH_2}$), 46.8 ($-\mathbf{CH}CH_3$), 55.3 ($-\mathbf{OCH_3}$), 61.4 ($O-\mathbf{CH_2}CH_3$), 105.6, 119.18, 126.28, 127.59, 128.99, 129.62, 133.81, 136.02, 157.77 (Ar–**C**), 169.85 (–C=O amide), 174.44 (–C=O ester). MS-ESI: $m/z$ 315.6 ($M^+$).

*3.2. Spectral Data for Pro-NP.* White powder; yield (90%), m.p. (72–75°C). Microwave condition: $T^\circ$ = 150°C, power = 200 wat, pressure = 247 psi, time = 5 minute. IR (KBr): $\nu_{max}/cm^{-1}$ 3311 (NH), 1730 (C=O, ester), 1640 (C=O, amide). $^1$H NMR: ($CDCl_3$) $\delta$ 0.91 (t, 3H, $J$ = 7.0 Hz, $-OCH_2CH_2\mathbf{CH_3}$), $\delta$ 1.59–1.65 (m, 5H, $-CH\mathbf{CH_3}$ and $-\mathbf{CH_2}CH_3$), $\delta$ 3.76–3.79 (q, 1H, $\mathbf{CH}CH_3$), 3.94–4.05 (m, 7H, $-\mathbf{OCH_3}$, $NH\mathbf{CH_2}$ and $-O\mathbf{CH_2}$), 5.9 (s, 1H, **NH**, $D_2O$-exchangeable) 7.14–7.19 (m, 2H, Ar**H**), 7.43 (d, 1H, $J$ = 8 Hz, Ar**H**), 7.72–7.76 (m, 3H, Ar**H**). $^{13}$C NMR: $\delta$ 10.25 ($OCH_2CH_2\mathbf{CH_3}$), 18.4 ($CH\mathbf{CH_3}$), 21.83 ($OCH_2\mathbf{CH_2}CH_3$), 41.48 ($\mathbf{CH_2}NH-$), 46.88 ($-\mathbf{CH}CH_3$), 55.34 ($-\mathbf{OCH_3}$), 67.00 ($O-\mathbf{CH_2}CH_3$), 105.65, 119.19, 126.29, 127.61, 129.00, 129.26, 133.82, 136.00, 157.78 (Ar–C), 169.92 (–C=O amide), 174.42 (–C=O ester). MS-ESI: $m/z$ 330.3 (M + 1). MS-EI: $m/z$ 330.3 (M + 1) (1%), 329.2 ($M^+$) (10%), 270 (1%), 242 (1%), 212 (2%), 186 (21%), 185 (100%), 171 (9%), 169 (12%), 155.1 (3%), 152 (8%), 140.9 (11%), 114.9 (7%), 74 (5%), 55.9 (2%).

*3.3. Results of Forced Degradation Stability Studies/Specificity.* During the forced degradation studies, the peaks of pro-NP were found to be pure and no significant degradation of drug substance was observed in thermal, photolytic, and oxidative conditions (Table 2). During the acidic hydrolysis analysis in 0.1 N HCl and after 24 h, a 21.5% degradation occurred (Figure 3). The guidelines [21, 22] consider that amount of degradation is enough to be representative of all possible acidic hydrolysis degradation. The two possible acidic degradation products are (2-(6-methoxynaphthalen-2-yl) propanoyl)glycine ($R_t$ = 2.4 min) and naproxen ($R_t$ = 3.6 min) (illustrated in Figure 5). Moreover, during the alkaline hydrolysis analysis in both 0.1 N and 0.01 N NaOH at zero time, complete degradation of the drug substance occurred (Figure 4). This can only be explained by the breakdown of the ester bond in pro-NP and the formation of a soluble sodium salt. Furthermore, alkaline hydrolysis of esters is irreversible.

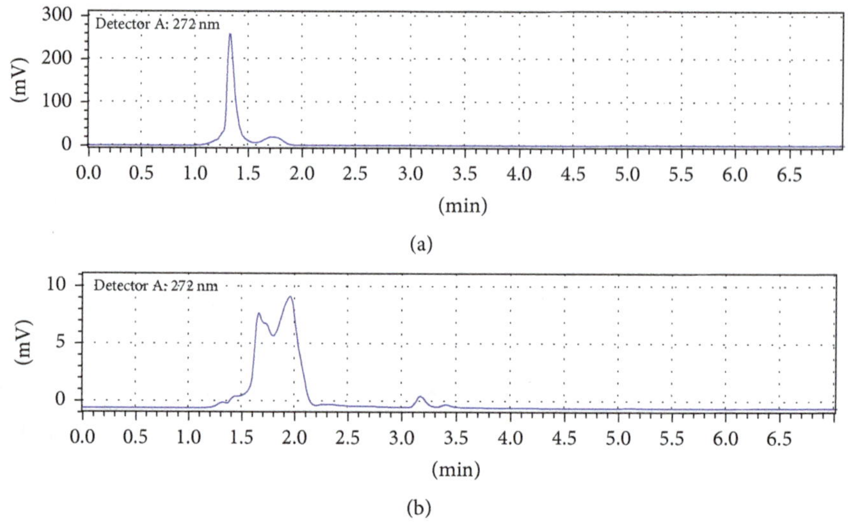


FIGURE 4: Chromatograms showing the complete degradation of pro-NP in (a) 0.1 N NaOH and (b) 0.01 N NaOH.

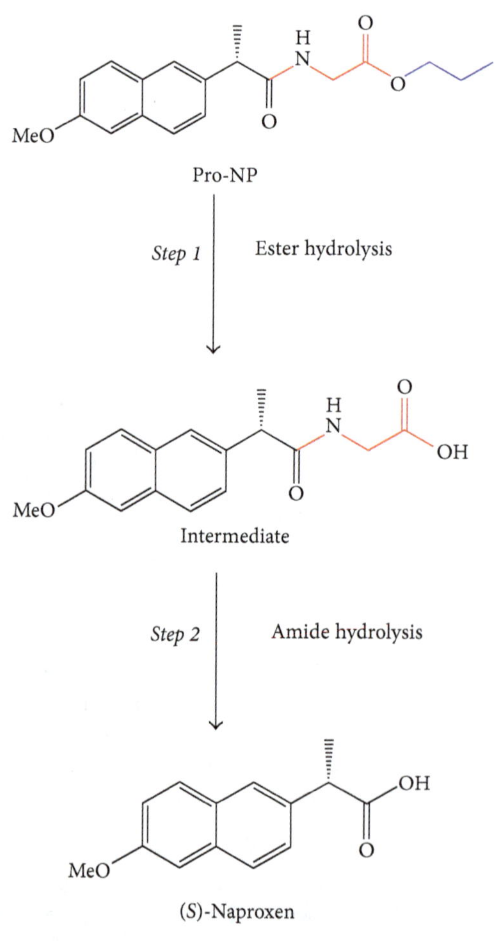


FIGURE 5: The degradation pathway of pro-NP.

*3.4. Results of Method Validation*

*3.4.1. Specificity (Selectivity).* NP is an expected degradation product of pro-NP. NP seems to be the peak that elutes the closest to the drug peak; therefore, a mixture of NP : pro-NP in the ratio of 1 : 2 was prepared and injected into the HPLC. Through visual inspection alone, the observer can determine that those two peaks are well resolved; $R_S > 1.5$ (Figure 6).

TABLE 4: Data from the precision evaluation.

| Experiment number | Peak areas | | | | % RSD |
|---|---|---|---|---|---|
| | Repeatability | Day 1 | Day 2 | Day 3 | |
| 1 | 482726 | 483990 | 494994 | 483664 | 1.322806 |
| 2 | 467668 | 486134 | 494258 | 495749 | 1.051686 |
| 3 | 473884 | 467312 | 487120 | 493997 | 2.869627 |
| 4 | 480740 | 468877 | 491463 | 486034 | 2.445337 |
| 5 | 469853 | 483700 | 488475 | 496976 | 1.373156 |
| 6 | 477899 | 481472 | 490697 | 484151 | 0.9776 |
| *% RSD* | *1.266338* | 1.728179 | 0.631699 | 1.250237 | *1.67*<br>*1.2* |

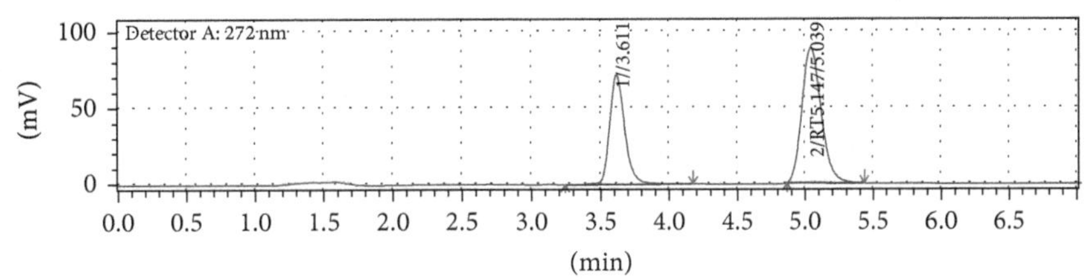


FIGURE 6: A chromatogram indicating the resolution between pro-NP and NP.

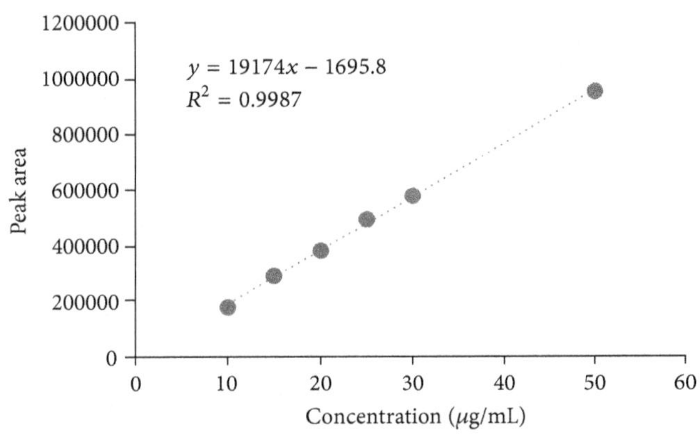


FIGURE 7: Calibration curve.

*3.4.2. Linearity.* The linearity of the method was determined at six concentration levels. A plot of peak area ($Y$) against concentration ($X$, $\mu$g/mL) was drawn for pro-NP standard solutions with 10, 15, 20, 25, 30, and 50 $\mu$g/mL concentrations. The results revealed an excellent correlation between the peak area and analyte concentration, as evident by the coefficient of determination, $r^2$, which is greater than 0.9987. The calibration curve and its results are presented in Figure 7 and Table 3, respectively.

*3.4.3. Precision.* The first step in establishing precision was repeatability. The % RSD of repeatability was 1.266. The % RSD of the intraday precision evaluation was 1.203, while that of interday precision was 1.673. The % RSD values of the repeatability, intraday, and interday studies were all <2, confirming that the method is precise. Calculations of the precision data are presented in Table 4.

*3.4.4. Robustness.* In all the deliberately varied chromatographic conditions (mobile phase pH and the detection wavelength), the developed method showed satisfactory resolution (% RSD < 2), illustrating the robustness of the method (Table 6).

*3.4.5. Accuracy.* The accuracy was accepted based on the results of the linearity, precision, and sensitivity studies. This approach is accepted and justified by the ICH guidelines [23].

*3.4.6. System Suitability.* System suitability data are calculated in Table 5.

TABLE 5: System suitability data.

| Parameter | Value |
|---|---|
| Theoretical plates | 5822.588 |
| Tailing factor | 1.211 |
| Resolution | 1.586 |
| Capacity factor | 5.039 |

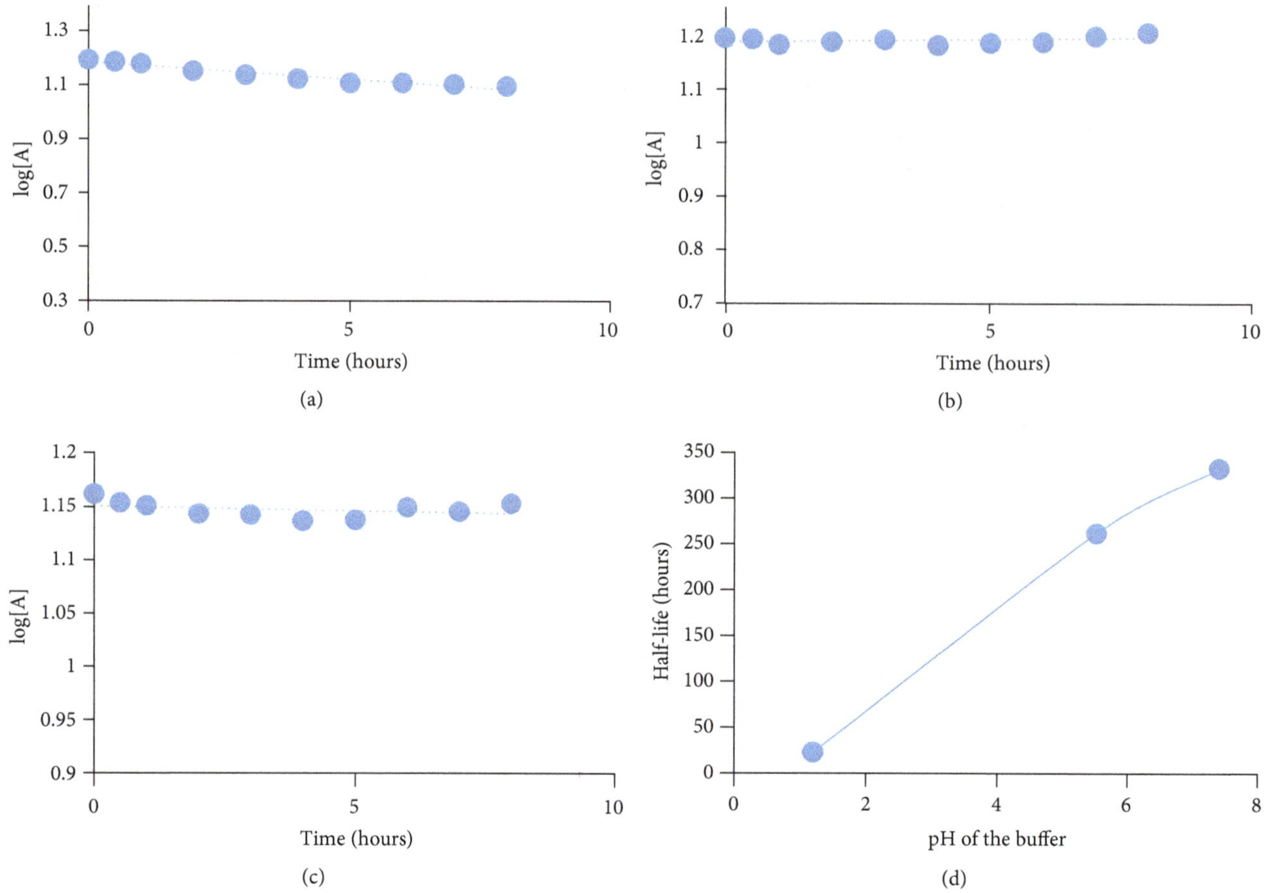


FIGURE 8: Hydrolysis of pro-NP in (a) SGF, (b) pH 7.4, and (c) pH 5.5. (d) is $t_{1/2}$ of pro-NP in different buffer solutions.

TABLE 6: Robustness analysis.

| Parameter | Peak areas | |
|---|---|---|
| | pH | Wavelength |
| Normal (pH 4.0, 272 nm detection) | 576794 | 576794 |
| Change 1 (pH 3.7, 270 nm detection) | 597818 | 588379 |
| | 596622 | 585239 |
| | 598154 | 602703 |
| Change 2 (pH 4.3, 274 nm detection) | 599370 | 577334 |
| | 598128 | 573709 |
| | 596365 | 576894 |
| *% RSD* | *1.342097* | *1.738497* |

*3.5. Results of the Hydrolysis Study.* The hydrolysis kinetics studies were carried out in aqueous buffer solutions (pH 1.2, 5.5, and 7.4) to establish the fate of the synthesized pro-NP. Under the experimental conditions, the targeted compound hydrolyzed to release the parent drug (NP) as evident by HPLC analysis. At constant pH and temperature, the reaction exhibited first-order kinetics, since plots of log concentration versus time resulted in straight lines (Figure 8). The rate constants ($K_{obs}$) and the corresponding half-lives ($t_{1/2}$) for the prodrug were calculated (Tables 7 and 8). The half-life was calculated using the equation $t_{1/2} = 0.693/K$, which derives from the first-order kinetic law. In SGF, more than 20% of pro-NP was hydrolyzed after 8 h. In pH 7.4, only 2% of the drug was hydrolyzed after 8 h. In pH 5.5, the drug was found intact after 8 h, indicating complete absence of hydrolysis at that pH. The pro-NP was found to be stable in all investigated GI fluids with half-lives more than 20 hours which confirm that this prodrug will pass intact through GIT; when reaching the systemic circulation, a reasonable degradation will take place by nonspecific esterases and amidases in the liver to liberate the parent drug NP.

## 4. Conclusion

The method developed is simple, accurate, precise, specific, and robust. Forced degradation and spiking with degradants showed that the degradants are well resolved and separated from the prodrug. Pro-NP is therefore suitable to attain the goals of oral administration since it is stable during its passage through GIT.

TABLE 7: The results from the hydrolysis kinetics.

| Time (hours) | pH 7.4 | | SGF | | pH 5.5 | |
|---|---|---|---|---|---|---|
| | [A*] | log [A] | [A] | log [A] | [A] | log [A] |
| 0 | 14.498 | 1.161308 | 15.912 | 1.201725 | 15.658 | 1.194736 |
| 0.5 | 14.223 | 1.152991 | 15.601 | 1.193152 | 15.586 | 1.192735 |
| 1 | 14.134 | 1.150265 | 15.41 | 1.187803 | 15.223 | 1.1825 |
| 2 | 13.895 | 1.142859 | 14.423 | 1.159056 | 15.409 | 1.187774 |
| 3 | 13.872 | 1.142139 | 13.975 | 1.145352 | 15.542 | 1.191507 |
| 4 | 13.697 | 1.136625 | 13.53 | 1.131298 | 15.171 | 1.181014 |
| 5 | 13.712 | 1.137101 | 13.052 | 1.115677 | 15.315 | 1.185117 |
| 6 | 14.099 | 1.149188 | 13.04 | 1.115278 | 15.367 | 1.186589 |
| 7 | 13.966 | 1.145072 | 12.88 | 1.109916 | 15.754 | 1.197391 |
| 8 | 14.208 | 1.152533 | 12.675 | 1.102948 | 15.998 | 1.204066 |

*A is for pro-NP.

TABLE 8: The calculated rate constants and half-lives.

| pH | $K_{obs}$ ($hour^{-1}$) | $T_{1/2}$ (hours) |
|---|---|---|
| 1.2 | $2.951 \times 10^{-2}$ | 23.479 |
| 5.5 | $2.6 \times 10^{-3}$ | 262.0067 |
| 7.4 | $2.075 \times 10^{-3}$ | 333.946 |

## Conflicts of Interest

The authors declare that there are no conflicts of interest regarding the publication of this paper.

## References


[1] V. Lohar, S. Singhal, and V. Arora, "Prodrug: approach to better drug delivery," *International Journal of Pharmaceutical Research*, vol. 4, no. 1, pp. 15–21, 2012.

[2] T. Elsaman, O. A. A. Aldeeb, T. Aboul-Fadl, and E. I. Hamedelneil, "Synthesis, characterization and pharmacological evaluation of certain enzymatically cleavable NSAIDs amide prodrugs," *Bioorganic Chemistry*, vol. 70, pp. 144–152, 2017.

[3] A. M. Qandil, "Prodrugs of nonsteroidal anti-inflammatory drugs (NSAIDs), more than meets the eye: A critical review," *International Journal of Molecular Sciences*, vol. 13, no. 12, pp. 17244–17274, 2012.

[4] P. K. Halen, P. R. Murumkar, R. Giridhar, and M. R. Yadav, "Prodrug designing of NSAIDs," *Mini-Reviews in Medicinal Chemistry*, vol. 9, no. 1, pp. 124–139, 2009.

[5] S. Kumar, D. K. Tyagi, and A. Gupta, "Synthesis, characterization and pharmcological activity of ester prodrugs of naproxen," *Asian Journal of Pharmaceutical and Clinical Research*, vol. 3, no. 3, pp. 208–211, 2010.

[6] P. C. Sharma, S. Yadav, R. Pahwa, D. Kaushik, and S. Jain, "Synthesis and evaluation of novel prodrugs of naproxen," *Medicinal Chemistry Research*, vol. 20, no. 5, pp. 648–655, 2011.

[7] T. Elsaman and M. M. Ali, "Nonsteroidal anti-inflammatory drugs ( NSAIDs ) derivatives with anti-cancer activity," *American Journal of Research Communication*, vol. 4, no. 4, 2016.

[8] L. F. Wang, H. N. Chiang, and W. B. Chen, "Synthesis and properties of a naproxen polymeric prodrug," *Journal of Pharmacy and Pharmacology*, vol. 54, no. 8, pp. 1129–1135, 2002.

[9] V. R. Shanbhag, A. M. Crider, R. Gokhale, A. Harpalani, and R. M. Dick, "Ester and amide prodrugs of ibuprofen and naproxen: Synthesis, anti-inflammatory activity, and gastrointestinal toxicity," *Journal of Pharmaceutical Sciences*, vol. 81, no. 2, pp. 149–154, 1992.

[10] J. Deb, J. Majumder, S. Bhattacharyya, and S. S. Jana, "A novel naproxen derivative capable of displaying anti-cancer and anti-migratory properties against human breast cancer cells," *BMC Cancer*, vol. 14, no. 1, article no. 567, 2014.

[11] H. Azizian, Z. Mousavi, H. Faraji et al., "Arylhydrazone derivatives of naproxen as new analgesic and anti-inflammatory agents: Design, synthesis and molecular docking studies," *Journal of Molecular Graphics and Modelling*, vol. 67, pp. 127–136, 2016.

[12] M. Thing, Y. Lu, L. Ågårdh et al., "Prolonged naproxen joint residence time after intra-articular injection of lipophilic solutions comprising a naproxen glycolamide ester prodrug in the rat," *International Journal of Pharmaceutics*, vol. 451, no. 1-2, pp. 34–40, 2013.

[13] J. Rautio, T. Nevalainen, H. Taipale et al., "Piperazinylalkyl prodrugs of naproxen improve in vitro skin permeation," *European Journal of Pharmaceutical Sciences*, vol. 11, no. 2, pp. 157–163, 2000.

[14] C. Larsen, "Quantitative determination of dextran-naproxen ester pro-drugs with varying molecular weights and degrees of substitution in biological media by means of high-performance size exclusion chromatography with fluorescence detection," *Journal of Pharmaceutical and Biomedical Analysis*, vol. 7, no. 10, pp. 1173–1181, 1989.

[15] M. Sheha, A. Khedr, and H. Elsherief, "Biological and metabolic study of naproxen-propyphenazone mutual prodrug," *European Journal of Pharmaceutical Sciences*, vol. 17, no. 3, pp. 121–130, 2002.

[16] M. H. Helal, S. Y. Abbas, M. A. Salem, A. A. Farag, and Y. A. Ammar, "Synthesis and characterization of new types of 2-(6-methoxy-2-naphthyl) propionamide derivatives as potential antibacterial and antifungal agents," *Medicinal Chemistry Research*, vol. 22, no. 11, pp. 5598–5609, 2013.

[17] M. Spadaro, A. Bucolo, C. Ronsisvale, and G. Pappalardo, "Design, synthesis and antiinflammatory activity of novel alfa -tocopherol," *Journal of Pharmaceutical Sciences and Research*, vol. 1, no. 4, pp. 88–95, 2009.

[18] G. Forte, I. Chiarotto, I. Giannicchi et al., "Characterization of naproxen-polymer conjugates for drug-delivery," *Journal of Biomaterials Science, Polymer Edition*, vol. 27, no. 1, pp. 69–85, 2016.

[19] M. Najlah, S. Freeman, D. Attwood, and A. D'Emanuele, "Synthesis, characterization and stability of dendrimer prodrugs," *International Journal of Pharmaceutics*, vol. 308, no. 1-2, pp. 175–182, 2006.

[20] T. Aboul-Fadl, T. Elsaman, O. A. Al-Deeb, H. A. Ghabbour, C. S. C. Kumar, and H.-K. Fun, "Some new NSAIDs prodrugs: an efficient synthesis, spectral characterization and X-ray crystal structure studies," *Asian Journal of Chemistry*, vol. 26, no. 16, pp. 5249–5254, 2014.

[21] ICH, Stability Testing of New Drug Substances and Products Q1A(R2), Int. Conf. Harmon, no. February, 2003.

[22] S. Niazi, *Handbook of Pharmaceutical Manufacturing Formulations*, CRC Press, 2009.

[23] ICH., ICH, Validation of Analytical Procedures: Text and Methodology, Int. Conf. Harmon, no. November, 1996.

# 9

# Determination of the Phenolic Composition and Antioxidant Activity of Pear Extracts

**Mindaugas Liaudanskas,[1] Kristina Zymonė,[1] Jonas Viškelis,[2] Almantas Klevinskas,[1] and Valdimaras Janulis[1]**

[1] *Department of Pharmacognosy, Faculty of Pharmacy, Lithuanian University of Health Sciences, Sukilėlių St. 13, LT-44307 Kaunas, Lithuania*

[2] *Institute of Horticulture, Lithuanian Research Centre for Agriculture and Forestry, Kauno St. 30, Babtai, LT-54333 Kaunas, Lithuania*

Correspondence should be addressed to Mindaugas Liaudanskas; m.liaudanskas@yahoo.com

Academic Editor: Paula G. De Pinho

The specific HPLC analytical procedure was developed and validated for the determination of phenolic compounds in pear samples of different popular cultivars "Conference," "Concordia," "Grabova," and "Patten." HPLC mobile phase consisted of 0.05% (v/v) trifluoroacetic acid in water and 100% (v/v) acetonitrile. The HPLC method was used to identify and confirm the specificity of 8 analytes: chlorogenic acid, rutin, hyperoside, isoquercitrin, isorhamnetin rutinoside, quercitrin, quercitrin malonyl glucoside, and isorhamnetin glucoside. Repeatability % RSD did not exceed 3.87%, and intermediate precision did not exceed 4.63%. The total content of phenolic compounds varied from 0.51 ± 0.001 mg/g (cv. "Concordia") to 1.11 ± 0.013 mg/g (cv. "Patten"). Chlorogenic acid was the major component in all the tested pear cultivars. The highest amount of chlorogenic acid (0.69 ± 0.033 mg/g) was found in pear samples of the cultivar "Grabova," and the highest amount of flavonol compounds (1.11 ± 0.013 mg/g) was found in pear samples of the cultivar "Concordia."

## 1. Introduction

Pears are among the most common fruit grown in Europe and worldwide. Approximately 3 million tonnes of pears per year are grown in the European Union alone [1], while almost 23 million tonnes of pears is grown per year in the world [2]. For this reason, pears are an important plant-based part of human diet both in Europe and in other continents. Various phenolic compounds have been detected in pear fruit—flavan-3-ols, flavonols, phenolic acids, anthocyanins, and so forth. Different groups of phenolic compounds as well as individual phenolic compounds have been found to have a specific effect on various physiological systems.

Phenolic compounds are an important group of plant-based biologically active compounds that strengthen the organism and prevent disease [3, 4]. Phenolic compounds have a particularly strong antioxidant effect [5–7], which is closely related to the anti-inflammatory [8, 9] and anticancer [10–12] effect. Multiple epidemiologic studies have proven a significant correlation between the consumption of phenolic compound-rich food and a decreased risk for developing cardiovascular and neurodegenerative diseases [13, 14]. Phenolic compounds have been proven to have a strong antimicrobial [15, 16] and antiviral [17, 18] effect. Based on abundant research data proving the beneficial effect of phenolic compounds in humans, it is appropriate to perform in-depth qualitative and quantitative analysis of the composition of phenolic compounds in pears.

A reliable analytical tool is required to achieve extensive results on phytochemical composition. HPLC methods for the determination of the qualitative and quantitative composition of phenolic compounds in pears were described in literature [19–21]. Based on these assumptions, extensive scientific research requires the redevelopment of sample preparation techniques and mobile phase gradients of those developed in other laboratories. Therefore, the objective was set to develop, optimise, and validate a suitable HPLC technique for the evaluation of phenolic compounds in pear samples grown in Lithuania.

## 2. Materials and Methods

*2.1. Pears.* The following pear cultivars were included in the study: "Conference"; "Concordia"; "Grabova"; and "Patten.". Pear samples were purchased in the supermarket. Each pear was cut into the slices of equal size (up to 1 cm in thickness), and the stalks and the seeds were removed. The pear slices were immediately dried in a drying oven at the temperature of 50°C. Loss on drying before the analysis was determined by drying the dried pears in a laboratory drying oven to the complete evaporation of water and volatile compounds (105°C temperature; a difference in weight between the measurements was up to 0.01 g) and by calculating the difference in raw material weight before and after drying [22]. The data were recalculated for absolute dry weight.

*2.2. Chemicals.* All the solvents, reagents, and standards used were of analytical grade. Acetonitrile and trifluoroacetic acids were obtained from Sigma-Aldrich GmbH (Buchs, Switzerland), and ethanol was obtained from Stumbras AB (Kaunas, Lithuania). Hyperoside (≥98%), isoquercitrin (≥99%), rutin (≥99%), quercitrin (≥98.5%), isorhamnetin-3-O-rutinoside (≥99%), isorhamnetin glucoside (≥95%), and chlorogenic acid (≥99%) standards were purchased from Extrasynthese (Genay, France). Quercetin-3-*O*-malonylglucoside (≥85%), DPPH, 1,1-diphenyl-2-picrylhydrazyl, radical (≤100%), ABTS, 2,2′-azinobis(3-ethylbenzothiazoline-6-sulphonic acid) (>98%), potassium persulphate (99.99%), sodium acetate trihydrate (≥99%), iron (III) chloride hexahydrate (≥99%), and TPTZ (2,4,6-tripyridyl-s-triazine) (≥98%) were obtained from Sigma-Aldrich GmbH (Steinheim, Germany). Deionised water produced by the Crystal E HPLC (Adrona SIA, Riga, Latvia) water purification system was used.

*2.3. Extraction.* 2.5 g of milled dried pear (exact weight) was weighed, added to 30 mL of ethanol (70%, v/v), and extracted in a Sonorex Digital 10 P ultrasonic bath (Bandelin Electronic GmbH & Co. KG, Berlin, Germany) for 15 minutes at 60°C. Ultrasound strength was 1100 W, and frequency was 80 kHz. The extract obtained was filtered through a paper filter; the residue on the filter was washed twice with 10 mL of ethanol (70%, v/v) in a 50 mL flask. The extract was filtered through a membrane filter with a pore size of 0.22 $\mu$m (Carl Roth GmbH, Karlsruhe, Germany).

*2.4. Spectrophotometric Studies.* All the spectrophotometric measurements were carried out with a double beam UV/VIS spectrophotometer M550 (Spectronic Camspec, Garforth, England, United Kingdom).

*2.4.1. Determination of Total Phenolic Content.* The total phenolic content (mg GAE/g DW) in the extracts of pears was determined by the Folin-Ciocalteu method [23], calculated from a gallic acid calibration curve, and expressed as mg/g gallic acid equivalent (GAE) per gram of absolutely dry weight (DW) (mg GAE/g DW).

*2.4.2. Evaluation of Antioxidant Activity*

*(1) DPPH$^\bullet$ Free Radical Scavenging Assay.* The DPPH free radical scavenging activity was evaluated using the method proposed by Brand-Williams et al. [24]. DPPH$^\bullet$ solution in 96.3% v/v ethanol (3 mL, $6 \times 10^{-5}$ M) was mixed with 10 $\mu$L of the ethanol extract of pear. A decrease in absorbance was determined at a wavelength of 515 nm after keeping the samples for 30 minutes in the dark.

*(2) ABTS$^{\bullet+}$ Radical Cation Decolourisation Assay.* An ABTS$^{\bullet+}$ radical cation decolourisation assay was applied according to the methodology described by Re et al. [25]. A volume of 3 mL of ABTS$^{\bullet+}$ solution (absorbance $0.800 \pm 0.02$) was mixed with 10 $\mu$L of the ethanol extract of pear. A decrease in absorbance was measured at the wavelength of 734 nm after keeping the samples for 30 minutes in the dark.

*(3) FRAP Assay.* The ferric reducing antioxidant power (FRAP) assay was carried out as described by Benzie and Strain [26]. The working FRAP solution included TPTZ (0.01 M dissolved in 0.04 M HCl), $FeCl_3 \times 6H_2O$ (0.02 M in water), and acetate buffer (0.3 M, pH 3.6) at the ratio of 1 : 1 : 10. A volume of 3 mL of a freshly prepared FRAP reagent was mixed with 10 $\mu$L of the pear extract. An increase in absorbance was recorded after 30 minutes at the wavelength of 593 nm.

*(4) Calculation of Antioxidant Activity of the Pear Extracts.* The antioxidant activity of the extracts was calculated from the Trolox calibration curve and expressed as $\mu$mol Trolox equivalent (TE) per gram of absolutely dry weight (DW). TE was calculated according to the following formula: TE $= c \times V/m$ ($\mu$mol/g), $c$ being the concentration of Trolox established from the calibration curve (in $\mu$M); $V$ being the volume of pear extract (in L); and $m$ being the weight (precise) of the raw plant material (in g).

*2.5. Instrumentation and Chromatographic Conditions.* A Waters 2695 chromatograph equipped with a Waters 2998 PDA detector (Waters, Milford, USA) was used for HPLC analysis. Chromatographic separation was managed, chromatograms were recorded, and data were processed with the Empower® v. 3.0 software (Waters, Milford, USA). Chromatographic separation was carried out by using an ACE (3 $\mu$m, C18, 150 × 4.6 mm i.d.) column equipped with an ACE (3 $\mu$m, $C_{18}$) precolumn (Advanced Chromatography Technologies, Aberdeen, Scotland). The column was operated at a constant temperature of 40°C. The volume of the extract being investigated was 10 $\mu$L. The flow rate was 1 mL/min, and gradient elution was used. The mobile phase consisted of 0.05% (v/v) trifluoroacetic acid in water (solvent A) and 100% (v/v) acetonitrile (solvent B). The following conditions of elution were applied: 0–38 minutes, 14–23% B; 38–39 minutes, 23–90% B; 39–44 minutes, 90% B (isocratic elution); and 44-45 minutes, 90–14% B. The identification of the chromatographic peaks was achieved by comparing the retention times and spectral characteristics ($\lambda$ = 200–400 nm) of the eluting peaks with those of the reference compounds. The compounds identified were confirmed by spiking the sample with the standard compound and monitoring the changes in the peak shape and spectral characteristics. For quantitative analysis, a calibration curve was obtained by injection of known concentrations of different standard compounds. The concentrations of phenolic compounds identified in the pear extracts were within the limits of the calibration curves. Chlorogenic acid was quantified at 325 nm, and flavonols were quantified at 350 nm.

*2.6. Statistical Data Analysis.* All the experiments were carried out in triplicate. Means and standard deviations were calculated using computer software Microsoft Office Excel 2013 (Microsoft, Redmond, Washington, USA) and SPSS 20.0 (Chicago, Illinois, USA). A single factor analysis of variance (ANOVA) along with the post hoc Tukey test was employed for statistical analysis. The correlation was evaluated by Spearman's analysis. Differences were considered to be significant at $p < 0.05$.

## 3. Results and Discussion

*3.1. Selection of the Extraction Technique.* The selection of the solvent and the conditions of extraction is an important analytical step in the development of the technique for the qualitative and quantitative analysis of the biologically active compounds in raw plant material. The solvent of the extraction is one of the main factors in the prognosis of the qualitative and quantitative composition of the isolated phenolic compounds [27–29].

During the first stage of the investigation for the selection of the conditions of extraction, we evaluated the influence of ethanol at various concentrations (20–96%) on the extraction of phenolic compounds from pear fruit samples. The obtained results showed that the greatest total amount of phenolic compounds (6.36 ± 0.23 mg GAE/g, $p < 0.05$) was obtained from pear fruit samples by applying extraction with 70% (v/v) ethanol, and thus this concentration of ethanol was selected for further studies (Figure 1).

During the second stage of the investigation for the selection of the conditions of extraction, we evaluated the influence of various temperatures (20–70°C) on the extraction of phenolic compounds from pear fruit samples. The greatest total amount of phenolic compounds (4.63 ± 0.10 mg GAE/g, $p < 0.05$) was obtained by extracting fruit samples at the temperature of 60°C (Figure 2), and thus this temperature was selected for further studies.

During the third stage of the investigation for the selection of the conditions of extraction, we evaluated the influence of the duration of the extraction on the extraction of phenolic compounds from pear fruit samples. The results of the experiments showed that the greatest total amount of phenolic compounds (5.18 ± 0.18 mg GAE/g, $p < 0.05$) was obtained by extracting fruit samples in an ultrasonic bath for

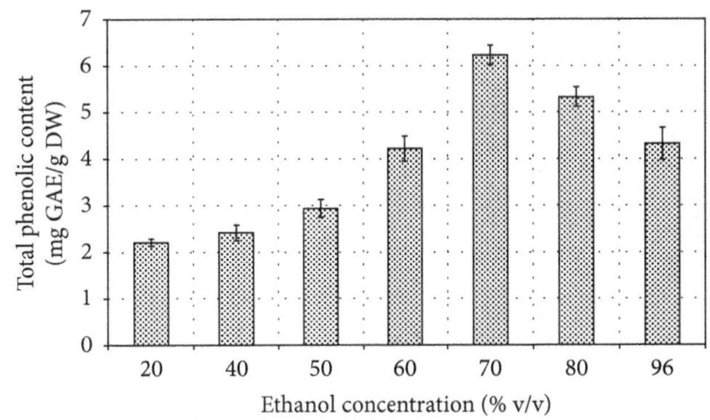


FIGURE 1: Effect of the concentration of the extractant on the extraction output of phenolic compounds from pear fruit samples.

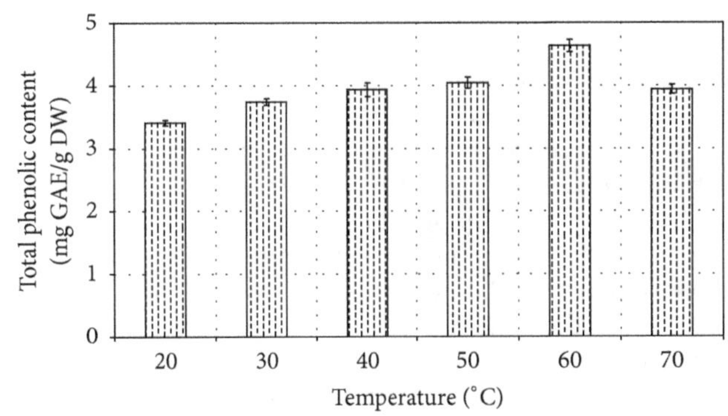


FIGURE 2: Effect of different temperatures on the extraction output of phenolic compounds from pear fruit samples.

15 min (Figure 3). For this reason, this duration of extraction was selected for further studies.

Summing up the results of the experiments for the selection of the extraction technique, the greatest output of phenolic compounds from pear fruit samples was obtained by extracting this raw plant material in 70% ethanol for 15 min in an ultrasonic bath at the temperature of 60°C. Based on these results, these conditions of extraction were selected for further studies.

*3.2. Development and Optimisation of Chromatographic Separation.* It is especially important to select the optimal analytical column for the analysis of the raw plant material. Currently, reversed-phase C18 columns are most common for the analysis of phenolic compounds [27, 30]. Reversed-phase HPLC uses C18 or other alkyl-chains [31]. Silicon dioxide or polyamide chains are also used [32]. Usually, the column length varies from 100 to 300 mm [31].

We performed tests to determine the best analyte peak resolution and the separation of the phenolic compound mixture. Selectivity and resolution of two differently sized chromatographic columns filled with different sorbents were evaluated: ACE C18 (150 mm × 4.6 mm, 3 $\mu$m) and YMC C18 (250 mm × 4.6 mm, 5 $\mu$m). Best selectivity and resolution were achieved using ACE C18 (150 mm × 4.6 mm, 3 $\mu$m); therefore, this column was selected for further studies.

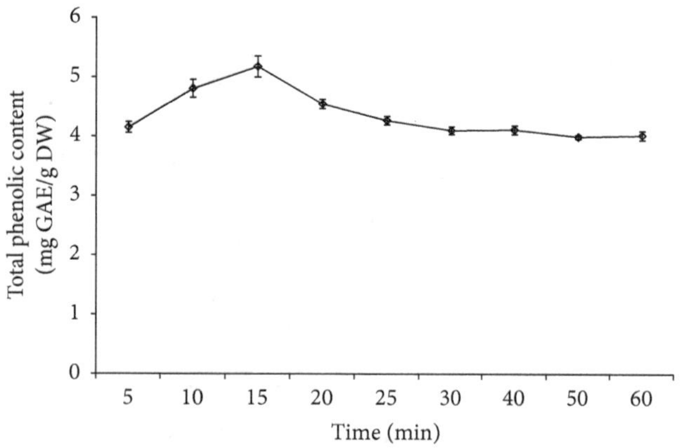


FIGURE 3: Effect of different duration of extraction on the extraction output of phenolic compounds from pear fruit samples.

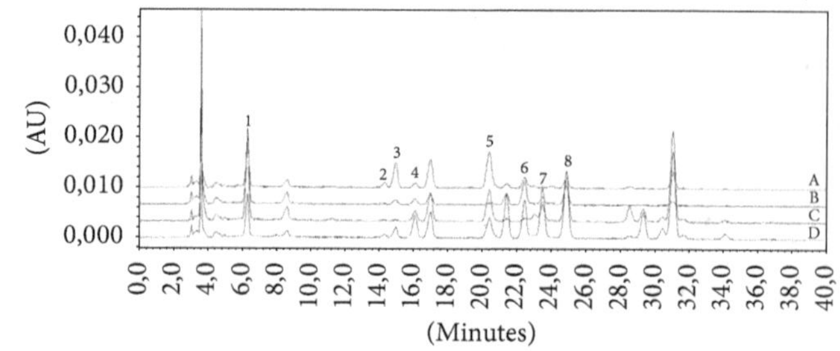


FIGURE 4: Chromatograms of the investigated pear sample extracts, $\lambda = 360$ nm. The uppercase letters denote different pear cultivars: A: "Patten"; B: "Conference"; C: "Grabova"; and D: "Concorde." Numbers indicate the peaks of analytes: 1: chlorogenic acid; 2: quercetin-3-*O*-rutinoside (rutin); 3: quercetin-3-*O*-galactoside (hyperoside); 4: quercetin-3-*O*-glucoside (isoquercitrin); 5: quercetin-3-*O*-malonylglucoside; 6: isorhamnetin rutinoside; 7: quercetin-3-*O*-rhamnoside (quercitrin); and 8: isorhamnetin glucoside.

In phenolic compound analysis, mobile phase usually consists of water and organic component solvent (usually acetonitrile or methanol) [31]. In order to reduce peak asymmetry (the spread of the analyte concentration zone), eluents may be acidified [27]. In order to achieve optimal analyte separation, we performed tests where the mobile phase was 0.05% trifluoroacetic acid and acetonitrile. The gradient elution method was optimised in order to achieve sufficient resolution for quantitative analysis (Figure S1, in Supplementary Material available online at https://doi.org/10.1155/2017/7856521). We performed experiments with different flow speed of the mobile phase (Figure S2) and different temperature (Figure S3), and we optimised the mobile phase gradient. Temperature affects separation, and, in order to ensure repeatability, it is recommended to maintain constant column temperature [27]; therefore, the column was placed in a thermostat during the analysis, and the impact of temperature on the efficacy of phenolic compound separation was evaluated. By applying conditions described in Section 2.4, we efficiently separated flavonoid glycoside and phenolic acid complex (quercetin and isorhamnetin glycoside and chlorogenic acid) from pear fruit extracts. The analyte peak resolution of the analysed extracts was higher than 2, which allows for performing qualitative and quantitative analysis of biologically active compounds. Salta et al. [19] described a shorter method (duration, 25 min), but it only allows for evaluating the quantitative composition of phenolic acids, flavan-3-ols. Escarpa and González [20, 33] developed a 25-minute HPLC method, but it allowed for separating 10 analytes. Our samples have a higher number of analytes, and this method failed to separate them properly.

The chromatogram of the ethanol extracts of pear samples is depicted in Figure 4. In our study, the elution order of the identified analytes was as follows: chlorogenic acid, quercetin-3-*O*-rutinoside (rutin), quercetin-3-*O*-galactoside (hyperoside), quercetin-3-*O*-glucoside (isoquercitrin), QMG, isorhamnetin rutinoside, quercetin-3-*O*-rhamnoside (quercitrin), and isorhamnetin glucoside.

*3.3. Validation of the Method.* Validation of the method was performed based on the ICH recommendations [34]. The following characteristics of validation were evaluated: the specificity of the method, precision (repeatability and intermediate precision), detection and quantitation limits of the analytes, and linearity.

The evaluation of the specificity of the method for peak identification and purity was based on the comparison of retention times and UV spectra of the analytes with those of the standard compounds. The compounds identified were confirmed by adding the standard compound into the analysed extract and monitoring the changes in the peak shape and spectral characteristics.

The quantitation limit of analytes was assessed by comparing the peak height to the baseline noise. The ratio of the peak height to baseline noise (signal-to-noise ratio) used for the estimation of the detection limit was 3 : 1, and the quantitation limit was 10 : 1. The determined limit of detection (LOD) varied from 0.021 $\mu$g/mL (chlorogenic acid) to 0.224 $\mu$g/mL (rutin). The limit of quantitation varied from 0.070 $\mu$g/mL (chlorogenic acid) to 0.747 $\mu$g/mL (rutin). The results obtained confirm that this method can be used for the quantitative analysis of phenolic compounds. The estimated determination coefficients ($R^2$) of the calibration curves were greater than 0.9989, and this proves the linearity of the quantitative evaluation method (Table 1).

Precision of the HPLC method was assessed based on two parameters: repeatability and intermediate precision. The obtained results describing the precision of our self-developed HPLC method are summarised in Table 2. The evaluation of result repeatability was based on the results of consecutive analyses performed within the same day (6 consecutive analyses of a mixture of standards at 3 different concentrations performed within the same day). The percent relative standard deviation (% RSD) of the method repeatability, determined according to the peak area, varied from 1.22% (chlorogenic acid) to 1.52% ((−)-epicatechin). Concentrations of reference compounds used in repeatability assessment are shown in Table S1. The intermediate precision of the results was assessed based on the results of analyses performed

TABLE 1: Characteristics of the quantitative evaluation of phenolic compounds.

| Compound | Confirmed linearity range (μg/mL) | LOD (μg/mL) | LOQ (μg/mL) | $R^2$ |
|---|---|---|---|---|
| Chlorogenic acid | 0.1–10 | 0.021 | 0.07 | 0.9998 |
| Rutin | 1.6–25 | 0.224 | 0.75 | 0.9999 |
| Hyperoside | 2.5–40 | 0.068 | 0.22 | 0.9999 |
| Isoquercitrin | 1.4–22 | 0.080 | 0.27 | 0.9989 |
| QMG | 1.1–17 | 0.119 | 0.400 | 0.9998 |
| Isorhamnetin rutinoside | 0.8–50 | 0.112 | 0.38 | 0.9999 |
| Quercitrin | 2.5–40 | 0.072 | 0.24 | 0.9999 |
| Isorhamnetin glucoside | 0.8–100 | 0.089 | 0.300 | 0.9999 |

TABLE 2: Precision of the HPLC method for the quantitative determination of phenolic compounds.

| Compound | Relative standard deviation (%) for the peak area | | | |
|---|---|---|---|---|
| | Intraday[a] | | | Interday[b] |
| | 1st concentration | 2nd concentration | 3rd concentration | |
| Chlorogenic acid | 0.33 | 0.89 | 3.39 | 4.48 |
| Rutin | 0.45 | 1.09 | 3.87 | 4.63 |
| Hyperoside | 0.33 | 0.4 | 0.46 | 3.4 |
| Isoquercitrin | 0.15 | 1.61 | 2.99 | 2.76 |
| QMG | 0.45 | 0.75 | 2.67 | 4.29 |
| Isorhamnetin rutinoside | 0.36 | 0.72 | 3.15 | 0.79 |
| Quercitrin | 0.65 | 0.52 | 2.53 | 1.16 |
| Isorhamnetin glucoside | 0.56 | 0.80 | 2.49 | 2.4 |

[a]Repeatability. [b]Intermediate precision.

within 3 different days (6 consecutive analyses of a mixture of standards at the mean concentration performed within the same day; 18 analyses in total), and its % RSD for the peak area varied from 0.79% (isorhamnetin rutinoside) to 4.63% (rutin).

*3.4. Identification of Phenolic Compounds in the Analysed Pear Samples.* In this study, the developed, optimised, and validated HPLC method was applied to investigate the ethanol extracts of pear samples. The content of phenolic compounds in the pear of cultivars "Conference"; "Concordia"; "Grabova"; and "Patten" was analysed qualitatively and quantitatively. The chromatograms of the tested pear cultivars are shown in Figure 4, and Table 3 summarises the characteristics of the quantitative content.

Chlorogenic acid was the major component of the identified phenolic compounds in all tested pear cultivars. Chlorogenic acid has properties that are beneficial for human health: besides antioxidant activity [35] and an anti-inflammatory effect [36], it lowers the risk of type II diabetes mellitus [37], improves cardiac and vascular function [38], and inhibits carcinogenesis processes [39, 40]. Other scientists [41, 42] also have detected chlorogenic acid in pear fruit.

The pears of the cv. "Grabova" contained the highest amount of chlorogenic acid (0.69 ± 0.033 mg/g), and this amount was 2.16-fold greater than the lowest amount of this acid (0.32 ± 0.005 mg/g) found in the pear samples of the cultivar "Patten" (Table 3). Our results confirm the findings of other studies indicating that chlorogenic acid is one of the most common phenolic compounds in pears [42–44]. Salta et al. [19] detected 4.3–62.4 mg/100 g FW of chlorogenic acid in fruit of the analysed pear cultivars.

In the investigated ethanol extracts of pear samples, flavonol compounds (quercetin and isorhamnetin glycosides) were the most abundant group of phenolic compounds. The consumption of products rich in quercetin glycosides reduces the risk of cancer and neurodegenerative diseases [45, 46]. Quercetin glycosides inhibit the angiotensin-converting enzyme and thus have an antihypertensive effect [47].

The following flavonol derivatives were identified and evaluated quantitatively: rutin, hyperoside, isoquercitrin, quercetin malonyl glucoside, quercitrin, isorhamnetin rutinoside, and isorhamnetin glucoside. Kolniak-Ostek and Oszmiański [41] also identified rutin, hyperoside, isoquercitrin, isorhamnetin rutinoside, and isorhamnetin glucoside in pear fruit. These scientists identified all three quercetin glycosides and isorhamnetin glucoside in pear peels, rutin, isoquercitrin, and isorhamnetin rutinoside in the pulp and hyperoside, isoquercitrin, and both isorhamnetin glycosides in seeds. Lin and Harnly [42] who analysed fruit peels of various pear cultivars have also identified quercetin malonyl glucoside in addition to rutin, hyperoside, and isoquercitrin. In their study, quercetin malonyl glucoside was not detected in all the analysed pear samples, as it was in our case as well. In our analysed samples, quercetin malonyl glucoside was not detected in pear samples of the cultivar "Conference,"

TABLE 3: Content of phenolic compounds in ethanol extracts obtained from pears (expressed for absolute dry lyophilisate weight).

| Compound | Amount of phenolic compounds in different cultivars[a], mg/g | | | |
|---|---|---|---|---|
| | "Conference" | "Concordia" | "Grabova" | "Patten" |
| Chlorogenic acid | 0.53 ± 0.007[b] | 0.66 ± 0.011[a] | 0.69 ± 0.033[a] | 0.32 ± 0.005[c] |
| Rutin | 0.01 ± 0.001[c] | 0.03 ± 0.007[b] | <LLOD | 0.09 ± 0.001[a] |
| Hyperoside | 0.02 ± 0.002[b] | 0.06 ± 0.009[a] | 0.03 ± 0.002[b] | 0.02 ± 0.0003[b] |
| Isoquercitrin | 0.04 ± 0.001[c] | 0.08 ± 0.004[a] | 0.07 ± 0.006[b] | 0.07 ± 0.001[b] |
| QMG | <LLOD | 0.07 ± 0.004[c] | 0.09 ± 0.012[b] | 0.12 ± 0.004[a] |
| Isorhamnetin rutinoside | 0.05 ± 0.001[b] | 0.08 ± 0.011[a] | 0.01 ± 0.001[c] | 0.02 ± 0.0001[c] |
| Quercitrin | 0.02 ± 0.004[c] | 0.09 ± 0.002[a] | 0.03 ± 0.003[b] | <LLOD |
| Isorhamnetin glucoside | 0.03 ± 0.004[c] | 0.12 ± 0.004[a] | 0.08 ± 0.009[b] | 0.02 ± 0.001[c] |
| Total | 0.71 ± 0.011[c] | 1.18 ± 0.013[a] | 1.00 ± 0.013[b] | 0.66 ± 0.001[d] |

[a]Values are means ($n = 3$). The different letters indicate significant differences between the values ($p < 0.05$).

and rutin was not detected in pear samples of the cultivar "Grabova." Other scientists also failed to detect rutin in any of the pear samples of the analysed cultivars [42]. The total amount of the identified flavonol compounds in pear samples of the analysed cultivars varied from 0.17 ± 0.004 mg/g (cv. "Conference") to 0.22 ± 0.002 mg/g ("Grabova"). The highest amounts of hyperoside, isoquercitrin, isorhamnetin rutinoside, quercitrin, and isorhamnetin glucoside were detected in pear samples of the cultivar "Concordia," and the highest amounts of rutin and quercetin malonyl glucoside were detected in pear samples of the cultivar "Patten."

Our analysis showed a wide variation in the phenolic content between different pear cultivars. The total content of phenolic compounds identified in the ethanol extracts of pear samples ranged from 0.51 ± 0.001 mg/g (cv. "Concordia") to 1.11 ± 0.013 mg/g (cv. "Patten").

The analysis of the qualitative and quantitative composition of pear cultivars showed that the cultivar itself as a factor had an impact on the parameters of qualitative and quantitative content of phenolic compounds in pears, and, depending on the pear cultivar, these parameters could vary considerably. The highest amount of the identified phenolic compounds was detected in pear samples of the cultivar "Concordia," and, therefore, upon completion of more in-depth agrotechnical studies, pears of this cultivar could potentially be beneficial for a wider cultivation in commercial and amateur gardens.

*3.5. Measurements of Antioxidant Activity in Pear Extracts.* After studying the qualitative and quantitative composition of phenolic compounds of pears of different cultivars, it was important to examine and assess the antiradical and reduction activity in the pear extracts. The results obtained during studies will be useful for the selection of pear cultivars in order to provide consumers with products rich in antioxidants and will allow for predicting the antioxidant effect of pear extracts *in vivo*.

In order to thoroughly evaluate the antioxidant activity of the pear extracts, different antioxidant capacity assays (DPPH, ABTS, and FRAP) were employed. The results of *in vitro* evaluation of antioxidant activity in the pear extracts (cultivars "Concorde," "Conference," "Grabova," and "Patten") are summarised in Figure 5.

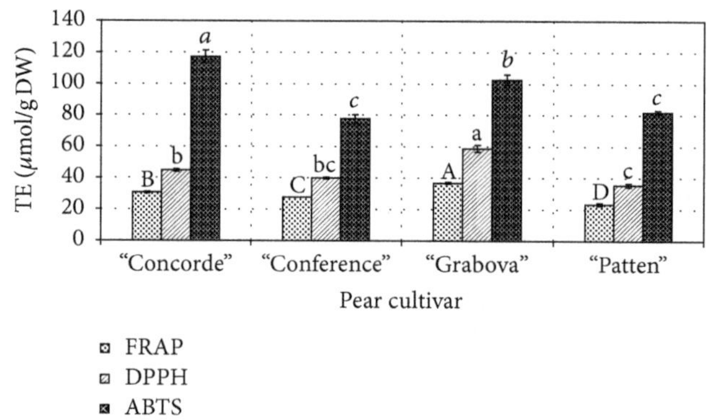


FIGURE 5: The antioxidant activity of pear extracts determined by using DPPH, ABTS, and FRAP assays. Values are means, and error bars indicate standard deviations ($n = 3$). The different letters indicate significant differences in the values between pear cultivars ($p < 0.05$).

After evaluating the antioxidant activity of pear extracts examined *in vitro* by FRAP and DPPH UV-VIS spectrophotometric assays, the strongest antioxidant activity was detected in pear extracts of the "Grabova" cultivar (resp., 36.86 ± 0.59 μmol/g and 58.75 ± 2.23, $p < 0.05$). Meanwhile, the ABTS method showed that the highest antiradical activity *in vitro* (117.29 ± 3.96 μmol/g, $p < 0.05$) was in the "Concorde" cultivar pear extracts.

To our knowledge, there are very few studies on the antioxidant activity of pears. For this reason—and due to the differences in the applied research methods—it is difficult to compare our results with those obtained by other scientists. Therefore, the findings from this study on the antioxidant activity of pear extracts *in vitro* were compared with those obtained in studies on other plants belonging to the *Rosaceae* family. The total reducing activity in apple extract samples was established by the UV-VIS spectrophotometric FRAP method *in vitro* and averaged 117.13 μmol/g DW [48], and it was by 3.2–5.0 times more potent than the reducing activity of pear fruit extract. Such differences could have been influenced by the differences in the qualitative and quantitative composition of pear and apple fruit, which was determined not only by different genus factors but also by various other factors (composition of the soil, the climatic-meteorological, cultivation, or storage conditions, orchard fertigation, etc.) that influence the chemical composition of pears. We did not evaluate them in this study. The extracts obtained from the fruit of different *Sorbus* species by DPPH and FRAP showed a higher antioxidant activity; that is, the TE values were higher than those determined in the present study: 86.9–181.1 $\mu$moL/g DW by the DPPH assay and 221.1–396.0 $\mu$moL/g DW by the FRAP assay. These differences could be due to the intermediate differences between the chemical composition of pears and rowan and the influence of other factors. The results of antiradical activity studies on different types of rowan berries were obtained using the ABTS method and are similar to our results. They varied from 65.3 $\mu$moL/g DW to 131.1 $\mu$moL/g DW [49].

After performing correlation analysis there was a very strong positive correlation between antioxidant activity and the content of chlorogenic acid (DPPH: 0.999, $p < 0.01$; FRAP: 0.999, $p < 0.01$). No more statistically significant correlation was found between individual identified phenolic compound and the total antioxidant activity of sampled extracts *in vitro.*

## 4. Conclusions

The HPLC-based analytical procedure for the detection and evaluation of phenolic compounds in pear samples of 4 popular cultivars was optimised and validated. The specificity, precision, and assay range confirmed the suitability of this method.

The distribution of 8 phenolic compounds was determined in the tested pear samples. Chlorogenic acid was the major component in the samples of all the tested cultivars. The highest amounts of chlorogenic acid were detected in pear samples of the cultivar "Grabova," and the highest amounts of flavonol compounds were detected in pear samples of the cultivar "Concordia."

The highest antioxidant activity *in vitro* measured using the DPPH and FRAP methods was found in pear extracts of the "Grabova" cultivar, and the highest antioxidant activity *in vitro* measured with the ABTS method was found in the "Concorde" cultivar extracts.

## Abbreviations

cv.: Cultivar
DW: Dry weight
GAE: Gallic acid equivalent
HPLC: High-performance liquid chromatography
LLOD: Lower than limit of detection
PDA: Photodiode array
QMG: Quercetin malonyl glucoside
% RSD: Percent relative standard deviation
v/v: Volume/volume.

## Conflicts of Interest

The authors declare that there are no conflicts of interest regarding the publication of this paper.

## References

[1] EU-27, "Fresh Deciduous Fruit Annual. Good Prospects for EU-27 Apple and Pear Production. Global Agriculture Information Network," GAIN Report Network: E60062.

[2] "Data from THE WORLD APPLE AND PEAR ASSOCIATION. Available from the internet," http://www.wapa-association.org/asp/page_1.asp.

[3] H. El Gharras, "Polyphenols: food sources, properties and applications—a review," *International Journal of Food Science & Technology*, vol. 44, no. 12, pp. 2512–2518, 2009.

[4] J. Sun, Y.-F. Chu, X. Wu, and R. H. Liu, "Antioxidant and antiproliferative activities of common fruits," *Journal of Agricultural and Food Chemistry*, vol. 50, no. 25, pp. 7449–7454, 2002.

[5] A. Scalbert, I. T. Johnson, and M. Saltmarsh, "Polyphenols: antioxidants and beyond," *American Journal of Clinical Nutrition*, vol. 81, no. 1, pp. 215S–217S, 2005.

[6] K. B. Pandey and S. I. Rizvi, "Plant polyphenols as dietary antioxidants in human health and disease," *Oxidative Medicine and Cellular Longevity*, vol. 2, no. 5, pp. 270–278, 2009.

[7] R. C. Hider, Z. D. Liu, and H. H. Khodr, "Metal chelation of polyphenols," *Methods in Enzymology*, vol. 335, pp. 190–203, 2001.

[8] A. W. Boots, L. C. Wilms, E. L. R. Swennen, J. C. S. Kleinjans, A. Bast, and G. R. M. M. Haenen, "In vitro and ex vivo anti-inflammatory activity of quercetin in healthy volunteers," *Nutrition Journal*, vol. 24, no. 7-8, pp. 703–710, 2008.

[9] S. Pastore, A. Potapovich, V. Kostyuk et al., "Plant polyphenols effectively protect HaCaT cells from ultraviolet C-triggered necrosis and suppress inflammatory chemokine expression," *Annals of the New York Academy of Sciences*, vol. 1171, pp. 305–313, 2009.

[10] C. Kandaswami, L. T. Lee, P. P. Lee et al., "The antitumor activities of flavonoids," *In Vivo*, vol. 19, no. 5, pp. 895–909, 2005.

[11] P. Lakhanpal and D. K. Rai, "Quercetin: A Versatile Flavonoid," *Internet Journal of Medical Update*, vol. 2, no. 2, pp. 22–37, 2007.

[12] M. López-Lázaro, "Flavonoids as anticancer agents: Structure-activity relationship study," *Current Medicinal Chemistry - Anti-Cancer Agents*, vol. 2, no. 6, pp. 691–714, 2002.

[13] E. Weichselbaum, L. Wyness, and S. Stanner, "Apple polyphenols and cardiovascular disease - a review of the evidence," *Nutrition Bulletin*, vol. 35, no. 2, pp. 92–101, 2010.

[14] J. P. E. Spencer, "The impact of fruit flavonoids on memory and cognition," *British Journal of Nutrition*, vol. 104, no. 3, pp. S40–S47, 2010.

[15] L. J. Nohynek, H.-L. Alakomi, M. P. Kähkönen et al., "Berry phenolics: antimicrobial properties and mechanisms of action against severe human pathogens," *Nutrition and Cancer*, vol. 54, no. 1, pp. 18–32, 2006.

[16] T. P. Cushnie and A. J. Lamb, "Antimicrobial activity of flavonoids," *International Journal of Antimicrobial Agents*, vol. 26, no. 5, pp. 343–356, 2005.

[17] J. H. Chávez, P. C. Leal, R. A. Yunes et al., "Evaluation of antiviral activity of phenolic compounds and derivatives against rabies virus," *Veterinary Microbiology*, vol. 116, no. 1-3, pp. 53–59, 2006.

[18] K. Likhitwitayawuid, B. Supudompol, B. Sritularak, V. Lipipun, K. Rapp, and R. F. Schinazi, "Phenolics with anti-HSV and anti-HIV activities from Artocarpus gomezianus, Mallotus pallidus, and Triphasia trifolia," *Pharmaceutical Biology*, vol. 43, no. 8, pp. 651–657, 2005.

[19] J. Salta, A. Martins, R. G. Santos et al., "Phenolic composition and antioxidant activity of Rocha pear and other pear cultivars - A comparative study," *Journal of Functional Foods*, vol. 2, no. 2, pp. 153–157, 2010.

[20] A. Escarpa and M. C. González, "Fast separation of (poly)phenolic compounds from apples and pears by high-performance liquid chromatography with diode-array detection," *Journal of Chromatography A*, vol. 830, no. 2, pp. 301–309, 1999.

[21] M. G. L. Hertog, E. J. M. Feskens, and P. C. H. Hollman, "Content of potentially anticarcinogenic flavonoids of 28 vegetables and 9 fruits commonly consumed in the Netherlands," *Journal of Agricultural and Food Chemistry*, vol. 40, no. 12, pp. 2379–2383, 1992.

[22] European pharmacopoeia, Council of Europe: Strasbourg, 7th Edition. Volume 1. 01/2008:2.2.32., pp. 51, 2010.

[23] R. Bobinait, P. Viškelis, and P. R. Venskutonis, "Variation of total phenolics, anthocyanins, ellagic acid and radical scavenging capacity in various raspberry (*Rubus* spp.) cultivars," *Food Chemistry*, vol. 132, no. 3, pp. 1495–1501, 2012.

[24] W. Brand-Williams, M. E. Cuvelier, and C. Berset, "Use of a free radical method to evaluate antioxidant activity," *LWT - Food Science and Technology*, vol. 28, no. 1, pp. 25–30, 1995.

[25] R. Re, N. Pellegrini, A. Proteggente, A. Pannala, M. Yang, and C. Rice-Evans, "Antioxidant activity applying an improved ABTS radical cation decolorization assay," *Free Radical Biology & Medicine*, vol. 26, no. 9-10, pp. 1231–1237, 1999.

[26] I. F. F. Benzie and J. J. Strain, "Ferric reducing/antioxidant power assay: direct measure of total antioxidant activity of biological fluids and modified version for simultaneous measurement of total antioxidant power and ascorbic acid concentration," *Methods in Enzymology*, vol. 299, pp. 15–27, 1999.

[27] J. Dai and R. J. Mumper, "Plant phenolics: extraction, analysis and their antioxidant and anticancer properties," *Molecules*, vol. 15, no. 10, pp. 7313–7352, 2010.

[28] J. Azmir, I. S. M. Zaidul, M. M. Rahman et al., "Techniques for extraction of bioactive compounds from plant materials: a review," *Journal of Food Engineering*, vol. 117, no. 4, pp. 426–436, 2013.

[29] T. V. Ngo, C. J. Scarlett, M. C. Bowyer, P. D. Ngo, and Q. V. Vuong, "Impact of different extraction solvents on bioactive compounds and antioxidant capacity from the root of Salacia chinensis L," *Journal of Food Quality*, vol. 2017, Article ID 9305047, 2017.

[30] N. Kumar, P. Bhandari, B. Singh, A. P. Gupta, and V. K. Kaul, "Reversed phase-HPLC for rapid determination of polyphenols in flowers of rose species," *Journal of Separation Science*, vol. 31, no. 2, pp. 262–267, 2008.

[31] K. Robards, "Strategies for the determination of bioactive phenols in plants, fruit and vegetables," *Journal of Chromatography A*, vol. 1000, no. 1-2, pp. 657–691, 2003.

[32] E. de Rijke, P. Out, W. M. A. Niessen, F. Ariese, C. Gooijer, and U. A. T. Brinkman, "Analytical separation and detection methods for flavonoids," *Journal of Chromatography A*, vol. 1112, no. 1-2, pp. 31–63, 2006.

[33] A. Escarpa and M. C. González, "Evaluation of high-performance liquid chromatography for determination of phenolic compounds in pear horticultural cultivars," *Chromatographia*, vol. 51, no. 1-2, pp. 37–43, 2000.

[34] ICH Harmonised Tripartite Guideline, "Validation of Analytical Procedures: Text and Methodology Q2 (R1)," in *Proceedings of the International Conference on Harmonisation of Technical Requirements for Registration of Pharmaceuticals for Human Use*, 1994.

[35] Y. Sato, S. Itagaki, T. Kurokawa et al., "In vitro and in vivo antioxidant properties of chlorogenic acid and caffeic acid," *International Journal of Pharmaceutics*, vol. 403, no. 1-2, pp. 136–138, 2011.

[36] X. Zhang, H. Huang, T. Yang et al., "Chlorogenic acid protects mice against lipopolysaccharide-induced acute lung injury," *Injury*, vol. 41, no. 7, pp. 746–752, 2010.

[37] T. Thomas and A. F. Pfeiffer, "Foods for the prevention of diabetes: How do they work?" *Diabetes/Metabolism Research and Reviews*, vol. 28, no. 1, pp. 25–49, 2012.

[38] Y. Kanno, R. Watanabe, H. Zempo, M. Ogawa, J. Suzuki, and M. Isobe, "Chlorogenic acid attenuates ventricular remodeling after myocardial infarction in mice," *International Heart Journal*, vol. 54, no. 3, pp. 176–180, 2013.

[39] N. Cinkilic, S. K. Cetintas, T. Zorlu et al., "Radioprotection by two phenolic compounds: chlorogenic and quinic acid, on X-ray induced DNA damage in human blood lymphocytes in vitro," *Food and Chemical Toxicology*, vol. 53, pp. 359–363, 2013.

[40] B. Stefanska, H. Karlic, F. Varga, K. Fabianowska-Majewska, and A. G. Haslberger, "Epigenetic mechanisms in anti-cancer actions of bioactive food components—the implications in cancer prevention," *British Journal of Pharmacology*, vol. 167, no. 2, pp. 279–297, 2012.

[41] J. Kolniak-Ostek and J. Oszmiański, "Characterization of phenolic compounds in different anatomical pear (Pyrus communis L.) parts by ultra-performance liquid chromatography photodiode detector-quadrupole/time of flight-mass spectrometry (UPLC-PDA-Q/TOF-MS)," *International Journal of Mass Spectrometry*, vol. 392, pp. 154–163, 2015.

[42] L.-Z. Lin and J. M. Harnly, "Phenolic compounds and chromatographic profiles of pear skins (Pyrus spp.)," *Journal of Agricultural and Food Chemistry*, vol. 56, no. 19, pp. 9094–9101, 2008.

[43] A. Schieber, P. Keller, and R. Carle, "Determination of phenolic acids and flavonoids of apple and pear by high-performance liquid chromatography," *Journal of Chromatography A*, vol. 910, no. 2, pp. 265–273, 2001.

[44] D. Tanrioven and A. Ekşi, "Phenolic compounds in pear juice from different cultivars," *Food Chemistry*, vol. 93, no. 1, pp. 89–93, 2005.

[45] F. Dajas, "Life or death: neuroprotective and anticancer effects of quercetin," *Journal of Ethnopharmacology*, vol. 143, no. 2, pp. 383–396, 2012.

[46] M. González-Vallinas, M. González-Castejón, A. Rodríguez-Casado, and A. Ramírez de Molina, "Dietary phytochemicals in cancer prevention and therapy: a complementary approach with promising perspectives," *Nutrition Reviews*, vol. 71, no. 9, pp. 585–599, 2013.

[47] N. Balasuriya and H. P. V. Rupasinghe, "Antihypertensive properties of flavonoid-rich apple peel extract," *Food Chemistry*, vol. 135, no. 4, pp. 2320–2325, 2012.

[48] L. Raudone, R. Raudonis, M. Liaudanskas, V. Janulis, and P. Viskelis, "Phenolic antioxidant profiles in the whole fruit, flesh and peel of apple cultivars grown in Lithuania," *Scientia Horticulturae*, vol. 216, pp. 186–192, 2017.

[49] M. A. Olszewska and P. Michel, "Antioxidant activity of inflorescences, leaves and fruits of three sorbus species in relation to their polyphenolic composition," *Natural Product Research (Formerly Natural Product Letters)*, vol. 23, no. 16, pp. 1507–1521, 2009.

# 10

# Comparison of an Electronic Nose Based on Ultrafast Gas Chromatography, Comprehensive Two-Dimensional Gas Chromatography, and Sensory Evaluation for an Analysis of Type of Whisky

**Paulina Wiśniewska, Magdalena Śliwińska, Tomasz Dymerski, Waldemar Wardencki, and Jacek Namieśnik**

*Department of Analytical Chemistry, Faculty of Chemistry, Gdańsk University of Technology, 11/12 Narutowicza St., 80-233 Gdańsk, Poland*

Correspondence should be addressed to Paulina Wiśniewska; p.m.wisniewska@gmail.com and Tomasz Dymerski; tomasz.dymerski@gmail.com

Academic Editor: Lee Y. Heng

Whisky is one of the most popular alcoholic beverages. There are many types of whisky, for example, Scotch, Irish, and American whisky (called bourbon). The whisky market is highly diversified, and, because of this, it is important to have a method which would enable rapid quality evaluation and authentication of the type of whisky. The aim of this work was to compare 3 methods: an electronic nose based on the technology of ultrafast gas chromatography (Fast-GC), comprehensive two-dimensional gas chromatography (GC × GC), and sensory evaluation. The selected whisky brands included 6 blended whiskies from Scotland, 4 blended whiskies from Ireland, and 4 bourbons produced in the USA. For data analysis, peak heights of chromatograms were used. The panelists who took part in sensory evaluations included 4 women and 4 men. The obtained data were analyzed by 2 chemometric methods: partial least squares discriminant analysis (PLS-DA) and discrimination function analysis (DFA). E-nose and GC × GC allowed for differentiation between whiskies by type. Sensory analysis did not allow for differentiation between whiskies by type, but it allowed giving consumer preferences.

## 1. Introduction

Whisky is an alcoholic beverage produced by the distillation of malted or fermented grain mash. Whisky is matured in oak barrels of a capacity not exceeding 700 l. Only water and pure caramel, a natural colorant, can be added to the distillate [1–3]. Scotch whisky is the most popular alcoholic drink of this type, which, in addition to all the above-mentioned requirements, has to be exclusively produced in Scottish distilleries. Scotch whisky is mainly produced from barley malt, water, and an admixture of some grains, for example, wheat and rye.

Irish whisky is made in Ireland and, as in the case of Scotch, it also consists of malted barley and some grains. The taste of Irish whisky is more delicate than that of Scotch, because the former is obtained by 3 separate distillations [4]. Besides those described above, one of the most popular types of American whisky should be mentioned, namely, bourbon. Bourbon must fulfill the following requirements established by the US Congress in 1964:

(i) It has to be prepared from a mix of grains containing at least 51% corn (maize).

(ii) The distillate must contain no more than 80% pure alcohol.

(iii) It has to be only made from natural ingredients.

(iv) It has to be rested in new oak barrels that are charred inside.

(v) It cannot be poured into the barrels if the content of pure alcohol in the distillate exceeds 62.5%.

(vi) It has to be produced in the United States of America.

In general, bourbon is made from a mash containing 70% corn, wheat or rye, and malted barley. The mash undergoes fermentation which is followed by distillation and maturation. Due to the fact that resting is a very important stage of bourbon production, most bourbon brands specify the duration of the aging period on the label of the product. During the resting period, whisky develops its color and woody flavor. The longer the resting period, the darker the color [5, 6].

In the case of raw material, whisky can be divided into 2 groups. One is made partly from corn, such as bourbon; the second is produced without corn, such as Irish or Scotch whisky. Apart from botanical origin, it is important to analyze a whisky for its type of production. Irish and Scotch whisky are produced from the same raw materials, but in different ways. The main difference is triplicate distillation in the case of Irish whisky.

The composition of whisky is most frequently analyzed by means of one-dimensional gas chromatography [7–9]. More advanced analyses are conducted by employing two-dimensional gas chromatography [10]. Besides the 2 aforementioned techniques, liquid chromatography [11], spectroscopy [12, 13], sensory evaluation [14], and sensor-based electronic nose [15] are also used for this purpose. Such broad diversity of applied analytical techniques is related to the complex composition of available whisky products. Whisky is also analyzed to control its composition and quality as well as authenticate the brand. Distinguishing between the whisky brands in relation to the production procedure, product origin, and quality can be provided by using an electronic nose based on the technology of ultrafast gas chromatography, comprehensive two-dimensional gas chromatography, and sensory analysis.

Comprehensive two-dimensional gas chromatography (GC × GC) is a technique used for the separation of complex mixtures of chemical compounds and their identification. In comparison to one-dimensional gas chromatography, it allows for the detection of a larger amount of compounds and reduction of the risk of coelution; unfortunately, it is also very expensive.

The electronic noses are analytical tools designed to detect and identify mixes of aromatic substances by mimicking the mode of action of the human olfactory organ. To this day, specific chemical sensors are mainly used. However, at present, e-noses based on mass spectrometry and fast gas chromatography are also applied [16]. The electronic nose, based on Fast-GC, combines the advantages of both the aforementioned chromatographic technique and an electronic nose. This allows for obtaining comprehensive information about the sample composition and the volatile fraction profile [17]. In comparison to GC × GC, analysis by using e-nose based on Fast-GC is cheaper and less time-consuming, but there is no possibility of detecting so many compounds included in the matrix [18].

Sensory evaluation is mainly used for the quality assessment of products. This type of analysis allows investigating the preferences of consumers and, in comparison to other techniques, it is the cheapest one. However, due to the presence of people, it can be considered as the least objective analysis.

In this work, the suitability of an electronic nose based on the technology of ultrafast gas chromatography, comprehensive two-dimensional gas chromatography, and sensory evaluation for the analysis of whisky was tested. The criteria of whisky selection were botanical origin and way of production. The aim of this work was to compare the 3 types of analysis. The whisky market is highly diversified, and because of this, it is important to invent a method which would enable a rapid quality evaluation and authentication of a whisky sample. For data analysis, we tested such chemometric methods like partial least squares with discriminant analysis (PLS-DA) and discriminant function analysis (DFA).

## 2. Materials and Methods

*2.1. Materials.* Only samples purchased in local stores in Gdańsk (Poland) were used in the study. The selected whisky brands included 6 blended whiskies from Scotland, 4 blended whiskies from Ireland, and 4 bourbons. All samples had been matured for 3 years. Among the samples were whiskies in different price ranges.

*2.2. Sample Preparation.* All samples used in the analysis by e-nose and GC × GC-TOFMS were processed according to the same analytical procedure. An aliquot of 1.75 ml of whisky was diluted with 6.25 ml of deionized water. Next, in the case of analysis by the electronic nose, the sample was mixed and incubated at 40°C for 20 min. A sample prepared in this manner was analyzed in triplicate. In the case of GC × GC-TOFMS, samples were prepared by using headspace solid phase microextraction (HS-SPME). A sample prepared in this manner was analyzed in duplicate. In the case of sensory analysis, each panelist received 20 ml of sample in a glass. The general scheme of procedures for sample preparation and analysis is shown in Figure 1.

*2.3. GC × GC-TOFMS Analysis.* For solid phase microextraction, divinylbenzene/carboxen/polydimethylsiloxane (DVB/CAR/PDMS) fiber was used. Thermostating was carried out for 5 min at 40°C. Extraction was carried out for 20 min at 40°C, and desorption was carried out for 20 min at 250°C.

The GC × GC system consisted of an Agilent 6890 gas chromatograph (Agilent Technologies, Palo Alto, CA, USA) equipped with a liquid nitrogen-based single-jet dual-stage cryogenic modulator and a split/splitless injector, coupled with a Pegasus III time-of-flight mass spectrometer (LECO Corp., St. Joseph, MI, USA). The column set consisted of a 30 m × 0.25 mm × 0.25 $\mu$m primary column with Equity 1 (Sigma-Aldrich, St. Louis, MI, USA) stationary phase and a 2 m × 0.1 mm × 0.1 $\mu$m secondary column with SolGel-Wax (SGE Analytical Science, Ringwood, Victoria, Australia) stationary phase. The modulation period was 6 s. The following optimized temperature program was used for the separation

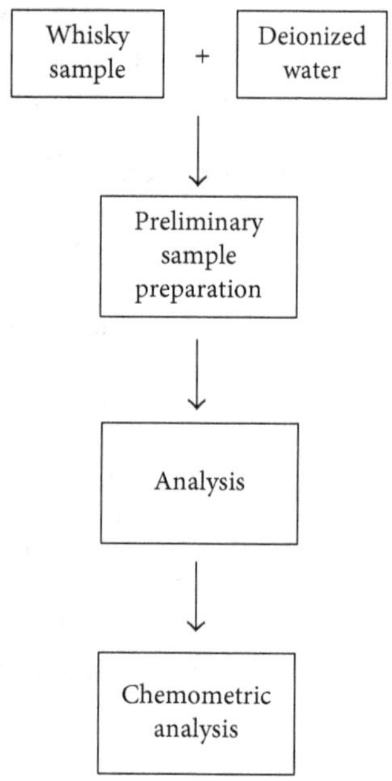


FIGURE 1: General scheme of procedures for sample preparation and analysis.

of the sample: initial temperature of 40°C maintained for 3.5 min and then ramped to 250°C and kept for 5 min. The total analysis time was 28 min 50 s. The injector was operated at 250°C in splitless mode. Helium was used as the carrier gas at a constant flow of 1 mL/min. The temperature of the ion source was 250°C, whereas the detector voltage was set to 1600 V. The analyzed $m/z$ range of ions was 40–400. The data acquisition rate was 200 spectra/s.

*2.4. Electronic Nose Analysis.* A HERACLES II Alpha MOS electronic nose based on the technology of ultrafast gas chromatography (Toulouse, France) was used in this study. The device was equipped with a sorption trap, an injection module for gas and liquid samples, HS100 dedicated autosampler, a set of independent columns of different polarity (10 m long nonpolar MXT-5 and medium polar MXT-1701 columns), 2 FID detectors, AlphaSoft V12 software package for HERACLES II with implemented modules for chromatographic and chemometric data analysis and chemical and sensory characterization of detected compounds, and AroChemBase V4 HERACLES V12 library. The autosampler loaded with samples was incubated at 40°C for 1200 s, while the syringe temperature was kept at 100°C. A 2.5 ml aliquot of sample was dispensed for 15 s at the injector temperature of 200°C and the flow rate of carrier gas equaled 30 ml/min. The sample was retained in the sorption trap for 20 s at 40°C. The initial temperature of analysis was set to 40°C and kept for 2 s. Next, the temperature was raised at a rate of 3°C/s up to 270°C and kept for 18 s. Hydrogen was used as the carrier gas.

*2.5. Sensory Analysis.* Out of the 20 respondents, 10 people were selected for preliminary tests of taste and smell. The general scheme of the election of members of the sensory panel is presented in Figure 2. Each person gets 5 mixtures to test the taste and 5 mixtures to test the smell. Mixtures used to test the taste include solutions of caffeine, sodium chloride, citric acid, starch, and monosodium glutamate, while mixtures to test the smell include the solution of citral, benzaldehyde, vanillin, eugenol, and 2-methylisoborneol. To qualify for the group of panelists, people had to give 80% correct answers in each test. From the group of 12 people, 8 people were chosen to take part in the sensory analysis of whisky. The group of panelists included 4 women and 4 men. Samples were analyzed based on smell, taste, and color. The panelists evaluated which type whisky belonged to.

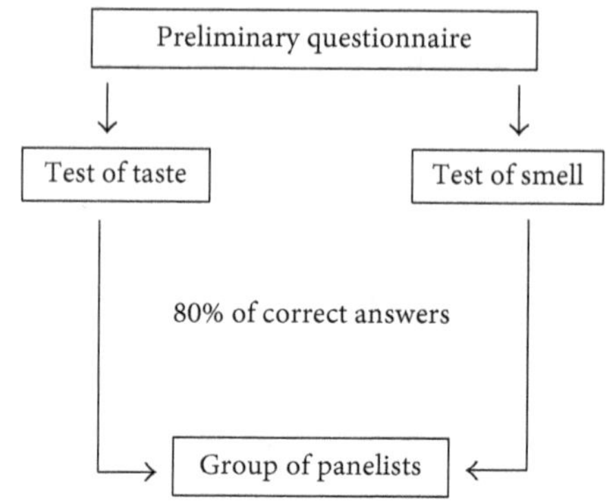


FIGURE 2: Scheme of the election of members of the sensory panel.

*2.6. Chemometric Analysis.* The area of compounds identified by the GC × GC analysis was used as data for chemometric analysis. Compounds present in the volatile fraction of whisky were identified by comparing their spectra with the spectra of fragment ions available in the NIST library. Then, a comparative analysis of different types of whisky was performed using the Matlab software v.7.8 (MathWorks, MA, USA) and the PLS Toolbox software v.6.2 (Eigenvector Research, Manson, WA, USA). The partial least squares with discriminant analysis (PLS-DA) model was created for the discrimination of whisky.

The data obtained from the e-nose analysis were analyzed by means of dedicated software which includes different chemometric methods. Discriminant function analysis (DFA) was used for discrimination of different types of whisky. To compare this technique to other techniques, partial least squares with discriminant analysis (PLS-DA) using Matlab software v.7.8 (MathWorks, MA, USA) and the PLS Toolbox software v.6.2 (Eigenvector Research, Manson, WA, USA) were used.

Results from sensory evaluation were analyzed by using Matlab software v.7.8 (MathWorks, MA, USA) and PLS Toolbox software v.6.2 (Eigenvector Research, Manson, WA, USA). The PLS-DA models were created for the discrimination of whisky.

## 3. Results and Discussion

Whiskies belong to a group of highly aromatic alcohols. This is due to a variety of raw materials used in their

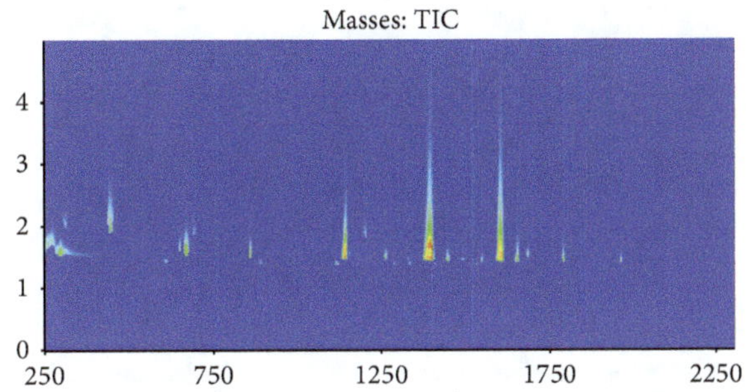


FIGURE 3: Exemplary chromatogram obtained from the analysis of whisky by GC × GC-TOFMS.

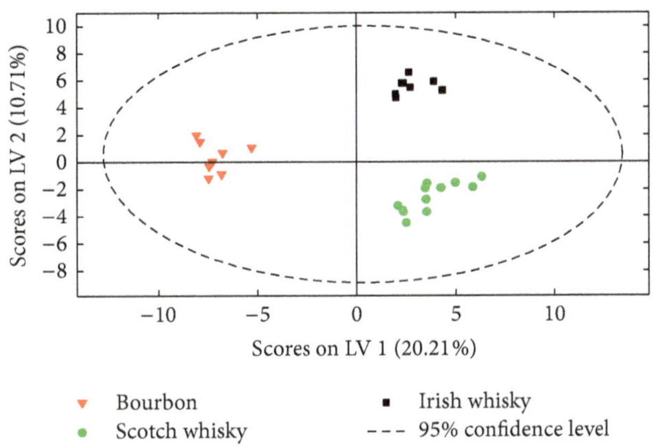


FIGURE 4: Scores plot of the first 2 factors of the PLS-DA model built with the area of the chosen peaks for GC × GC-TOFMS analysis.

production and to the aging of alcohol in barrels. Whisky includes a variety of compounds from many different groups. Fourteen samples in duplicate were analyzed by using GC × GC-TOFMS, which allows for the detection of hundreds of compounds. Such large number of compounds can be detected due to the use of two columns of different polarity and cryogenic modulator. Thanks to this, the mixtures of aromatic compounds are separated in two dimensions, thus avoiding the coelution of some compounds. An exemplary chromatogram is shown in Figure 3. Among the samples were 4 bourbons, 6 types of Scotch whisky, and 4 Irish whiskies. From hundreds of compounds, 119 compounds were used for data analysis (Table 1). To select such compounds, the retention times and spectra of fragment ions of all detected compounds were compared with the NIST library. In each case, the compounds were considered to have been identified if the similarity of their spectra to the spectra found in NIST library was at least 85% and if the signal-to-noise (s/n) ratio was higher than 100. For data, PLS-DA (Figure 4) was used. Autoscale and SNV algorithm were used for preprocessing, and a Venetian blind was used for cross-validation. PLS-DA allows for the discrimination of samples by type, which includes different botanical origins and types of production. As can be seen in Table 2, showing confusion matrices and confusion tables, both for calibration and for cross-validation data, all samples were classified correctly.

A quick tool for assessing the quality of the whisky may be an electronic nose. An e-nose based on Fast-GC was used for the discrimination of whisky according to the type. Six samples of Scotch whisky, four samples of Irish whisky, and four samples of bourbon from four different batches were analyzed in triplicate. For data analysis, the area of the peaks was used as input data. In the case of Fast-GC analysis, a smaller number of compounds can be detected than in the case of GC × GC analysis. This number is connected with the use of shorter column and with the coelution of some compounds, which cannot be separated in one dimension. An exemplary chromatogram is shown in Figure 5. Due to the presence of the implemented software, DFA was used for data analysis. All groups were discriminated by type. In Table 3, DFA was used for data analysis (Figure 6). All four groups of whisky were discriminated according to botanical origin and the way of production by DFA analysis. The verification of this method was obtained by calculation of the Euclidean distances between groups and pattern discrimination indices which are shown in Table 3. The smallest distance (10.59) is between the Irish and Scotch whisky. This is because the same type of raw material is used for the production of both types of whisky. The variables used as input data for DFA analysis, namely, the most discriminant peak areas for specific compounds, were presented in Figure 7 and the names of selected chemical compounds were presented in Table 4.

Table 1: Selected compounds identified in a volatile fraction of whisky used as input data in chemometric analysis.

| Name | Fragment ions | Irish whisky 1/2/3/4 | | | | Scotch whisky 1/2/3/4/5/6 | | | | | | Bourbon 1/2/3/4 | | | |
|---|---|---|---|---|---|---|---|---|---|---|---|---|---|---|---|
| 1,1-Diethoxypropanal | 47, 75, 103 | + | − | − | − | − | − | − | + | − | + | + | + | + | + |
| 2-Methyl-1-butanol | 57, 41, 70 | + | + | + | + | + | + | + | + | + | + | + | + | + | + |
| 3-Methyl-1-butanol | 55, 70, 42 | + | + | + | + | + | + | + | + | + | + | + | + | + | + |
| 3-Methyl-1-butanol acetate | 43, 70, 55 | + | + | + | + | + | + | + | + | + | + | + | + | + | + |
| 1-Decanol | 56, 70, 83 | + | + | + | + | + | + | + | + | + | + | + | + | + | + |
| 1-Dodecanol | 55, 69, 83 | + | + | + | + | + | + | + | + | + | + | + | + | + | + |
| 1-Heptanol | 41, 56, 70 | − | − | − | − | − | − | − | − | − | − | − | + | + | − |
| 1-Hexanol | 56, 43, 69 | − | − | − | + | − | + | − | − | − | − | + | + | + | + |
| 2-Ethyl-1-hexanol | 57, 70, 83 | − | − | − | − | + | − | − | + | − | − | − | + | + | + |
| 1-Nonanol | 56, 41, 70 | − | − | − | + | − | − | − | − | − | − | + | + | + | + |
| 1-Nonene | 41, 56, 69 | − | − | − | − | − | + | + | − | − | − | + | + | + | + |
| 1-Octanol | 56, 41, 70, | − | − | − | − | − | − | − | − | − | − | + | + | + | + |
| 2-Butyl-1-octanol | 57, 43, 69 | + | + | + | + | + | + | + | + | + | + | + | + | + | + |
| 1-Octene | 43, 55, 70 | − | − | − | − | − | + | − | − | − | − | + | + | + | + |
| 1-Pentanol | 42, 55, 70 | − | − | − | + | − | − | − | − | − | − | + | + | + | + |
| 2-Methyl-1-propanol | 43, 74, 55 | + | + | + | + | + | + | + | + | + | + | + | + | + | + |
| 1-Tetradecyl acetate | 43, 55, 69 | − | + | + | + | − | + | − | − | − | + | − | − | − | − |
| 2-Dodecanol | 45, 55, 69 | − | − | − | − | − | − | − | − | − | − | + | − | + | − |
| 2-Furancarboxaldehyde | 96, 67, 51 | − | − | − | + | − | + | + | − | − | − | + | − | + | + |
| 2-Heptanone | 43, 58, 71 | − | − | − | + | − | + | − | − | + | − | − | + | − | + |
| 2-Nonanone | 43, 58, 71 | + | − | − | + | + | + | − | + | − | + | + | + | + | + |
| 2-Octanone | 43, 58, 71 | − | − | − | − | + | + | + | − | − | − | + | + | + | + |
| 2-Pentadecanone | 43, 58, 71 | − | − | + | + | − | + | + | − | − | − | − | − | − | − |
| 2-Undecanol | 45, 41, 55 | + | − | − | + | − | − | − | + | + | − | − | − | − | − |
| 2-Undecanone | 43, 58, 71 | + | + | + | + | + | + | + | + | + | + | + | + | + | + |
| 3-Methylbutyl dodecanoate | 70, 43, 55 | − | + | + | − | + | + | − | + | − | − | + | − | + | − |
| 3-Nonenoic acid, ethyl ester | 69, 55, 41 | + | + | + | − | + | + | + | + | + | + | + | + | + | + |
| 3,7,11-Trimethyl-6,10-dodecadien-1-ol | 69, 81, 123 | + | − | − | + | + | + | + | − | − | − | − | − | − | − |
| Acetaldehyde | 44 | + | + | + | + | + | + | + | + | + | + | + | + | + | + |
| Acetic acid, 2-methylpropyl ester | 43, 56, 73 | + | + | − | + | + | + | + | − | + | + | + | + | + | + |
| Acetic acid, 2-phenylethyl ester | 104, 43, 91 | + | + | + | + | + | + | + | + | + | + | + | + | + | + |
| Acetic acid, heptyl ester | 43, 56, 70 | + | + | + | + | + | + | + | − | + | + | + | + | + | + |
| Acetic acid, hexyl ester | 43, 56, 61 | + | + | + | + | + | + | + | + | + | + | + | + | + | + |
| Acetic acid, nonyl ester | 43, 56, 70 | + | + | + | + | + | − | + | − | + | + | + | − | + | + |
| $\alpha$-Farnesene | 93, 107, 79 | − | − | − | − | − | + | + | + | − | + | − | − | − | − |
| $\alpha$-Ionone | 121, 93, 43 | − | − | − | − | − | − | − | − | − | − | − | + | + | − |
| $\alpha$-Muurolene | 105, 91, 161 | + | − | − | + | − | − | + | − | − | + | − | + | + | − |
| $\alpha$-Pinene | 93, 77, 39 | + | + | + | + | + | + | + | + | + | + | + | + | + | + |
| Benzaldehyde | 77, 105, 51 | + | − | + | + | + | + | + | + | + | + | + | − | + | + |

Table 1: Continued.

| Name | Fragment ions | Irish whisky 1/2/3/4 | | | | Scotch whisky 1/2/3/4/5/6 | | | | | | Bourbon 1/2/3/4 | | | |
|---|---|---|---|---|---|---|---|---|---|---|---|---|---|---|---|
| 1,1-Dimethylpropylbenzene | 119, 91, 148 | − | − | − | − | − | + | + | − | − | − | + | + | − | − |
| 1-Ethylnonylbenzene | 91, 119, 41 | + | − | − | − | − | + | − | + | + | − | − | − | − | − |
| 1-Methyldecylbenzene | 105, 91, 79 | + | − | + | − | − | + | − | + | + | − | − | − | − | − |
| 1-Methylethylbenzene | 105, 120, 77 | + | − | − | + | + | + | + | − | + | − | + | + | + | + |
| 1-Pentylheptylbenzene | 91, 105, 119 | − | − | − | − | − | + | − | + | + | − | − | − | − | − |
| 1-Propyloctylbenzene | 91, 133, 105 | − | − | − | − | − | + | − | + | + | − | − | − | − | − |
| Benzene, 1-(1,5-Dimethyl-4-hexenyl)-4-methylbenzene | 132, 119, 202 | + | + | + | − | + | + | − | − | − | + | − | + | + | + |
| 1,2,3,4-Tetramethylbenzene | 119, 134, 91 | + | + | + | − | + | − | + | − | − | + | − | − | − | − |
| 1-Ethyl-2,4-dimethylbenzene | 119, 134, 91 | − | + | − | − | − | − | + | − | + | − | + | + | − | − |
| 1-Ethyl-2-methylbenzene | 105, 120, 77 | − | + | + | − | − | − | + | − | + | + | + | + | + | + |
| 1-Ethyl-4-methylbenzene | 105, 120, 77 | − | + | − | − | − | + | − | + | − | + | − | + | − | + |
| 1-Methyl-2-propylbenzene | 105, 134, 77 | − | − | − | − | − | − | + | − | − | − | + | + | − | − |
| 1-Methyl-3-(1-methylethyl)benzene | 119, 91, 134 | − | − | + | + | + | + | + | + | − | + | + | + | − | − |
| 1-Methyl-4-(1-methylethenyl)benzene | 117, 132, 91 | − | − | − | − | − | + | + | − | − | + | + | + | + | + |
| Ethylbenzene | 91, 106, 51 | + | + | + | + | + | + | + | − | + | + | + | + | + | + |
| Hexylbenzene | 91, 162, 65 | − | − | − | − | − | + | − | − | − | − | + | − | + | − |
| Methylbenzene | 91, 65, 39 | − | − | − | − | − | + | + | + | + | + | − | − | − | + |
| Pentylbenzene | 91, 148, 65 | + | − | − | − | − | + | + | − | − | − | + | + | + | + |
| Propylbenzene | 91, 120, 65 | − | − | − | − | − | + | + | − | − | − | + | + | + | + |
| Benzeneacetic acid, ethyl ester | 91, 65, 164 | + | + | + | − | + | + | + | + | + | + | + | + | + | + |
| Benzofuran | 118, 89, 63 | − | − | − | − | + | + | + | − | − | − | − | − | − | − |
| Benzoic acid, ethyl ester | 105, 77, 122 | + | + | + | + | + | + | + | + | + | + | + | + | + | + |
| Biphenyl | 154, 76, 65 | + | + | + | + | − | + | + | + | + | + | − | − | − | − |
| 1,1-Diethoxy-3-methyl-butane | 103, 47, 71 | − | − | − | − | + | + | + | − | − | − | + | − | − | − |
| Butanedioic acid, diethyl ester | 101, 129, 55 | + | + | − | + | − | + | − | + | + | + | + | + | + | + |
| 2-Methylbutanoic acid, ethyl ester | 57, 102, 41 | − | − | + | − | + | + | + | − | − | − | + | + | + | + |
| 3-Methylbutanoic acid, 3-methylbutyl ester | 85, 70, 57 | + | − | + | − | − | − | − | − | − | − | − | − | − | + |
| Butanoic acid, 3-methylbutyl ester | 43, 71, 55 | − | + | + | − | − | − | + | − | − | − | + | − | − | + |
| Butanoic acid, ethyl ester | 71, 43, 88 | + | + | + | + | + | + | + | + | + | + | + | + | + | + |
| Camphene | 93, 121, 41 | − | − | + | − | − | − | − | + | − | + | + | + | + | + |

TABLE 1: Continued.

| Name | Fragment ions | Irish whisky 1/2/3/4 | | | | Scotch whisky 1/2/3/4/5/6 | | | | | | Bourbon 1/2/3/4 | | | |
|---|---|---|---|---|---|---|---|---|---|---|---|---|---|---|---|
| γ-Terpinene | 93, 121, 136 | − | − | − | − | + | + | + | + | + | + | + | + | − | + |
| Decanal | 41, 57, 82 | + | + | + | + | + | + | + | + | + | + | + | + | + | + |
| Decanoic acid, ethyl ester | 88, 101, 43 | + | + | + | + | + | + | + | + | + | + | + | + | + | + |
| Decanoic acid, methyl ester | 74, 87, 143 | + | + | + | + | + | + | + | + | + | + | + | + | + | − |
| Dodecane | 43, 57, 71 | − | − | + | + | + | + | + | + | + | + | + | + | + | + |
| Dodecanoic acid, ethyl ester | 88, 101, 73 | + | + | + | + | + | + | + | + | + | + | + | + | + | + |
| Dodecanoic acid, methyl ester | 74, 87, 55 | + | + | + | + | + | + | + | − | + | + | + | + | + | + |
| 1,1-Diethoxy-ethane (CAS) | 45, 73, 103 | + | − | − | + | + | + | + | + | + | + | + | + | + | + |
| 1-(2-Furanyl)-ethanone | 95, 110, 43 | − | − | − | − | + | + | − | − | − | + | + | + | − | − |
| Ethyl 9-hexadecenoate | 55, 69, 88 | + | − | − | − | − | + | + | + | + | + | − | − | − | − |
| Ethyl acetate | 43, 61, 70 | + | + | − | + | + | + | + | + | + | + | + | + | − | + |
| Ethyl tiglate | 55, 83, 113 | − | − | − | − | + | + | + | − | − | − | + | + | + | + |
| 2,3-Dihydrofuran | 41, 70 | + | + | − | − | + | + | + | + | + | + | − | − | − | + |
| Heptadecane | 57, 43, 71 | − | − | + | − | − | + | − | + | + | + | + | + | − | + |
| Heptanoic acid, ethyl ester | 88, 113, 43 | + | + | + | + | + | + | + | + | + | + | + | + | + | + |
| Hexadecane | 43, 57, 71 | − | − | − | − | − | + | + | + | + | + | + | + | − | − |
| Hexadecanoic acid, ethyl ester | 88, 101, 43 | + | + | − | + | + | + | + | + | + | + | + | + | + | + |
| Hexadecanoic acid, methyl ester | 74, 87, 43 | − | + | − | − | − | − | + | + | − | − | + | − | + | − |
| Hexanal | 44, 56, 72 | − | − | − | − | + | + | + | − | − | − | + | + | + | + |
| Hexanoic acid, ethyl ester | 88, 99, 70 | + | + | + | + | + | + | + | + | + | + | + | + | + | + |
| Hexyl octanoate | 43, 57, 145 | − | − | − | − | − | − | − | − | − | − | + | + | + | − |
| Isoamyl hexanoate | 43, 70, 99 | + | + | + | + | + | + | + | − | + | + | − | − | − | + |
| Isobutyl caprate | 56, 41, 155 | + | + | + | + | + | + | + | + | − | + | + | + | + | + |
| Isobutyl caprylate | 57, 127, 41 | + | + | + | + | + | + | + | − | + | + | + | + | + | + |
| Isobutyl laurate | 56, 41, 183 | + | − | − | − | + | + | + | − | + | + | + | + | − | − |
| Isopentyl hexanoate | 70, 43, 99 | − | − | − | − | − | − | − | − | − | − | + | + | − | − |
| Methyl laurate | 91, 106, 77 | − | + | − | − | − | + | + | + | + | + | − | + | + | + |
| Nerolidol | 69, 41, 81 | − | − | − | + | + | + | + | + | + | + | − | − | − | + |
| Nonadecanoic acid, ethyl ester | 88, 101, 43 | − | + | + | − | − | − | − | − | − | − | − | − | + | − |
| Nonanal | 57, 70, 82 | − | − | − | + | − | − | − | + | + | − | + | + | + | + |
| Nonanoic acid, ethyl ester | 88, 101, 73 | + | + | + | + | + | + | + | + | + | + | + | + | + | + |
| Octadecanal | 43, 57, 82 | + | − | − | − | − | + | − | − | − | − | − | + | + | + |
| Octanoic acid, ethyl ester | 88, 57, 101 | + | + | + | + | + | + | + | + | + | + | + | + | + | + |
| Octanoic acid, methyl ester | 74, 87, 55 | + | − | + | + | − | + | − | − | − | − | + | − | + | + |
| Oxalic acid, nonyl 2-phenylethyl ester | 104, 43, 55 | − | + | + | − | − | + | − | + | + | + | − | − | − | − |
| Pentadecane | 57, 71, 85 | + | + | + | + | + | + | + | + | + | + | + | + | + | + |

Table 1: Continued.

| Name | Fragment ions | Irish whisky 1/2/3/4 | | | | Scotch whisky 1/2/3/4/5/6 | | | | | | Bourbon 1/2/3/4 | | | |
|---|---|---|---|---|---|---|---|---|---|---|---|---|---|---|---|
| Pentadecanoic acid, ethyl ester | 88, 101, 70 | + | + | + | + | + | + | + | + | + | + | + | + | + | + |
| 3-Methylpentanoic acid, ethyl ester | 88, 101, 43 | – | + | – | + | – | – | – | – | – | – | + | + | + | + |
| Pentanoic acid, ethyl ester | 88, 57, 41 | – | – | – | – | + | + | + | – | + | + | + | + | + | + |
| Phenylethyl alcohol | 91, 122, 65 | – | – | – | – | + | + | + | + | + | – | + | + | + | + |
| 2-Methylpropanal | 43, 72, 57 | – | – | + | – | + | – | + | – | – | + | – | – | – | + |
| 1,1-Diethoxy-2-methylpropane | 47, 103, 73 | – | – | – | – | + | + | + | – | – | – | + | – | – | + |
| 2-Methylpropanoic acid, ethyl ester | 43, 71, 88 | – | – | + | + | + | + | + | – | – | + | + | + | + | + |
| Propyl dodecanoate | 61, 43, 55 | + | + | + | + | + | – | + | – | + | + | – | – | – | + |
| Propyl hexanoate | 43, 99, 117 | – | – | – | – | – | + | + | – | – | – | + | + | – | + |
| Styrene | 104, 78, 51 | – | – | – | – | + | + | + | – | + | + | + | + | + | + |
| Tetradecane | 43, 57, 71 | + | + | + | + | + | + | + | + | + | + | + | + | + | + |
| Tetradecanoic acid, ethyl ester | 88, 101, 55 | + | + | + | + | + | + | + | + | + | + | + | + | + | + |
| Undecanal | 41, 57, 82 | + | + | + | + | + | + | + | + | + | + | – | + | + | + |
| Undecanoic acid, ethyl ester | 88, 101, 73 | + | + | + | + | + | + | + | – | + | – | + | + | + | + |

"+" means that the compound was considered to be identified if its spectra were at least 85% similar to the spectra included in NIST library and if the signal-to-noise (s/n) ratio was higher than 100
"–" means that the compound did not fulfill the above-mentioned requirements.

TABLE 2: Confusion matrices and confusion tables for results obtained by the use of GC × GC-TOFMS, e-nose, and sensory analysis.

| | GC × GC-TOFMS | | | | E-nose | | | | Sensory analysis | | | |
|---|---|---|---|---|---|---|---|---|---|---|---|---|
| **Model results** | | | | | | | | | | | | |
| **Confusion matrix** | | | | | | | | | | | | |
| Class | TP | FP | TN | FN | TP | FP | TN | FN | TP | FP | TN | FN |
| Bourbon | 1.000 | 0.000 | 1.000 | 0.000 | 1.000 | 0.000 | 1.000 | 0.000 | 1.000 | 0.000 | 1.000 | 0.000 |
| Scotch | 1.000 | 0.000 | 1.000 | 0.000 | 1.000 | 0.000 | 1.000 | 0.000 | 1.000 | 0.000 | 1.000 | 0.000 |
| Irish | 1.000 | 0.000 | 1.000 | 0.000 | 1.000 | 0.000 | 1.000 | 0.000 | 1.000 | 0.000 | 1.000 | 0.000 |
| **Confusion table** | | | | | | | | | | | | |
| Predicted as | Bourbon | | Scotch | Irish | Bourbon | | Scotch | Irish | Bourbon | | Scotch | Irish |
| Bourbon | 8 | | 0 | 0 | 48 | | 0 | 0 | 4 | | 0 | 0 |
| Scotch | 0 | | 12 | 0 | 0 | | 72 | 0 | 0 | | 6 | 0 |
| Irish | 0 | | 0 | 8 | 0 | | 0 | 48 | 0 | | 0 | 4 |
| **Cross-validation results** | | | | | | | | | | | | |
| **Confusion matrix** | | | | | | | | | | | | |
| Class | TP | FP | TN | FN | TP | FP | TN | FN | TP | FP | TN | FN |
| Bourbon | 1.000 | 0.000 | 1.000 | 0.000 | 1.000 | 0.000 | 1.000 | 0.000 | 0.250 | 0.300 | 0.700 | 0.000 |
| Scotch | 1.000 | 0.000 | 1.000 | 0.000 | 1.000 | 0.000 | 1.000 | 0.000 | 0.500 | 0.375 | 0.725 | 0.000 |
| Irish | 1.000 | 0.000 | 1.000 | 0.000 | 1.000 | 0.000 | 1.000 | 0.000 | 0.250 | 0.300 | 0.700 | 0.000 |
| **Confusion table** | | | | | | | | | | | | |
| Predicted as | Bourbon | | Scotch | Irish | Bourbon | | Scotch | Irish | Bourbon | | Scotch | Irish |
| Bourbon | 8 | | 0 | 0 | 48 | | 0 | 0 | 1 | | 1 | 2 |
| Scotch | 0 | | 12 | 0 | 0 | | 72 | 0 | 2 | | 3 | 1 |
| Irish | 0 | | 0 | 8 | 0 | | 0 | 48 | 1 | | 2 | 1 |

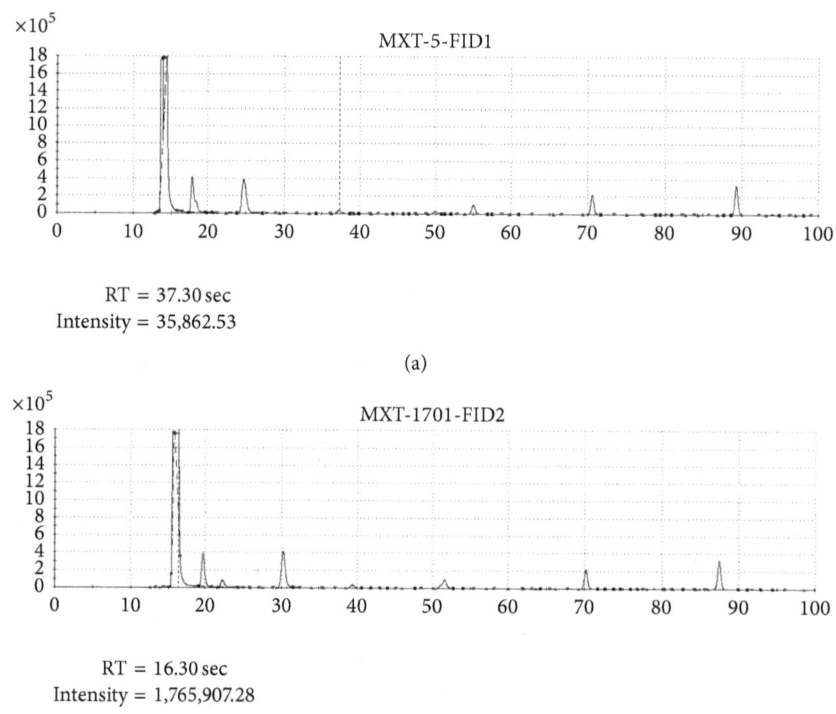


FIGURE 5: Exemplary chromatogram obtained from the analysis of whisky by e-nose using Fast-GC-A.

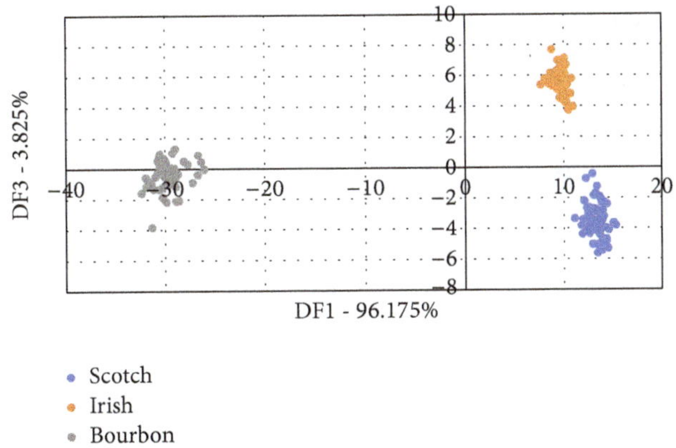


FIGURE 6: Scores plot of the first 2 factors of the DFA model for e-nose analysis of whiskies according to their type.

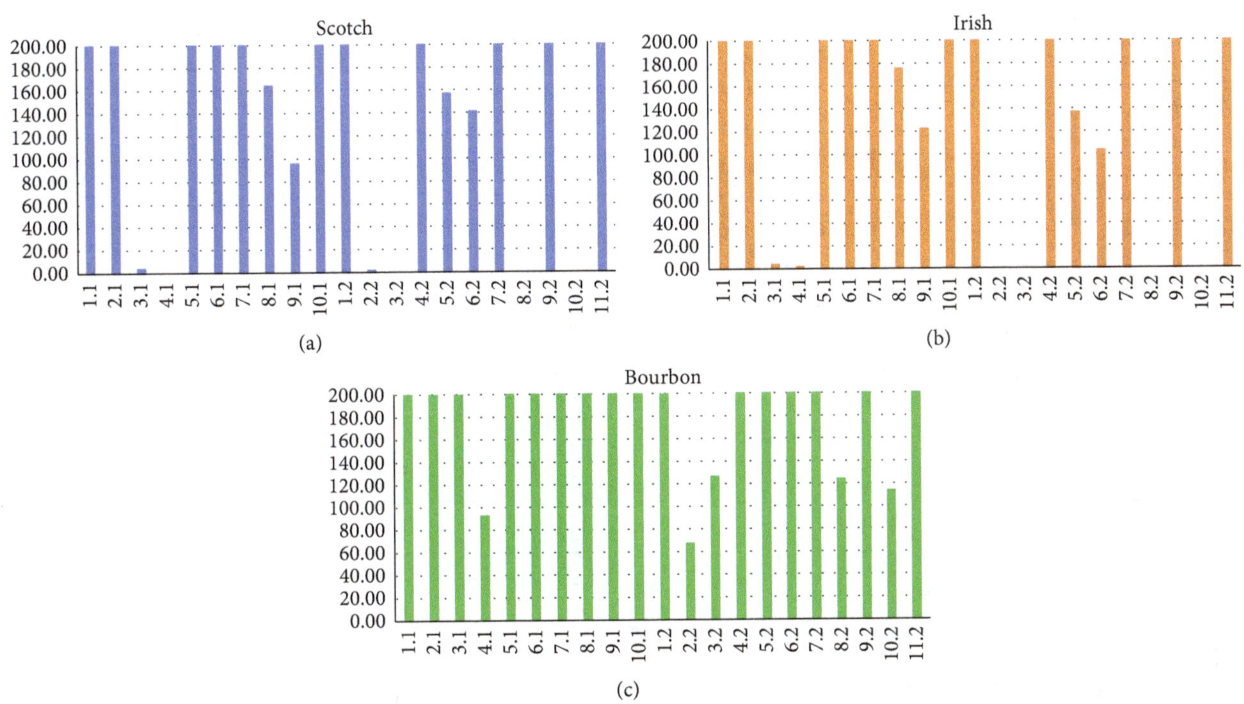


FIGURE 7: Mean bar graphs of selected peak areas used as raw data in chemometrics representing key chemical compounds, which are important for discrimination of (a) Scotch whisky, (b) Irish whisky, and (c) bourbon.

TABLE 3: Calculated values of Euclidean distances and pattern discrimination indices between groups of whisky samples according to DFA results.

| Relation between groups | | Distances | Pattern discrimination index (%) |
|---|---|---|---|
| Scotch | Bourbon | 421800.88 | 94.59 |
| Scotch | Irish | 48167.83 | 10.59 |
| Irish | Bourbon | 419440.47 | 89.00 |

Substances present in the sample were identified on the basis of comparison of calculated linear temperature-programmed retention indices (LTPRI) with indices included in the database. To compare this technique to the others, PLS-DA was used for data analysis (Figure 8). As input data, the areas of all detected peaks were used as variables. Preprocessing and cross-validation methods were the same as in the case of GC × GC analysis. As can be seen in Table 2, all samples were classified correctly.

During the sensory analysis, the panelists evaluated which type the tested whisky belonged to. PLS-DA was used for data analysis (Figure 9). Preprocessing and cross-validation methods were the same as in the case of other analyses. None of the groups was discriminated correctly, which can be seen in Table 2. However, this type of analysis

Table 4: Key chemical compounds (variables), which are important for sample discrimination.

| Number of variables-column number | Name of the compound | Scotch | Irish | Bourbon |
|---|---|---|---|---|
| 1-1 | 2-Propanol | + | + | + |
| 2-1 | n-Butanol | + | + | + |
| 3-1 | Furfural | + | + | + |
| 4-1 | Ethyl isovalerate | − | + | + |
| 5-1 | Alpha-pinene | + | + | + |
| 6-1 | Decanal | + | + | + |
| 7-1 | Eugenol | + | + | + |
| 8-1 | Geosmin | + | + | + |
| 9-1 | Gamma-decalactone | + | + | + |
| 10-1 | 12-Methyltridecanal | + | + | + |
| 1-2 | n-Butanol | + | + | + |
| 2-2 | Furfural | + | − | + |
| 3-2 | 1-Octanol | − | − | + |
| 4-2 | Decanal | + | + | + |
| 5-2 | 2,4-Nonadienal | + | + | + |
| 6-2 | Phenylethyl acetate | + | + | + |
| 7-2 | (E,E)-2,3-Decadienal | + | + | + |
| 8-2 | 4-Vinylguaiacol | − | − | + |
| 9-2 | Geosmin | + | + | + |
| 10-2 | Gamma-nonalactone | − | − | + |
| 11-2 | Gamma-decalatone | + | + | + |

"+" means that the compound was considered as variable for the chosen group.
"−" means that the compound was not considered as variable for the chosen group.

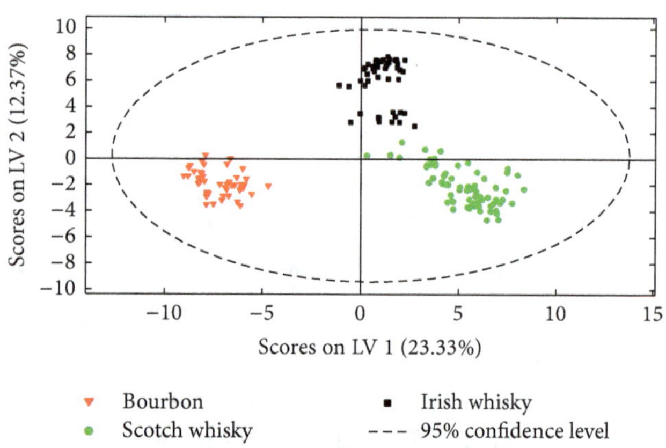


Figure 8: Scores plot of the first 2 factors of the PLS-DA model built with the area of the chosen peaks for e-nose analysis.

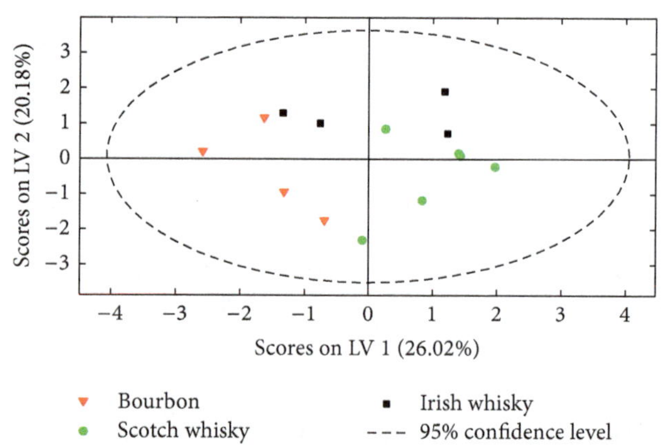


Figure 9: Scores plot of the first 2 factors of the PLS-DA model built for sensory analysis.

showed the preferences of consumers, who evaluated the taste and smell of Scotch whisky as the most desirable.

The classification results, both for calibration and for cross-validation data for all techniques, are compared in Table 5. Sensitivity is defined as the proportion of samples belonging to a given class that are correctly classified in that class. Specificity is the proportion of samples not belonging to a given class that are correctly rejected in that class. Classification error is the proportion of samples incorrectly classified.

GC × GC and e-nose analysis allow for discrimination of whisky due to the botanical origin and the way of production in contrast to sensory analysis. The quickest analysis was obtained by using e-nose. The most complex analysis of

TABLE 5: Classification results for Scotch whisky, bourbon, and Irish whisky.

| Modeled class | GC × GC | | | E-nose | | | Sensory analysis | | |
|---|---|---|---|---|---|---|---|---|---|
| | Bourbon | Scotch | Irish | Bourbon | Scotch | Irish | Bourbon | Scotch | Irish |
| Sensitivity (Cal) | 1.000 | 1.000 | 1.000 | 1.000 | 1.000 | 1.000 | 1.000 | 1.000 | 1.000 |
| Specificity (Cal) | 1.000 | 1.000 | 1.000 | 1.000 | 0.990 | 1.000 | 0.900 | 0.875 | 1.000 |
| Class. err. (Cal) | 0.000 | 0.000 | 0.000 | 0.000 | 0.005 | 0.000 | 0.050 | 0.062 | 0.000 |
| Sensitivity (CV) | 1.000 | 1.000 | 1.000 | 1.000 | 1.000 | 1.000 | 0.250 | 0.500 | 0.500 |
| Specificity (CV) | 1.000 | 1.000 | 1.000 | 1.000 | 0.990 | 0.992 | 0.700 | 0.625 | 0.800 |
| Class. err. (CV) | 0.000 | 0.000 | 0.000 | 0.000 | 0.005 | 0.004 | 0.525 | 0.438 | 0.350 |

TABLE 6: Comparison of GC × GC-TOFMS e-nose based on Fast-GC and sensory analysis for whisky samples discrimination.

| Type of analysis | Discrimination of whisky due to the type | Advantages | Disadvantages |
|---|---|---|---|
| GC × GC-TOFMS | Correct | It allows for precise examination of the composition of the matrix. | It is time-consuming and expensive. |
| E-nose based on Fast-GC | Correct | It allows for a partial analysis of the composition of the matrix. It is quick and relatively cheap. The software includes chemometric analysis. | It does not allow for precise examination of the composition of the matrix. |
| Sensory analysis | Incorrect | It allows for quality assessment due to consumer preferences. It is relatively cheap. | It is subjective and does not allow for discrimination of whisky based on the type. |

matrices was obtained by using GC × GC analysis. However, GC × GC analysis is too expensive to use for regular analysis of whisky. Sensory analysis was too subjective to obtain a correct classification of whisky by type, but it was the only one allowing for an evaluation of consumer preferences. The comparison of GC × GC-TOFMS, e-nose based on Fast-GC, and sensory analysis is shown in Table 6.

## 4. Conclusion

To sum up, the electronic nose based on the technology of ultrafast gas chromatography, comprehensive two-dimensional gas chromatography, and sensory evaluation was applied for distinguishing between selected whiskies of different botanical origins and different ways of production. GC × GC-TOFMS allowed for discrimination of whisky according to type. Apart from this, it allowed for identification of hundreds of compounds. Unfortunately, analysis of whisky by means of this technique is too expensive and it will take too long for it to be a routine evaluation of whisky. Sensory evaluation allowed access to people's preferences. It is a good tool for quality assessment but it does not allow for discrimination of whisky by type. Because of people, it can be also too subjective to ensure correct evaluation. An e-nose based on Fast-GC allows for good discrimination of whisky by type. Besides, it is not as subjective as sensory evaluation and the analysis is cheaper and less time-consuming than in the case of gas chromatography. This technique could be used for rapid and routine classification in industrial production, which will be the subject of further studies.

## Competing Interests

The authors declare that there are no competing interests regarding the publication of this paper.

## Acknowledgments

The authors acknowledge the financial support for this study by Grant no. 2012/05/B/ST4/01984 from the National Science Centre of Poland. The received funding did not lead to any competing interests regarding the publication of this manuscript.

## References

[1] Regulation (EC) No 110/2008 of the European Parliament and the Council of 15 January 2008 on the definition, description, presentation, labelling and protection of geographical indications of spirit drinks and repealing Council Regulation (EEC) No 1576/89.

[2] N. Christoph and C. Bauer-Christoph, "Vodka," in *Flavours and Fragrances Chemistry, Bioprocessing and Sustainability*, R. G. Berger, Ed., p. 231, Springer, Berlin, Germany, 2007.

[3] C. N. Rhodes, K. Heaton, I. Goodall, and P. A. Brereton, "Gas chromatography carbon isotope ratio mass spectrometry

applied to the detection of neutral alcohol in Scotch whisky: an internal reference approach," *Food Chemistry*, vol. 114, no. 2, pp. 697–701, 2009.

[4] C. Bamforth, "Whisky analysis," in *Whisky: Technology, Production and Marketing*, I. Russell and G. Stewart, Eds., pp. 275–306, Elsevier, London, UK, 2003.

[5] R. Ralph, "Production of American whiskies: bourbon, corn, rye and Tennessee," in *The Alcohol Textbook*, K. A. Jacques, T. P. Lyons, and D. R. Kelsall, Eds., pp. 275–285, Nottingham University Press, Nottingham, UK, 2003.

[6] S. Reigler, *Kentucky Bourbon Country: The Essential Travel Guide*, The University Press of Kentucky, Lexington, Ky, USA, 2013.

[7] P. J. S. Barbeira and N. R. Stradiotto, "Anodic stripping voltammetric determination of Zn, Pb and Cu traces in whisky samples," *Fresenius' Journal of Analytical Chemistry*, vol. 361, no. 5, pp. 507–509, 1998.

[8] M. P. Martí, O. Busto, J. Guasch, and R. Boqué, "Electronic noses in the quality control of alcoholic beverages," *TrAC - Trends in Analytical Chemistry*, vol. 24, no. 1, pp. 57–66, 2005.

[9] J. S. Garcia, B. G. Vaz, Y. E. Corilo et al., "Whisky analysis by electrospray ionization-Fourier transform mass spectrometry," *Food Research International*, vol. 51, no. 1, pp. 98–106, 2013.

[10] A. G. Mignani, L. Ciaccheri, B. Gordillo et al., "Identifying the production region of single-malt Scotch whiskies using optical spectroscopy and pattern recognition techniques," *Sensors and Actuators, B: Chemical*, vol. 171-172, pp. 458–462, 2012.

[11] P. C. Ashok, B. B. Praveen, and K. Dholakia, "Near infrared spectroscopic analysis of single malt Scotch whisky on an optofluidic chip," *Optics Express*, vol. 19, no. 23, pp. 22982–22992, 2011.

[12] R. I. Aylott, A. H. Clyne, A. P. Fox, and D. A. Walker, "Analytical strategies to confirm Scotch whisky authenticity," *Analyst*, vol. 119, no. 8, pp. 1741–1746, 1994.

[13] T. Adam, E. Duthie, and J. Feldmann, "Investigations into the use of copper and other metals as indicators for the authenticity of scotch whiskies," *Journal of the Institute of Brewing*, vol. 108, no. 4, pp. 459–464, 2002.

[14] I. G. Parker, S. D. Kelly, M. Sharman, M. J. Dennis, and D. Howie, "Investigation into the use of carbon isotope ratios (13C/12C) of Scotch whisky congeners to establish brand authenticity using gas chromatography-combustion-isotope ratio mass spectrometry," *Food Chemistry*, vol. 63, no. 3, pp. 423–428, 1998.

[15] K.-Y. M. Lee, A. Paterson, J. R. Piggott, and G. D. Richardson, "Sensory discrimination of blended Scotch whiskies of different product categories," *Food Quality and Preference*, vol. 12, no. 2, pp. 109–117, 2001.

[16] M. Śliwińska, P. Wiśniewska, T. Dymerski, J. Namieśnik, and W. Wardencki, "Food analysis using artificial senses," *Journal of Agricultural and Food Chemistry*, vol. 62, no. 7, pp. 1423–1448, 2014.

[17] Alpha MOS, http://www.alpha-mos.com/pdf/en/instruments/electronic_nose/Alpha-MOS_HERACLES_Electronic_Nose.pdf.

[18] P. Wiśniewska, M. Śliwińska, T. Dymerski, W. Wardencki, and J. Namieśnik, "Qualitative characteristics and comparison of volatile fraction of vodkas made from different botanical materials by comprehensive two-dimensional gas chromatography and the electronic nose based on the technology of ultra-fast gas chromatography," *Journal of the Science of Food and Agriculture*, vol. 97, no. 4, pp. 1316–1325, 2017.

# 11

# Analytical Methods for the Determination of the Distribution of Total Petroleum Hydrocarbons in the Water and Sediment of Aquatic Systems

**A. O. Adeniji,[1,2] O. O. Okoh,[1,2] and A. I. Okoh[1,3]**

[1]*SAMRC Microbial Water Quality Monitoring Centre, University of Fort Hare, Alice 5700, South Africa*
[2]*Department of Chemistry, University of Fort Hare, Alice 5700, South Africa*
[3]*Applied and Environmental Microbiology Research Group, Department of Biochemistry and Microbiology, University of Fort Hare, Alice 5700, South Africa*

Correspondence should be addressed to A. O. Adeniji; adenijigoke@gmail.com

Academic Editor: Ana Moldes

Several methods of extraction and analytical determination for total petroleum hydrocarbons (TPHCs) in aqueous and solid samples are reviewed. Infrared spectroscopy is one of the efficient methods that are being replaced today pursuant to getting rid of some halogenated solvents classified as ozone-depleting substances. The gravimetric method which uses n-hexane as an extraction solvent for the determination of oil and grease, as well as the nonpolar materials, has become a preferred choice, despite being not suitable for volatiles because of the mandatory evaporation step. Other frequently used methods include gas chromatography with a flame ionization detector (FID) or mass spectrometric detector (MSD) which has the capacity to reveal the type of hydrocarbons present and is applicable to both volatile and semivolatile samples. Ultraviolet fluorescence is another method that is available both as a portable field device and as off-site laboratory equipment. Each of the methods has its own advantages and disadvantages; hence, the choice of method is guided by the type of data needed as discussed in detail in this review. The distribution of TPHC in water and sediments across the globe and the factors influencing the distribution were also reviewed.

## 1. Introduction

Petroleum hydrocarbons (PHCs) are prominent among organic contaminants which are frequently disposed into the marine environment in lower concentrations in the form of urban runoffs, automobile wastes, stormwater, industrial effluents, or domestic wastes [1–3] and occasionally in higher concentrations as a major oil spill [4, 5]. The main constituents of these PHCs are degraded crude oils, combusted fossil fuels, and normal alkanes [6, 7]. They are generally less soluble in water but are readily adsorbed onto particulate matter and are subsequently scavenged to the bottom sediment which has become a reservoir for several hydrophobic contaminants. The danger in the sedimentation of these organic pollutants is their tendency to accumulate and bioconcentrate over time in aquatic organisms [8–10].

Total petroleum hydrocarbons (TPHCs) is the term generally used to describe the amount of petroleum-based hydrocarbon extracted and quantified by a particular method in an environmental matrix [11, 12]. TPHC analytical methods vary in scope to a great extent, with each providing results within a particular range, while some are nonspecific. Hence, understanding how the analysis was carried out is very crucial for correct interpretation of the results obtained. Methods with different extraction efficiencies could produce different TPHC concentrations for the same sample [13, 14]. Hence, results generated using different procedures are always difficult to compare because different calibration standards and extracting solvents are involved [6].

Petroleum hydrocarbons consist majorly of three groups of compounds, namely, alkanes, alkenes, and aromatics [15]. Alkanes are the key components of many refined petroleum products (e.g., gasoline, diesel fuel, kerosene, jet fuel, and heating oil) and are subdivided into several classes which include the following: linear alkanes (n-alkanes), branched alkanes, isoprenoids, cycloalkanes (e.g., steranes and triterpanes), and unresolved complex mixtures (UCM) which are difficult to resolve by capillary column gas chromatography [6, 16]. The presence of UCM in chromatograms is usually an indication of petroleum degradation. They are commonly found in the high-molecular-weight range of the n-alkanes but are sometimes present in the low molecular range as well [17, 18].

TPHC is sometimes determined as the sum of concentrations of n-alkanes and UCM [19] or as the total concentration of all hydrocarbons eluting from n-nonane (C9) to n-hexatriacontane (C36) excluding the priority polycyclic aromatic hydrocarbons [20]. They are mostly found in the higher-molecular-weight range of the hydrocarbons and sometimes also in the low-molecular-weight range [21]. TPHC could also be measured as the sum of aliphatic and aromatic hydrocarbons, especially those in the following groups: C9 to C18 (aliphatic), C19 to C36 (aliphatic), and C11 to C22 (aromatic) [22–24].

Total petroleum hydrocarbons (TPHCs) themselves may not be a direct pointer of hazard to humans or the environment [13, 25], but their presence indicates the health status of the water body and is also used for source tracking of the contaminants in the coastal waters and sediments [26, 27]. In this paper, we review the existing methods of extraction for petroleum hydrocarbon in water and solid samples, compare methods of final determination, and assess the levels of petroleum hydrocarbons in the aquatic environments across the globe and the available tools for source tracking using n-alkanes as reference compounds.

## 2. Physicochemical Properties

*2.1. Boiling Points.* Generally, the boiling point of hydrocarbons (HCs) increases as the number of carbon atoms increases. For n-alkanes, an increase of approximately 20°C is observed for every addition of new carbon atom to the chain. PAHCs are characterized by higher melting points and boiling points than their corresponding n-alkanes and therefore they are solid in nature [28].

*2.2. Solubility, Volatility, and Persistence.* The extent of interaction of hydrocarbon structures with water depends on their degree of polarity. Aliphatic hydrocarbons are generally less polar and are therefore less soluble in water compared to their corresponding aromatic counterparts. For instance, the solubility of aliphatic hydrocarbons in the range of C5-C6 is 36 mg/L while aromatic hydrocarbons in the same range have solubility as high as 220 mg/L. The most soluble components of the aromatic hydrocarbons are the lower-molecular-weight compounds. Therefore, solubility decreases as the number of the aromatic rings and molecular-weight increase. On the other hand, the aliphatic HCs are relatively more volatile than the aromatic HCs with a similar number of carbon atoms [29]. Nearly all the normal alkanes with carbon number less than 15 are lost by volatilization in the early days of spillage while those with carbon number greater than 20 are lost some weeks later [13, 30]. Lower weight hydrocarbons are less persistent in the environment unlike the high-molecular-weight hydrocarbons. Therefore, persistence of petroleum hydrocarbons rises as the boiling point of the compound increases [28].

*2.3. Toxicity and Microbial Degradation.* Aromatic hydrocarbons are considered to be relatively more toxic and more resistant to microbial degradation than the corresponding normal alkanes, isoalkanes, cycloalkanes, and the unsaturated aliphatic hydrocarbons [5, 31]. The degree of toxicity of contaminants in water will to a large extent depend on their level of solubility [28]. Microbial degradation of hydrocarbons is a function of many factors which include the level of dissolved oxygen, pH, and microbial population in the environment. The population of the degrading bacteria can be lower in an unpolluted environment but can drastically increase when in contact with spilled petroleum substances [32–36].

The rate of susceptibility of hydrocarbons to microbial degradation differs from one class to another [37, 38] and bacteria can degrade aliphatic, alicyclic, and aromatic hydrocarbons of low to moderate weights at a higher rate when the environmental conditions are very ideal. However, the rate of biodegradation decreases as the resistance due to the increase in the molecular weight of the compounds increases [39]. The typical decreasing order of biodegradation is given as follows: n-alkanes > branched alkanes > low-molecular-weight aromatics > cyclic alkanes > high-molecular-weight aromatics. Aromatic hydrocarbons are generally less biodegradable than the corresponding aliphatic hydrocarbons [5, 37, 38].

Oxygenases can convert alkanes to alcohol by the process of degradation. The alcohol can be oxidized to aldehyde and then to fatty acids which can subsequently be utilized in the $\beta$-oxidation of the aliphatic chain. In a similar manner, cycloalkanes can be degraded to cyclic alcohol by oxidases and then dehydrogenated to ketone. Similarly, aromatic hydrocarbons can be transformed to an epoxide through the action of oxygenases. The product can then be hydrated to *trans*-dihydrodiols which are later oxidized to catechols. Catechol can later be oxidized further to muconic acid or 2-hydroxymuconic semialdehyde [40] which can finally be degraded to yield organic acids that are used by microorganisms for cell synthesis and energy production [41].

## 3. Sources of Petroleum Hydrocarbons in the Marine Environment

Petroleum hydrocarbons in the marine environment come from natural (biogenic or diagenetic) and anthropogenic (petrogenic or pyrogenic) sources [42]. Biogenic source hydrocarbons refer to those hydrocarbons produced by living organisms (e.g., planktons, algae, and bacteria) or through biological processes involving terrestrial plants. Examples include C23–C33 aliphatic hydrocarbons which are largely produced by the terrestrial plant detritus and odd-numbered n-alkanes (C15, C17, and C19) from biogenic marine input which are usually available in low quantity [43]. Hydrocarbons from diagenetic sources are those produced as a result of physical and geochemical changes gone through by a sediment after its initial deposition as it changes into rock. They enter marine water mostly by natural seepage [44].

The anthropogenic hydrocarbons are usually more in the environment than the natural ones. They are commonly generated from various human activities like industrialization, urbanization, transportation, oil utilization, operations and storage, shipping, and fishing [45, 46]. Abundance of C16, C18, and C20 in most cases suggests oil spillage [18] from tanks and oil pipelines constructed to transfer crude oil from one place to another [44]. For anthropogenic derived hydrocarbons, the level of odd-numbered alkanes is almost equal to that of the even-numbered ones, usually in the ratio 1 : 1 [43]. In the early 1980s, it was concluded that approximately 10% of the crude oil in the marine environment comes from natural oil seeps, and another 27% comes from oil production, refining, and transportation, while the balance of 63% originates from urban and river runoff, atmospheric emission, and municipal and industrial discharges, among others [44].

## 4. Sample Collection and Preservation

Sample collection and preparation strategies are very crucial in the analysis of petroleum hydrocarbons in the environment [47]. The sample must be collected without contamination and must be wholly representative of the environmental medium where it was taken [11, 13]. The analyst must ensure that no bias or error is introduced into the samples during and after the collection. The sampling equipment must be thoroughly cleaned in between sampling points to avoid cross-contamination and the sample must be homogeneous. A minimum of one duplicate field sample is required in good practice for aqueous samples and space should be left for air above the sample to prevent overflow [47, 48].

The use of 1-liter amber glass bottles with Teflon-lined screw cap for aqueous sample collection is recommended. Samples are to be preserved at the time of collection with concentrated HCl or $H_2SO_4$ to pH < 2 and held at 4°C immediately after collection until extracted within 14 days of collection. However, unpreserved samples must be extracted within 7 days [23, 48, 49]. Soil or sediment samples are to be collected with a grab or core sampler or any other equipment as found appropriate in 120 mL wide-mouth glass jar or vessel of any other sizes, depending on the quantity of samples needed for the analysis [47]. The samples must be cooled to 4°C and extracted within 14 days of collection. Both the aqueous and the soil/sediment extracts must be analyzed within 40 days of extraction [11, 49]. A volume of 400 mL is the smallest aqueous sample size allowed for this analysis provided that the solvent/sample ratio is upheld [23].

Soil or sediment samples can be otherwise preserved at 4 ± 2°C at the time of collection and frozen thereafter at −10°C within 48 hours. The sample must be extracted not later than 14 days when thawed and analyzed within the next 40 days. The frozen samples can be held for up to a year before analysis if need be but must be extracted within 24 hours of defrosting [48].

## 5. Extraction and Cleanup Methods for TPHC in Water and Sediments

The most commonly used extraction methods for the isolation of the semivolatile compounds from water, soil, and sediment samples prior to analytical determination are discussed below.

*5.1. Water Samples.* The most commonly used extraction method for TPHC in aqueous samples is the EPA 3510 liquid-liquid extraction [5, 50, 51]. Other available methods include continuous liquid-liquid extraction [11], solid phase extraction [51], vortex extraction [52], and microextraction [53].

The liquid-liquid extraction (LLE) technique involves the use of a separatory funnel for multiple extraction of the sample with an organic solvent of choice with the intent of achieving good recovery of the desired analytes. The technique requires strong shaking of the liquid sample with an organic solvent in a funnel shaped piece of glassware for a couple of minutes, followed by the separation of the organic layer (extract) from the aqueous phase. The extract is then filtered through a desiccant to remove moisture before a final concentration prior to the analytical procedure. The method is however labor-intensive and solvent consuming [11, 54]. Another method is the EPA 3520, the continuous liquid-liquid extraction (CLLE) which involves continuous recycling of fresh organic solvent in the form of droplets which are passed through the water sample. It is very efficient, requires less attention of the operator, and consumes very little volume of solvent. CLLE is very good for samples containing emulsion-forming solids but has few disadvantages. It consumes more time (18–24 hours) than the conventional LLE (EPA 3510) and has a tendency to degrade the thermally unstable compounds, as well as lose some volatile ones in the process [11] (http://www.chemistry.sc.chula.ac.th/course_info/2302548/Wk3.pdf).

Solid phase extraction (SPE, EPA Method 3535) involves passing of a water sample through a cartridge containing an adsorbent like silica or alumina in which analytes are retained and afterwards eluted with a small portion of organic solvent [11, 55]. It has better extraction efficiency and uses lesser solvent and glassware compared to LLE and CLLE. Sample transfer in SPE is very minimal, guaranteeing little or no cross-contamination [56]. Many laboratories are still using LLE and SPE today, even though they are time-consuming, because they are relatively inexpensive, simple, and highly efficient [57].

*5.2. Solvent Selection.* The choice of solvent used for organic extraction in relation to the type of material extracted determines to a very large extent the efficiency of such an extraction process. A good extracting solvent should have low boiling point, low miscibility with water, and high solubility for the desired analytes [58]. Other criteria to be considered in the choice of solvents include cost, toxicity, spectral qualities, availability, method regulations, and extraction efficiency [11]. Organic contaminants were in time past extracted with halogenated solvents like Freon 113 (trichlorotrifluoroethane) which were less toxic to humans and possess good infrared spectral qualities. However, they are presently being phased out globally because of their implication in ozone layer depletion. Possible replacements for these solvents include dichloromethane, pentane, n-hexane, tetrachloroethene, and carbon tetrachloride [11, 13, 59].

Extraction efficiency of the organic analytes in aqueous samples is frequently monitored by the addition of surrogate standard solution (e.g., 1-chlorooctadecane for the n-alkanes; 2-fluorobiphenyl for the PAHs) to the sample matrix before extraction. Many laboratories are also performing matrix spikes with the intent of determining whether the compounds of interest are retained in the water, soil, or sediment matrices [60].

*5.3. Solid Samples.* The most frequently used methods for the extraction of soil and sediment samples prior to their analytical determination include conventional Soxhlet extraction [27, 61], ultrasonication [52, 62], vortex/mechanical shaking [63], accelerated solvent extraction [5, 64], and supercritical fluid extraction. Extraction efficiency for all these methods solely depends on the sample and solvent matrix [11].

Mechanical shaking is quite easy and suitable for petroleum hydrocarbon extraction from various environmental samples [65, 66]. It involves shaking the sample with a solvent. Desiccant (e.g., $Na_2SO_4$) is sometimes added to remove humidity and facilitate the breaking of the sample, thereby increasing its surface area for excellent contact with the extracting solvent. However, it is not an approved standard method by the USEPA because its efficiency depends on the type of sample [11]. Soxhlet extraction (EPA Method SW-846 3540) is the most commonly used and it is quite efficient for the extraction of the semivolatiles. It involves heating and refluxing of the sample with a solvent continuously for a long period of time (16–48 hours or more) but generates a large volume of extract for concentration and it is less appropriate for the volatiles [67, 68].

Ultrasonic extraction (EPA Method SW-846 3550) method is also very appropriate for the semivolatiles. It uses sound waves (ultrasonic vibrations) to enhance the transfer of the desired compounds from the sample to the solvent. It is a faster method than Soxhlet extraction and requires a relatively lesser amount of solvent [11]. However, it is labor-intensive in that it requires multiple extractions, decanting, and filtration for every sample processed. Accelerated solvent extraction (ASE) is another method for the extraction of semivolatile organic compounds. It is otherwise referred to as subcritical fluid extraction and has reportedly met the requirements for EPA Method 3545 (pressurized fluid extraction) in that it has wider coverage in its application and a good extraction efficiency equivalent to Soxhlet extraction method [67].

ASE uses organic solvents like dichloromethane for its operation. The solvent is heated and pressurized, before being passed through the sample for quick and efficient removal of the desired compounds from the sample matrix [11, 67]. The method requires a shorter extraction period and smaller volume of solvent compared to the conventional methods and, thus, it is a good alternative to both Soxhlet and ultrasonic extraction [23, 68]. However, it is an expensive technique and has the tendency of extracting many interferences together with the desired analytes because of the high temperature it uses for its operation. Therefore, column cleanup may be required after the extraction process before the analytical determination. Moreover, the extraction process is though shorter compared to Soxhlet extraction method but the preparation time for the extraction cells is long and rinsing consumes a lot of solvent [68, 69].

Supercritical fluid extraction (SFE) is a standard EPA method for total recoverable petroleum hydrocarbons (3540) and polycyclic aromatic hydrocarbons (3561). It involves heating and pressuring of the mobile phase (e.g., $CO_2$) to supercritical levels, whereby it possesses the properties of both gas and liquid (supercritical fluid). The fluid is then passed through the sample for the concentration of the analytes on a particular sorbent, after which they are eluted with a solvent and analyzed conventionally [11].

*5.4. Column Cleanup and Separation.* Column cleanup is a process used to separate organic analytes from their interfering compounds of different polarity which were possibly coextracted along with the analytes [70]. It is a very important aspect of infrared-based and gravimetric methods because they are very sensitive to nonpetroleum interferences. It is however not commonly utilized in gas chromatography-based methods because the presence of interfering compounds is not difficult to identify by experienced analysts [11].

Such interferences include the following:

(i) Natural organics that are not related to petroleum, for example, animal and vegetable derived hydrocarbons.

(ii) Oxygenated metabolites from biodegraded petroleum origin, for example, organic acids, alcohols, ketones, aldehydes, and phenols.

(iii) Artifacts from laboratory or sampling equipment, for example, phthalates.

(iv) Non-petroleum-related chemicals, for example, pesticides and chlorinated solvents [70].

In the infrared-based and gravimetric category, a small quantity of silica gel is added to the extract vial and shaken to remove the polar compounds before the analysis whereas, in the GC methods, the adsorbent (e.g., silica gel) is packed into a glass column. The sample extract is then run through the column which retains the polar substances while the compounds of interest are eluted with an appropriate solvent. This is followed by reconcentration of the eluates and analysis [70]. Aside from the cleaning of the extract, silica gel column also fractionates the petroleum hydrocarbon extract into aliphatic and aromatic fractions for separate determination [13, 48].

## 6. Analytical Procedures for TPHC in Water and Sediments

There are many conventional analytical methods available for the measurement of TPHC concentrations in various environmental samples. Some are more applicable for field screening while others are suitable for laboratory uses [11, 13, 71]. The most commonly used include infrared (IR) spectroscopy (EPA Method 418.1), gravimetry (EPA Method 1664A), gas chromatography with flame ionization detection (GC/FID) (EPA Method 8015), gas chromatography with mass spectrometric detection (GC/MSD) (EPA Methods 8270 and 625), ultraviolet spectrophotometry, immunoassay (IMA) (EPA Methods 4030 and 4035), Raman spectroscopy, and fluorescence spectroscopy [8, 52, 72, 73]. Liquid samples are mostly prepared for analytical determination by liquid-liquid extraction (LLE) or solid phase extraction (SPE) while solid samples are commonly made ready by Soxhlet extraction method [74]. This section will therefore present the strengths and weaknesses of some of the methods.

*6.1. Gravimetric Method.* There are basically two gravimetric EPA methods. These include EPA 413.1 for oil and grease and EPA 1664A for silica gel n-hexane extractable material (also regarded as total petroleum hydrocarbons). They both involve solvent extraction, evaporation, and weighing to a constant weight [52] but the oil and grease method does not differentiate the polar organic matter from the petroleum hydrocarbon portion [74]. EPA 1664A was developed to replace EPA 413.1 that uses Freon 113 (a Class I ozone-depleting substance) as its extracting solvent. The new method performs extraction with n-hexane which is more environmentally friendly but flammable and could determine total petroleum hydrocarbon concentration in the same procedure for the oil and grease analysis [75].

EPA 1664A procedure begins with adjustment of the aqueous sample pH to a value less than 2 with either $H_2SO_4$ or HCl. The sample is then extracted with n-hexane using LLE (or SPE) method and the organic portion is filtered through a funnel containing anhydrous $Na_2SO_4$, after which the solvent is distilled off from the flask at 85°C in a water bath. The flask is then dried and weighed to constant weight for the estimation of the oil and grease concentration [75]. For the TPHC determination, the oil and grease in the flask is dissolved in fresh portion of n-hexane. A suitable amount of silica gel is added to remove the polar substances (e.g., fats, dyes, oils, waxes, and some heavier organic molecules from animals and plants origin) in the extract and the mixture is stirred using a magnetic stirrer, filtered, and gravimetrically determined as nonpolar material (NPM), also known as silica gel treated n-hexane extractable material (SGT-HEM) [74, 76].

Solid samples (sludge, soil, and sediment) are catered for in the EPA SW-846 Test Method 9071B in which a wet sample is to be acidified with concentrated HCl. Acidified sludge sample is dried with either anhydrous sodium sulphate or magnesium sulphate and soil or sediment sample with sodium sulphate prior to the Soxhlet extraction with n-hexane. The NPM and SGT-HEM in the sample are thereafter gravimetrically determined as earlier described in the EPA 1664 procedures [77]. Other gravimetric methods reported for solid samples involved the addition of 3 g KOH (to enhance the extraction of the humic substances) and 100 $cm^3$ methanol to about 100 g air-dried samples prior to the Soxhlet extraction process. The subsequent steps involving gravimetric measurement of the polar and nonpolar compounds are as discussed above [78–80].

Gravimetric determination of TPHC in environmental samples is quite simple and affordable and has thereby gained a very huge patronage over the years because it does not require any special equipment. However, it is more applicable to samples with high concentration of analytes because of its low sensitivity. It is time-consuming, especially at the solvent evaporation stage, and has great tendency of losing the low-molecular-weight compounds, thereby introducing errors into the measurement [58, 73]. The extraction efficiency of the solvent is relatively low for the higher-molecular-weight compounds but can be reasonably enhanced with the addition of salt (sodium chloride). The coextraction of some organic matter with many polar functional groups is also encouraged by the low polarity of the solvent [81, 82].

*6.2. Infrared Spectrophotometry.* Similarly, two infrared (IR) based methods are in use for petroleum hydrocarbon related measurements. They include EPA 413.2 for total recoverable oil and grease (TOG) and EPA 418.1 for total recoverable petroleum hydrocarbons (TRPHCs) [13]. The oil and grease method is limited in application unlike EPA 418.1 or the modified EPA 801.5 which was widely accepted as a standard method for the assessment of hydrocarbon levels in the petroleum contaminated sites [13, 46]. The method is capable of determining different classes of hydrocarbon as either oil and grease (OG), gasoline range organics (GRO), or total petroleum hydrocarbons (TPHCs) [46]. It uses light absorption of the $CH_2$ bond at 2930 $cm^{-1}$ which is characteristic of the oil and grease for its measurement [58].

EPA 418.1 involves acidification of the water sample to pH $< 2$, extraction of the analytes from the sample matrix using 1,1,2-trichloro-1,2,2-trifluoroethane (Freon 113) or tetrachloromethane ($CCl_4$), and filtration of the extract through anhydrous sodium sulphate for moisture removal. The filtered extract is then made up to 100 mL with the extraction solvent and then stirred with a sufficient quantity of silica gel for the removal of the fatty matter before the absorbance measurement from 3200 $cm^{-1}$ to 2700 $cm^{-1}$. The final reading is then compared with the absorbance of a reference oil standard [13, 75]. The advantages of the method include elimination of the solvent evaporation step, making it more appropriate for the volatile components of petroleum [58, 79]. The measurement is very reproducible and more accurate, saves time, and is relatively easier [82]. Generally, IR methods give higher results than gravimetric methods, possibly because of the loss of volatile components due to the evaporation step in the latter [58].

However, IR readings sometimes may not suggest petroleum contamination in water or sediment sample because some compounds from nonpetroleum origin (e.g., humic or fulvic acid) may generate strong absorbance in the infrared range, yielding a false positive result especially when the cleanup step is omitted [52, 59]. It is also limited in its application as no adequate information is provided on the type of hydrocarbon present. The halogenated solvents (Freon, $CCl_4$) used for extraction in this method are though relatively nontoxic and nonflammable but are primary sources of ozone-layer-depleting substances which were required to be phased out in the Clean Air Act Amendments of 1990 [46]. EPA 418.1 is therefore being gradually replaced with EPA gravimetric (EPA 1664) and/or gas chromatographic (e.g., ISO 9377-2 or 8015D) method [52, 76, 82].

*6.3. Gas Chromatography.* The revised EPA 8015D called "Non-Halogenated Organics Using GC/FID" was specifically developed to take care of volatile and semivolatile organics separately as gasoline range organics (GRO) in the boiling point range of 60°C–170°C (C6–C10) and diesel range organics (DRO) in range of 170°C–430°C (C10 to C28). The method has the ability to provide some necessary information on the product type unlike the gravimetric and IR methods [13]. The volatile samples are usually introduced into the gas chromatograph (GC) by purge and trap method (EPA 5030 for aqueous or EPA 5035 for solid sample) [52, 83].

Aqueous semivolatile samples are mostly prepared by liquid-liquid extraction (EPA 3510) technique, although solid phase extraction (EPA 3535) may as well be used. Solid samples are usually extracted with Soxhlet device (EPA 3540). The extracts are subjected to either alumina (EPA 3611) or silica gel (EPA 3630) column cleanup with which the fatty matter is removed on the basis of polarity and petroleum hydrocarbons are fractionated into aliphatic and aromatic fractions. Each fraction is then injected directly into a gas chromatograph equipped with a nonpolar capillary column for analytical determination [11, 83]. Identification of analytes is done by comparing the retention time of an individual compound to that of a reference standard. Calibration standards are used to prepare calibration curves with which the compounds of interest are quantified [11].

The aliphatic and aromatic fractions are analyzed separately on the gas chromatograph using a flame ionization detector (GC-FID) which is mostly preferred for its relative sensitivity and selectivity for hydrocarbons. The two fractions are usually divided into carbon number ranges (GRO and DRO) on the basis of the PAH and n-alkane markers, respectively [52, 84]. The areas covered by the resolved and unresolved (UCM) components are most commonly estimated to determine the total petroleum hydrocarbon (TPHC) level in the sample [84–86]. Unfractionated extract can also be quantified as the oil and grease, yielding a concentration greater than the sum of the aliphatic and aromatic fractions [84].

Total hydrocarbons are sometimes measured with FID while the aromatic fraction is determined with a photoionization detector (PID). The aliphatic fraction is therefore estimated by subtracting the aromatic from the total hydrocarbon result. However, PID is not entirely selective for aromatic compounds and therefore can overestimate [13]. Alternatively, GC coupled with mass spectrometric detector (MSD), especially in the selected ion monitoring (SIM) mode, can produce a better result in the analysis of aromatic compounds. This is because it has the capacity to detect a compound present at an extremely low concentration in a complex matrix, provided that the desired compound has a prominent fragment ion at a mass that differentiates it from other hydrocarbon compounds and can also offer more comprehensive information about the identity of the compound [52, 72].

With GC methods, the identity of the fresh to mildly degraded oil can be unraveled in the environmental samples for pattern recognition [86]. The profile of unresolved complex mixtures (UCM) in the contaminated samples can be adequately characterized and resolved [87]. But the cost of analysis, operational time required, and sample matrix effects may be higher [88]. Overlap of carbon number ranges could be another challenge because a section of the GRO may be reported as DRO and vice versa [13, 52].

*6.4. Ultraviolet Fluorescence Spectrophotometry.* Ultraviolet fluorescence (UVF) spectrophotometry is a fast and valuable screening method for TPHC analysis in the environmental samples. It involves solvent extraction and optical measurement of the hydrocarbon molecules using an ultraviolet fluorescence spectrometer. It is always used in the field for aqueous samples without solvent extraction [58, 74, 89]. The aromatic hydrocarbons fluoresce when their $\pi$-bonds are excited by ultraviolet light and the hydrocarbon concentration in the sample extracts is measured using a fluorometer [74]. The technique has greater potential to distinguish low-molecular-weight aromatic hydrocarbons with 2 to 3 rings from the high-molecular-weight compounds [11, 58]. North Carolina in the United States has recently approved the method for monitoring spillage from leaking underground storage tanks as a replacement for the GC/FID (EPA 8015D) method [90]. However, aliphatic alkanes do not respond to ultraviolet radiation because they do not have *$\pi$-bonds*. Therefore, samples containing paraffinic oil cannot be detected by this method [74].

For reliable analytical results, a calibration curve must be prepared using specific hydrocarbon contaminants. Analytical recovery of UVF was reportedly good with n-hexane (132%) as an extraction solvent compared with methanol (40%) [74] but the accuracy and reproducibility at a low concentration range were found to be relatively low [82]. It is not a suitable method for fingerprint analysis because nonhydrocarbon compounds can also emit at the same wavelength as the analytes, thus causing interference [89].

## 7. Levels of Total Petroleum Hydrocarbons in the Environment

The total petroleum hydrocarbon results published from different countries across the world are presented in Tables 1 and 2. The levels reported in water matrix ranged from "not detected" (ND) to 260,000 $\mu$g/L. The highest values (60,000–260,000 $\mu$g/L) were recorded in the Deepwater Horizon, Gulf of Mexico [91], while the lowest values (0.05–4.13 $\mu$g/L) were found in the Gulf of Thailand and East Coast of Peninsular Malaysia [6]. The first world immense accidental blowout occurred in June 1979 at the Ixtoc-I most productive oil well in the Gulf of Mexico where more than 3.4 million barrels were spilled [92]. This might be responsible for the higher level of TPHC measured from the

Table 1: Levels of TPHC in water samples around the world (μg/L).

| Sample source | Concentrations reported | Analytical method | References |
|---|---|---|---|
| | *Africa* | | |
| Groundwater samples collected in some communities in Rivers State, Nigeria | 33,076 (max) | GC/FID | [63] |
| Ubeji surface water, Warri, Nigeria | 73,500 | Gravimetry | [96] |
| Surface water from the vicinity of Nigerian National Petroleum Corporations (NNPC) Oil Depot in Apata, Ibadan Metropolis, Nigeria | 20,340–27,400 | Gravimetry | [110] |
| Groundwater in communities in Niger Delta region of Nigeria | 1,352–12,110 | GC-FID | [111] |
| | *Asia* | | |
| Strait of Johor, in Peninsular Malaysia | 25–2,795 | — | [112] |
| Bohai Bay, China | 24–508 | — | [109] |
| Main Outfall Drain in Al-Nassiriya City, Southern Iraq | 0.01–8.76 | UVF | [113] |
| Setiu Wetlands, Terengganu, Malaysia | 4–121 | UVF | [114] |
| Coastal waters from Malaysian west coast | 5–386 | — | [112] |
| Dungun River basin water, Malaysia | 12–41 | UVF | [50] |
| Gulf of Thailand and East Coast of Peninsular Malaysia | 0.05–4.13 | UVF | [6] |
| Levantine Basin, Israeli Coastline | 19–88 | — | [115] |
| Northwestern Arabian Gulf | 1.2–546 | — | [116] |
| | *America* | | |
| Jamaica Bay, New York Metropolitan Area | 10–20 | GC/FID, UVF & GC/MS | [117] |
| Deepwater Horizon, Gulf of Mexico | 60,000–260,000 | — | [91] |
| Ixtoc-I Blowout, Gulf of Mexico | 5–106,000 | — | [93] |
| | *Europe* | | |
| Seawater of North Cape (Norway) | 483 (mean) | GC/MS | [72] |
| Romanian Black Sea Sector | 29–271 | UVF | [118] |
| Kara Sea | 8–408 | — | [119] |
| Pechora Sea | 2–87 | — | [120] |
| | *Australia* | | |
| Yellow Water, Kakadu National Park | ND–200 | GC/FID | [121] |

site some years later [93]. Another spill in the region from an oil well in the Deepwater Horizon popularly referred to as "the worst environmental disaster in the history of the US" claimed not less than 11 lives in April 2010. About 4.2 million barrels were reportedly spilled in the ugly incidence that lasted for about 87 days in which British Petroleum (BP) was implicated, hence the reason for higher TPHC level in the study conducted at the site about three years later [91, 94].

The European Union's target value for TPHC in estuary and harbor water is 300 μg/L. This value was greatly exceeded in many water bodies across the world, signifying anthropogenic input of hydrocarbons into the environments [95]. With reference to the collated data as shown in Table 1, TPHC levels were exceptionally higher in the Gulf of Mexico and the Niger Delta region of Nigeria, where petroleum exploration activities are being undertaken, than in other regions [63, 91, 93, 96].

Table 2 showed that the TPHC levels in the sediment samples analyzed across the world varied from ND (not detected) to 16,500 mg/kg. The lowest values (ND–1.71 mg/kg) were recorded from the coastline and mangroves of the Northern Persian Gulf [97] and the highest ones (300–16,500 mg/kg) were recorded from the Pearl River Estuary, China [98]. The minimum value of 10 mg/kg stipulated by the United Nations Environment Programme was exceeded in most countries, suggesting increasing petroleum hydrocarbon pollution in the fresh and marine aquatic environments across the globe. Urban-industrial activities can increase the concentration to between 10 and 1,000 mg/kg while levels >3,000 mg/kg will usually be found in oil spill sites [4, 99, 100]. This indicates a greater influence of industrial activities as well as oil spillage on the petroleum hydrocarbon levels in most of the coastal sediments all over the world. The levels were extremely higher in some estuaries in China [56, 98, 101–103], Ceuta Harbor in North Africa [104], São Gonçalo Channel in Pelotas, Rio Grande do Sul, Brazil [22], and Clyde Estuary, United Kingdom [105]. TPHC was generally much higher in the sediment than in the water column.

TABLE 2: Levels of TPHC in sediment samples around the world (mg/kg).

| Sample source | Concentrations reported | Analytical method | References |
|---|---|---|---|
| | *Africa* | | |
| Ceuta Harbor, North Africa | 496–6,972 | Gravimetry | [104] |
| Soil in communities in Niger Delta region of Nigeria | 1,242–5,200 | GC-FID | [111] |
| Bizerte Lagoon, Tunisia | 0.05–20 | — | [122, 123] |
| | *Asia* | | |
| Arabian Gulf | 444 (mean) | UVF | [124] |
| Bohai Bay, China | 6.3–535 | UVF | [125] |
| Arabian Gulf sediment, Kuwait | 50–1,122 | UVF | [78] |
| Musa Bay sediments | 17–97 | UVF & GC/MS | [126] |
| Khowr-e Musa Bay | 18–89 | UVF & GC/MS | [127] |
| Coastal Area of Putatan and Papar, Sabah | 0.26–4.59 | UV/VIS | [128] |
| Main Outfall Drain in Al-Nassiriya City, Southern Iraq | 0.09–5.70 | UVF | [113] |
| Coastline and mangroves of the Northern Persian Gulf | ND–1.71 | UVF | [97] |
| Southeast Coast of India | 2.2–5.6 | UVF | [8] |
| Koh Sichang-Sriracha, Thailand | 1.1–153.4 | UVF | [99] |
| Saudi Arabia | 62–1,400 | | [129] |
| Changjiang Estuary, China | 2,200–11,820 | — | [56, 101] |
| Pearl River Estuary, China | 300–16,500 | — | [98, 102, 103] |
| | *America* | | |
| São Gonçalo Channel, Pelotas, Rio Grande do Sul, Brazil | 177–5,893 | Gravimetry/GC/MS | [22] |
| Barnegat-Bay-Little Egg Harbor Estuary, USA | 47–1,003 | Iatroscan Mk6s/FID | [130] |
| Galveston Bay, USA | 4.2–1,814 | — | [131, 132] |
| Todos os Santos Bay, Brazil | 1.6–11 | GC/MS | [133] |
| Trinidad Coast | 3.0–1,824 | UVF | [134] |
| | *Europe* | | |
| Clyde Estuary, UK | 34–4,386 | Iatroscan Mk6s/FID | [105] |

General assessment of the analytical methods employed by researchers for the determination of total petroleum hydrocarbons in both water and sediment matrices as shown in Tables 1 and 2 revealed greater use of UVF which was quicker and easier than other existing methods [106]. Although the equipment is mostly available as a portable field device, any data obtained should often be validated with off-site laboratory equipment for better quality and reliability [107]. The IR method (EPA 418.1) is gradually fading away because of the mandated halogenated solvents which are environmentally unfriendly, as gas chromatographic and gravimetric (EPA 1664) procedures are now increasing in patronage [108]. Total petroleum hydrocarbon is more determined in the Asian environment than in other continents of the world probably because of the great abundance of crude oil in many countries in the region. It is very obvious that spillage of oil, industrial effluent discharge, and urban runoff are the major reasons for higher TPHC levels in marine environment and therefore steps should be taken to minimize these to the barest minimum across the globe [109].

## 8. Conclusion

Although solid phase extraction has a better extraction efficiency than LLE, its application in the petroleum hydrocarbon extraction from the water matrix is limited because it is yet to be validated by the United States Environmental Protection Agency for this particular analysis. Therefore, LLE remains the most commonly used method because of its affordability, simplicity, and relative efficiency. Similarly, the solid samples are more commonly extracted with Soxhlet extractor which is time-consuming compared to the accelerated solvent extraction because it is easy to use and quite cheaper. Aliphatic hydrocarbons which are the primary components of petroleum hydrocarbons are mostly more determined in the marine sediment than in the water column because they have relatively low solubility in water and thereby concentrate in the bottom sediment.

Many analytical methods are available for the TPHC measurement in the environmental samples. Some measure TPHC on the basis of the aliphatic or aromatic fraction alone while few others combine the two fractions for the measurement. The results taken using different methods are therefore in most cases incomparable owing to their different background. However, off-site laboratory equipment is more preferred for the analytical determination of the analytes than the portable field devices because it generates good data which are much more reliable. The TPHC concentrations were found to be extremely higher in the spillage sites and greatly industrialized areas. Hence, there is a greater need for compliance with environmental laws as regards the oil exploration and transportation, as well as waste discharge into the marine environment in order to ensure safety of aquatic and human lives.

## Conflicts of Interest

The authors declare that there are no conflicts of interest regarding the publication of this paper.

## References


[1] V. Hatje, T. E. Payne, D. M. Hill, G. McOrist, G. F. Birch, and R. Szymczak, "Kinetics of trace element uptake and release by particles in estuarine waters: effects of pH, salinity, and particle loading," *Environment International*, vol. 29, no. 5, pp. 619–629, 2003.

[2] C. Olver, *Report on Issues Relating to the Potential Relocation of the Port Elizabeth Manganese Terminal and Tank Farm to the Port of Ngqura*, Port Elizabeth Regional Chamber of Commerce, Port Elizabeth, South Africa, 2008.

[3] River Health Programme (RHP), *State-of-Rivers Report: Buffalo River System*, Department of Water Affairs and Forestry, Pretoria, South Africa, 2004.

[4] G. M. M. Tehrani, R. Hshim, S. H. Halim, S. B. Tavakoly Sany, R. Khani jazani, and Z. Monazami Tehrani, "Assessment of contamination by petroleum hydrocarbons in sediments of Musa Bay, Northwest of the Persian Gulf-Iran," *International Proceedings of Chemical, Biological and Environmental Engineering*, vol. 33, pp. 75–80, 2012.

[5] O. E. Ahmed, S. A. Mahmoud, and A. E. M. Mousa, "Aliphatic and poly-aromatic hydrocarbons pollution at the drainage basin of Suez Oil Refinery Company," *Current Science International*, vol. 4, no. 1, pp. 27–44, 2015.

[6] P. Wongnapapan, G. Wattayakorn, and A. Snidvongs, "Petroleum hydrocarbon in seawater and some sediments of the South China Sea, Area I: Gulf of Thailand and East Coast of Peninsular Malaysia," in *Proceedings of the 1st Technical Seminar on Marine Fishery Resources Survey in the South China Sea*, pp. 105–110, 1999.

[7] G. Wattayakorn, "Petroleum pollution in the Gulf of Thailand: a historical review," *Coastal Marine Science*, vol. 35, no. 1, pp. 234–245, 2012.

[8] A. Muthukumar, G. Idayachandiran, S. Kumaresan, T. A. Kumar, and T. Balasubramanian, "Petroleum hydrocarbons (PHC) in sediments of three different ecosystems from Southeast Coast of India," *International Journal of Pharmaceutical & Biological Archives*, vol. 4, no. 3, pp. 543–549, 2013.

[9] J. W. Farrington and H. Takada, "Persistent organic pollutants (POPs), polycyclic aromatic hydrocarbons (PAHs), and plastics: examples of the status, trend, and cycling of organic chemicals of environmental concern in the ocean," *Oceanography*, vol. 27, no. 1, pp. 196–213, 2014.

[10] S. A. M. Ali, C. Payus, and M. M. Ali, "Surface sediment analysis on petroleum hydrocarbon and total organic carbon from coastal area of papar to Tuaran, Sabah," *Malaysian Journal of Analytical Sciences*, vol. 19, no. 2, pp. 318–324, 2015.

[11] W. Weisman, Ed., *Analysis of Petroleum Hydrocarbons in Environmental Media (Total Petroleum Hydrocarbon Criteria Working Group Series)*, vol. 1, Amherst Scientific Publishers, Amherst, Mass, USA, 1998.

[12] Institute for Environmental and Coastal Management (IECM), "Port of Ngqura: marine biomonitoring programme. Annual Report 2002-2003," Tech. Rep. C86, Institute for Environmental and Coastal Management, Port Elizabeth, South Africa, 2003.

[13] Agency for Toxic Substances and Disease Registry (ATSDR), *Toxicological Profile for Total Petroleum Hydrocarbon*, US Department of Health and Human Services, Public Health Service, Atlanta, Ga, USA, 1999.

[14] J. G. Speight, *The Chemistry and Technology of Petroleum*, vol. 137 of *Chemical Industries*, CRC Press, 2014.

[15] P. E. Wigger and B. E. Torkelson, "Petroleum hydrocarbon fingerprinting: numerical interpretation developments," in *Proceedings of the International Petroleum Environmental Conference*, San Antonio, Tex, USA, 1997.

[16] M. A. de Abreu-Mota, C. A. de Moura Barboza, M. C. Bícego, and C. C. Martins, "Sedimentary biomarkers along a contamination gradient in a human-impacted sub-estuary in Southern Brazil: a multi-parameter approach based on spatial and seasonal variability," *Chemosphere*, vol. 103, pp. 156–163, 2014.

[17] A. T. Revill, M. R. Carr, and S. J. Rowland, "Use of oxidative degradation followed by capillary gas chromatography-mass spectrometry and multi-dimensional scaling analysis to fingerprint unresolved complex mixtures of hydrocarbons," *Journal of Chromatography A*, vol. 589, no. 1-2, pp. 281–286, 1992.

[18] M. Sakari, B. S. Hsia, R. Tahir et al., "Estuary and sea-associated wetlands as final sink for organic pollutants: a case study in Sabah, Malaysia," *International Journal of Environment and Bioenergy*, vol. 9, no. 1, pp. 1–16, 2014.

[19] R. R. Harji, A. Yvenat, and N. B. Bhosle, "Sources of hydrocarbons in sediments of the Mandovi estuary and the Marmugoa harbour, west coast of India," *Environment International*, vol. 34, no. 7, pp. 959–965, 2008.

[20] R. Klee, *Selection of Analytical Methods to Characterize Petroleum Releases Guidance Document*, Connecticut Department of Energy & Environmental Protection, 2015.

[21] M. A. Gouch, M. M. Rhead, and S. J. Rowland, "Biodegradation studies of unresolved complex mixtures of hydrocarbons: model UCM hydrocarbons and the aliphatic UCM," *Organic Geochemistry*, vol. 18, no. 1, pp. 17–22, 1992.

[22] P. J. Sanches Filho, E. M. Böhm, G. M. Böhm, G. O. Montenegro, L. A. Silveira, and G. R. Betemps, "Determination of hydrocarbons transported by urban runoff in sediments of São Gonçalo Channel (Pelotas—RS, Brazil)," *Marine Pollution Bulletin*, vol. 114, no. 2, pp. 1088–1095, 2017.

[23] Washington State Department of Ecology (WSDE), "Analytical methods for petroleum hydrocarbons," Publication no. ECY 97-602, Department of Ecology Publications Distribution Center, Olympia, Wash, USA, 1997.

[24] J. M. Guerra-García, F. J. González-Vila, and J. C. García-Gómez, "Aliphatic hydrocarbon pollution and macrobenthic assemblages in Ceuta harbour: a multivariate approach," *Marine Ecology Progress Series*, vol. 263, pp. 127–138, 2003.

[25] United States Department of Health and Human Services (USDHHS), *Toxicological Profile for Total Petroleum Hydrocarbons (TPH)*, Public Health Service Agency for Toxic Substances and Disease Registry, 1999.

[26] H. Zaghden, M. Kallel, A. Louati, B. Elleuch, J. Oudot, and A. Saliot, "Hydrocarbons in surface sediments from the Sfax coastal zone, (Tunisia) Mediterranean Sea," *Marine Pollution Bulletin*, vol. 50, no. 11, pp. 1287–1294, 2005.

[27] I. Zrafi, L. Hizem, H. Chalghmi, A. Ghrabi, M. Rouabhia, and D. Saidane-Mosbahi, "Aliphatic and aromatic biomarkers for petroleum hydrocarbon investigation in marine sediment," *Journal of Petroleum Science Research*, vol. 2, no. 4, pp. 145–155, 2013.

[28] World Health Organization (WHO), "Petroleum products in drinking-water: background document for development of WHO guidelines for drinking-water quality," Tech. Rep. WHO/SDE/WSH/05.08/123, 2005.

[29] H. I. Abdel-Shafy and M. S. M. Mansour, "A review on polycyclic aromatic hydrocarbons: source, environmental impact, effect on human health and remediation," *Egyptian Journal of Petroleum*, vol. 25, pp. 107–123, 2016.

[30] K. Scally, *The use of forensic polycyclic aromatic hydrocarbon signatures and compound ratio analysis techniques (CORAT) for the source characterisation of petrogenic/pyrogenic environmental releases [Master of Science Thesis]*, Higher Education and Training, Galway Mayo Institute of Technology, 2005.

[31] S. F. Mehanna and F. I. El-Gammal, "Gulf of Suez fisheries: current status, assessment and management," *Journal of King Abdulaziz University (Marine Sciences)*, vol. 18, pp. 3–18, 2007.

[32] A. I. Okoh, "Biodegradation of Bonny light crude oil in soil microcosm by some bacterial strains isolated from crude oil flow stations saver pits in Nigeria," *African Journal of Biotechnology*, vol. 2, no. 5, pp. 1014–1108, 2003.

[33] A. I. Okoh, A. O. Olaniran, and P. Golyshin, "Dechlorination of 1,2- Dichloroethane by *Pseudomonas aeruginosa* OK1 isolated from a waste dumpsite in Nigeria," *African Journal of Biotechnology*, vol. 3, no. 10, pp. 508–511, 2004.

[34] A. I. Okoh, "Biodegradation alternative in the cleanup of petroleum hydrocarbon pollutants," *Biotechnology and Molecular Biology Reviews*, vol. 1, no. 2, pp. 38–50, 2006.

[35] A. I. Okoh and M. R. Trejo-Hernandez, "Remediation of petroleum hydrocarbon polluted systems: exploiting the bioremediation strategies," *African Journal of Biotechnology*, vol. 5, no. 25, pp. 2520–2525, 2006.

[36] S. D. Williams, D. E. Ladd, and J. J. Farmer, "Fate and transport of petroleum hydrocarbons in soil and groundwater at big south fork National River and recreation area, Tennessee and Kentucky, 2002-2003," U.S. Geological Survey Scientific Investigations Report 2005-5104, 2006.

[37] P. Fusey and J. Oudot, "Relative influence of physical removal and biodegradation in the depuration of petroleum-contaminated seashore sediments," *Marine Pollution Bulletin*, vol. 15, no. 4, pp. 136–141, 1984.

[38] J. G. Leahy and R. R. Colwell, "Microbial degradation of hydrocarbons in the environment," *Microbiological Reviews*, vol. 54, no. 3, pp. 305–315, 1990.

[39] T. H. Wiedemeier, J. T. Wilson, D. H. Kampbell, R. N. Miller, and J. E. Hansen, *Technical Protocol for Implementing the Intrinsic Remediation with Long-term Monitoring Option for Natural Attenuation of Dissolved-phase Fuel Contamination in Groundwater*, U.S. Air Force Center for Environmental Excellence, Brooks Air Force Base, San Antonio, Tex, USA, 1995.

[40] C. E. Cerniglia, "Microbial transformation of aromatic hydrocarbons," in *Petroleum Microbiology*, R. M. Atlas, Ed., pp. 99–128, Macmillan Publishing Co, New York, NY, USA, 1984.

[41] V. Kothari, M. Panchal, and N. Srivastava, *Microbial Degradation of Hydrocarbons*, Institute of Science, Nirma University, Ahmedabad, India, 2013.

[42] Y. Wu, J. Zhang, T.-Z. Mi, and B. Li, "Occurrence of n-alkanes and polycyclic aromatic hydrocarbons in the core sediments of the Yellow Sea," *Marine Chemistry*, vol. 76, no. 1-2, pp. 1–15, 2001.

[43] M. Sakari, L. S. Ting, L. Y. Houng et al., "Urban effluent discharge into rivers; a forensic chemistry approach to evaluate the environmental deterioration," *World Applied Sciences Journal*, vol. 20, no. 9, pp. 1227–1235, 2012.

[44] K. A. Kvenvolden and C. K. Cooper, "Natural seepage of crude oil into the marine environment," *Geo-Marine Letters*, vol. 23, no. 3-4, pp. 140–146, 2003.

[45] M. Sakari, M. P. Zakaria, M. Junos et al., "Spatial distribution of petroleum hydrocarbon in sediments of major rivers from east coast of Peninsular Malaysia," *Coastal Marine Science*, vol. 32, pp. 1–10, 2008.

[46] T. Strother, S. Lowry, and B. Bravo, "Measurement of dispersed oil in water using an infrared analysis method," Application Note 52439, Thermo Fisher Scientific Inc, 2013.

[47] S. Buckland, *Draft Sampling Protocols and Analytical Methods for Determining Petroleum Products in Soil and Water*, Oil Industry Environmental Working Group, 1999.

[48] Kansas Department of Health and Environment (KDHE), *Kansas Method for the Determination of Mid-Range Hydrocarbons (MRH) and High-Range Hydrocarbons (HRH)*, Revision 1.0, 2015.

[49] ESS Laboratory, "Total petroleum hydrocarbons (SW 846 Method 8100 modified)," SOP No: 60_8100-mod. Revision 4: 23, 2008.

[50] S. Suratman, "Distribution of total petrogenic hydrocarbon in Dungun River basin, Malaysia," *Oriental Journal of Chemistry*, vol. 29, no. 1, pp. 77–80, 2013.

[51] J. P. Morris, *Residual diesel range organics and selected frothers in process waters from fine coal flotation [Master of Science Thesis in Mining and Minerals Engineering]*, Virginia Polytechnic Institute and State University, 2013.

[52] J. E. Cortes, A. Suspes, S. Roa, C. González, and H. E. Castro, "Total petroleum hydrocarbons by gas chromatography in Colombian waters and soils," *American Journal of Environmental Sciences*, vol. 8, no. 4, pp. 396–402, 2012.

[53] J. Ilavsky and J. Hrivnak, "Determination of petroleum hydrocarbons in water by microextraction and capillary gas chromatogarphy," *Slovakia Journal of Civil Engineering*, vol. 1, pp. 13–17, 2004.

[54] B. Johnson, "Determining trace amounts of contaminants in water," Water Conditioning and Purification, 2016, http://www.horizontechinc.com/PDF/WCP_article_2011.pdf.

[55] D. Gallagher, "Determining Diesel and Oil Range Organics in Water Using Automated Solid Phase Extraction," Horizon Technology, Inc. AN064-100625: 1–4, 2002.

[56] W. Zhang, H. Feng, J. Chang, J. Qu, H. Xie, and L. Yu, "Heavy metal contamination in surface sediments of Yangtze River intertidal zone: an assessment from different indexes," *Environmental Pollution*, vol. 157, no. 5, pp. 1533–1543, 2009.

[57] M. Fingas, Ed., *Handbook of Oil Spill Science and Technology*, John Wiley & Sons, Hoboken, NJ, USA, 2015.

[58] M. K. Stenstrom, S. Fam, and G. S. Silverman, "Analytical methods for quantitative and qualitative determination of hydrocarbons and oil and grease in water and wastewater," *Environmental Technology Letters*, vol. 7, no. 1-12, pp. 625–636, 1986.

[59] American Water Works Association (AWWA), *American Water Works Association*, American Water Works Association (AWWA), Denver, Colo, USA, 1st edition, 1998.

[60] J. G. Speight, *Lange's Handbook of Chemistry*, McGraw-Hill, New York, NY, USA, 6th edition, 2005.

[61] S. Filho, P. J. Luz, G. R. Betemps, M. D. R. G. Silva, and E. B. Caramao, "Studies of n-alkanes in the sediments of colony Z3 (Pelotas-RS-Brazil)," *Brazilian Journal of Aquatic Science and Technology*, vol. 17, no. 1, pp. 27–33, 2013.

[62] O. L. G. Maioli, K. C. Rodrigues, B. A. Knoppers, and D. A. Azevedo, "Distribution and sources of aliphatic and polycyclic aromatic hydrocarbons in suspended particulate matter in water from two Brazilian estuarine systems," *Continental Shelf Research*, vol. 31, no. 10, pp. 1116–1127, 2011.

[63] I. J. Alinnor and M. A. Nwachukwu, "Determination of total petroleum hydrocarbon in soil and groundwater samples in some communities in Rivers State, Nigeria," *Journal of Environmental Chemistry and Ecotoxicology*, vol. 5, no. 11, pp. 292–297, 2013.

[64] F. Galbiati and L. Teli, *Determination of Total Petroleum Hydrocarbons in Rubble and Soils by Accelerated Solvent Extraction and GC-FID*, Thermo Fisher Scientific, 2015.

[65] E. V. Lau, S. Gan, and H. K. Ng, "Extraction techniques for polycyclic aromatic hydrocarbons in soils," *International Journal of Analytical Chemistry*, vol. 2010, Article ID 398381, 9 pages, 2010.

[66] I. J. Okop and S. C. Ekpo, "Determination of total hydrocarbon content in soil after petroleum spillage," in *Proceedings of the World Congress on Engineering (WCE '12)*, vol. 3, London , UK, 2012.

[67] Dionex Corporation, "Extraction of total petroleum hydrocarbon contaminants (diesel and waste oil) in soils by accelerated solvent extraction (ASE)," Dionex Application Note 338, 2011.

[68] I. De Dobbeleer, F. Galbiati, and L. Teli, "Determination of total petroleum hydrocarbons in rubble and soils by accelerated solvent extraction and GC-FID," Poster Note 64829, Thermo Fisher Scientific, 2016.

[69] H. Giergielewicz-Możajska, Ł. Dąbrowski, and J. Namieśnik, "Accelerated solvent extraction (ASE) in the analysis of environmental solid samples—some aspects of theory and practice," *Critical Reviews in Analytical Chemistry*, vol. 31, no. 3, pp. 149–165, 2001.

[70] D. A. Zemo, K. A. Synowiec, R. I. Magaw, and R. E. Mohler, "Comparison of shake and column silica gel cleanup methods for groundwater extracts to be analyzed for TPHd/DRO," *Groundwater Monitoring and Remediation*, vol. 33, no. 4, pp. 108–112, 2013.

[71] R. Saddler and D. Connell, "Analytical methods for the determination of total petroleum hydrocarbons in soil," in *Proceedings of the 5th National Workshop on the Assessment of Site Contamination*, pp. 133–150, National Environment Protection Council Service Corporation, 2003.

[72] C. M. Reddy and J. G. Quinn, "GC-MS analysis of total petroleum hydrocarbons and polycyclic aromatic hydrocarbons in seawater samples after the North Cape oil spill," *Marine Pollution Bulletin*, vol. 38, no. 2, pp. 126–135, 1999.

[73] R. N. Okparanma and A. M. Mouazen, "Determination of total petroleum hydrocarbon (TPH) and polycyclic aromatic hydrocarbon (PAH) in soils: a review of spectroscopic and nonspectroscopic techniques," *Applied Spectroscopy Reviews*, vol. 48, no. 6, pp. 458–486, 2013.

[74] M. Matthew, *A comparison study of gravimetric and ultraviolet fluorescence methods for the analysis of total petroleum hydrocarbons in surface water [M.S. thesis]*, Northeastern University, Boston, Mass, USA, 2009.

[75] United States Environmental Protection Agency (USEPA), "Analytical method guidance for EPA method 1664a implementation and use (40 CFR part 136)," Office of Water 4303, 2000.

[76] United States Environmental Protection Agency (USEPA), "Method 1664, Revision A: n-Hexane extractable material (HEM; oil and grease) and silica gel treated n-hexane extractable material (SGTHEM; non-polar material) by extraction and gravimetry," Tech. Rep. EPA-821-R-98-002. PB99-121949, 1999.

[77] United States Environmental Protection Agency (USEPA), *Method 9071B: N-hexane Extractable Material (HEM) for Sludge, Sediment, and Solid Samples*, Revision 2, 1998.

[78] M. S. Massoud, F. Al-Abdali, A. N. Al-Ghadban, and M. Al-Sarawi, "Bottom sediments of the Arabian Gulf—II. TPH and TOC contents as indicators of oil pollution and implications for the effect and fate of the Kuwait oil slick," *Environmental Pollution*, vol. 93, no. 3, pp. 271–284, 1996.

[79] P. C. Onianwa, "Petroleum hydrocarbon pollution of urban topsoil in Ibadan City, Nigeria," *Environment International*, vol. 21, no. 3, pp. 341–343, 1995.

[80] A. A. Adeniyi and J. A. Afolabi, "Determination of total petroleum hydrocarbons and heavy metals in soils within the vicinity of facilities handling refined petroleum products in Lagos metropolis," *Environment International*, vol. 28, no. 1-2, pp. 79–82, 2002.

[81] M. E. Essington, *Soil and Water Chemistry: An Integrative Approach*, CRC Press, Boca Raton, Fla, USA, 2004.

[82] M. Martl and W. Ritter, "Brings back CFC-free IR measurement for oil-in-water," ASTM Standard D7678-11, QuantaRed Technologies GmbH, 2013, https://www.petro-online.com/.

[83] United States Environmental Protection Agency (USEPA), *Method 8015D: Non-Halogenated Organics Using GC/FID. Revision 4*, United States Environmental Protection Agency, Washington, DC, USA, 2003.

[84] British Colombia Ministry of Environment (BCME), "Aliphatic/aromatic separation of extractable petroleum hydrocarbons in solids or water by silica gel fractionation," in *Analytical Method 7 for Contaminated Sites, Version 2.1*, pp. 3–11, 2001.

[85] G. S. Douglas, K. J. McCarthy, D. T. Dahlen et al., "The use of hydrocarbon analyses for environmental assessment and remediation," *Journal of Soil Contamination*, vol. 1, no. 3, pp. 197–216, 1992.

[86] Z. Wang and M. F. Fingas, "Development of oil hydrocarbon fingerprinting and identification techniques," *Marine Pollution Bulletin*, vol. 47, no. 9-12, pp. 423–452, 2003.

[87] G. S. Frysinger, R. B. Gaines, L. Xu, and C. M. Reddy, "Resolving the unresolved complex mixture in petroleum-contaminated sediments," *Environmental Science and Technology*, vol. 37, no. 8, pp. 1653–1662, 2003.

[88] J. Krupcík, P. Oswald, D. Oktavec, and D. W. Armstrong, "Calibration of GC-FID and IR spectrometric methods for

determination of high boiling petroleum hydrocarbons in environmental samples," *Water, Air, and Soil Pollution*, vol. 153, no. 1–4, pp. 329–341, 2004.

[89] The International Tanker Owners Pollution Federation Limited (ITOPFL), "Sampling and monitoring of marine oil spills," Technical Information Paper 14, 2012.

[90] Quick Results on Site (QROS), *Hydrocarbon Analysis with QED®*, 2012, http://www.qros.co.uk/hydrocarbon_analysis.html.

[91] P. W. Sammarco, S. R. Kolian, R. A. Warby, J. L. Bouldin, W. A. Subra, and S. A. Porter, "Distribution and concentrations of petroleum hydrocarbons associated with the BP/Deepwater Horizon Oil Spill, Gulf of 108 Mexico," *Marine Pollution Bulletin*, vol. 73, pp. 129–143, 2013.

[92] L. A. Soto, A. V. Botello, S. Licea-Durán, M. L. Lizárraga-Partida, and A. Yáñez-Arancibia, "The environmental legacy of the Ixtoc-I oil spill in Campeche Sound, southwestern Gulf of Mexico," *Frontiers in Marine Science*, vol. 1, no. 57, pp. 1–9, 2014.

[93] P. D. Boehm and D. L. Fiest, "Subsurface distributions of petroleum from an offshore well blowout. The Ixtoc I blowout, Bay of Campeche," *Environmental Science and Technology*, vol. 16, no. 2, pp. 67–74, 1982.

[94] D. Griffin, N. Black, and C. Devine, "5 years after the Gulf oil spill: what we do (and don't) know. 2010 Gulf Oil Spills," CNN, International Edition. Updated 1707 GMT (0107 HKT), 2015, http://edition.cnn.com/2015/04/14/us/gulf-oil-spill-unknowns/.

[95] J. A. Sciortino and R. Ravikumar, *Fishery Harbour Manual on the Prevention of Pollution*, BOBP/MAG/22, Hay of Bengal Programme, Madras, India, 1990.

[96] G. O. Adewuyi, O. T. Etchie, and O. T. Ademoyegun, "Determination of total petroleum hydrocarbons and heavy metals in surface water and sediment of Ubeji River, Warri, Nigeria," *Bioremediation, Biodiversity and Bioavailability*, vol. 5, no. 1, pp. 46–51, 2011.

[97] S. L. Mohebbi-Nozar, M. P. Zakaria, W. R. Ismail et al., "Total petroleum hydrocarbons in sediments from the coastline and mangroves of the northern Persian Gulf," *Marine Pollution Bulletin*, vol. 95, no. 1, pp. 407–411, 2015.

[98] X. Peng, G. Zhang, B. Mai, J. Hu, K. Li, and Z. Wang, "Tracing anthropogenic contamination in the Pearl River estuarine and marine environment of South China Sea using sterols and other organic molecular markers," *Marine Pollution Bulletin*, vol. 50, no. 8, pp. 856–865, 2005.

[99] G. Wattayakorn and S. Rungsupa, "Petroelum hydrocarbon residues in the marine environment of Koh Sichang-Sriracha, Thailand," *Coastal Marine Science*, vol. 35, no. 1, pp. 122–128, 2012.

[100] United Nations Environment Programme (UNEP), "Determination of petroleum hydrocarbons in sediments," *Reference Methods for Marine Pollution Studies*, vol. 20, pp. 33–75, 1992.

[101] I. Bouloubassi, J. Fillaux, and A. Saliot, "Hydrocarbons in surface sediments from there Changjiang (Yangtze River) Estuary, East China Sea," *Marine Pollution Bulletin*, vol. 42, no. 12, pp. 1335–1346, 2001.

[102] J. Li, B. Zhou, J. Shao, Q. Yang, Y. Liu, and W. Cai, "Influence of the presence of heavy metals and surface-active compounds on the sorption of bisphenol A to sediment," *Chemosphere*, vol. 68, no. 7, pp. 1298–1303, 2007.

[103] Q. Li, Z. Wu, B. Chu, N. Zhang, S. Cai, and J. Fang, "Heavy metals in coastal wetland sediments of the Pearl River Estuary, China," *Environmental Pollution*, vol. 149, no. 2, pp. 158–164, 2007.

[104] J. M. Guerra-García and J. C. García-Gómez, "Assessing pollution levels in sediments of a harbour with two opposing entrances. Environmental implications," *Journal of Environmental Management*, vol. 77, no. 1, pp. 1–11, 2005.

[105] C. H. Vane, S. R. Chenery, I. Harrison, A. W. Kim, V. Moss-Hayes, and D. G. Jones, "Chemical signatures of the Anthropocene in the Clyde estuary, UK: sediment-hosted Pb, 207/206Pb, total petroleum hydrocarbon, polyaromatic hydrocarbon and polychlorinated biphenyl pollution records," *Philosophical Transactions of the Royal Society A: Mathematical, Physical and Engineering Sciences*, vol. 369, no. 1938, pp. 1085–1111, 2011.

[106] S. Greason, *Field Screening Petroleum Hydrocarbons Using Ultraviolet Fluorescence Technology*, Sitelab Corporation, West Newbury, Mass, USA, 2009.

[107] Environment Agency (EA), *EA (UK Environment Agency) Framework for the Use of Rapid Measurement Techniques (RMT) in the Risk Management of Land Contamination*, Environment Agency, Bristol, UK, 2009.

[108] R. Becker, M. Koch, S. Wachholz, and T. Win, "Quantification of total petrol hydrocarbons (TPH) in soil by IR-spectrometry and gas chromatography—conclusions from three proficiency testing rounds," *Accreditation and Quality Assurance*, vol. 7, no. 7, pp. 286–289, 2002.

[109] Y. Li, Y. Zhao, S. Peng, Q. Zhou, and L. Q. Ma, "Temporal and spatial trends of total petroleum hydrocarbons in the seawater of Bohai Bay, China from 1996 to 2005," *Marine Pollution Bulletin*, vol. 60, no. 2, pp. 238–243, 2010.

[110] G. O. Adewuyi and R. A. Olowu, "Assessment of oil and grease, total petroleum hydrocarbons and some heavy metals in surface and groundwater within the vicinity of NNPC Oil Depot in Apata, Ibadan metropolis, Nigeria," *International Journal of Research and Revies in Applied Sciences*, vol. 13, no. 1, pp. 166–174, 2012.

[111] I. J. Alinnor, C. E. Ogukwe, and N. C. Nwagbo, "Characteristic level of total petroleum hydrocarbon in soil and groundwater of oil impacted area in the Niger Delta region, Nigeria," *Journal of Environment and Earth Science*, vol. 4, no. 23, pp. 188–194, 2014.

[112] A. R. Abdullah, W. C. Woon, and R. A. Bakar, "Distribution of oil and grease and petroleum hydrocarbons in the Straits of Johor, Peninsular Malaysia," *Bulletin of Environmental Contamination and Toxicology*, vol. 57, no. 1, pp. 155–162, 1996.

[113] A. A. Maktoof, B. Y. ALKhafaji, and Z. Z. Al-Janabi, "Evaluation of total hydrocarbons levels and traces metals in water and sediment from main outfall drain in Al-Nassiriya City/Southern Iraq," *Natural Resources*, vol. 5, no. 13, pp. 795–803, 2014.

[114] S. Suratman, N. M. Tahir, and M. T. Latif, "A Preliminary study of total petrogenic hydrocarbon distribution in setiu wetland, southerzn South China Sea (Malaysia)," *Bulletin of Environmental Contamination and Toxicology*, vol. 88, no. 5, pp. 755–758, 2012.

[115] Shemen Oil and Gas Resources Israel (SOGRLI), *Final Report: Yam-3 Environmental Monitoring Post-Drill Survey Report*, CSA Ocean Sciences, Stuart, Fla, USA, 2014.

[116] M. I. El Samra, H. I. Emara, and F. Shunbo, "Dissolved petroleum hydrocarbon in the northwestern Arabian Gulf," *Marine Pollution Bulletin*, vol. 17, no. 2, pp. 65–68, 1986.

[117] J. T. Tanacredi, "Petroleum hydrocarbons from effluents: detection in marine environment," *Journal of the Water Pollution Control Federation*, vol. 49, no. 2, pp. 216–226, 1977.

[118] D. Ţigănuş, V. Coatu, L. Lazăr, and A. Oros, "Present level of petroleum hydrocarbons in seawater associated with offshore

exploration activities from the Romanian Black Sea sector," *Cercetări Marine*, vol. 46, pp. 98–108, 2016.

[119] A. Korshenko, I. Matveichuk, T. Plotnikova, and V. Luchkov, *Marine Water Pollution, Annual Report 2003*, Hydrometeoizdat, St. Petersburg, Russia, 2005 (Russian).

[120] A. Korshenko, I. Matveichuk, T. Plotnikova, and V. Luchkov, "Marine waters quality on hydrochemical parameters," Annual Report, Hydrometeoizdat, St. Petersburg, Russia, 2003 (Russian).

[121] R. A. Van Dam, C. Camilleri, and C. Turley, "Preliminary assessment of petroleum hydrocarbons in water and sediment at Yellow Water, Kakadu National Park," Internal Report 283, Department of the Environment and Energy, Parkes, Australia, 1998.

[122] N. Mzoughi, M. Dachraoui, and J.-P. Villeneuve, "Evaluation of aromatic hydrocarbons by spectrofluorometry in marine sediments and biological matrix: what reference should be considered?" *Comptes Rendus Chimie*, vol. 8, no. 1, pp. 97–102, 2005.

[123] A. B. Garali, M. Ouakad, and M. Gueddari, "Contamination of superficial sediments by heavy metals and iron in the Bizerte lagoon, northern Tunisia," *Arabian Journal of Geosciences*, vol. 3, no. 3, pp. 295–306, 2010.

[124] M. S. Massoud, F. Al-Abdali, and A. N. Al-Ghadban, "The status of oil pollution in the Arabian Gulf by the end of 1993," *Environment International*, vol. 24, no. 1-2, pp. 11–22, 1998.

[125] R. Zhou, X. Qin, S. Peng, and S. Deng, "Total petroleum hydrocarbons and heavy metals in the surface sediments of Bohai Bay, China: long-term variations in pollution status and adverse biological risk," *Marine Pollution Bulletin*, vol. 83, no. 1, pp. 290–297, 2014.

[126] G. M. Tehrani, R. Hashim, A. H. Sulaiman et al., "Distribution of total petroleum hydrocarbons and polycyclic aromatic hydrocarbons in musa bay sediments (Northwest of the Persian Gulf)," *Environment Protection Engineering*, vol. 39, no. 1, pp. 115–128, 2013.

[127] G. H. Monazami Tehrani, H. Rosli, A. H. Sulaiman et al., "Petroleum hydrocarbon assessment in the wastewaters of petrochemical special economic zone and sediment benchmark calculation of the coastal area—northwest of the Persian Gulf," *Iranian Journal of Fisheries Sciences*, vol. 13, no. 1, pp. 119–134, 2014.

[128] S. A. Mohd Ali, R. Tair, S. Z. Yang, and M. Mohd Ali, "Petroleum hydrocarbon in surface sediment from coastal area of Putatan and Papar, Sabah," *Malaysian Journal of Analytical Sciences*, vol. 17, no. 2, pp. 286–290, 2013.

[129] S. W. Fowler, J. W. Readman, B. Oregioni, J.-P. Villeneuve, and K. McKay, "Petroleum hydrocarbons and trace metals in nearshore Gulf sediments and biota before and after the 1991 war: an assessment of temporal and spatial trends," *Marine Pollution Bulletin*, vol. 27, pp. 171–182, 1993.

[130] C. H. Vane, I. Harrison, A. W. Kim, V. Moss-Hayes, B. P. Vickers, and B. P. Horton, "Status of organic pollutants in surface sediments of Barnegat Bay-Little Egg Harbor Estuary, New Jersey, USA," *Marine Pollution Bulletin*, vol. 56, no. 10, pp. 1802–1808, 2008.

[131] L. P. Rozas, T. J. Minello, and C. B. Henry, "An assessment of potential oil spill damage to salt marsh habitats and fishery resources in Galveston Bay, Texas," *Marine Pollution Bulletin*, vol. 40, no. 12, pp. 1148–1160, 2000.

[132] P. H. Santschi, B. J. Presley, T. L. Wade, B. Garcia-Romero, and M. Baskaran, "Historical contamination of PAHs, PCBs, DDTs, and heavy metals in Mississippi River Delta, Galveston Bay and Tampa Bay sediment cores," *Marine Environmental Research*, vol. 52, no. 1, pp. 51–79, 2001.

[133] J. J. Celino, O. M. C. de Oliveira, G. M. Hadlich, de. Souza, A. F. Queiroz, and K. S. Garcia, "Assessment of contamination by trace metals and petroleum hydrocarbons in sediments from the tropical estuary of Todos os Santos Bay, Brazil," *Revista Brasileira de Geociências*, vol. 38, no. 4, pp. 753–760, 2008.

[134] J. B. R. Agard, M. Boodoosingh, and J. Gobin, "Petroleum residues in surficial sediments from the Gulf of Paris, Trinidad," *Marine Pollution Bulletin*, vol. 19, no. 5, pp. 231–233, 1988.

# 12

# Flavonoids in Different Parts of *Lysimachia clethroides* Duby Extracted by Ionic Liquid: Analysis by HPLC and Antioxidant Activity Assay

**Jin-feng Wei,[1,2] Zhi-juan Zhang,[1] Li-li Cui,[1] and Wen-yi Kang[1,2]**

[1] *Institute of Chinese Materia Medica, Henan University, Kaifeng 475004, China*
[2] *Kaifeng Key Laboratory of Functional Components in Health Food, Kaifeng 475004, China*

Correspondence should be addressed to Wen-yi Kang; kangweny@hotmail.com

Academic Editor: Yoshihiro Kudo

To establish methods for simultaneous determination of isoquercitrin, astragalin in leaves, quercetin, and kaempferol in flowers of *Lysimachia clethroides* Duby, respectively, the methods were ultrasound-assisted extraction combined with RP-HPLC, and ionic liquid was used as the extraction solvent. Meanwhile, the antioxidant activity of the different extracts of *L. clethroides* was evaluated. Purospher STAR RP-$C_{18}$ column (4.6 mm × 250 mm, 5 $\mu$m) was used for analysis. The flow rate was 0.6 mL·min$^{-1}$, and the column temperature was 25°C. The detection wavelength was 360 nm. The mobile phases a and b consisted of acetonitrile-0.4% phosphoric acid (18 : 82, v/v), methanol (A), and 0.4% phosphoric acid (B), respectively. Linear ranges were 0.068~1.64, 0.060~1.44, 0.0080~0.19, and 0.0077~0.18 $\mu$g for isoquercitrin, astragalin, quercetin, and kaempferol, respectively. The average recoveries of the four constituents were 99.17%, 98.39%, 100.68%, and 98.81%, respectively. The antioxidant activity of the extracts was detected by DPPH, ABTS, and FRAP. Under the optimized conditions, all the test solutions showed a certain antioxidant activity and the ionic liquid extracts were better than that of extract of methanol. Ionic liquid used as the extraction solvent had the potential to extract active ingredients efficiently from *L. clethroides*, and this method improved the antioxidant activity with accurate and reliable results.

## 1. Introduction

*Lysimachia clethroides* Duby belongs to the Primulaceae family and is abundant and widely distributed in China. The root or the whole plant of *L. clethroides* was used as traditional Chinese herbal medicine with many health benefits in the treatment of edema, jaundice, dysentery, rheumatic fever, amenorrhea, bruises, fractures, traumatic bleeding, mastitis, boils, and snakebite [1]. Furthermore, its tender stems and leaves serve as popular vegetable in some areas [2]. Pharmacological research showed that *L. clethroides* had many pharmacological activities such as antitumor [3, 4], antioxidant [5], hypoglycemic [6], and hepatoprotective effect [7]. The main chemical components of *L. clethroides* were flavonoids, triterpenes, and organic acids [8–11]. The flavonoids are responsible for the bioactivity of *L. clethroides* [12, 13].

Ionic liquids (ILs) are solely composed of cations and anions as liquid near room temperature [14]. ILs own a large amount of excellent properties, including negligible vapour pressure, good thermal stability, wide liquid range, good dissolving, and high electric conductivity [15]. ILs are expected to be greener alternatives to traditional volatile organic solvents with the characteristic of being environment-friendly. ILs are promising solvents for extracting the active ingredients from traditional Chinese medicine, which are not only environment-friendly but also highly efficient [16, 17]. However, it unavoidably involved some organic solvents during the extraction. ILs in traditional Chinese medicine extraction are applied in the initial stage; it is necessary to conduct in-depth research.

Our group had reported the content of four flavonoids in *L. clethroides* [18], and on the basis of other studies [19–21], the contents of isoquercitrin, astragalin, quercetin, and kaempferol in different parts of *L. clethroides* using high-performance liquid chromatography (HPLC) were compared and a rapid, effective, and environment-friendly ionic liquid-based ultrasonic-assisted approach was established to improve the extraction yield. In addition, the antioxidant activities of the different extracts of *L. clethroides* were measured in order to prove the efficient advantage of ionic liquid in vitro.

## 2. Materials and Methods

*2.1. Reagents and Materials.* Isoquercitrin, astragalin, quercetin, and kaempferol of purity greater than 98% were purchased from Chengdu Pufei De Biotech Co., Ltd. Acetonitrile used for HPLC was purchased from Avantor Performance Materials, Inc. (USA). Methanol used for HPLC was purchased from Tianjin Shield Fine Chemicals Co., Ltd. The water was Wahaha pure water.

1-Butyl-3-methylimidazolium tetrafluoroborate ($[BMIM]BF_4$), 1-butyl-3-methylimidazolium hexafluorophosphate ($[BMIM]PF_6$), 1-hexyl-3-methylimidazolium hexafluorophosphate ($[HMIM]PF_6$), and 1-butyl-3-methylimidazolium bromide ([BMIM]Br) were obtained from Merck KGaA.

1,1-Diphenyl-2-picrylhydrazyl (DPPH) was obtained from Tokyo Chemical Industry Co., Ltd. 2,2′-Azino-bis(3-ethylbenzothiazoline-6-sulfonic acid) (ABTS) was obtained from Fluka. 2,4,6-Tri(2-pyridinyl)-1,3,5-triazine (TPTZ), butylated hydroxytoluene (BHT), butyl-p-hydroxyanisole (BHA), and gallic acid propyl (PG) were obtained from Sigma. 6-Hydroxy-2,5,7,8-tetramethylchroman-2-carboxylic acid (Trolox) was obtained from Aldrich Chemical Co., Ltd.

Air-dried plant was collected from Huaxi District, Guizhou Province of China, in June 2014. The plant was identified as *L. clethroides* of Primulaceae family by Professor Changqin Li (Henan University, Kaifeng, China). The voucher specimens were deposited in the Institute of Natural Products, Henan University (number 20140612).

*2.2. Preparation of Standard Solution.* The four standard solutions of isoquercitrin, astragalin, quercetin, and kaempferol were prepared in methanol at a concentration of 1365, 1200, 160, and 154 $\mu g\cdot mL^{-1}$ and stored at 4°C, respectively. The standard working solutions were prepared by appropriate dilution of the four stock standard solutions with methanol to the required concentrations. Standard working solutions (a and b) were prepared by diluting isoquercitrin and astragalin and quercetin and kaempferol stock solutions at proper concentrations, respectively.

*2.3. Preparation of Test Sample Solution.* Leaves (0.50 g) and flowers (0.25 g) of *L. clethroides* were mixed with 5 mL of $[BMIM]BF_4$ methanol solution ($0.5\ mol\cdot L^{-1}$) in glass vessel, respectively. The samples of leaves and flowers were extracted by ultrasonic-assisted extraction (20 and 45 min, 500 W) and then centrifuged (4000 and 5000 $r\cdot min^{-1}$, resp.). Then, the extract was filtrated through a 0.22 $\mu m$ microporous membrane. The subsequent filtrate was the test sample solutions a and b (the optimum extraction solvent, concentration of selected extraction solvent, mesh, solid-liquid ratio, ultrasonic time and power, and centrifugation speed were systematically investigated in the study).

*2.4. HPLC Analysis.* LC-20AT high-performance liquid chromatography (Shimadzu) was equipped with a degasser, a quaternary gradient low pressure pump, the CTO-20A column oven, a SPD-M20A UV-detector, and a SIL-20A autosampler. The data were acquired and processed using LC-Solution chromatography data processing system.

The two chromatographic separations were all performed on a Purospher STAR RP-C18 column (4.6 mm × 250 mm, 5 $\mu m$) at a column temperature of 25°C. The flow rate of mobile phase was 0.6 $mL\cdot min^{-1}$. The UV detection wavelength was set at 360 nm. All injection volumes were 10 $\mu L$. Mobile phase a consists of acetonitrile-0.4% phosphoric acid (18 : 82, v/v). Mobile phase b consists of methanol (A) and 0.4% phosphoric acid (B), and the gradient program was as follows: 0~25 min, 52% A; 25~30 min, 52%~65% A; 30~46 min, 65% A. The HPLC chromatograms of the standard solution and the extract of sample were shown in Figure 1.

*2.5. Experimental Procedure.* The schematic plot of the experimental procedure was shown in Figure 2.

*2.6. Antioxidant Activity In Vitro*

*2.6.1. Preparation of the Sample Solution.* According to the optimized conditions, methanol and ionic liquid sample solutions were prepared and concentrated to 1 mL, respectively.

*2.6.2. DPPH Assay.* The sample was diluted into a range of concentrations with methanol. 10 $\mu L$ of sample solution (or methanol) and 175 $\mu L$ of DPPH methanol solution (200 $\mu M$) were added to 96-well microplate and mixed. The solution was measured at 515 nm using a Multiskan GO microplate reader after 20 min in the dark place, with PG, BHA, and BHT as positive controls. All the experiments were performed in triplicate. The percentage of radical scavenging rate was calculated using formula (1). $IC_{50}$ value was calculated from the concentration-effect linear regression curve [22].

$$\text{DPPH radical scavenging rate (\%)} = \left[\frac{(A_0 - A_1)}{A_0}\right] \times 100\%. \tag{1}$$

*Note.* $A_0$ was the absorbance of the DPPH itself; $A_1$ was the absorbance of sample and the positive control. $IC_{50}$ represents the concentration of sample in $\mu g\cdot mL^{-1}$ which inhibits DPPH radical by 50%.

*2.6.3. ABTS Assay.* According to literature [23], the ABTS radical working solution was prepared. Then, the sample was diluted into a range of concentrations with methanol. 10 $\mu L$ of sample solution (or methanol) and 200 $\mu L$ of ABTS solution

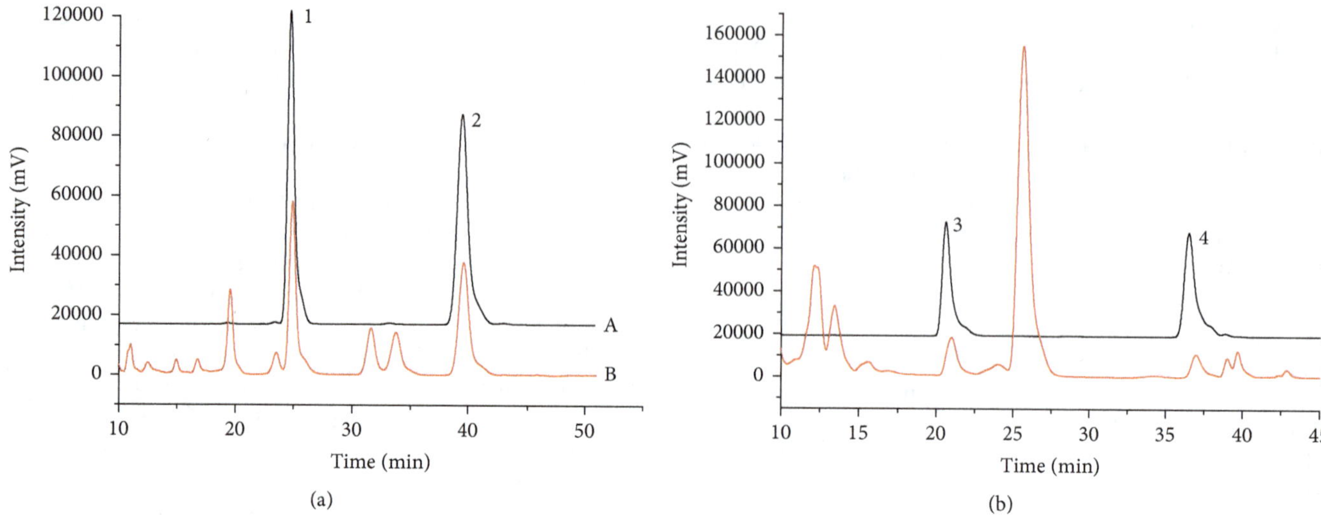


FIGURE 1: HPLC chromatograms of the standard solution (A) and the sample solution (B): (1) isoquercitrin, (2) astragalin, (3) quercetin, and (4) kaempferol.

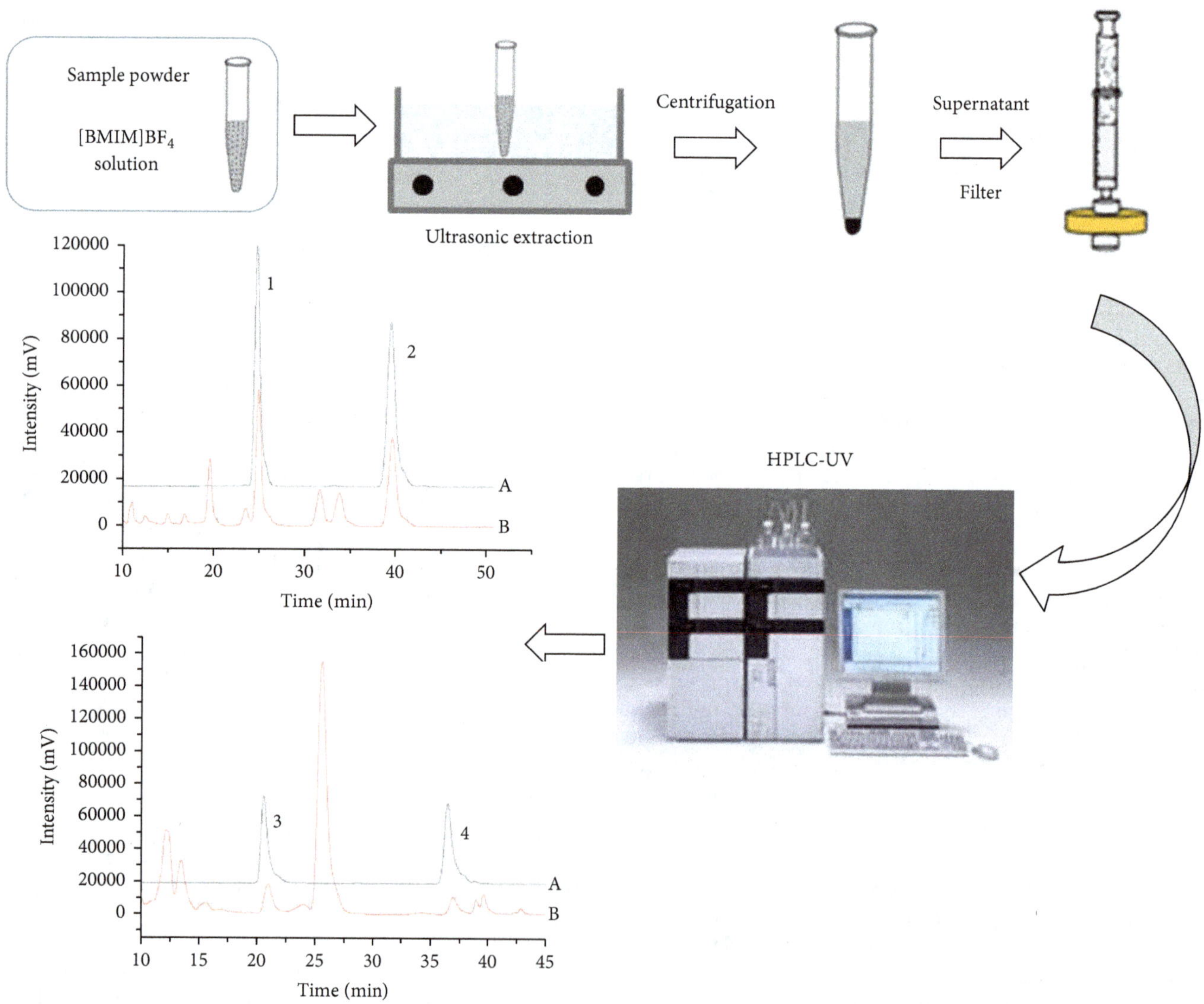


FIGURE 2: Schematic plot of the experimental procedure.

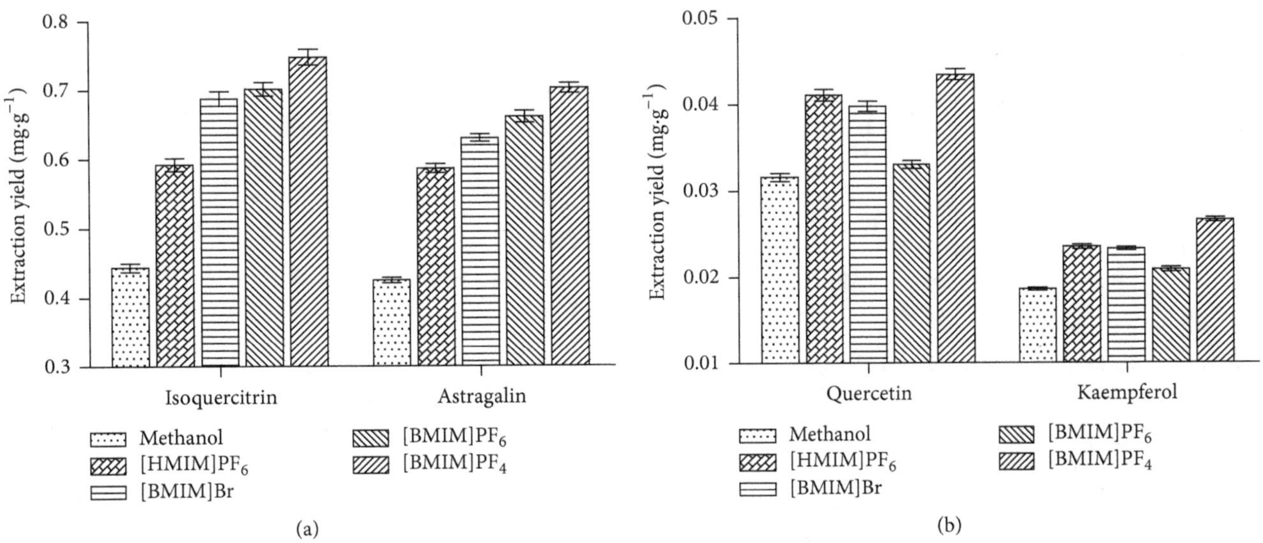


FIGURE 3: Effect of type of ILs on extraction yields. Concentration of the four ILs in methanol solution: 0.5 mol·L$^{-1}$; the particle size: 50 mesh; solid-liquid ratio: 1 : 10 g·mL$^{-1}$; ultrasound time: 30 min; ultrasound power: 500 W; centrifugation speed: 5000 r·min$^{-1}$.

were added to 96-well microplate and mixed. The solution was measured at 405 nm after 20 min in the dark place. All the experiments were performed in triplicate. Calculate the percentage of radical scavenging rate and $IC_{50}$ value.

*2.6.4. FRAP Assay.* According to literature [24], the TPTZ working solution was prepared. Then, the sample was diluted into a range of concentrations with methanol. 10 $\mu$L of sample solution (or methanol) and 200 $\mu$L of TPTZ solution were added to 96-well microplate and mixed. The solution was measured at 593 nm after 20 min at 37°C. All reactions were carried out with three replications. Calculate the reducing ability of $Fe^{3+}$ of the sample. The results were expressed in $\mu$mol Trolox equivalents (TE)/g sample. The standard curve was linear when the concentration of Trolox was between 50 and 1600 $\mu$M. Further dilution of the sample was needed if the FRAP value measured exceeded the linear range of the standard curve.

*2.7. Statistical Analysis.* All the grouped data were statistically evaluated with SPSS 19.0 software. All results are expressed as mean ± standard deviation (SD).

## 3. Results and Discussion

*3.1. Optimization of Extraction Conditions.* The contents of isoquercitrin, astragalin, quercetin, and kaempferol in the roots, stems, leaves, and flowers of *L. clethroides* were determined according to the method of literature [18]. The results showed that the contents of the four flavonoids were different, and the contents of isoquercitrin, astragalin in leaves, quercetin, and kaempferol in flowers were the highest, respectively. Therefore, the extraction conditions of isoquercitrin, astragalin in leaves, quercetin, and kaempferol in flowers of *L. clethroides* were chosen to optimize the conditions of extraction.

*3.1.1. Type of ILs.* The structure of IL has a significant effect on its physicochemical properties including solubility in water, the viscosity, and extraction capacity, which might have great impact on the extraction efficiency of target compounds [25]. In our study, four kinds of ILs were used as the extraction solvents, including water-soluble and water-insoluble types. In addition, methanol was selected as extraction solvent for evaluating the extraction yield compared with the developed method. In Figure 3, the addition of ILs to methanol obviously improved the extraction efficiency of the four flavonoid compounds compared with methanol as solvent. These might be due to the stronger multi-interactions between the analytes and ILs including hydrogen bonding and van der Waals interaction energy [26]. Among the four types of ILs, the extraction efficiency of $[BMIM]BF_4$ was the highest, but others did not show a certain regularity. The reason for the above phenomenon may be the fact that other ingredients in leaves and flowers had an influence on the extraction of target analytes. Therefore, $[BMIM]BF_4$ was chosen as the component of extraction solvent in the following studies.

*3.1.2. $[BMIM]BF_4$ Concentration.* The concentration effect of $[BMIM]BF_4$ on the extraction was shown in Figure 4. The extraction yields continued to rise when the concentration increased from 0.1 to 0.5 mol·L$^{-1}$ and then different degrees of decrease appeared with the further increase of the concentration. This was because the solubility of the ionic liquid solution increased with the increasing concentration, but the diffusion and transfer capability of extraction solvent decreased, which led to declining of the solvent's penetration power into the interior of sample matrix [27, 28].

*3.1.3. Particle Size.* The effect of the particle sizes of sample on extraction efficiency was investigated. In Figure 5, the highest extraction yields of isoquercitrin, astragalin in leaves, quercetin, and kaempferol in flowers were achieved at 24 and

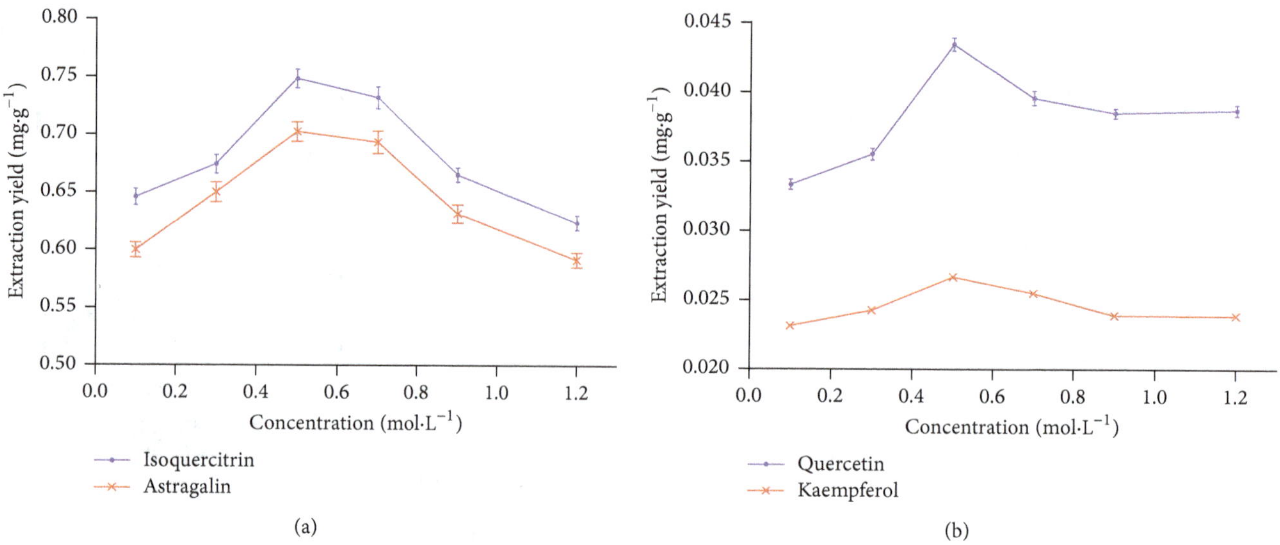


FIGURE 4: Effect of the concentration of IL on extraction yields. Type of IL: [BMIM]$BF_4$; the particle size: 50 mesh; solid-liquid ratio: 1:10 g·mL$^{-1}$; ultrasound time: 30 min; ultrasound power: 500 W; centrifugation speed: 5000 r·min$^{-1}$.

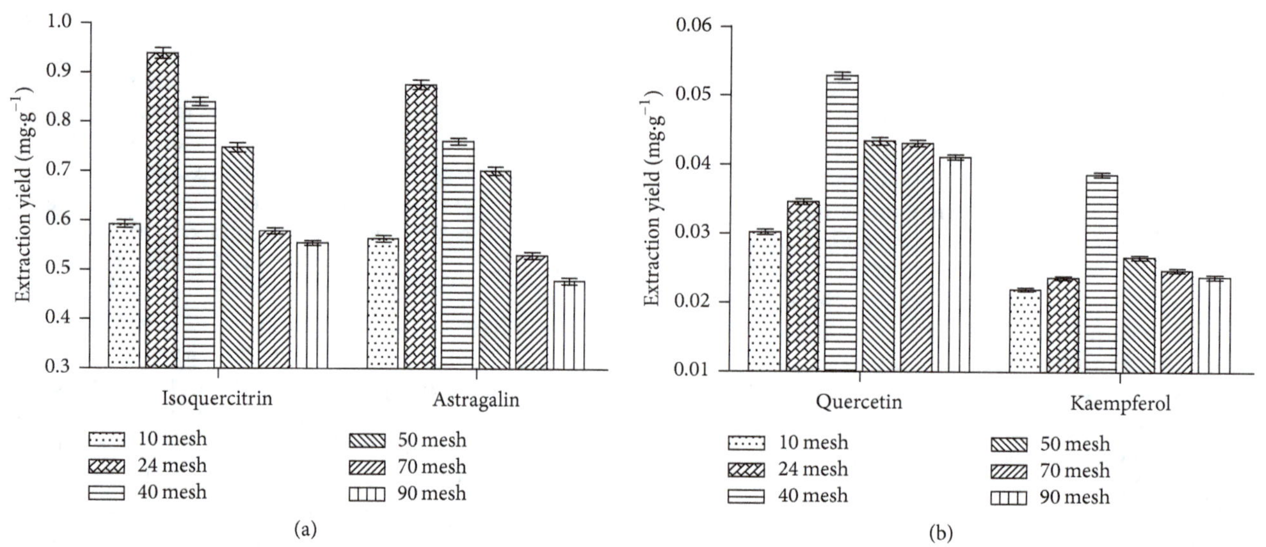


FIGURE 5: Effect of the particle size on extraction yields. Type of IL: [BMIM]$BF_4$; concentration: 0.5 mol·L$^{-1}$; solid-liquid ratio: 1:10 g·mL$^{-1}$; ultrasound time: 30 min; ultrasound power: 500 W; centrifugation speed: 5000 r·min$^{-1}$.

40 mesh, respectively. It might be the smaller the particle size, the easier the extraction, but the samples of leaves and flowers can easily be conglomerated by ionic liquid when the particle size was too small [29]. In addition, the different optimum particle sizes of leaves and flowers might be caused by the difference of organizational structures.

*3.1.4. Solid-Liquid Ratio.* Six different ratios of solid-liquid (1:5, 1:10, 1:20, 1:30, 1:50, and 1:80 g·mL$^{-1}$) were chosen to evaluate the effect of solid-liquid ratio on the extraction yield. The results in Figure 6 indicated that the maximum extraction rates of isoquercitrin, astragalin in leaves, quercetin, and kaempferol in flowers were obtained at 1:10 and 1:20 g·mL$^{-1}$, respectively. Results showed that the target analytes in the sample had basically been extracted. When the proportion of extraction agent was increased, the extraction efficiencies were constant or decreased slightly; this finding might be related to the viscosity of ionic liquid.

*3.1.5. Ultrasound Time.* To investigate the influence of ultrasound time on the extraction efficiency, six time points ranging from 10 to 60 min were chosen. In Figure 7, the maximum extraction efficiencies of isoquercitrin, astragalin in leaves, quercetin, and kaempferol in flowers were obtained at 20 and 45 min, respectively. Nevertheless, the ultrasound time was further extended; the extraction yields of target

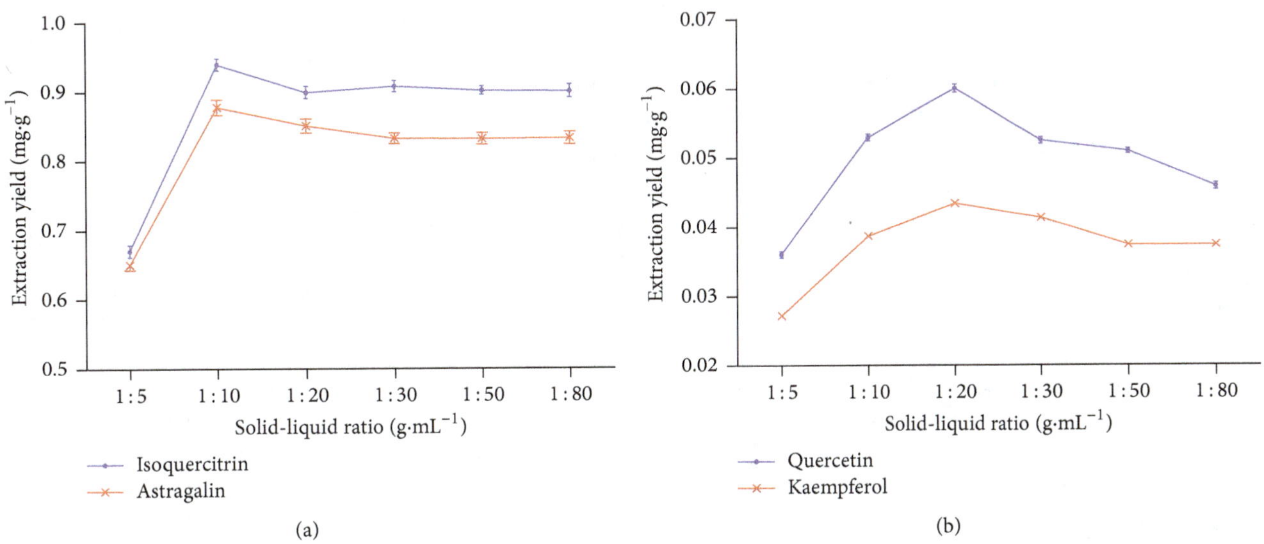


FIGURE 6: Effect of solid-liquid ratio on extraction yields. Type of IL: [BMIM]$BF_4$; concentration: 0.5 mol·L$^{-1}$; the particle size: 24 mesh (a) and 40 mesh (b); ultrasound time: 30 min; ultrasound power: 500 W; centrifugation speed: 5000 r·min$^{-1}$.

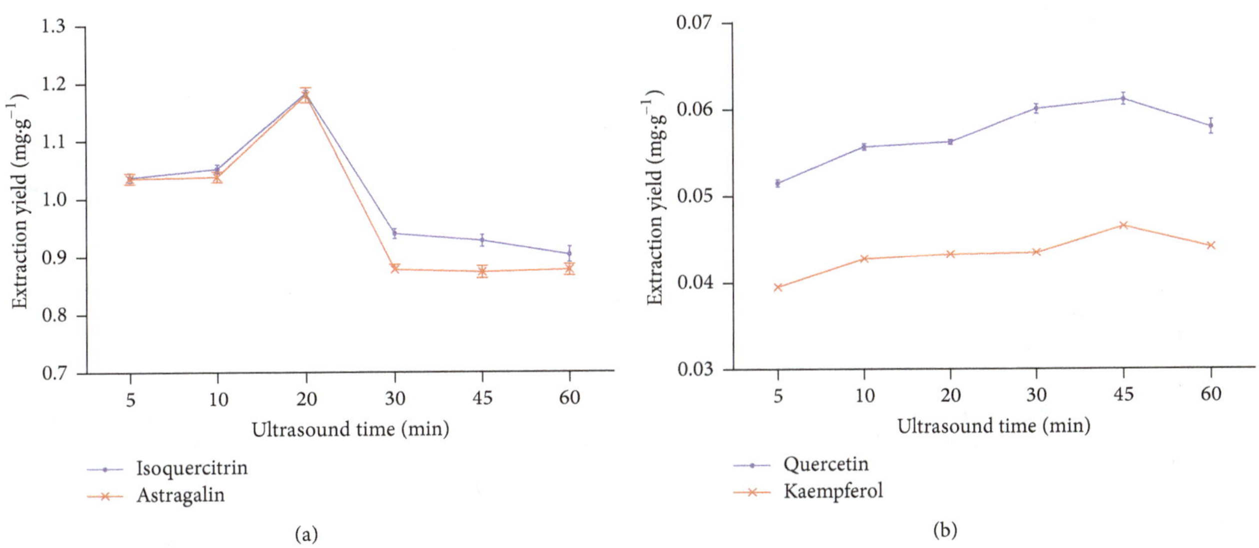


FIGURE 7: Effect of ultrasound time on extraction yields. Type of IL: [BMIM]$BF_4$; concentration: 0.5 mol·L$^{-1}$; the particle size: 24 mesh (a) and 40 mesh (b); solid-liquid ratio: 1 : 10 g·mL$^{-1}$ (a) and 1 : 20 g·mL$^{-1}$ (b); ultrasound power: 500 W; centrifugation speed: 5000 r·min$^{-1}$.

analytes had a certain decrease; the reason for the above phenomenon may be the fact that abundant sonication heat destroys the structure of the analytes [30].

*3.1.6. Ultrasonic Power.* The ultrasonic power had a significant effect on the extraction of the analytes. In our study, the effect of different ultrasonic powers on extraction efficiency was investigated from 200 to 500 W (in Figure 8). Results showed that the extraction efficiencies were increased with the increase of ultrasonic power. Therefore, 500 W was chosen as ultrasonic power in this study.

*3.1.7. Centrifugation Speed.* The influence of centrifugation speed on the extraction yield of the target analyte was investigated. The centrifugation time was set at 5 min, which was a reasonable time for productivity of the method [31]. In Figure 9, results showed that the maximum extraction yields of isoquercitrin, astragalin in leaves, quercetin, and kaempferol in flowers were obtained at 4000 and 5000 r·min$^{-1}$, respectively. The reason might be that the higher centrifugal speed affects the solubility of the analytes, so that the solubility of the analytes is reduced.

### *3.2. Method Validation*

*3.2.1. System Suitability Test (SST).* System suitability test (SST) of HPLC includes the theoretical plate number ($n$), the resolution (Rs), the tailing factor ($T$), and the repeatability.

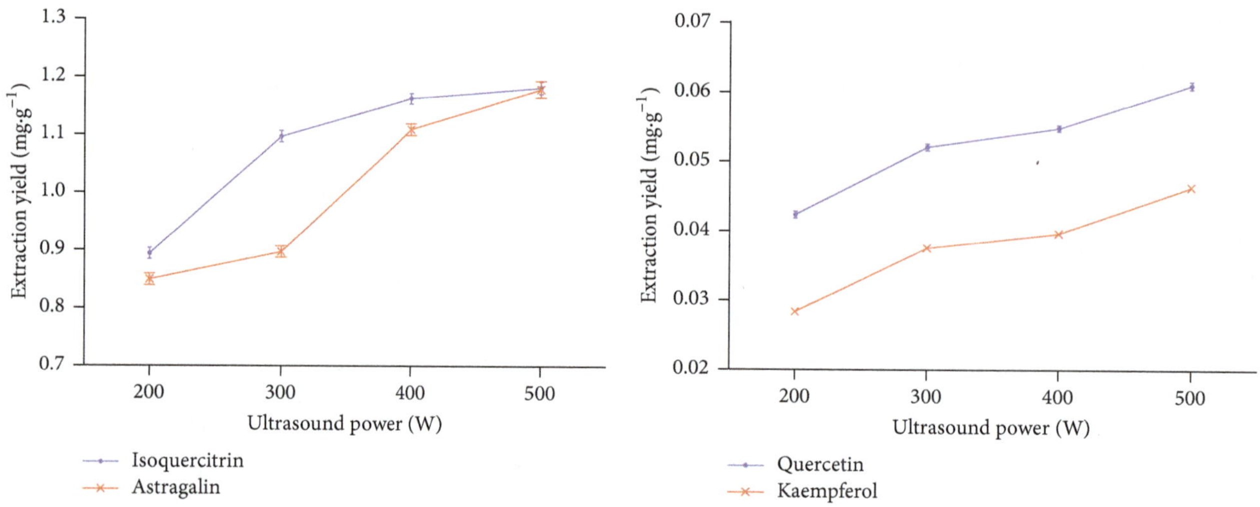


(a) (b)

FIGURE 8: Effect of ultrasound power on extraction yields. Type of IL: $[BMIM]BF_4$; concentration: 0.5 mol·L$^{-1}$; the particle size: 24 mesh (a) and 40 mesh (b); solid-liquid ratio: 1 : 10 g·mL$^{-1}$ (a) and 1 : 20 g·mL$^{-1}$ (b); ultrasound time: 20 min (a) and 45 min (b); centrifugation speed: 5000 r·min$^{-1}$.

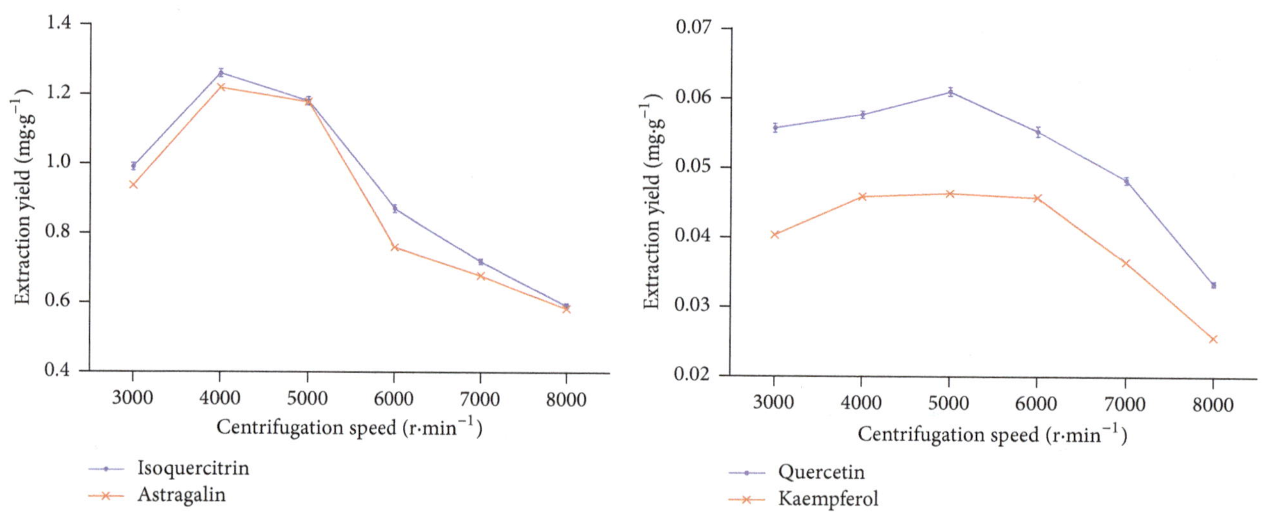


(a) (b)

FIGURE 9: Effect of centrifugation speed on extraction yields. Type of IL: $[BMIM]BF_4$; concentration: 0.5 mol·L$^{-1}$; the particle size: 24 mesh (a) and 40 mesh (b); solid-liquid ratio: 1 : 10 g·mL$^{-1}$ (a) and 1 : 20 g·mL$^{-1}$ (b); ultrasound time: 20 min (a) and 45 min (b); ultrasound power: 500 W.

The repeatability was represented by the relative standard deviation of the peak area (RSD%). The results of SST were shown in Table 1, which shows that the chromatographic systems had good suitability.

*3.2.2. Linearity.* The stock standard solutions a and b were accurately injected 0.5, 1, 2, 4, 6, 8, 10, and 12 $\mu$L to chromatographic instrument for the construction of calibration curves, respectively. The calibration curves were constructed by plotting the peak areas versus the injection volume of each compound. The results were demonstrated in Table 2. $r$ values were in the range from 0.9996 to 1.0000, which indicated that the methods had good linearity.

*3.2.3. Precision.* The stock standard solutions a and b were accurately injected 10 $\mu$L to chromatographic instrument, respectively. And they were analyzed in six replicates within one day for determining the precision of the developed assay. The RSDs of peak areas for isoquercitrin, astragalin, quercetin, and kaempferol were 0.46%, 0.83%, 0.23%, and 0.24%, respectively. The results indicated that the methods were precise for quantitative analysis of isoquercitrin, astragalin, quercetin, and kaempferol.

*3.2.4. Stability.* The sample solutions were prepared under the optimum extraction conditions and placed at room temperature and then were injected 10 $\mu$L to chromatographic

TABLE 1: System suitability test (SST) of HPLC.

| Compound | Theoretical plate number | Resolution | Tailing factor | Repeatability |
|---|---|---|---|---|
| Isoquercitrin | 8392 | 1.52 | 1.31 | 0.87 |
| Astragalin | 8834 | 10.95 | 1.18 | 0.64 |
| Quercetin | 7071 | 5.50 | 1.25 | 1.13 |
| Kaempferol | 13956 | 8.54 | 1.35 | 1.02 |

TABLE 2: Regression equations and linear ranges for four compounds.

| Compound | Regressive equation | $r$ | Linear range/$\mu g$ |
|---|---|---|---|
| Isoquercitrin | $Y = 3.19 \times 10^6 X - 2.92 \times 10^4$ | 0.9996 | 0.068~1.64 |
| Astragalin | $Y = 3.00 \times 10^6 X - 2.32 \times 10^4$ | 0.9997 | 0.060~1.44 |
| Quercetin | $Y = 7.42 \times 10^6 X - 9.58 \times 10^3$ | 1.0000 | 0.0080~0.19 |
| Kaempferol | $Y = 8.90 \times 10^6 X - 1.49 \times 10^4$ | 1.0000 | 0.0077~0.18 |

$Y$: peak area; $X$: injection amount ($\mu$g).

TABLE 3: Antioxidant activity of the different extracts of *L. clethroides*.

| Sample | DPPH | | ABTS | | FRAP |
|---|---|---|---|---|---|
| | Radical scavenging rate/% | $IC_{50}/\mu g{\cdot}mL^{-1}$ | Radical scavenging rate/% | $IC_{50}/\mu g{\cdot}mL^{-1}$ | $TEAC/\mu mol{\cdot}g^{-1}$ |
| Leaf's methanol extract | 93.91 | 1403 ± 80 | 97.01 | 144 ± 5 | 38.58 ± 0.72 |
| Leaf's ionic liquid extract | 88.48 | 638 ± 60 | 93.23 | 50 ± 2 | 105.79 ± 7.02 |
| Flower's methanol extract | 93.92 | 1004 ± 12 | 98.04 | 151 ± 1 | 50.69 ± 1.10 |
| Flower's ionic liquid extract | 89.77 | 460 ± 36 | 96.36 | 51 ± 2 | 105.83 ± 1.25 |
| $[BMIM]BF_4$ | −4.43 | NT | 0.03 | NT | NT |
| BHA | 76.00 | 15 ± 1.46 | 99.94 | 2.06 ± 0.08 | 5919.11 ± 14.10 |
| BHT | 73.06 | 12 ± 0.48 | 99.42 | 6.56 ± 0.10 | 589.20 ± 59.20 |
| PG | 95.61 | 4.01 ± 0.08 | 95.80 | 0.69 ± 0.05 | 9441.33 ± 69.27 |

*Note.* NT means not available because of low activity; BHA, BHT, and PG are positive controls.

instrument at 0, 3, 6, 9, 12, and 24 h, respectively. The RSDs of peak areas for isoquercitrin, astragalin, quercetin, and kaempferol were 0.67%, 0.83%, 1.65%, and 1.74%. The results indicated that the sample solution was basically stable at room temperature within 24 h.

*3.2.5. Repeatability.* Six test sample solutions a and b were processed under the optimum extraction conditions and then injected 10 $\mu$L to chromatographic instrument for analysis. The RSDs of peak areas for the four compounds were 2.03%, 1.84%, 1.15%, and 1.40%, indicating that the analytical methods had an acceptable level of repeatability.

*3.2.6. Recovery.* Nine batches of leaves and flowers samples of *L. clethroides* were prepared and divided into three groups, respectively. Then the four standard substances at three different amounts were added to the leaves and flowers samples. The spiked samples were prepared according to the optimum extraction conditions. All the calculated recovery values of the analytes ranged from 98.39 to 100.68% and the RSDs were less than 2.20%. The results demonstrated that the methods were reasonable and feasible.

*3.3. Antioxidant Activity of the Different Extracts of L. clethroides In Vitro.* The antioxidant activities of methanol and ionic liquid sample solutions were evaluated by the DPPH, ABTS, and FRAP assay comprehensively, and the results were shown in Table 3. Results showed that the ionic liquid extract had stronger antioxidant activity than that of the methanol extract, and the ionic liquid itself had no antioxidant activity. Results indicated that the ionic liquid used as extraction solvent had more advantages for extracting ingredients.

*3.4. Sample Analysis.* Ionic liquid was successfully applied for the determination of isoquercitrin, astragalin in leaves, quercetin, and kaempferol in flowers of *L. clethroides* by the optimized conditions, respectively. The mass fractions of isoquercitrin, astragalin in leaves, quercetin, and kaempferol in flowers were 1.262, 1.219, 0.061, and 0.046 $mg{\cdot}g^{-1}$ ($n = 3$), respectively.

## 4. Conclusion

The method ionic liquid-based ultrasonic-assisted extraction had been developed for the extraction of isoquercitrin, astragalin in leaves, quercetin, and kaempferol in flowers of *L. clethroides*. Compared with the conventional extraction with methanol, the proposed approaches provided higher extraction efficiencies. Results demonstrated that the methods were simple, rapid, green, and effective. ILs can be recycled by some methods, such as vacuum distillation, membrane filtration, salting out, and liquid-liquid extraction [32]. Considering the unique properties of ILs, the methods have a great promising prospect in sample preparation of Chinese herbal medicine.

## Conflicts of Interest

The authors declare that there are no conflicts of interest.

## Authors' Contributions

Wen-yi Kang and Jin-feng Wei conceived the research idea. Zhi-juan Zhang and Li-li Cui conducted the experiments, collected the plant specimens, analyzed and interpreted the data, and prepared the first draft. Wen-yi Kang, Zhi-juan Zhang, and Jin-feng Wei critically read and revised the paper. All the authors read and approved the paper before its final submission.

## Acknowledgments

This work was supported by Henan Province University Science and Technology Innovation Team (16IRTSTHN019), Science and Technology Innovation Team of Kaifeng City (2016-124), Kaifeng City Science and Technology Innovation Talent (1509010), and Key project in Science and Technology Agency of Kaifeng (1603111).

## References

[1] X. M. Hu, W. K. Zhang, and Q. S. Zhu, The Chinese Materia Medica. Shanghai: Shanghai Science and Technology Press, 1999.

[2] Z.-J. Zhang, Z.-Y. Xia, J.-M. Wang, X.-T. Song, J.-F. Wei, and W.-Y. Kang, "Effects of flavonoids in Lysimachia clethroides duby on the activities of cytochrome P450 CYP2E1 and CYP3A4 in rat liver microsomes," *Molecules*, vol. 21, no. 6, article no. 738, 2016.

[3] L. H. Tang, X. Y. Xu, B. G. You, W. Zhang, and Y. Q. Wang, "Anti-tumor effect and its mechanism of total flavones of *Lysimachia* clethroides Duby," *Shanghai Journal of Traditional Chinese Medicine*, vol. 41, no. 5, pp. 74–76, 2007.

[4] Q.-M. Xu, Y.-L. Liu, Y.-L. Feng, L.-H. Tang, and S.-L. Yang, "A new e-ring $\gamma$-lactone pentacyclic triterpene from lysimachia clethroides and its cytotoxic activities," *Chemistry of Natural Compounds*, vol. 48, no. 4, pp. 597–600, 2012.

[5] C. F. Li, Y. L. Song, Y. X. Liu, and W. Y. Kang, "Antioxidant activity of extracts from *Lysimachia clethroides*," *Fine Chemicals*, vol. 25, pp. 1191–1193, 2008.

[6] J. F. Wei, "Study on Chemical Constitutens, Hepatoprotective and Antihyperglycemic Effect of *Lysimachia paridiformis* var. *stenophylla* and *Lysimachia* clethroides," *MSD thesis*, 2012, Henan University, Kaifeng.

[7] D. Liang, "Study on the chemical composition and biological activity of *Lysimachia clethroides* and chemical conversion of synthetic," *MSD thesis*, 2012, Peking Union Medical College, Beijing.

[8] S. M. Yue, B. Q. Chen, P. F. Yuan, W. H. Cui, and W. Y. Kang, "Chemical Constituents of *Lysimachia clethroides* Duby," *Chinese Phramaceutical Journal*, vol. 46, pp. 341–343, 2011.

[9] L. F. Ding, Y. D. Guo, X. D. Wu, and Y. H. Ma, "Chemical constituents, of flavonoids in *Lysimachia clethroides*," *Chinese Traditional Patent Medicine*, vol. 32, no. 05, pp. 827–830, 2010.

[10] Q. M. Xu, L. H. Tang, X. Li, L. L. Hao, X. R. Li, and S. L. Yang, "Isolation and Identification of 3-O-p-Coumaroyloxyl Pentacyclic Triterpenoids from *Lysimachia clethroides* Duby," *Chin Pharm J*, vol. 45, pp. 825–828, 2010.

[11] H. Y. Zou and P. F. Tu, "Studies on the chemical constituents of *Lysimachia clethroides* Duby," *Chin Tradit Herb Drugs*, vol. 40, pp. 704–708, 2009.

[12] L. H. Tang, X. Y. Xu, B. G. You, Y. Q. Wang, and W. Zhang, "Effect of flavones from *Lysimachia clethroides* on gene expression of HL-60 cell," *Pharm Clin Res*, vol. 15, pp. 21–24, 2007.

[13] Y. Q. Wang, L. H. Tang, Z. Q. Liang, and Z. L. Gu, "Primary studies on the anti-uterine cervix cancer effects of the extract ZE4 from *Lysimachia clethroides* Duby," *Chin Pharmacol Bull*, vol. 23, pp. 925–929, 2007.

[14] A. A. Aerov, A. R. Khokhlov, and I. I. Potemkin, "Why ionic liquids can possess extra solvent power," *Journal of Physical Chemistry B*, vol. 110, no. 33, pp. 16205–16207, 2006.

[15] J. F. Wei, P. R. Cao, J. M. Wang, and W. Y. Kang, "Analysis of tilianin and acacetin in Agastache rugosa by high-performance liquid chromatography with ionic liquids-ultrasound based extraction," *Chemistry Central Journal*, vol. 10, p. 76, 2016.

[16] W. Y. Ma, Y. B. Lu, R. L. Hu, J. H. Chen, Z. Z. Zhang, and Y. J. Pan, "Application of ionic liquids based microwave-assisted extraction of three alkaloids N-nornuciferine, O-nornuciferine, and nuciferine from lotus leaf," *Talanta*, vol. 80, no. 3, pp. 1292–1297, 2010.

[17] W. Liu, D. D. Li, H. S. Yang et al., "Determination of oleanic acid and paeoniflorin in *Paeonia lactiflora* by ultrasound-assisted ionic liquid-reversed phase liquid chromatography," *China J Chin Mater Med*, vol. 40, pp. 443–449, 2015.

[18] J. F. Wei, Z. J. Zhang, D. D. Li, W. Liu, and W. Y. Kang, "Analysis of four flavonoids in *Lysimachia clethroides* using ionic liquid-assisted extraction," *China J Chin Mater Med*, vol. 40, pp. 1305–1310, 2015.

[19] J. F. Wei, Z. H. Yin, and F. D. Shang, "Volatiles in the *Lysimachia clethroides* Duby by head space solid phase microextraction coupled with gas chromatography-mass spectrometry (HS-SPME-GC-MS)," *African Journal of Pharmacy and Pharmacology*, vol. 6, pp. 2484–2487, 2012.

[20] J. F. Wei, Y. Y. Li, Z. H. Yin, F. Gong, and F. D. Shang, "Antioxidant activities *in vitro*and hepatoprotective effects of *Lysimachia clethroides* Duby on $CCl_4$-induced acute liver injury in mice," *Afr J Pharm Pharmacol*, vol. 6, pp. 743–750, 2012.

[21] J. F. Wei, X. Chang, W. Wang, and W. Y. Kang, "Chemical constituents from *Lysimachia circaeoides*," *Journal of Chinese Medicine Material*, vol. 36, pp. 1441–1443, 2013.

[22] W.-Y. Kang, Y.-L. Song, and L. Zhang, "$\alpha$-Glucosidase inhibitory and antioxidant properties and antidiabetic activity of Hypericum ascyron L.," *Medicinal Chemistry Research*, vol. 20, no. 7, pp. 809–816, 2011.

[23] W.-Y. Kang, C.-F. Li, and Y.-X. Liu, "Antioxidant phenolic compounds and flavonoids of Mitragyna rotundifolia (Roxb.) Kuntze in vitro," *Medicinal Chemistry Research*, vol. 19, no. 9, pp. 1222–1232, 2010.

[24] W. Kang and J. Wang, "In vitro antioxidant properties and in vivo lowering blood lipid of Forsythia suspense leaves," *Medicinal Chemistry Research*, vol. 19, no. 7, pp. 617–628, 2010.

[25] C. F. Poole and S. K. Poole, "Extraction of organic compounds with room temperature ionic liquids," *Journal of Chromatography A*, vol. 1217, no. 16, pp. 2268–2286, 2010.

[26] Z. Guo, B. Lue, K. Thomasen, A. S. Meyer, and X. Xu, "Predictions of flavonoid solubility in ionic liquids by COSMO-RS: experimental verification, structural elucidation, and solvation characterization," *Green Chemistry*, vol. 9, no. 12, pp. 1362–1373, 2007.

[27] S. Y. Dong, Q. Hu, J. Li, Y. L. Fu, Y. Q. Xing, and T. L. Huang, "Ionic liquid-based microwave-assisted extraction followed high-performance liquid chromatography for the simultaneous determination of five organic oxygen pesticides in soil samples," *Chinese Journal of Analysis Laboratory*, vol. 32, pp. 48–52, 2013.

[28] H. Wu, M. Chen, Y. Fan, F. Elsebaei, and Y. Zhu, "Determination of rutin and quercetin in Chinese herbal medicine by ionic liquid-based pressurized liquid extraction-liquid chromatography- chemiluminescence detection," *Talanta*, vol. 88, pp. 222–229, 2012.

[29] R. Jin, L. Fan, and X. An, "Ionic liquid-assisted extraction of paeonol from cynanchum paniculatum," *Chromatographia*, vol. 73, no. 7-8, pp. 787–792, 2011.

[30] P. Liang, F. Wang, and Q. Wan, "Ionic liquid-based ultrasound-assisted emulsification microextraction coupled with high performance liquid chromatography for the determination of four fungicides in environmental water samples," *Talanta*, vol. 105, pp. 57–62, 2013.

[31] C. K. Zacharis, C. Christophoridis, and K. Fytianos, "Vortex-assisted liquid-liquid microextraction combined with gas chromatography-mass spectrometry for the determination of organophosphate pesticides in environmental water samples and wines," *Journal of Separation Science*, vol. 35, no. 18, pp. 2422–2429, 2012.

[32] H. Liu, X. Y. Wei, J. H. Li, T. Li, and F. Wang, "Review of ionic liquids recycling," *Journal of Cellulose Science Technology*, vol. 21, pp. 63–69, 2013.

13

# Equilibrium Measurements of the $NH_3$-$CO_2$-$H_2O$ System: Speciation Based on Raman Spectroscopy and Multivariate Modeling

**Maths Halstensen,[1] Henrik Jilvero,[2] Wathsala N. Jinadasa,[1] and Klaus-J. Jens[1]**

[1]*Applied Chemometrics Research Group (ACRG), University College of Southeast Norway, Porsgrunn, Norway*
[2]*Department of Energy and Environment, Chalmers University of Technology, 412 96 Göteborg, Sweden*

Correspondence should be addressed to Maths Halstensen; maths.halstensen@usn.no

Academic Editor: Adriano A. Gomes

Liquid speciation is important for reliable process design and optimization of gas-liquid absorption process. Liquid-phase speciation methods are currently available, although they involve tedious and time-consuming laboratory work. Raman spectroscopy is well suited for in situ monitoring of aqueous chemical reactions. Here, we report on the development of a method for speciation of the $CO_2$-$NH_3$-$H_2O$ equilibrium using Raman spectroscopy and PLS-R modeling. The quantification methodology presented here offers a novel approach to provide rapid and reliable predictions of the carbon distribution of the $CO_2$-$NH_3$-$H_2O$ system, which may be used for process control and optimization. Validation of the reported speciation method which is based on independent, known, $NH_3$-$CO_2$-$H_2O$ solutions shows estimated prediction uncertainties for carbonate, bicarbonate, and carbamate of 6.45 mmol/kg $H_2O$, 34.39 mmol/kg $H_2O$, and 100.9 mmol/kg $H_2O$, respectively.

## 1. Introduction

The rapid increase in the level of $CO_2$ in the earth's atmosphere is recognized as the single most important environmental challenge facing our global society [1]. All climate change mitigation plans rely on carbon dioxide capture and storage as a near-term "immediate response" technology [2]. Despite the various global $CO_2$ capture research and development initiatives, the well-established gas-liquid absorption process is expected to be the technology of choice for early, large-scale deployment [3], with the chilled ammonia process (CAP) [4] being one of the currently demonstrated technologies.

Out of several postcombustion techniques available to capture $CO_2$ from coal power plants, amine solutions have been commonly tested and used. The disadvantages of amine technology are that it requires large amount of energy in the stripping process and has thermal and oxidative degradation and corrosion problems [5]. Ammonia technology is an alternative to overcome these drawbacks. This process requires less energy for stripping and the heat of reaction is much lower than amine process. There is less maintenance cost than amine process as there are no degradation or corrosion issues. $CO_2$ reacted ammonia can be used to produce fertilizer. The process of $CO_2$ capture by ammonia can be twofold depending on the temperature of $CO_2$ absorption. If the absorption is performed under 2–10°C, which is called the chilled ammonia process, there can be precipitations of ammonium carbonate compounds while if the absorption is increased to 25–40°C, the precipitation problem is eliminated [6].

If chemical speciation data could be generated concomitant with the determination of the physical and chemical solvent properties, the development and accuracy of the thermodynamic process model would be greatly facilitated. Improvement and optimization of commercial processes that target needed cost reductions require access to rigorous thermodynamic models that build on liquid-phase speciation data. Such data are currently acquired through tedious and time-consuming laboratory work. The standard wet chemistry titration to determine liquid phase is the most popular method. This method takes time and errors can be propagated during sampling, chemical preparation, weighing, and titration [7]. Various analytical techniques can be used to determine the species present in $CO_2$ reacted ammonia solution. $^{13}C$ nuclear magnetic resonance (NMR) spectroscopy has been used [8] to determine ionic species in a range of ammonia concentration from 0.69 to 8.95 mol/L and $CO_2$ loading (total $CO_2$ moles/initial solvent moles) concentrations from 0.33 to 0.72. Fourier transform infrared spectroscopy (FT-IR) and X-ray diffraction (XRD) were used to qualitatively distinguish ammonium bicarbonate from ammonium carbonate and ammonium carbamate, while CHN elemental analysis and near-infrared (NIR) spectroscopy were used to quantify ammonium bicarbonate based on multivariate regression methods as reported in [9]. Wen and Brooker [10], Zhao et al. [11], and Kim et al. [12] suggested methods to determine carbon species in $CO_2$-$NH_3$-$H_2O$ system based on factor analysis where they assumed that the corresponding Raman intensities were directly related to the concentrations of species.

This work reports on a method that combines Raman spectroscopy and partial least-squares regression (PLS-R) for in situ solvent speciation in the chilled ammonia process and which does not rely on calibration by an independent analysis method [13], for example, NMR. It describes the development, validation, and application of the method for determination of the liquid phase composition of the $NH_3$-$CO_2$-$H_2O$ system, while comparisons of the speciation results obtained here and those from established $CO_2$-$NH_3$-$H_2O$ thermodynamic equilibrium models are described in a separate publication [14].

In the thermodynamic modeling of the $CO_2$-$NH_3$-$H_2O$ system, a vapor-liquid equilibrium (VLE) is assumed to exist for water, ammonia, and $CO_2$. The liquid phase can be described by reactions (1)–(5). The chemical composition of aqueous ammonia solution is described by

$$NH_3 + H_2O \rightleftarrows NH_4^+ + OH^- \qquad (1)$$

The reactions of dissolved $CO_2$ in aqueous ammonia solution are given by reactions (2)–(5). Within the present system, $CO_2$ is bound as the anion species of carbonate $[CO_3^{2-}]$, bicarbonate $[HCO_3^-]$, and carbamate $[NH_2COO^-]$.

$$CO_{2(aq)} + H_2O \rightleftarrows H^+ + HCO_3^- \qquad (2)$$

$$HCO_3^- \rightleftarrows H^+ + CO_3^{2-} \qquad (3)$$

$$CO_{2(aq)} + 2NH_3 \rightleftarrows NH_4^+ + NH_2COO^- \qquad (4)$$

$$NH_4^+ + NH_2COO^- + H_2O \rightleftarrows NH_4^+ + HCO_3^- + NH_3 \qquad (5)$$

During $CO_2$ absorption by the aqueous $NH_3$ solution, reactions (1)–(5) will equilibrate according to the $CO_2$ concentration, pressure, and temperature; the carbon distribution in the solvent is given by the concentrations of the anions in reactions (2)–(5). In addition to these reactions and depending on the reaction conditions, various other compounds may precipitate, such as ammonium bicarbonate ($NH_4HCO_3$), ammonium carbonate $[(NH_4)_2CO_3]$, ammonium carbamate ($NH_4NH_2CO_2$), and ammonium sesquicarbonate $[(NH4)_2CO_3 \cdot NH_4HCO_3]$. Even though the precipitation is promoted by low temperature, the high water content in the solvent reduces much of this possibility.

Spectroscopy, Raman spectroscopy in particular, is well known for in situ monitoring of the chemical reactions of aqueous solutions [16, 17]. The water molecule shows only weak Raman scattering; hence Raman spectroscopy has potential advantage over IR spectroscopy [18] for aqueous phase analysis such as required for the aqueous chilled ammonia $CO_2$ capture solvent. The Raman band envelopes of aqueous solutions of ammonium carbonate, ammonium bicarbonate, and ammonium carbamate have been identified and analyzed [10] and form the basis for the previously reported speciation studies of the $CO_2$-$NH_3$-$H_2O$ system [11, 12]. However, evaluation of the Raman spectra in these previous studies was based on univariate, single-band analysis. The use of superior multivariate partial least-squares regression (PLS-R) [19] methods, which exploit the multivariate information in the spectra, has, to the best of our knowledge, not been reported previously. Furthermore, the method developed in the present study is not limited to laboratory applications but can also be used to monitor continuously a reactive process, which presents an opportunity for its implementation as an on-line process analytical technology (PAT) [20] in carbon capture plants for optimization and efficient operation.

## 2. Materials and Methods

The development of the Raman spectroscopy-based method for speciation of the $CO_2$-$NH_3$-$H_2O$ system is based on PLS-R analysis of a series of samples of known composition. The procedure involves

(1) preparation of aqueous solutions that contain known concentrations of $[CO_3^{2-}]$, $[HCO_3^-]$, and $[NH_2COO^-]$: one set of solutions was prepared for PLS-R model calibration and a second independent set of solutions was used for validation;

(2) determination of the Raman spectrum of each sample solution;

(3) preprocessing of the Raman spectra by cropping each spectrum to cover the range 450–2300 $cm^{-1}$, with subsequent elimination of the inconsistently varying baseline from spectrum-to-spectrum and centering of each wavelength by subtraction of the mean;

(4) calibration of the PLS-R models for the anionic species $[CO_3^{2-}]$, $[HCO_3^-]$, and $[NH_2COO^-]$ using the Raman spectra of the solutions prepared for model calibration in the previous step (Step (3));

(5) validation of the PLS-R models using the Raman spectra of the solutions prepared for model validation in Step (3).

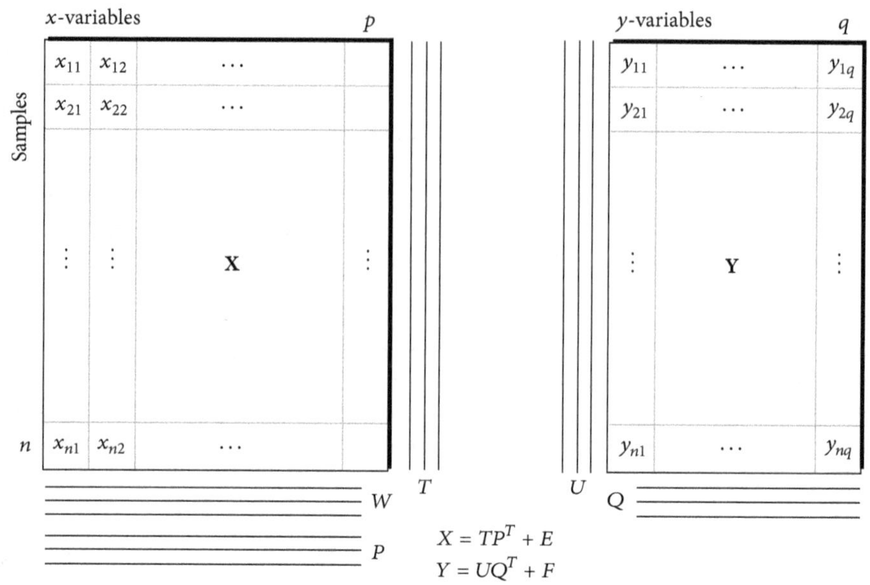


FIGURE 1: $X$-$Y$ data and models in the PLS-R modeling approach [15]. The parameters that emanate from PLS-R calibration include $P$ ($X$ loadings); $W$ ($X$ loading weights); $T$ ($X$ scores); $U$ ($Y$ scores); and $Q$ ($Y$ loadings). $E$ and $F$ are the residuals for $X$ and $Y$, respectively.

The procedures required to achieve reliable speciation using the proposed method are described in the following sections.

*2.1. Raman Spectroscopy.* The Raman phenomenon is based on quantized vibrational changes that are associated with electromagnetic radiation absorption. An important advantage of Raman spectroscopy over infrared spectroscopy is that the water molecule shows very weak Raman scattering, whereas, in infrared spectroscopy, the water molecule shows strong absorption across an important part of the infrared spectrum.

The Raman instrument used in the investigations reported in the present paper was the RXN2 portable multichannel Raman spectrometer (Kaiser Optical Systems Inc.). Four fiber optic probes can be connected and utilized through an automatic sequential scanning system that is integrated into the instrument. The specifications of the RXN2 Raman spectrometer are listed in Table 1.

Raman spectra of aqueous solutions can be acquired using either a noncontact probe optic, whereby the sample solutions are not in direct contact with the probe optic, or an immersion probe optic, whereby the sample solution is in direct contact with the probe optic. The Raman spectra were acquired using a short-focus (200 $\mu$m), sapphire window, Hastelloy immersion probe (Kaiser Optical Systems Inc.).

TABLE 1: Specifications of the RXN2 Raman spectrometer.

| Name | Description |
|---|---|
| Excitation laser wavelength (nm) | 785 |
| Spectral range (cm$^{-1}$) | 100–3425 |
| Spectral resolution (cm$^{-1}$) | 4 |
| Operating temperature range (°C) | 15–30 |
| Number of channels | 4 |
| Laser type | Invictus™ NIR diode laser |
| Spectrograph | f/1.8 Holographic imaging spectrograph |
| Grating | Holographic transmission grating |
| Detector | TE cooled, 1024 CCD Detector |
| Multichannel scanning | 4-Channel sequential operation |
| Cal-Check™ | Automatic analyzer monitor |
| Auto-Cal™ | Automated calibration of axis and laser wavelength |
| Immersion probe optic | 200 $\mu$m (short focus)/Hastelloy/sapphire window |

*2.2. Partial Least-Squares Regression Modeling.* PLS-R is an empirical data-driven modeling approach that requires both representative input data ($X$) and output data ($Y$). A detailed description of PLS-R and validation can be found in literature [15, 19, 21]. In this study PLS-R is used in combination with Raman spectroscopy. The $X$ matrix contains Raman spectra that represent different concentrations and the $Y$ vector contains known reference concentrations from the sample preparation step. All models reported in this article are validated based on independent test data (test set validation) [21].

The concentration range spanned by $X$ and $Y$ should reflect the concentrations to be predicted. The overall aim of PLS-R is to model simultaneously the multivariate input data ($X$) and the output response ($Y$) (Figure 1). PLS-R avoids many of the problems associated with the traditional multiple linear regression (MLR) and principal component regression (PCR) [15] methods. PLS-R is advantageous in cases where the $X$ matrix contains collinear data. Colinearity is in most cases unavoidable in spectroscopy, as there are significantly more variables (waveshifts) than samples (observations). When the number of variables is significantly higher than the number of samples, colinearity is guaranteed. When collinear data are used, MLR fails or becomes unstable. Although PCR can deal with colinearity, it has other issues related to the way in which the $X$-matrix is decomposed without utilizing information on $Y$, often leading to models with a higher number of components than the PLS-R case. Another advantage of applying PLS-R is that the feature parameters that emerge from the model calibration stage, that is, $X$ scores ($T$), $X$ loadings ($P$), $X$ loading weights ($W$), and $Y$ scores ($U$), can be plotted and interpreted to support the calibration procedure (see Figure 1).

An important aspect of PLS-R modeling is model validation [21], which is important to determine the complexity of the model. The correct complexity of a PLS-R model is defined as the optimal number of PLS components, which can only be determined by proper validation of the model using an independent dataset that is acquired and used exclusively for this purpose. To determine the optimal number of components in a PLS-R model, a criterion based on the so-called root mean square error of prediction (RMSEP) [15] is calculated and minimized. RMSEP is based on predictions of $Y$, for example, concentrations from the validation dataset. Using several models with different numbers of components, the predictions are compared to the reference values of $Y$ in the RMSEP calculation. The model that has the optimal number of components is defined as the one that ends up with the lowest RMSEP value. The RMSEP values are derived using (6):

$$\text{RMSEP} = \sqrt{\frac{\sum_{i=1}^{I}\left(\mathbf{y}_{\text{predicted}} - \mathbf{y}_{\text{reference}}\right)^2}{I}}, \quad (6)$$

where $\mathbf{y}_{\text{predicted}}$ is the predicted value from the PLS-R model, which is compared to the reference value $\mathbf{y}_{\text{reference}}$. The sum of the squared prediction errors is divided by $I$, which is the number of samples in the validation dataset.

*2.3. Preparation of the Calibration and Validation Samples.* The ranges of concentrations, expressed in moles of each anion per kilogram of $H_2O$ (molality), of the calibration and validation set were predetermined to reflect the concentrations expected in samples that would in the future be subjected to the developed speciation method. The calibration and validation sets covered the same concentration range. Analytical grade chemicals and Milli-Q water (18.2 MΩ·cm) were used to prepare the samples. The ammonia solution (25 wt%) was supplied by Merck KGaA (Darmstadt, Germany). Sodium hydrogen bicarbonate (99.7%), sodium carbonate (99.9%), and ammonium carbamate (98%) were supplied by Sigma-Aldrich (Steinheim, Germany). All chemicals were used as received. All solutions were prepared gravimetrically using a Mettler Toledo balance (±0.1 mg). Forty solutions of each of $Na_2CO_3$, $NaHCO_3$, and $NH_4NH_2CO_2$, spanning the concentration ranges of 0–0.7 mol/kg $H_2O$, 0–0.96 mol/kg $H_2O$, and 0–2.56 mol/kg $H_2O$, respectively, were prepared for calibration and validation purposes.

The selection of concentration range for calibration and validation range is important especially when the model is used for analyzing future samples. Three factors were considered during this selection which are the solubility of the chemicals, expected species concentrations, and Raman instrument performance. As reported in [22], solubility of $Na_2CO_3$ is 0.7 g at 0°C and 1.25 g at 10°C per kg $H_2O$. $NaHCO_3$ solubility in water is 0.69 g at 0°C and 0.815 g at 10°C per kg $H_2O$. Ammonium carbamate is freely soluble in water. Holmes et al. [8] report a comparison of equilibrium measurements of ionic system in $CO_2$-$NH_3$-$H_2O$ systems for different initial concentrations of $CO_2$ loading. The comparison is based on his experimental work on $^{13}C$ NMR measurements with three thermodynamic models of Pitzer model [23], NRTL model [24], and TIDES model [25]. This comparison gives an indication of expected species concentration for a given $CO_2$ loading and $NH_3$ concentration. For the demonstration of the proposed method in this study, different samples prepared using 5 wt% ammonia in the $CO_2$ loading range from 0 to 0.6 mol $CO_2$/mol $NH_3$ were used and the expected species concentration reasonably falls in the calibration and validation range based on the reported work by Holmes et al. [8]. The limitation of the Raman instrument was also considered when selecting the concentration range. An overview of the sample solutions used for the calibration and validation of the PLS-R models, including the respective concentrations, can be found in Appendix.

*2.4. Acquisition of Raman Spectra.* The Raman spectra were measured with the Kaiser RXN2 Raman spectrometer using a laser power of 400 mW and a total exposure time of 60 seconds with six scans of 10 seconds each being applied to achieve a good signal-to-noise ratio. To maintain a consistent temperature in the spectrometer, the instrument was stabilized for 30 minutes before each measurement series. The short-focus immersion optic was fitted onto the fiber optic probe head and cleaned with acetone. The immersion probe was then positioned vertically using a stand, with the optical window facing down. A glass container that contained the sample solution was positioned under the immersion optic, which was then carefully immersed in the solution. The tip of the optic was positioned in the center of the solution, approximately 20 mm from the bottom of the glass container. The sample and probe optic were protected from external light sources (such as fluorescent light) using aluminum foil. The Raman spectrum was obtained by initiating a scan in the instrument software. In the intervals between sample measurements, the probe was cleaned in acetone to avoid cross-contamination of sample solutions.

*2.5. Preprocessing of Raman Spectra.* Figure 2 presents an example of the preprocessing of the Raman spectra (calibration and validation samples for bicarbonate). During preprocessing, all the Raman spectra were cropped so as to cover the range of 450–2300 cm$^{-1}$, since wavelengths outside

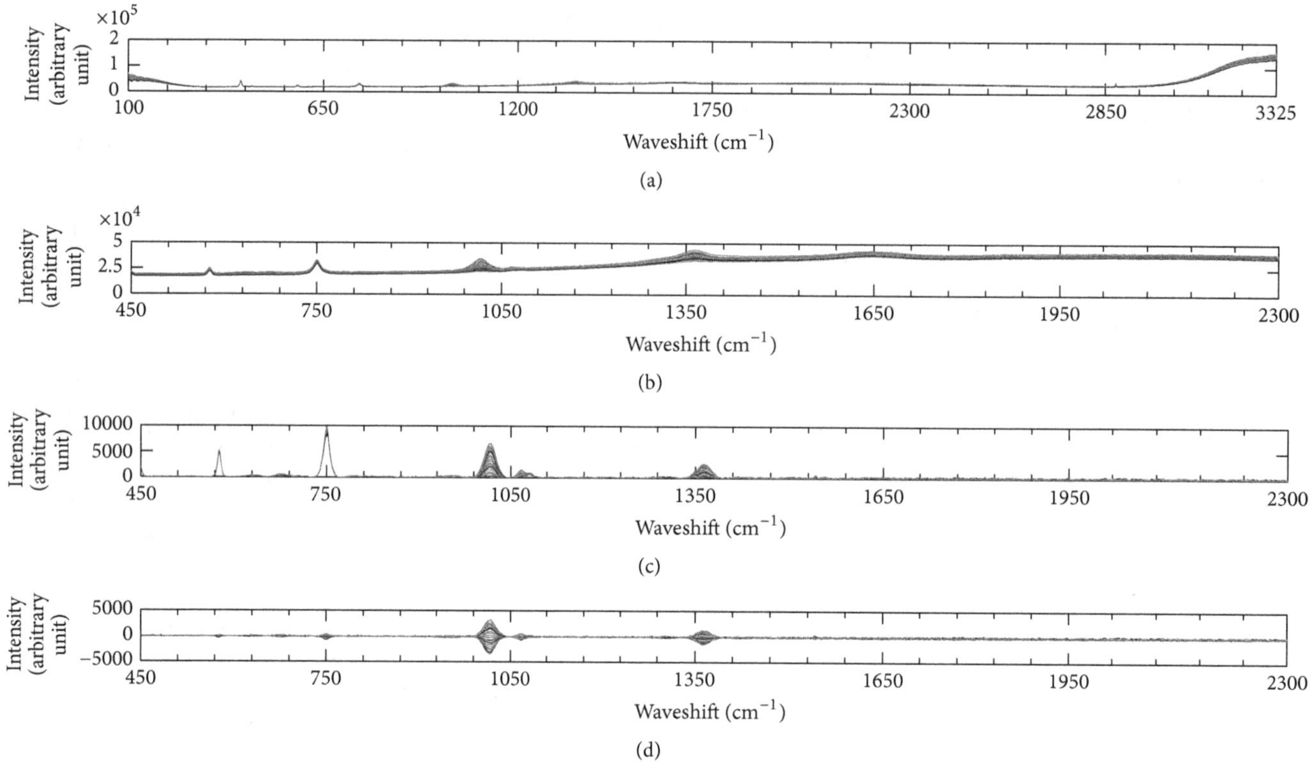


FIGURE 2: Preprocessing steps of the calibration and validation Raman spectra of bicarbonate. (a) Raw spectra for the range of 100–3325 $cm^{-1}$. (b) Spectra cropped to cover the range of 450–2300 $cm^{-1}$. (c) Spectra baseline corrected using the Whittaker filter (lambda = 100; $P$ = 0.001). (d) Mean-centered spectra for PLS-R model calibration.

this range were not useful for the PLS-R analysis. Figures 2(a) and 2(b) show the bicarbonate spectra before and after cropping, respectively. In this range, the Raman spectra show an inconsistent baseline, which is particularly evident at the highest Raman shifts (above 2000 $cm^{-1}$). Raman spectra with variable baselines often generate regression models that have a higher number of components than is necessary, since the model also needs to model the baseline drift. To ensure a less complex model, a baseline correction is performed to decrease or remove the baseline drift in each Raman spectrum. Several baseline correction methods are available [26]. Spectroscopic data from a Raman spectrometer can be decomposed into three parts: (1) the analytical signal; (2) the baseline; and (3) noise. Noise was not a problem in the present study, whereas the baseline required some attention. The goal of all the baseline correction methods is to estimate the baseline in order to remove it. The MATLAB® 2012b software (MathWorks Inc.) in combination with PLS Toolbox 7.31 (Eigenvector Research Inc.) was used to find a suitable baseline correction in the present case. The lowest RMSEP was gained using the Whittaker filtering method [27] included in PLS Toolbox 7.31, which preserves the original shape of the signal part of the Raman spectrum. The Whittaker filter was applied with the following parameters: lambda = 100 and $P$ = 0.001. The lambda parameter defines how much curvature is allowed, while the $P$ parameter holds information about asymmetry in the spectra. Figure 2(c) shows the Raman spectra of bicarbonate after baseline correction using the Whittaker filter. Finally, the spectra were centered by subtracting the mean from each variable (waveshift) in the Raman spectra. Centering [19] is applied prior to PLS-R calibration to avoid the need for an additional PLS-component to describe the mean of the data, which would entail a more complex model. Figure 2(d) shows the data after mean-centering. To summarize, the spectra were cropped to lie within the range of 450–2300 $cm^{-1}$, the baseline was corrected using the Whittaker filter with lambda value of 100 and $P$ value of 0.001, and the spectra were centered on the mean prior to calibration of the PLS-R models of the three species (carbonate, bicarbonate, and carbamate).

*2.6. PLS-R Modeling of Carbonate, Bicarbonate, and Carbamate.* Individual models for carbonate, bicarbonate, and carbamate were calibrated based on the obtained Raman spectra of aqueous solutions that contained known concentrations in the ranges of 0–0.7 mol/kg, 0–0.96 mol/kg, and 0–2.56 mol/kg, respectively. Since the aim was to develop PLS-R models that could be used for speciation of a real $CO_2$-$NH_3$-$H_2O$ system, a selection of variables (waveshifts) to be included in each model was carried out. The waveshifts to be included in the model of each respective anion were chosen based on backward selection, whereby only the wavelengths related to ammonia and the other two $CO_2$ anion species were omitted. In the model for the prediction of carbonate, the Raman wavelengths associated with ammonia,

TABLE 2: Vibrational assignments of $(NH_4)_2CO_3$, $NH_4HCO_3$, and $H_2NCOONH_4$ aqueous solutions, and omitted frequencies in the respective PLS-R models. Adapted from the data of Wen and Brooker [10].

| Frequency [$cm^{-1}$] | Assignment | Omitted frequencies for the PLS-R anion model | | |
|---|---|---|---|---|
| | | Carbonate | Bicarbonate | Carbamate |
| 3390 | Antisymmetric N-H stretch of $NH_3$ | X | X | X |
| 3310 | Symmetric N-H stretch of $NH_3$ | X | X | X |
| 3220 | Fermi resonance with N-H symmetry stretch of $NH_3$ | X | X | X |
| 3050 | Symmetric N-H stretch of $NH_4^+$ | X | X | X |
| 2850 | A combination of fundamentals of $NH_4^+$ | X | X | X |
| 1690 | $NH_2$ deformation of $NH_4^+$ | X | X | X |
| 1645 | Antisymmetric deformation of $NH_3$ | | | X |
| 1630 | C-O antisymmetric stretch of $HCO_3^-$ | | | X |
| 1550 | Antisymmetric $CO_2$ stretch of $H_2NCOO^-$ | | | X |
| 1436, 1380 | CO antisymmetric stretch of $CO_3^{2-}$ | X, X | X, X | X, X |
| 1430 | Antisymmetric $NH_2$ deformation of $NH_4^+$ | X | X | X |
| 1405 | Symmetric $CO_2$ stretch of $H_2NCOO^-$ | X | X | X |
| 1360 | CO symmetric stretch of $HCO_3^-$ | X | | X |
| 1302 | C-OH bend of $HCO_3^-$ | X | X | X |
| 1120 | CN stretch of $H_2NCOO^-$ | X | X | X |
| 1065 | CO symmetric stretch of $CO_3^{2-}$ | | X | X |
| 1034 | $NH_2$ wag of $H_2NCOO^-$ | X | X | |
| 1017 | C-OH stretch of $HCO_3^-$ | X | X | X |
| 680 | $CO_2$ antisymmetric deformation of $CO_3^{2-}$ | X | X | X |
| 640 | (OH)-CO bend of $HCO_3^-$ | X | X | X |
| 570 | Torsion about $CO_2$ skeleton of $H_2NCOO^-$ | X | X | X |

bicarbonate, and carbamate were omitted. The wavelength ranges used in the modeling of bicarbonate analysis were selected based on the same principle, while the carbamate model was based on a more limited range of wavelengths. Since PLS-R modeling of carbamate is more challenging than modeling of the other two species, forward selection was used to define the wavelength ranges in this model. Fine-tuning of the wavelength selection was made based on prediction results, in which the model with a combination of variables that resulted in the lowest prediction uncertainty was used. The carbonate, bicarbonate, and carbamate models were based on the respective frequency ranges of [450–520, 750–950, 1062–1100, 1140–1200, 1520–1650 and 1760–2300] $cm^{-1}$; [450–520, 750–950, 1140–1200, 1350–1390, 1520–1650 and 1760–2300] $cm^{-1}$; and [1033–1043] $cm^{-1}$. Table 2 lists the frequencies omitted from each PLS-R anion model.

The PLS-R models for carbonate, bicarbonate, and carbamate were all based on 20 calibration spectra, in addition to the 20 independent spectra that were used exclusively for the validation. Figure 3 shows the spectra used to calibrate and validate the carbonate, bicarbonate, and carbamate prediction models.

Dissolution of carbonate in water will lead to the following equilibrium state as given in reaction (7).

$$CO_3^{2-} + H_2O \rightleftarrows OH^- + HCO_3^- \quad (7)$$

However, given the detection limit of the Raman spectrometer, only the carbonate band was observed. Therefore, this reaction was neglected in the carbonate PLS-R model development. However, in the cases of bicarbonate dissolution, reaction (8) and for carbamate dissolution reactions (7)–(13) were observed.

$$HCO_3^- \rightleftarrows H^+ + CO_3^{2-} \quad (8)$$

$$CO_3^{2-}{}_{(aq)} + NH_4^+{}_{(aq)} \rightleftarrows HCO_3^-{}_{(aq)} + NH_{3(aq)} \quad (9)$$

$$NH_3 + H_2O \rightleftarrows NH_4^+ + OH^- \quad (10)$$

$$NH_4^+ \rightleftarrows H^+ + NH_3 \quad (11)$$

$$HCO_3^-{}_{(aq)} + NH_{3(aq)} \rightleftarrows H_2NCOO^-{}_{(aq)} + H_2O \quad (12)$$

$$H_2NCOO^-{}_{(aq)} + H_2O \rightleftarrows CO_3^{2-}{}_{(aq)} + NH_4^+{}_{(aq)} \quad (13)$$

Thus, the quantitative PLS-R models were developed according to the following steps:

(1) Calibrate and validate the carbonate model.

(2) Predict carbonate in the bicarbonate calibration and validation datasets, and correct the bicarbonate references accordingly.

(3) Calibrate and validate the bicarbonate model based on the corrected dataset obtained in Step (2).

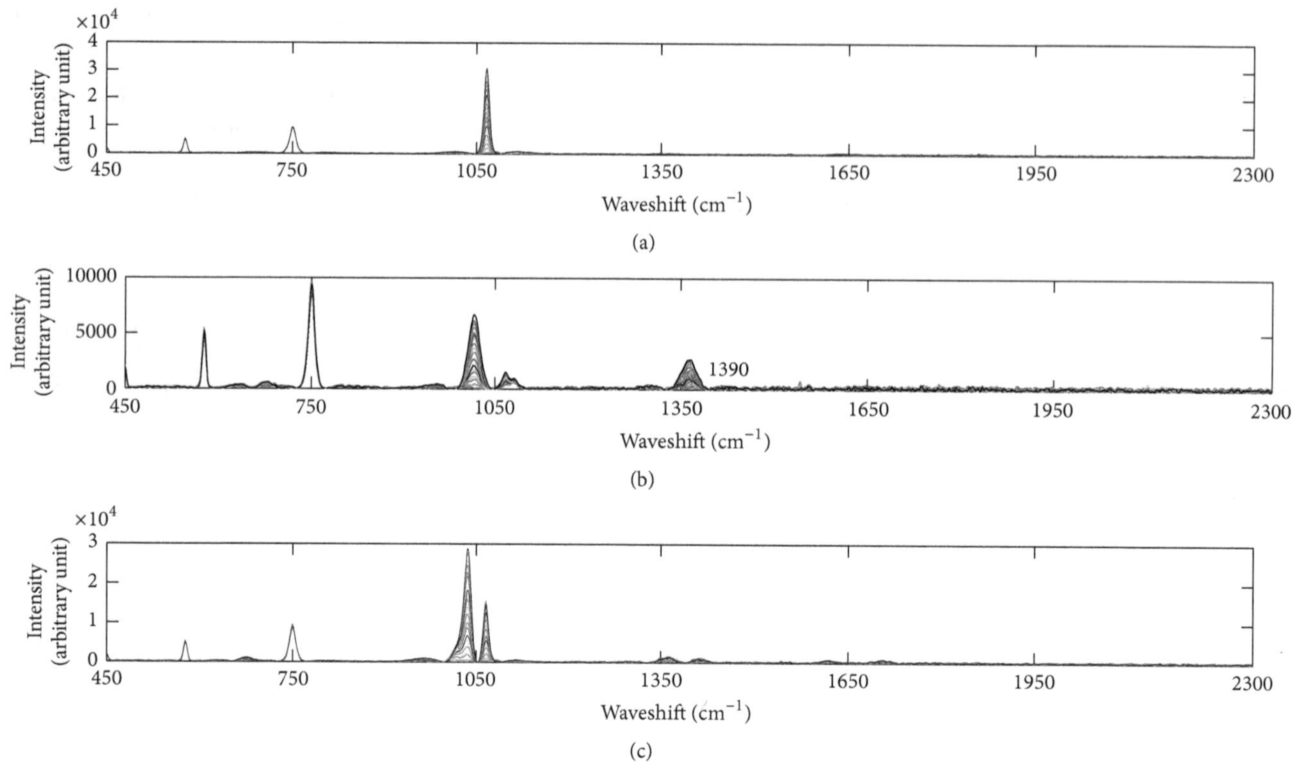


FIGURE 3: Raman spectra used in the PLS-R model calibration and validation. (a) Raman spectra of carbonate; (b) Raman spectra of bicarbonate; (c) Raman spectra of carbamate.

(4) Apply the carbonate model from Step (1) and bicarbonate model from Step (3) to predict carbonate and bicarbonate in the carbamate calibration and validation datasets.

(5) Correct the carbamate references in the calibration and validation datasets based on the predictions of carbonate and bicarbonate made in Step (4).

(6) Calibrate and validate the carbamate model.

All the PLS-R models developed in the present study were developed using the Unscrambler X ver. 10.3 software. Finally, the speciation method was demonstrated based on realistic aqueous solutions that contained 5 wt% ammonia loaded with $CO_2$. Since reference concentrations for carbonate, bicarbonate, and carbamate are not available for these datasets, the prediction results were assessed based on the calculated prediction uncertainties provided by Unscrambler X ver. 10.3.

## 3. Results and Discussion

The calibration and validation results for each of the respective PLS-R models for carbonate, bicarbonate, and carbamate are presented in the respective subsections below. Outliers were detected in the validation and calibration sets for carbonate using the $t_1 - u_1$ scatter plots (data not shown). An outlier may result from an air bubble sticking to the sapphire window in the tip of the probe optic or from the probe optic being inserted so far down into the solution that the measurement is influenced by the glass container.

### 3.1. PLS-R Validation Results for Carbonate, Bicarbonate, and Carbamate.

Figures 4–6 show the validation results for the PLS-R models of carbonate, bicarbonate, and carbamate, respectively. The results include plots of the scores ($t_1 - t_2$), regression coefficients ($B$), residual validation variances, and predicted concentrations versus the reference concentrations. The score plots are used to visualize how the calibration spectra compare to the validation spectra. The regression coefficients show the weight that each wavelength is assigned in the prediction. The residual validation variance plot shows the size of the residual for models with an increasing number of components. The predicted versus measured plots show how the predicted concentrations from the validation dataset in comparison with the references calculated from the sample preparation stage. Prediction performance is evaluated from an interpretation of the statistical parameters of merit, which include $r^2$, RMSEP, and the slope of the regression line.

For the carbonate series, one outlier, identified according to the definition provided previously [19], was removed from the validation dataset. No outliers were detected in the calibration dataset. The outlier was not considered thereafter. Figure 4 shows selected plots from the calibration and validation of the PLS-R model for carbonate. In bicarbonate series no outliers were detected in the calibration and validation datasets. Figure 5 shows selected plots from the calibration

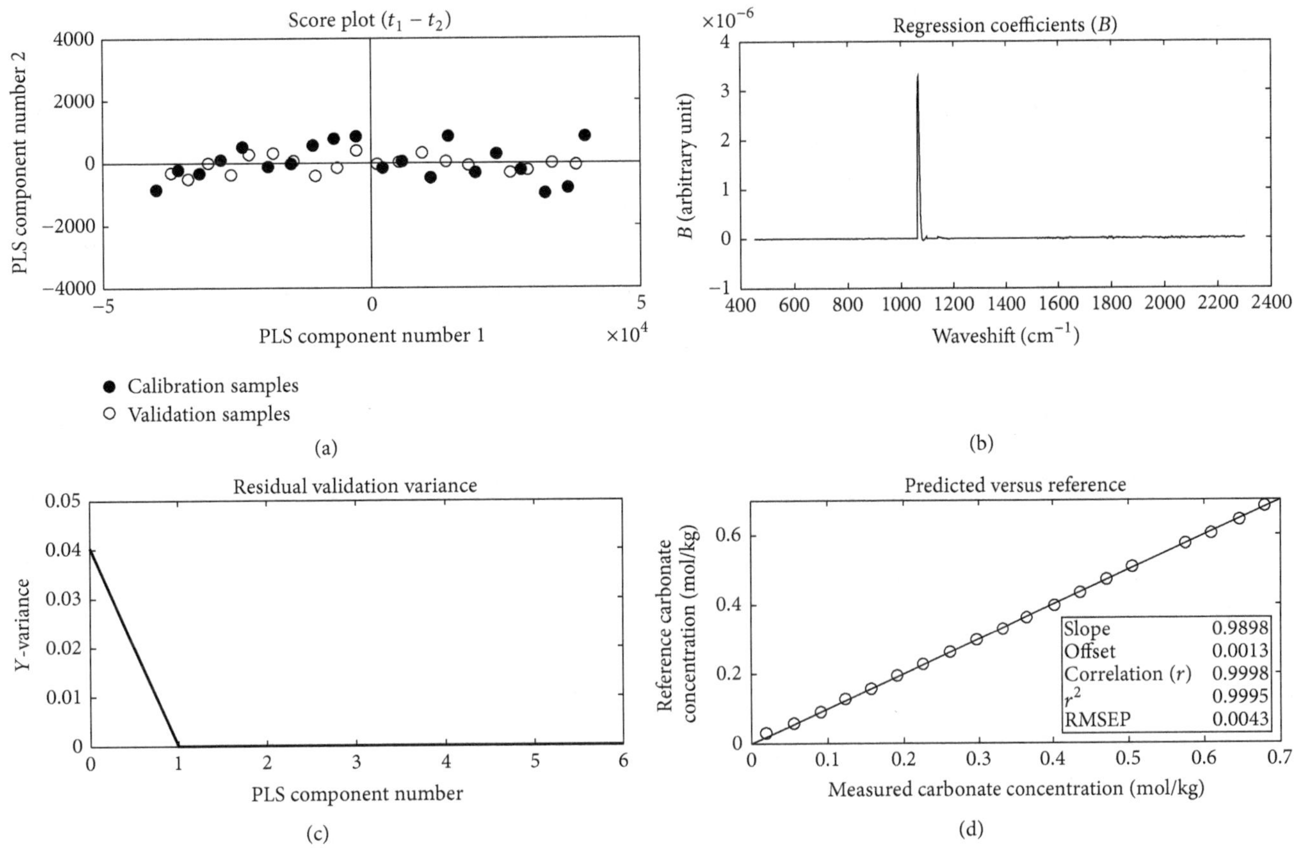


FIGURE 4: PLS-R model for carbonate. (a) Score plot of PLS components 1 versus 2 showing the calibration samples and validation samples. (b) Regression coefficients based on a one-component PLS-R model. (c) The residual validation variance shows that one PLS-component is optimal. (d) Predicted carbonate concentrations based on the one-component PLS-R model versus the reference concentrations obtained from the solute ion preparation stage.

TABLE 3: PLS-R modeling of carbonate, bicarbonate, and carbamate.

| Model | Carbonate model | Bicarbonate model | Carbamate model |
|---|---|---|---|
| Wavelength ranges included [cm$^{-1}$] | 450–520, 750–950, 1062–1100, 1140–1200, 1520–1650 and 1760–2300 | 450–520, 750–950, 1140–1200, 1350–1390, 1520–1650 and 1760–2300 | 1033–1043 |
| Slope | 0.9898 | 0.9636 | 1.0246 |
| $r^2$ | 0.9995 | 0.9932 | 0.9968 |
| RMSEP [mmol/kg] | 4.3 | 24.1 | 49.9 |
| Number of components | 1 | 1 | 1 |
| Number of outliers | 1 (in the validation set) | 0 | 0 |

and validation of the PLS-R model for bicarbonate. For carbamate series also no outliers were detected in the calibration or validation datasets. Figure 6 shows selected plots from the calibration and validation of the PLS-R model for carbamate.

The wavelength ranges used in the PLS-R models for the predictions of carbonate, bicarbonate, and carbamate are presented in Table 3. Some variables (waveshifts) with regression coefficients close to zero were also included, as it was found that the prediction uncertainties were slightly improved when these wavelengths were included. The values for the slope, $r^2$, RMSEP, the number of PLS components, and outliers in all three models are listed in Table 3.

Multivariate analysis has been proven to overcome many challenges in univariate method. Univariate method is simple and samples for calibration can be prepared using one chemical when there are no interferences from other constituents. There are many instances when the property of interest

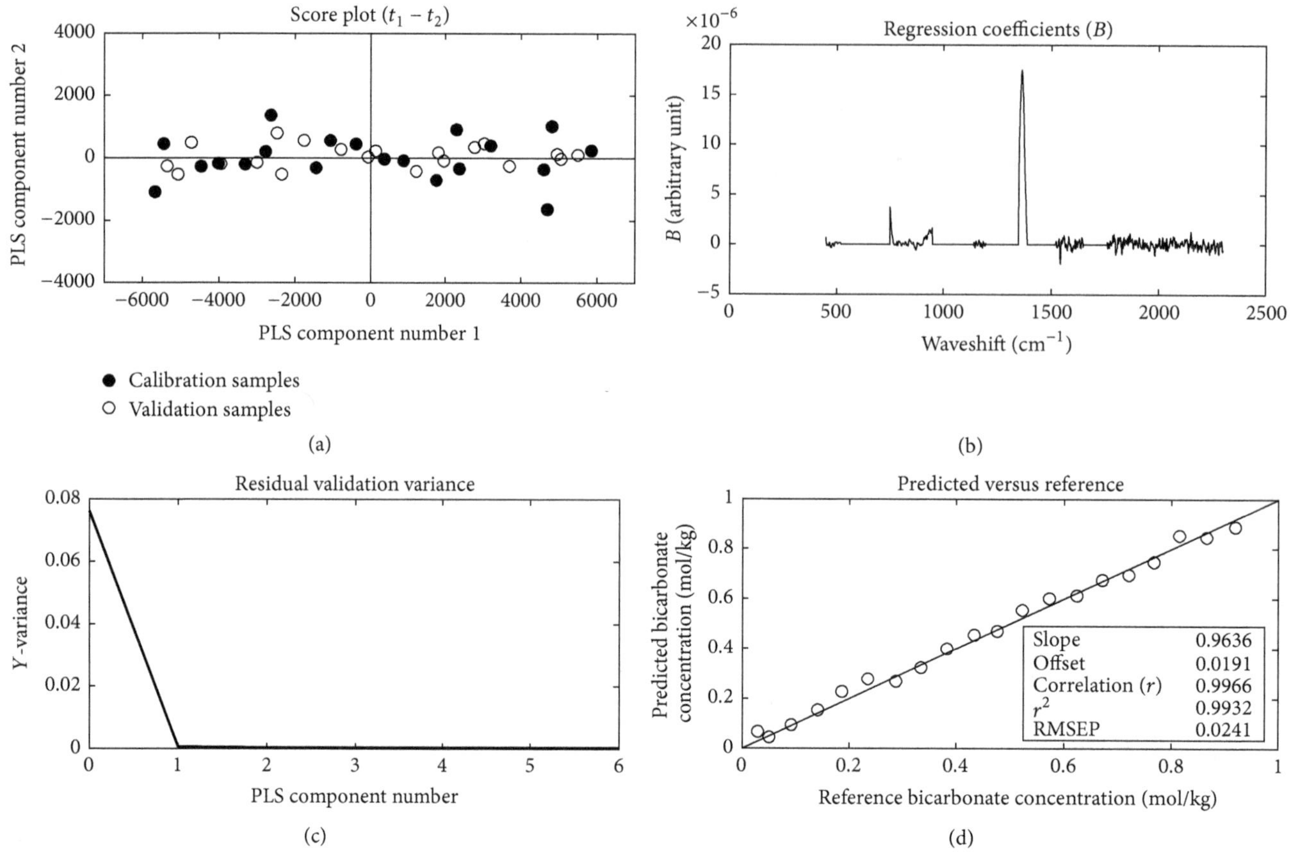


FIGURE 5: PLS-R model for bicarbonate. (a) Score plot of PLS components 1 versus 2 showing the calibration samples and validation samples. (b) Regression coefficients based on a one-component PLS-R model. (c) The residual validation variance shows that one PLS-component is optimal. (d) Predicted bicarbonate concentrations based on the one-component PLS-R model versus the reference concentrations obtained from the solution preparation stage.

cannot be described by one peak. According to Table 2, vibrational modes assigned to carbonate, bicarbonate, and carbamate fall closer to each other in the finger print area from 1000 to 1420 $cm^{-1}$. The $NH_3$-$CO_2$-$H_2O$ system consists of several equilibrium reactions and the compositions of carbonate, bicarbonate, and carbamate are influenced by the concentrations of each other. It can be seen from the calibration spectrum that there are completely visible, independent nicely shaped peak assigned to each other; however this may be not true when it comes to the real $CO_2$ loaded ammonia system. Two scenarios of misuse of the multivariate regression in spectroscopic applications have been explained by Esbensen et al. [28]. They are assigning individual peaks for regression which are identified by the preprocessed data or by regression coefficients. The regression coefficients are used to calculate the response value from the X-measurements. The size of the coefficients gives an indication of which variables have an important impact on the response variables. Assigning wavelengths selected during calibration development for regression must be done with caution because more than one wavelengths are associated with the functional group to some degree [28]. Many factors including scatter effect affect the wavelength position and only by using a wavelength region can the robustness of the calibration model be increased.

The score plots shown in Figures 4–6 reveal that the calibration and validation datasets in all three cases span the same score space, which indicates similarity of the datasets. The most important wavelengths are those with regression coefficients that show the largest deviation from zero. The regression coefficients listed in Figure 4 (carbonate), Figure 5 (bicarbonate), and Figure 6 (carbamate) show that the most important wavelength ranges are 1060–1070 $cm^{-1}$, 1350–1380 $cm^{-1}$, and 1033–1043 $cm^{-1}$, respectively. In all three models, only a small contribution is gained from wavelengths outside these ranges. The residual validation variance plot shows that one-component PLS-R is sufficient for all three models. The slope of the regression line is 0.96–1.02 and the $r^2$ is 0.0.993–0.999, which is close to the optimal value of 1.0. The average prediction errors, that is, RMSEP values, for carbonate, bicarbonate, and carbamate, were 4.3 mmol/kg $H_2O$, 24.1 mmol/kg $H_2O$, and 49.9 mmol/kg $H_2O$, respectively.

*3.2. Demonstration of the Method.* The proposed speciation method was demonstrated using aqueous solutions of 5 wt% ammonia loaded with $CO_2$. Since reference concentrations

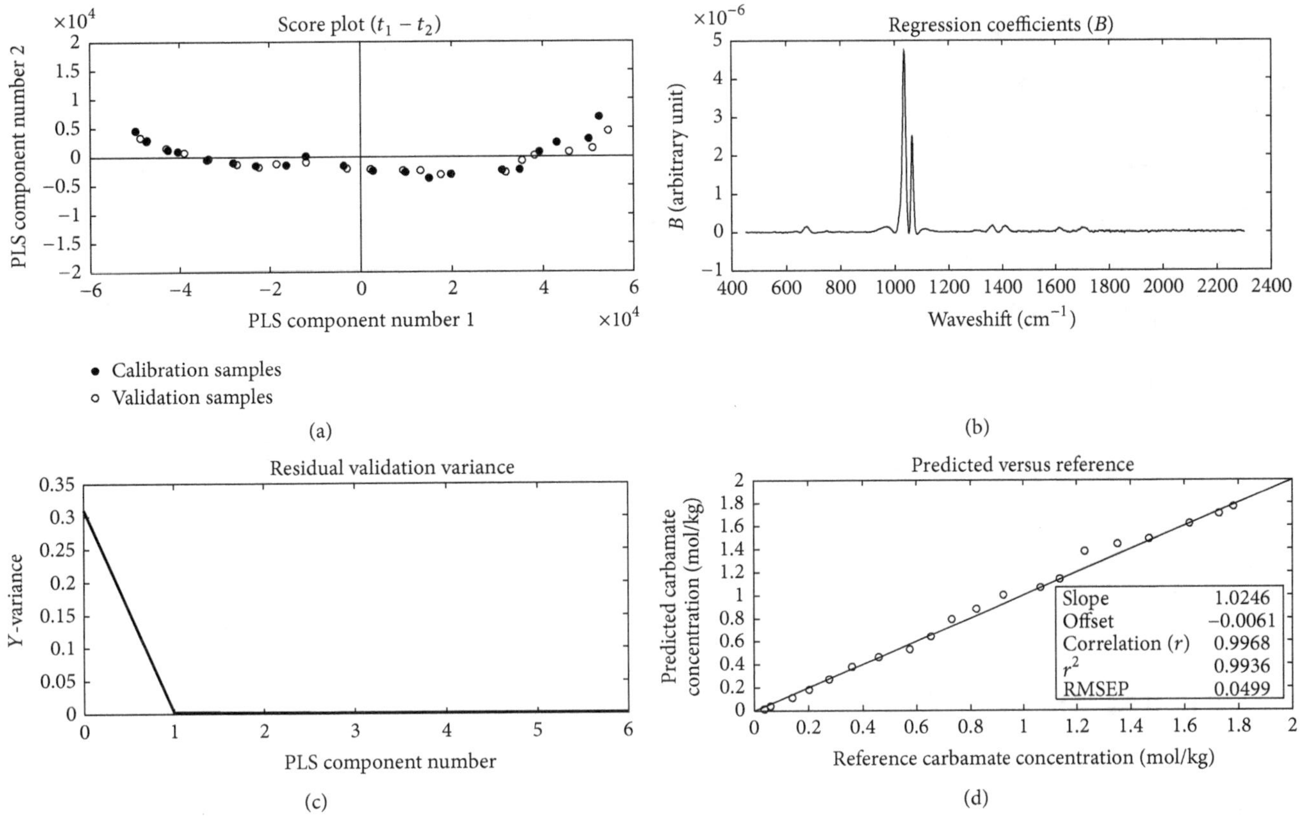


FIGURE 6: PLS-R model for carbamate. (a) Score plot of PLS components 1 versus 2 showing the calibration samples (filled circles) and validation samples (open circles). (b) Regression coefficients based on a one-component PLS-R model. (c) The residual validation variance shows that one PLS-component is optimal. (d) Predicted carbamate concentrations based on the one-component PLS-R model versus the reference concentrations obtained from the solution preparation stage.

of carbonate, bicarbonate, and carbamate were not available, the prediction performance level of each model was assessed based on the calculated prediction uncertainties, as defined previously [29]. The demonstration dataset was based on 14 solutions with different loadings (mole-ratio $CO_2/NH_3$) of $CO_2$, which were measured three times (42 samples in total). The solvent $CO_2$ loading is one of the primary process parameters in the operation of systems for the chemical absorption of $CO_2$.

Figure 7 shows the predicted concentrations of the measured species. The average prediction uncertainties for carbonate, bicarbonate, and carbamate were 6.45 mmol/kg $H_2O$, 34.39 mmol/kg $H_2O$, and 100.9 mmol/kg $H_2O$, respectively. Overall, the predictions of the sample solutions with corresponding uncertainties give a satisfactory outcome regarding speciation of all the anions. The last three predictions for carbamate shown in Figure 7(c) reveal greater uncertainties than those noted for the other predictions. This is due to either the precipitation of solids or the presence of a slightly higher concentration of ammonia in this sample. While a precipitate was not visible in this sample, the conditions were close to those for which precipitation is expected. Thus, the influence of precipitation could not be ruled out. In the accompanying report [14], the method presented here is compared to the experimental data of the same samples with the precipitation-titration method, with good agreement. In the present study, the difference observed between the demonstration dataset and the carbamate calibration is that even though all the same species are present, the $CO_2$ loading differs (see reactions (12) and (13)). In the demonstration dataset, the amounts of ammonia and water are roughly constant, and the amount of $CO_2$ increases continuously with increasing sample number. In the demonstration dataset, the $CO_2$ loading is increased from 0 to 0.6. Thus, the present method can relate the $CO_2$ loading to the liquid carbon distribution through reactions (2)–(5).

*3.3. Comparison of the Model with Literature.* Three models developed in this study have been used in the study of VLE data of chilled ammonia system [14]. This study shows the model predictability which has been compared with two thermodynamic models of Darde et al. [6] and Que and Chen [24] and experimental work carried out by Wen and Brooker [10], Holmes et al. [8], Zhao et al. [11], and Ahn et al. [30]. The composition analysis performed by Zhao et al. [11] for $CO_2$-$NH_3$-$H_2O$ system is based on univariate analysis of Raman measurements. They record results for different initial ammonia concentrations in the

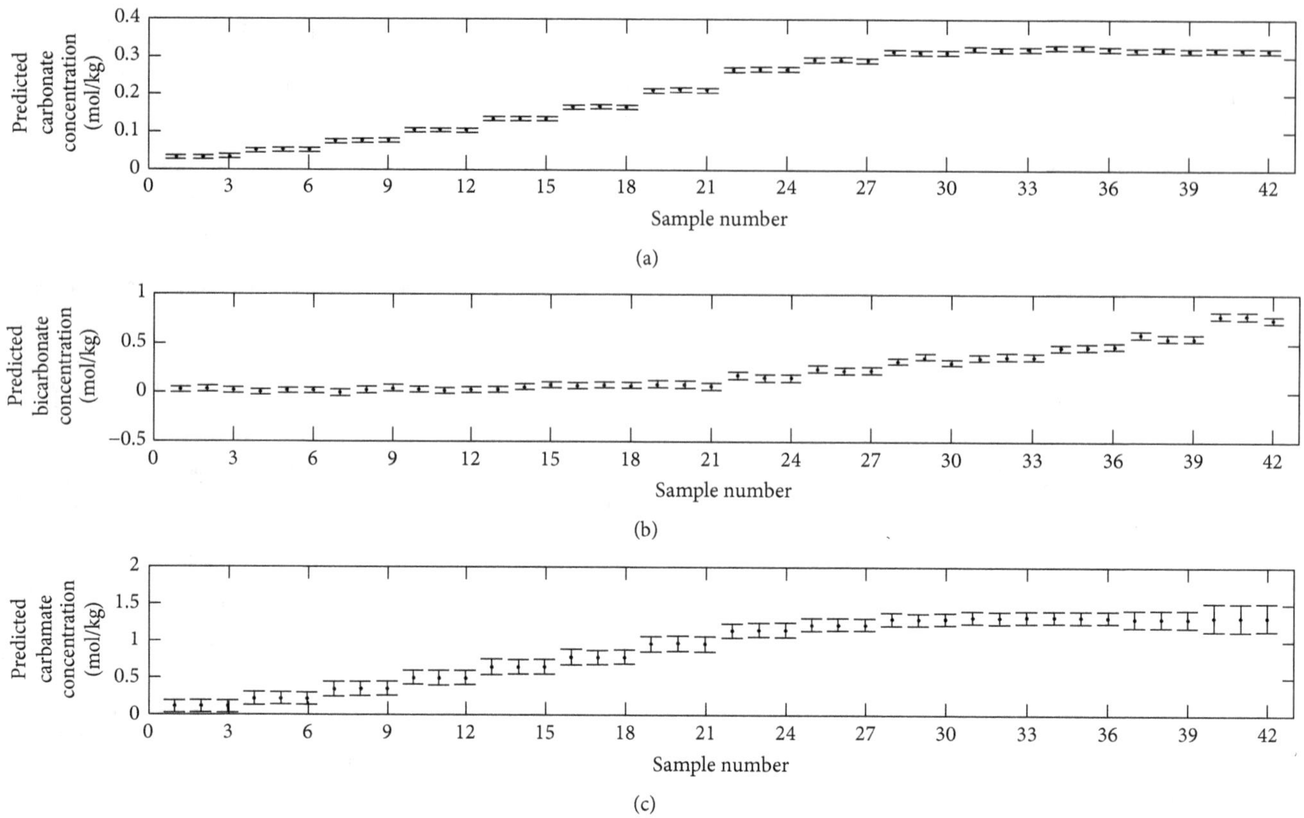


FIGURE 7: Predicted concentrations [mol/kg] of (a) carbonate; (b) bicarbonate; and (c) carbamate.

range of 0.69 mol/L to 2.10 mol/L and $CO_2$ concentration range from 0.18 to 0.67 mol/L. The work is related to low concentrations of ammonia. The proposed method in this study is independent of ammonia concentration and is based on three calibration sets each with 20 measurements which were followed by validation using independent data set spanning in the calibration range for each species. Zhao's method includes calculating molar scattering intensity (J) of each carbon species by preparation of series of solutions of sodium carbonate and sodium bicarbonate to calculate J of carbonate and bicarbonate while J of carbamate was calculated by carbon conservation balance. Therefore the molar scattering intensity of carbamate was dependent on those of other 2 components.

## 4. Conclusion

A method for speciation of the $CO_2$-$NH_3$-$H_2O$ system is proposed. The proposed method can be applied without the need for additional analytical calibration methods. Speciation is achieved based on a combination of Raman spectroscopy and multivariate PLS-R modeling, wherein the so-called full spectrum calibration method is applied to extract information from the entire spectrum. The concentrations of carbonate, bicarbonate, and carbamate were predicted with an average prediction error (RMSEP) as being 4.3 mmol/kg $H_2O$, 24.1 mmol/kg $H_2O$, and 49.9 mmol/kg $H_2O$, respectively. For the method demonstration case, which lacked reference concentrations, the prediction uncertainties for carbonate, bicarbonate, and carbamate were 6.45 mmol/kg $H_2O$, 34.39 mmol/kg $H_2O$, and 100.9 mmol/kg $H_2O$, respectively.

## Appendix

## Concentrations of the Carbonate, Bicarbonate, and Carbamate Species in the Solutions Used for Model Calibration and Validation

See Table 4.

## Conflicts of Interest

The authors declare no conflicts of interest regarding the publication of this research paper.

## Acknowledgments

The financial support provided by the Norwegian Research Council and the industrial partners in the KMB project (no. 199905/S60) and the Ph.D. scholarship financed by the Ph.D. programme in Process Energy and Automation at University College in Southeast Norway are gratefully acknowledged.

TABLE 4: Overview of sample solutions reported as mol/kg $H_2O$.

| Sample number | Carbonate [mol/kg $H_2O$] | | Bicarbonate [mol/kg $H_2O$] | | Carbamate [mol/kg $H_2O$] | |
|---|---|---|---|---|---|---|
| | Calibration | Validation | Calibration | Validation | Calibration | Validation |
| (1) | 0.00000 | 0.01974 | 0.00000 | 0.03441 | 0.00000 | 0.07923 |
| (2) | 0.03840 | 0.05585 | 0.05671 | 0.05704 | 0.14939 | 0.14652 |
| (3) | 0.07280 | 0.09137 | 0.10508 | 0.10159 | 0.29518 | 0.27757 |
| (4) | 0.11073 | 0.12401 | 0.15180 | 0.15504 | 0.39826 | 0.40627 |
| (5) | 0.14710 | 0.15777 | 0.20941 | 0.20138 | 0.53542 | 0.55145 |
| (6) | 0.18544 | 0.19187 | 0.25647 | 0.25214 | 0.67098 | 0.67585 |
| (7) | 0.22126 | 0.22658 | 0.30250 | 0.30090 | 0.80054 | 0.80525 |
| (8) | 0.25856 | 0.26310 | 0.35346 | 0.35263 | 0.93483 | 0.94422 |
| (9) | 0.29493 | 0.29814 | 0.40131 | 0.40289 | 1.08200 | 1.05871 |
| (10) | 0.33113 | 0.33291 | 0.46320 | 0.45270 | 1.21012 | 1.21201 |
| (11) | 0.36895 | 0.36560 | 0.49901 | 0.49946 | 1.33941 | 1.33618 |
| (12) | 0.39913 | 0.40255 | 0.54737 | 0.54932 | 1.47757 | 1.47498 |
| (13) | 0.44205 | 0.43683 | 0.59693 | 0.60006 | 1.61910 | 1.60257 |
| (14) | 0.47772 | 0.47227 | 0.65191 | 0.65196 | 1.73472 | 1.73898 |
| (15) | 0.51599 | 0.50607 | 0.70127 | 0.70248 | 1.87207 | 1.87232 |
| (16) | 0.55162 | 0.54010 | 0.74870 | 0.75143 | 2.02088 | 2.01094 |
| (17) | 0.59048 | 0.57554 | 0.79937 | 0.79809 | 2.12369 | 2.13626 |
| (18) | 0.62521 | 0.61002 | 0.84721 | 0.84682 | 2.27083 | 2.28208 |
| (19) | 0.66210 | 0.64656 | 0.90173 | 0.89758 | 2.41909 | 2.41723 |
| (20) | 0.69948 | 0.68035 | 0.95353 | 0.94859 | 2.55658 | 2.44528 |

## References


[1] S. Solomon, D. Qin, M. Manning et al., “Climate Change 2007: The Physical Science Basis,” Contribution of Working Group I to the Fourth Assessment Report of the Intergovernmental Panel on Climate Change, Cambridge University Press, Cambridge, UK, 2007.

[2] IEA, *Technology Roadmap: Carbon Capture and Storage*, International Energy Agency, Paris, France, 2013.

[3] G. T. Rochelle, “amine scrubbing for CO2 capture,” *Science*, vol. 325, no. 5948, pp. 1652–1654, 2009.

[4] G. d. Koeijer, Y. O. Enge, C. Thebault, S. Berg, J. Lindland, and S. J. Overå, “European CO2 test centre mongstad–testing, verification and demonstration of post-combustion technologies,” *Energy Procedia*, vol. 1, no. 1, pp. 1321–1326, 2009.

[5] O. Lawal, A. Bello, and R. Idem, “The role of methyl diethanolamine (MDEA) in preventing the oxidative degradation of CO2 loaded and concentrated aqueous monoethanolamine (MEA)-MDEA blends during CO2 absorption from flue gases,” *Industrial and Engineering Chemistry Research*, vol. 44, no. 6, pp. 1874–1896, 2005.

[6] V. Darde, K. Thomsen, W. J. M. van Well, and E. H. Stenby, “Chilled ammonia process for CO2 capture,” *International Journal of Greenhouse Gas Control*, vol. 4, no. 2, pp. 131–136, 2010.

[7] J. T. Yeh, H. W. Pennline, K. P. Resnik, and K. Rygle, *Absorption And Regeneration Studies for CO2 Capture by Aqueous Ammonia*, Pittsburgh, PA, USA, 2004.

[8] P. E. Holmes, M. Naaz, and B. E. Poling, “Ion concentrations in the CO2-NH3-H2O system from 13C NMR spectroscopy,” *Industrial and Engineering Chemistry Research*, vol. 37, no. 8, pp. 3281–3287, 1998.

[9] L. Meng, S. Burris, H. Bui, and W.-P. Pan, “Development of an analytical method for distinguishing ammonium bicarbonate from the products of an aqueous ammonia CO2 scrubber,” *Analytical Chemistry*, vol. 77, no. 18, pp. 5947–5952, 2005.

[10] N. Wen and M. H. Brooker, “Ammonium carbonate, ammonium bicarbonate, and ammonium carbamate equilibria: a raman study,” *Journal of Physical Chemistry®*, vol. 99, no. 1, pp. 359–368, 1995.

[11] Q. Zhao, S. Wang, F. Qin, and C. Chen, “Composition analysis of CO2-NH3-H2O system based on Raman spectra,” *Industrial and Engineering Chemistry Research*, vol. 50, no. 9, pp. 5316–5325, 2011.

[12] Y. J. Kim, J. K. You, W. H. Hong, K. B. Yi, C. H. Ko, and J.-N. Kim, “Characteristics of CO2 absorption into aqueous ammonia,” *Separation Science and Technology*, vol. 43, no. 4, pp. 766–777, 2008.

[13] P. A. G. L. Samarakoon, N. H. Andersen, C. Perinu, and K.-J. Jens, “Equilibria of MEA, DEA and AMP with bicarbonate and carbamate: a Raman study,” *Energy Procedia*, vol. 37, pp. 2002–2010, 2013.

[14] H. Jilvero, K.-J. Jens, F. Normann et al., “Equilibrium measurements of the NH3-CO2-H2O system - measurement and evaluation of vapor-liquid equilibrium data at low temperatures,” *Fluid Phase Equilibria*, vol. 385, pp. 237–247, 2015.

[15] K. H. Esbensen, D. Guyot, F. Westad, and L. P. Houmoller, *Multivariate Data Analysis: in Practice*, CAMO Software, Oslo, Norway, 2010.

[16] U. Lichtfers and B. Rumpf, “Infrared spectroscopic studies of the determination of species concentrations in aqueous solutions containing ammonia and carbon dioxide,” *Chemie Ingenieur Technik*, vol. 72, pp. 1526–1530, 2000.

[17] U. Lichtfers and B. Rumpf, "An infrared spectroscopic investigation on the species distribution in the system NH3 + CO2 + H2O," *Thermodynamic Properties of Complex Fluid Mixtures*, pp. 92–119, 2004.

[18] J. R. Ferraro, K. Nakamoto, and C. W. Brown, *Introductory Raman Spectroscopy*, Academic Press, San Diego , Calif, USA, 2 edition, 2008.

[19] H. Martens and T. Naes, *Multivariate calibration*, John Wiley & Sons, London, UK, 1989.

[20] K. A. Bakeev, *Process analytical technology: spectroscopic tools and implementation strategies for the chemical and pharmaceutical industries*, John Wiley & Sons, 2010.

[21] K. H. Esbensen and P. Geladi, "Principles of proper validation: use and abuse of re-sampling for validation," *Journal of Chemometrics*, vol. 24, no. 3-4, pp. 168–187, 2010.

[22] A. Seidell, *Solubilities of inorganic and organic compounds; a compilation of quantitative solubility data from the periodical literature*, D. Van Nostrand Company, New York, NY, USA, 2 edition, 1919.

[23] K. S. Pitzer, "Thermodynamics of electrolytes. I. theoretical basis and general equations," *Journal of Physical Chemistry*, vol. 77, no. 2, pp. 268–277, 1973.

[24] H. Que and C.-C. Chen, "Thermodynamic modeling of the NH3-CO2-H2O system with electrolyte NRTL model," *Industrial and Engineering Chemistry Research*, vol. 50, no. 19, pp. 11406–11421, 2011.

[25] E. M. Pawlikowski, J. Newman, and J. M. Prausnitz, "Phase equilibria for aqueous solutions of ammonia and carbon dioxide," *Industrial Engineering Chemistry Research Process Design and Development*, vol. 21, pp. 764–770, 1982.

[26] K. H. Liland, T. Almøy, and B.-H. Mevik, "Optimal choice of baseline correction for multivariate calibration of spectra," *Applied Spectroscopy*, vol. 64, no. 9, pp. 1007–1016, 2010.

[27] P. H. C. Eilers, "A perfect smoother," *Analytical Chemistry*, vol. 75, no. 14, pp. 3631–3636, 2003.

[28] K. Esbensen, P. Geladi, A. Larsen, and P. Williams, "Myth-busters: one can always assign a wavelength/wavelength region specific for one's application," *NIR News*, vol. 26, no. 4, pp. 15–17, 2015.

[29] M. Høy, K. Steen, and H. Martens, "Review of partial least squares regression prediction error in Unscrambler," *Chemometrics and Intelligent Laboratory Systems*, vol. 44, no. 1-2, pp. 123–133, 1998.

[30] C. K. Ahn, H. W. Lee, Y. S. Chang et al., "Characterization of ammonia-based CO2 capture process using ion speciation," *International Journal of Greenhouse Gas Control*, vol. 5, no. 6, pp. 1606–1613, 2011.

# 14

# Adsorptive Removal of Methylene Blue and Crystal Violet onto Micro-Mesoporous $Zr_3O$/Activated Carbon Composite: A Joint Experimental and Statistical Modeling Considerations

**H. Ait Ahsaine,[1] Z. Anfar,[1] M. Zbair,[2] M. Ezahri,[1] and N. El Alem[1]**

[1]*Materials and Environment Laboratory, Department of Chemistry, Ibn Zohr University, Agadir, Morocco*
[2]*Catalysis and Materials Corrosion Laboratory, Department of Chemistry, Chouaib Doukkali University, El Jadida, Morocco*

Correspondence should be addressed to H. Ait Ahsaine; a.hassan@uiz.ac.ma

Academic Editor: Luqman C. Abdullah

Zirconium oxide/activated carbon ($Zr_3O$/AC) composite was synthesized to remove methylene blue (MB) and crystal violet (CV) from the aqueous medium. The $Zr_3O$/AC sample was characterized by X-ray diffraction (XRD), scanning electron microscopy (SEM), energy dispersive X-ray analyses (EDS), Raman spectroscopy (RS), BET surface area, and Fourier transform infrared spectroscopy (FTIR). XRD profiles confirmed the successful synthesis of the zirconium oxide/activated carbon composite. SEM images showed multideveloped walls with irregular particle size with channel arrays. The nitrogen physisorption combines I and IV types with a calculated BET surface area of 1095 $m^2/g$. Raman spectrum illustrated a disorder of both crystalline structure and the graphitic structure. The adsorption was better fitted to the pseudo-second-order (PSO) kinetic model. Langmuir model fitted better the experimental results of MB adsorption, whereas the CV was better consistent with the Freundlich model. The obtained results suggested that the MB and CV adsorption might be influenced by the mass transfer that involves multiple diffusion steps. The maximum adsorption capacities are 208.33 and 204.12 mg/g for MB and CV, respectively. The MB and CV removal mechanisms were proposed, and statistical optimization was performed using central composite design combined with the response surface methodology.

## 1. Introduction

Nowadays, many industrial activities such as coloring, textile, cosmetic, paper, food, paint, printing, and pharmaceuticals are considered a big concern from the environmental point of view as many hazardous synthetic dyes are generally produced [1]. These pollutants are regarded as reactive, acid, and basic and are very harmful to environment including vegetation, humans, and animals [2]. Methylene blue (MB) is a common basic dye that is used for dyeing clothes in textile industries [3]. MB can cause the adverse effects such as cyanosis, vomiting, diarrhea, tachycardia, and jaundice on human beings. [3, 4]. Crystal violet is another type of toxic pollutant with a poor degradability that is widely used in textile industries because of its low cost and high solubility in water [5, 6].

Various processes are effectively investigated for the removal and elimination of pollutants like chemical precipitation, adsorption, membrane filtration, biological treatment, and photocatalytic degradation [7, 8]. Among these techniques, adsorption is the most effective and broadly utilized approach due to its many advantages being not only the relatively low cost and ease implementation of the operation process [9] but also the effectiveness of the adsorption process that primarily depends on the performance of the adsorbent mass transfer and thermodynamics. Nevertheless, traditional adsorbents including carbonaceous materials, metal nanoparticle oxides, and polymer resins suffer from either low adsorption capacities or low efficiency [10–16]. Therefore, in the general framework of environmental and waste management, there is a need to synthesize environmental recyclable and friendly materials with better uptake and thermodynamic and kinetic properties. Recently, $ZrO_2$ or zirconium-based composites was used as an adsorbent in the gas and liquid phase. More precisely, zirconia $ZrO_2$ was used for $CO_2$ adsorption and hydrogen adsorption to understand the relation to the dehydrogenation of hexane [17, 18]. While in recent works, zirconium and zirconia was used with activated carbon (AC) as composites for the removal of pollutants in water. For instance, Mullick and Neogi [19] have impregnated zirconium on powdered activated carbon (AC) using ultrasound as the tool for synthesis and applying it for fluoride adsorption from water. However, Xiong et al. [20] have used iron-zirconium-modified activated carbon nanofiber for the adsorption of phosphates. Similar works have used these types of composite for the removal of other pollutants [21, 22].

In the present work, commercial suboxide $Zr_3O$ nanoparticles have been fixed onto AC to improve the sorption capacity of MB and CV dyes. This suboxide is unique because of close-packed Zr layers in H stacking and O in octahedral voids, and additionally, the $OZr_6$ octahedra share vertices to form a 3D framework. This 3D framework is very beneficial if it is fixed within the prepared adsorbent leading to high density interactions. The effect of various parameters such as mass loading of $Zr_3O$/AC, pH, contact time, dye concentration, temperature, and regeneration was studied. In addition, the mechanism of MB and CV onto $Zr_3O$/AC and adsorptions experiments optimization were investigated using FTIR analysis after adsorption and RSM-CCD method. It should be noted that the optimization of the adsorption system using classic adsorption and experimental protocol based on changing each single variable does not give useful information about variables. Hence, RSM coupled with the CCD method was considered as a new approach and helpful allowing changing more than a variable at one time and studying the interaction effects between operational parameters [8, 23–29].

To the best of our knowledge, this work reports an effective methodology to study the adsorption of CV and MB, a straightforward study to calculate the optimum parameters through surface modeling, and a sorption mechanism based on different techniques before and after adsorption.

## 2. Materials and Methods

*2.1. Synthesis of the Activated Carbon.* The chemicals used in this study were of analytical reagent grade and used without purification. The activated carbon-based almond shell was prepared by chemical activation using KOH. For this purpose, the almond shell was mixed with KOH solution (Sigma-Aldrich, 98%), 10 g of KOH (Sigma-Aldrich, 90%), with 20 g of almond shell and 100 mL of distilled water to obtain a solution which was heated at 60°C for 12 h and then dried at 110°C. Then, the dried sample was pyrolyzed under $N_2$ flow (200 $cm^3$/min) at 300°C for 2 h and then at 800°C for 3 h at a heating rate of 10°C/min. The $Zr_3O$/AC composite was prepared by impregnation of 20 wt.% of zirconium oxynitrate $ZrO(NO_3)_2{\cdot}xH_2O$ (Sigma-Aldrich, 99%) on the prepared activated carbon after 6 hours of stirring at room temperature. Finally, the impregnated almond shell was dried at 70°C for 12 h and then pyrolyzed at 800°C for 1 hour.

*2.2. Adsorption Experiments.* The adsorption batch system tests were conducted in glass beakers (150 mL) containing 50 mL of CV or MB solutions with a constant stirring rate. Then, the influence of different key parameters was fully investigated. After a certain time, the pollutants concentration was determined using a UV-Vis spectrophotometer (UV 2300) at their respective maximum wavelengths of 664 nm for MB and 582 nm for CV. The adsorbed amount and removal percentage at different conditions were calculated using the following equations:

$$Q_{ads} = (C_0 - C_e) * \frac{V}{m}\ (\text{mg/g}),$$
$$\%\ \text{removal} = \left(\frac{(C_0 - C_e)}{C_0}\right) * 100, \tag{1}$$

where $C_0$ (mg/L) and $C_e$ (mg/L) are the initial and equilibrium liquid-phase concentrations, $m$ (g) is the mass of $Zr_3O$/activated carbon, and $V$ (L) is the volume of the MB/CV solution.

Langmuir [30] and Freundlich [31] were used to analyze the equilibrium data, and the adsorption isotherm models were also applied to simulate information about the distribution of adsorbate molecules at the solid-liquid interface. Table 1 summarizes the mathematical parameters of these models. However, the equations of the selected kinetics models to fit the experimental data are listed in Table 2.

## 3. Results and Discussion

*3.1. Structural and Textural Characterizations.* Figure 1 shows the XRD patterns of the synthesized zirconia-AC powder, and the diffractogram shows two aspects: amorphous aspect in the range of 20–30° and amorphous aspect in the range of 40–50°, which are characteristics to carbon/two-dimensional reflections of turbostratic layer stacks [32] and to graphitic structure of activated carbon. The intense diffraction profiles are associated with the crystalline hexagonal zirconium oxide $Zr_3O$ (JCPDS card no 03-065-7450), the calculated lattice parameters for zirconia by Le Bail fit are $a = b = 5.613$ (2) and $c = 5.192$ (3), and these parameters are very close to the standard $Zr_3O$ JCPDS zirconia file.

Thermal analyses, illustrated in Figure 1(b), show the thermal decomposition of zirconia/AC composite as a function of temperature. This experiment was performed under air. Two major weight losses were recorded: the first loss of 15% below 100°C is attributed to the surface-adsorbed $H_2O$ [33], while the second weight loss of 60% in the range of 300–480°C might be originated from thermal decomposition organic matters such as lignin, cellulose, and hemicelluloses of the almond shell. An intense endothermic peak was recorded for this loss showing the remarkable amount of decomposed products [34].

Figure 2 illustrates the morphology and elemental dispersive spectroscopy of $Zr_3O$/AC composite. In the SEM

TABLE 1: Isotherm models.

| Isotherm | Equation | Description |
|---|---|---|
| Langmuir | $1/Q_e = 1/Q_m + 1/(K_L Q_m C_e)$ | $C_e$: concentration at equilibrium; $Q_e$ = uptake at equilibrium; $Q_m$ = maximum adsorption capacity; $K_L$= energy of adsorption |
| Freundlich | $\ln(Q_e) = \ln K_F + 1/n \ln(C_e)$ | $K_F$ = adsorption capacity; $n$ = intensity of adsorption; $1/n = 0$ irreversible; $1/n >$ unfavourable; $0 < 1$ favourable |

TABLE 2: Kinetic models used to fit the experimental data.

| Kinetic models | Equation | Description |
|---|---|---|
| Pseudo-first order | $\ln(Q_e - Q_t) = \log(Q_e) - K_1 t/2.303$ | $Q_t$ = uptake at time $t$; $Q_e$ = uptake at equilibrium; $K_1$ = rate constant |
| Pseudo-second order | $t/Q_t = 1/(K_2 Q_t^2) + 1/Q_e$ | $K_2$ = rate constant |
| Intraparticle diffusion | $Q_t = K_{di} t^{0.5} + C$ | $K_{di}$ = rate constant |

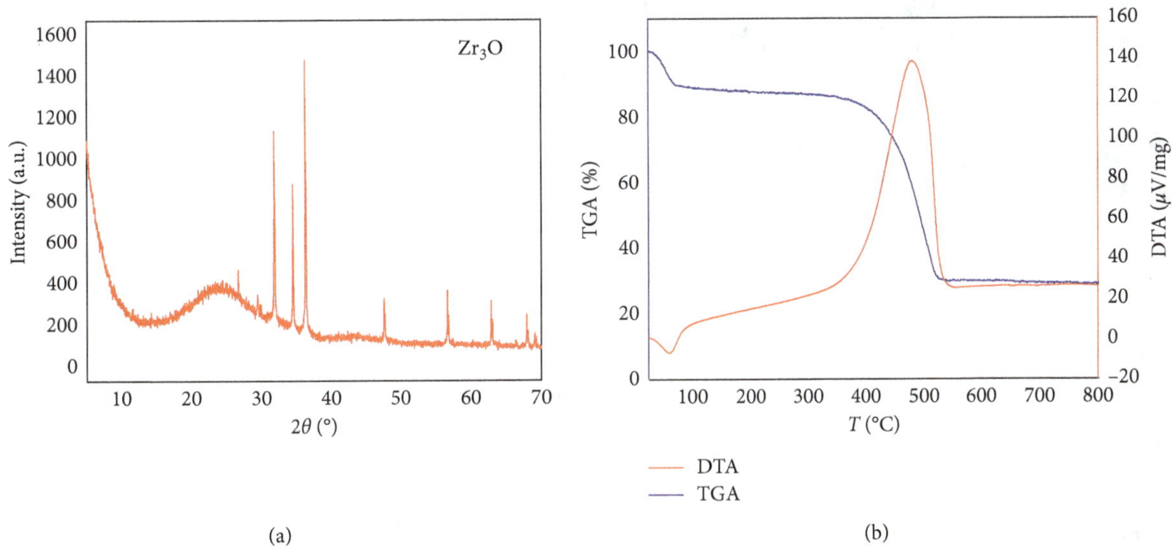


FIGURE 1: (a) X-Ray diffraction of $Zr_3O$/AC composite and (b) TGA/DTA of the $Zr_3O$/AC.

micrographs, multideveloped walls were observed, and pores can be seen with irregular particles and channel arrays in the range of few micrometers to few hundred micrometers. Clearly, knowing that almond shell contains micropores, it is inferred that the AC preserved the original structure of the almond after even after carbonization. The results of elemental analysis indicate the presence of three main atoms: carbon, oxygen, and zirconium.

Raman analysis is a very important tool for the characterization of active modes in the activated carbons adsorbents [35, 36]. Clearly, all the samples have two obvious D and G bands, which are assigned to the chaos and disorder of the crystalline structure and the graphitic structure of carbon materials, respectively [37, 38]. The synthesized $Zr_3O$/AC and AC were studied by Raman analysis, and the results are presented in Figure 3(a). The spectra displayed the typical spectra of AC with the characteristic D and G bands at 1362 and 1604 $cm^{-1}$, respectively. Furthermore, the intensity of $Zr_3O$/AC decreased indicating that the preparation process of $Zr_3O$/AC slightly changed the structure of activated carbon. Moreover, this means that the degree of graphitization of $Zr_3O$/AC decreases because zirconia occupies the pore channel [39].

The $N_2$ physisorption isotherm is illustrated in Figure 3(b). It combines types I and IV characteristics, meaning the coexistence of microporous and mesoporous structures in AC and $Zr_3O$/AC. Indeed, the distributions of pores size confirm the presence of micropores and mesopores in AC and $Zr_3O$/AC, and pore size distributions shown in Figure 3(c) are bimodal with a peak at around 0.5 nm for microporous structures and around 2 nm for mesoporous structures. The $Zr_3O$/AC BET surface area is 1095 $m^2/g$ with total pore volume of 0.66 $cm^3/g$ (0.16 $cm^3/g$ of microporous volume and 0.56 $cm^3/g$ of mesoporous volume).

The FTIR analysis of $Zr_3O$/AC composite is displayed in Figure 3(d). In the spectrum, a broad and strong absorption band at approximately 3449 $cm^{-1}$ attributed to the O-H stretching vibration of water molecules adsorbed on the surface $Zr_3O$/AC [40]. The spectrum of $Zr_3O$/AC presents a peak for Zr-O stretching vibration at 620 $cm^{-1}$ [41].

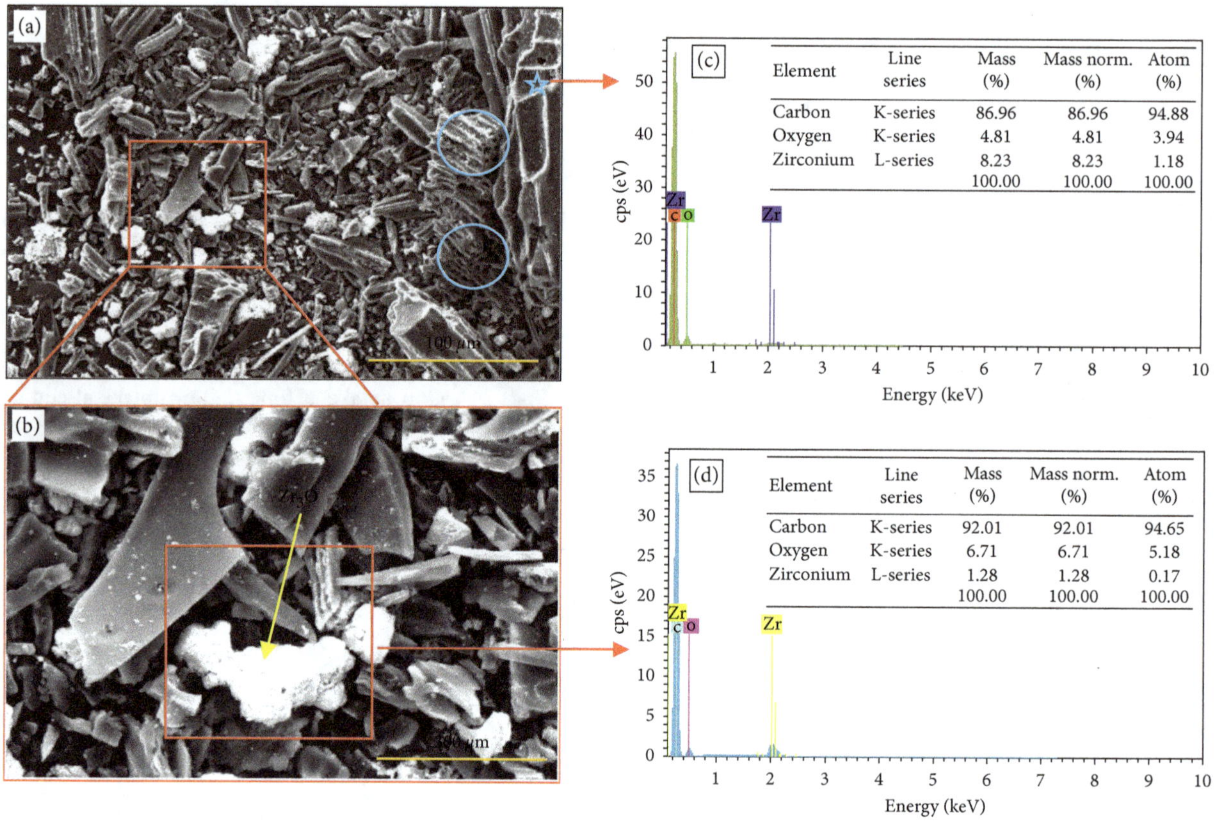


Figure 2: (a-b) SEM photographs of the $Zr_3O$/AC and (c-d) elemental EDS analyses.

However, the band centered on 1622 cm$^{-1}$ indicates enhancement in the aromatic C=C groups [27, 42].

### 3.2. Adsorption Results

*3.2.1. Influence of Operational Parameters.* Binding sites and surface promoted by $Zr_3O$/AC doses are one of the main parameters influencing the adsorption process. Consequently, several experiments of the dyes removal with different $Zr_3O$/AC doses have been conducted to evaluate the effect of these key factors. These experiments have been performed at room temperature unless mentioned. The influence of $Zr_3O$/AC doses on MB and CV dyes removal by $Zr_3O$/AC is presented in Figure 4; As the $Zr_3O$/AC composite weight increased from 5 mg to 16 mg, the percentage removal increased as well from 84.43% to 100% for MB removal and from 75.41% to 98% for CV removal. The increase in MB and CV removal at higher dosages explain (i) the increase in number of active binding sites and (ii) bulk surface area [43].

Additionally, up to 20 mg, the dyes removal was not significantly affected by further increase in the $Zr_3O$/AC amount due to the possibility of aggregation, which limits the active surfaces for adsorption. The optimum $Zr_3O$/AC dosage was found to be 16 mg for both dyes.

Figure 5(b) illustrates the removal of MB and CV dyes as a function of solution pH. Knowing that pH can play a substantial role in the adsorbent surface charges and interfacial solid-liquid transport [44], the pH-dependence experiment has been performed for both MB and CV. As the pH increases, adsorption of both dyes on $Zr_3O$/AC was found to be increasing; this is mainly due to elimination of the positive proton in the aqueous media thereby promoting activities of electrostatic attraction between the positive charges of both dye and the composite $Zr_3O$/AC. Indeed, according to the point of zero charge (Figure 5(a)), it is confirmed that the surface of the composite can be negatively charged at pH higher than 7.5. While at an acidic pH, the functional groups of activated carbon become protonated, which are mainly the carboxylic (-COOH-), phenolic (-OH-), and chromonic group. At pH 3, the surface charge of $Zr_3O$/AC becomes more positively charged, which enhances MB and CV electrostatic repulsion: the number of positively charged sites on $Zr_3O$/AC increased and negatively charged sites on $Zr_3O$/AC decreased. Therefore, the electrostatic repulsion between the $Zr_3O$/AC sites and the positively charged CV and MB dye molecules increased which inhibit the adsorption pathway.

Figure 6 describes the influence of contact time on MB and CV adsorption by $Zr_3O$/AC. It was observed that the adsorption process of both dyes takes place in two different stages:

(i) The first stage was rapid which infers strong electrostatics between cationic dyes and the $Zr_3O$/AC positive surface.

(ii) The second stage was with no significant improvement of the adsorbed dyes, and this equilibrium

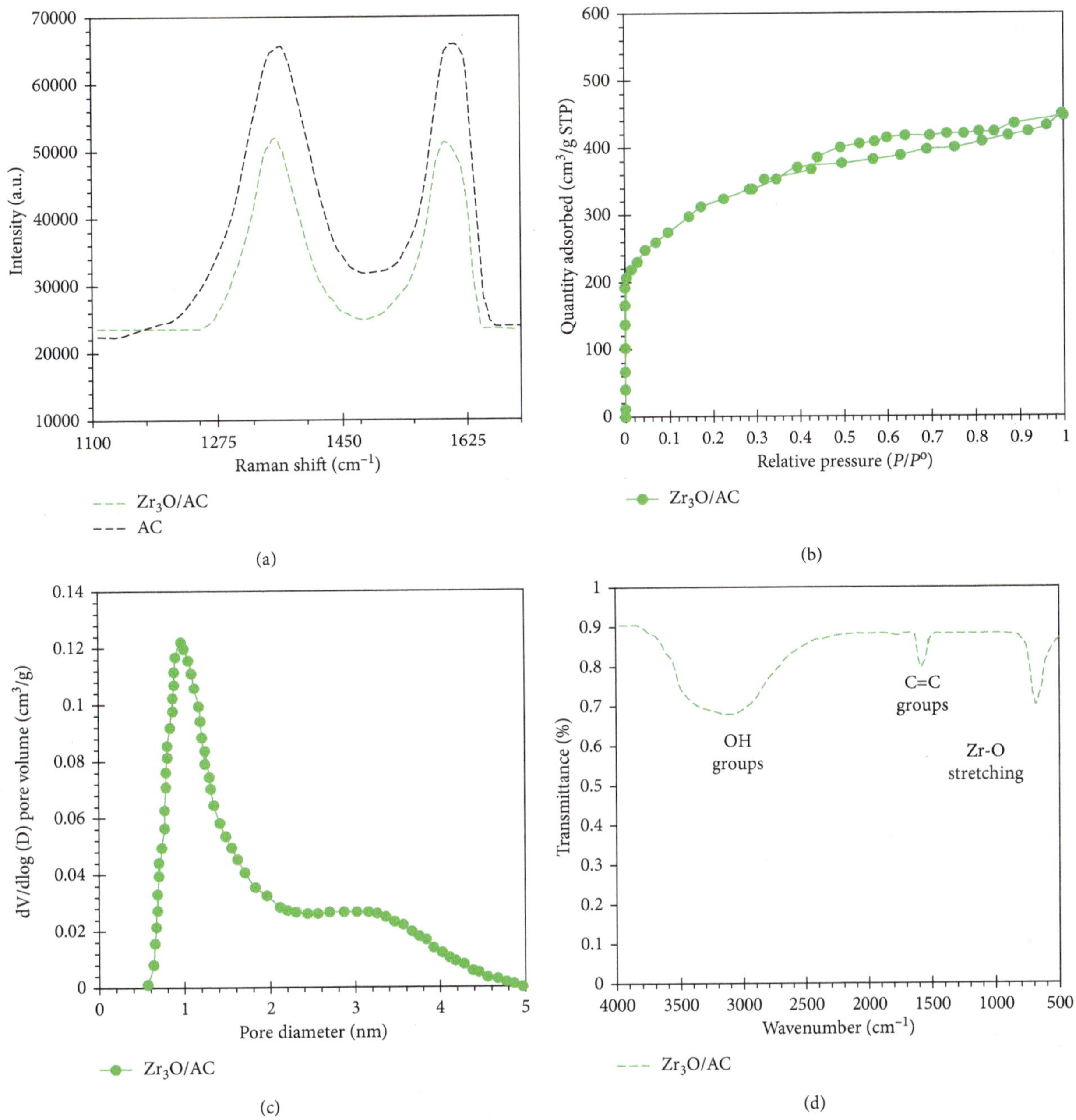


FIGURE 3: (a) Raman spectroscopy of the AC and $Zr_3O$/AC; (b) $N_2$ physisorption analyses; (c) pore size distribution curve; (d) FTIR spectroscopy of the $Zr_3O$/AC adsorbent.

stage could be due to the fast saturation of the binding sites in the $Zr_3O$/AC surface.

*3.2.2. Kinetics and Isotherm Adsorption.* In this section, pseudo-first-order (PFO), pseudo-second-order (PSO), and intraparticle diffusion (IPD) kinetic models were used to fit the experimental data of MB and CV adsorption on $Zr_3O$/AC. It is well established that the pseudo-first-order kinetic model has been widely applied for the adsorption of solid/liquid systems [45–47].

Table 3 lists different calculated parameters using PFO and PSO models. Figure S1 (Supporting information) shows the fitting of experimental data to the PSO model at different initial concentrations of both MB and CV dyes, and the correlation coefficient is more significant than those of the PFO model. Moreover, a large difference of $Q_e$ between the experiment ($Q_{e,exp}$) and calculation ($Q_{e,cal}$) is observed [48]. These data support the fact that the adsorption data are well represented by the PSO kinetics model for the entire adsorption period of MB and CV dyes onto $Zr_3O$/AC [49].

In order to gain more insights into the kinetics of the adsorption process, the IPD model was studied. The plot ($Q_t$ vs $t_{1/2}$) (Figure 7) shows an initial curved portion followed by a linear portion and a plateau.

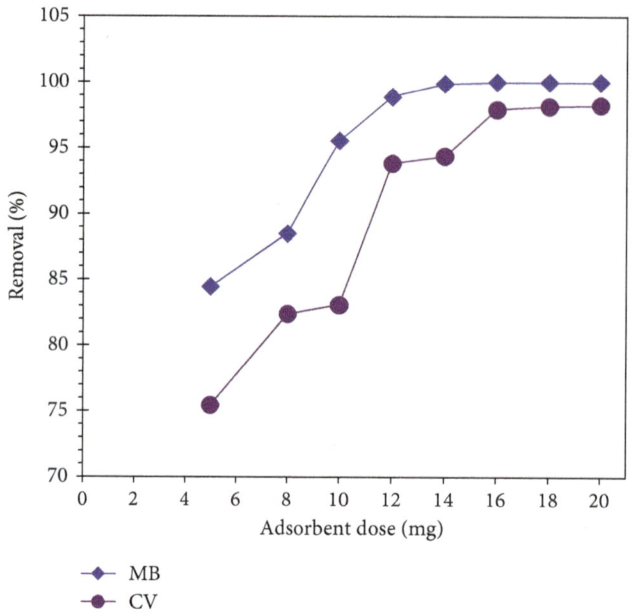


FIGURE 4: Effect of adsorbent dose on MB and CV dyes removal by $Zr_3O/AC$.

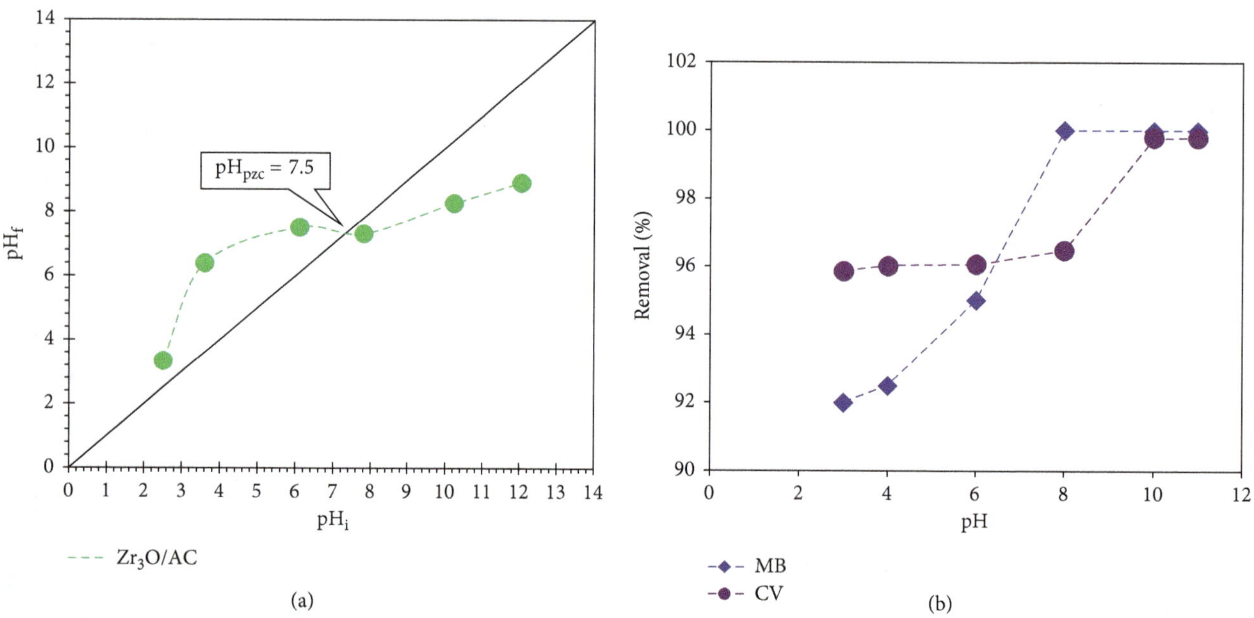


FIGURE 5: (a) Point of zero charge of the composite; (b) effect of initial pH solutions on MB and CV dyes removal by $Zr_3O/AC$.

It is clear that the adsorption processes of both MB and CV dyes can be roughly divided into three steps, as indicated by the guidelines for each portion: The first portion is attributed to the diffusion of the MB and CV molecule through solution to the external surface of the $Zr_3O/AC$ composite. This is a fast process that mainly depends on the surface area of the $Zr_3O/AC$. The second portion is attributed to gradual equilibrium with IPD as the rate-determining step. The last step is due to an adsorption/desorption equilibrium which is established for both MB and CV dyes [8]. As can be seen from the Figure 7, the $Zr_3O/AC$ exhibits faster adsorption kinetics at the initial adsorption stage due to its much higher surface area, as inferred in physisorption measurement.

The adsorption isotherm models were used to investigate the interaction between the MB and CV dyes and the $Zr_3O/AC$ when the adsorption process reaches equilibrium. Figure S2 shows plots of adsorption data of MB and CV dyes fitted to Langmuir and Freundlich isotherm models. Table 4 summarizes the isotherm parameters for this study. It is obvious that the Langmuir model provides a better fit for MB and Freundlich models for CV. A good fit with the Langmuir model indicate monolayer adsorption. However, the Freundlich model is indicating multilayer adsorption.

We have provided an extensive comparison of $Zr_3O/AC$ to the best performing materials reported to date in Table 5.

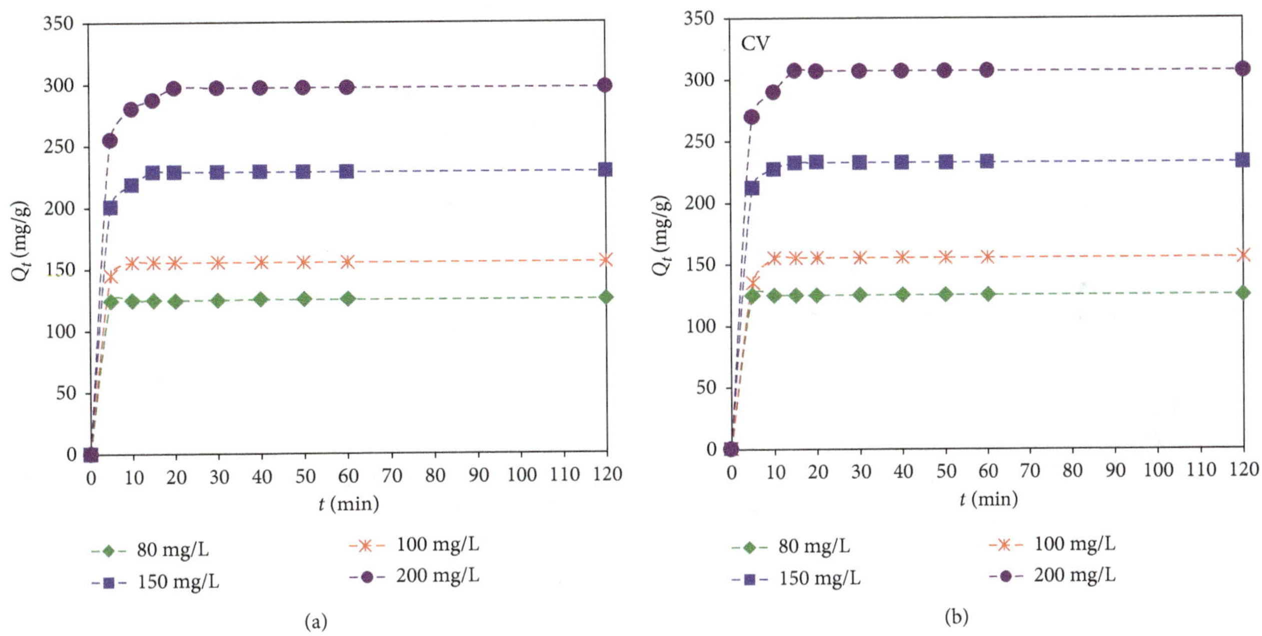


Figure 6: Effect of contact time on the removal of MB (a) and CV (b) onto $Zr_3O$/AC at 25°C.

Table 3: PFO and PSO fitting kinetic parameters of the experimental data.

| | $C_{intial}$ (mg/L) | $Q_{e,exp}$ (mg/g) | PFO | | | PSO | | |
|---|---|---|---|---|---|---|---|---|
| | | | $Q_{e,cal}$ (mg/g) | $K_1$ ($min^{-1}$) | $R^2$ | $Q_{e,cal}$ (mg/g) | $K_2$ (g/mg·min) | $R^2$ |
| MB | 80 | 125.000 | 0.752 | 0.005 | 0.054 | 125.000 | 0.128 | 0.999 |
| | 100 | 155.035 | 1.553 | 0.014 | 0.087 | 156.250 | 0.102 | 0.999 |
| | 150 | 228.507 | 2.837 | 0.032 | 0.163 | 227.273 | 0.022 | 0.999 |
| | 200 | 296.937 | 4.871 | 0.046 | 0.234 | 294.118 | 0.010 | 0.999 |
| CV | 80 | 125.000 | 0.760 | 0.005 | 0.049 | 125.000 | 0.107 | 0.999 |
| | 100 | 155.078 | 1.774 | 0.018 | 0.087 | 156.250 | 0.046 | 0.999 |
| | 150 | 232.510 | 2.355 | 0.026 | 0.156 | 232.558 | 0.031 | 0.999 |
| | 200 | 307.002 | 3.286 | 0.036 | 0.165 | 312.500 | 0.013 | 0.999 |

Our prepared material is one of the best adsorbents for the removal of both MB and CV.

*3.2.3. Regeneration of $Zr_3O$/AC.* Reusability was carried out by regenerating $Zr_3O$/AC with 1% sodium hydroxide at room temperature for 3 hours. The regenerated $Zr_3O$/AC was then washed with Milli-Q water and dried in a vacuum oven.

The $Zr_3O$/AC was separated and washed with distilled water. Recycling efficiency of $Zr_3O$/AC was investigated for the removal of MB and CV for 3 cycles. The results show that the adsorption efficiency of $Zr_3O$/AC was reduced to 88% for MB and to 82% for CV dye (Figure 8). We should note that, after every cycle, NaOH was used as a desorption medium to remove adsorbed MB ions from the $Zr_3O$/AC surface.

*3.2.4. Statistical Analysis and Model Fitting.* The significance of the independent variables and their interactions was tested with various statistical analyses such as analysis of variance (ANOVA), correlation coefficient ($R^2$), and adjusted correlation coefficient ($R^2_{adj}$) [58]. The Box-Behnken matrix used in this work is presented in Table 1S. The analysis of variance results (Table 6) suggests that the developed models (Equations (2) and (3)) are highly significant with a very low $P$ value for all models:

$$Y_{(R\%,\text{MB onto } ZrO_2\text{–AC})} = 99.498 + 0.379X_{AD} - 0.084X_{IC} + 0.112X_{CT} - 0.304X_{AD}^2 + 0.145X_{AD}X_{IC}, \quad (2)$$

$$Y_{(R\%,\text{VC onto } ZrO_2\text{–AC})} = 92.253 + 5.299X_{AD} - 0.686X_{IC} - 0.177X_{CT} - 3.287X_{AD}^2 + 3.058X_{IC}^2 + 2.956X_{CT}^2 + 0.555X_{AD}X_{IC}, \quad (3)$$

where $X_{AD}$, $X_{IC}$, and $X_{CT}$ represent the adsorbent dose, initial dyes concentration, and contact time, respectively.

The $R^2$ values of full second-order polynomial regression models were found to be 0.951 ($R^2_{adj}$ 0.895) and 0.972 ($R^2_{adj}$

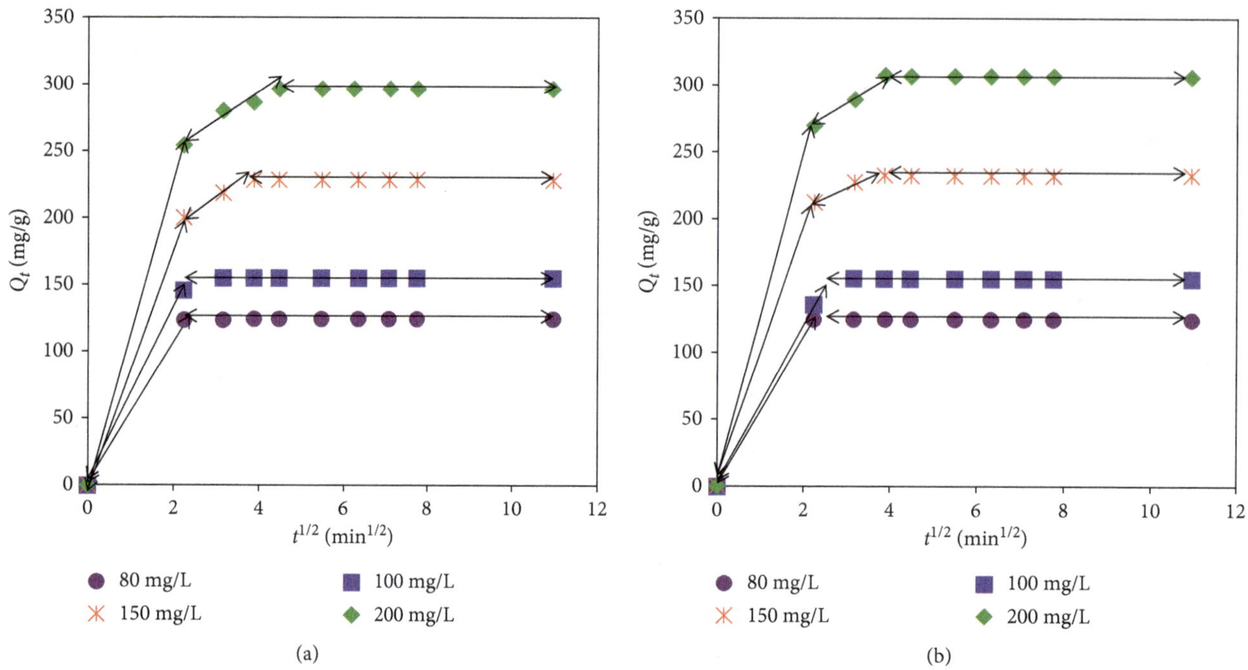


FIGURE 7: The plots of the intraparticle diffusion model of MB (a) and CV (b).

TABLE 4: Constants of Langmuir and Freundlich simulations of the MB and CV adsorption isotherms.

| | Langmuir | | | | Freundlich | | |
|---|---|---|---|---|---|---|---|
| | $Q_{max}$ (mg/g) | $K_L$ (L·min$^{-1}$) | $R_L$ | $R^2$ | $K_F$ (mg/g) | $n$ | $R^2$ |
| MB | 208.33 | 16.0 | 0.001 | 0.998 | 242.60 | 2.929 | 0.225 |
| CV | 169.49 | 11.80 | 0.001 | 0.492 | 204.12 | 0.910 | 0.915 |

TABLE 5: Comparison between different reported activated carbons and $Zr_3O$/AC composite.

| AC adsorbent | $Q_{max}$ (mg/g) | Ref. |
|---|---|---|
| *MB* | | |
| Karanj fruit hull | 154.8 | [50] |
| Macadamia nut endocarp | 194.70 | [51] |
| Wood | 60.97 | [52] |
| *Ficus carica* bast | 47.62 | [53] |
| Sawdust | 416.7 | [42] |
| This work | 208.33 | — |
| *CV* | | |
| Bamboo leaves powder | 393.16 | [54] |
| Male flowers of coconut tree | 85.84 | [55] |
| Male flowers of coconut tree | 60.42 | [55] |
| Waste apricot | 57.80 | [56] |
| Tomato plant root | 94.34 | [57] |
| This work | 204.12 | — |

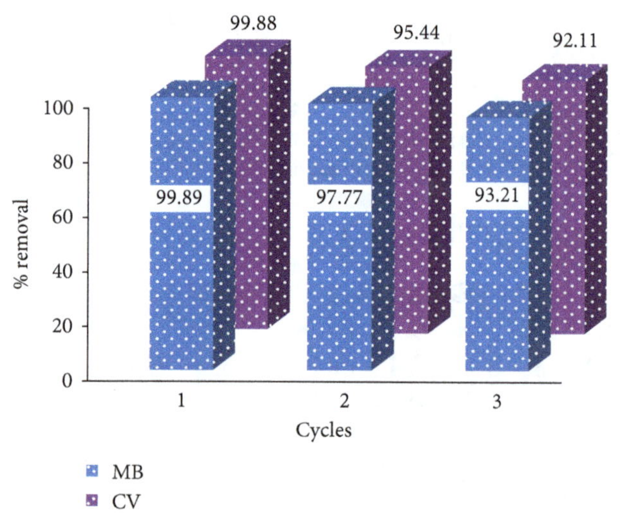


FIGURE 8: Regeneration efficiency of $Zr_3O$/AC for MB and CV removal.

0.952) for MB and CV adsorption on $Zr_3O$/AC, respectively [8]. Moreover, the control test was used to determine the significance of coefficients of the parameters [59]. The linear, quadratic, and interaction effects of the parameters are given in Table 7. The results confirm that most of the coefficients models were significant ($P$ values are less than 0.05) except quadratics effects of contact time, initial concentration, and the interactions effects of CT-IC and AD-CT, which are statistically insignificant ($P$ values are more than 0.05). The same conclusion was observed in the case of CV adsorption onto $Zr_3O$/AC.

*3.2.5. Optimization of CV and MB Using RSM.* Pareto diagrams (Figure S3) indicate the percent influence of different independent variables on MB and CV adsorption onto $Zr_3O$/AC. An observation of the Pareto diagram makes it possible to say that the $Zr_3O$/AC dose and initial

TABLE 6: Analysis of variance and coefficients of determination.

| | Sum of squares | Degrees of freedom | Mean square | $F_{obs}$ | $P$ value |
|---|---|---|---|---|---|
| *MB adsorption onto $Zr_3O/AC$* | | | | | |
| Regression | 1.8019 | 9 | 0.2002 | 189.4786 | <0.0001 |
| Residue | 0.0933 | 8 | 0.0117 | | |
| Lack of fit | 0.0880 | 3 | 0.0293 | 27.7681 | 0.00152 |
| Error | 0.0053 | 5 | 0.0011 | | |
| Total | 1.8952 | 17 | | | |
| $R^2$ | 0.951 | | | | |
| $R^2_A$ | 0.895 | | | | |
| *CV adsorption onto $Zr_3O/AC$* | | | | | |
| Regression | 348.6312 | 9 | 38.7368 | 6027.5101 | <0.0001 |
| Residue | 8.0777 | 8 | 1.0097 | | |
| Lack of fit | 8.0455 | 3 | 2.6818 | 417.2990 | <0.0001 |
| Error | 0.0321 | 5 | 0.0064 | | |
| Total | 356.7088 | 17 | | | |
| $R^2$ | 0.977 | | | | |
| $R^2_A$ | 0.952 | | | | |

TABLE 7: Analysis of the variance-reduced model.

| Source | Coefficient | $t$ exp | $P$ value |
|---|---|---|---|
| *MB adsorption onto $Zr_3O/AC$* | | | |
| $a_0$ | 99.498 | 7498 | <0.0001 |
| $a_{AD}$ | 0.379 | 32.96 | <0.0001 |
| $a_{IC}$ | −0.084 | −7.29 | 0.0007 |
| $a_{CT}$ | 0.112 | 9.79 | 0.00018 |
| $a^2_{AD}$ | −0.304 | −19.55 | <0.0001 |
| $a^2_{IC}$ | 0.006 | 0.37 | 0.723 |
| $a^2_{CT}$ | 0.013 | 0.86 | 0.431 |
| $a_{AD-IC}$ | 0.145 | 8.92 | 0.000295 |
| $a_{AD-CT}$ | −0.037 | −2.31 | 0.069 |
| $a_{CT-IC}$ | −0.008 | −0.46 | 0.664 |
| *CV adsorption onto $Zr_3O/AC$* | | | |
| $a_0$ | 92.253 | 2819 | <0.0001 |
| $a_{AD}$ | 5.299 | 186.95 | <0.0001 |
| $a_{IC}$ | −0.686 | −24.21 | <0.0001 |
| $a_{CT}$ | −0.177 | −4.15 | 0.0089 |
| $a^2_{AD}$ | −3.287 | −85.64 | <0.0001 |
| $a^2_{IC}$ | 3.058 | 79.69 | <0.0001 |
| $a^2_{CT}$ | 2.956 | 77.02 | <0.0001 |
| $a_{AD-IC}$ | 0.555 | 13.85 | <0.0001 |
| $a_{AD-CT}$ | −0.088 | −2.18 | 0.081 |
| $a_{CT-IC}$ | −0.003 | −0.06 | 0.953 |

concentration factors account more than 84.81% of the variation of the response in the case of MB adsorption onto $Zr_3O/AC$. For the adsorption of CV onto $Zr_3O/AC$, the $Zr_3O/AC$ dose and the quadratic effects of $Zr_3O/AC$ dose, contact time, and initial concentration are the most influential parameters on the adsorption efficiency, with an influence percentage of 48.59, 18.70, 16.18, and 15.12%, respectively.

From this study, it can be clearly seen that the $Zr_3O/AC$ dose (with a positive influence) and the initial concentration (with a negative influence) are the first factors influencing the MB and CV adsorption onto $Zr_3O/AC$ or $Zr_3O/AC$. This probably returns to the availability of active sites, when the $Zr_3O/AC$ mass of the support increases [60].

To achieve the maximum adsorption of the CV and MB dyes onto $Zr_3O/AC$ and AC, the RSM modeling was used (Figures 9(a) and 9(b)). The optimum values of different independent variables (initial dye concentration, $Zr_3O/AC$ dose, and contact time) are determined as shown in Table 8. It shows that the model-predicted values are in very good agreement with the experimentally determined values which confirm that the proposed model combined with RSM is an effective approach for modeling the adsorption process and to understand the relationships between the independent variables and response.

*3.2.6. Adsorption Mechanism.* From the FTIR, Raman, and EDS analyses, a large number of hydroxyl functional groups and carbonyl functional group on the surface of $Zr_3O/AC$ were observed. Knowing that the prepared material has a very high surface area, the functional groups are participating in the adsorption mechanism of CV and MB as a function of solution pH and the load of $Zr_3O/AC$ on the surface. After CV adsorption, the FTIR spectrum of the $Zr_3O/AC$, presented in Figure 10(a), shows characteristic changes due to the adsorption of CV onto $Zr_3O/AC$: the intensity of the peak representing the stretching vibrations of the O-H group shifted and the peak representing the stretching vibrations of the C-C group diminished. This might be due to the formation of chemical bonds between C-C of $Zr_3O/AC$ and electrophilic $N^+$ of CV. Furthermore, the Zr-O peak shifted to higher wavenumbers suggesting a formation of chemical bonds between Zr-O and $N^+$ of CV dye. The same behavior was observed in the case of MB adsorption on $Zr_3O/AC$. As shown in the FTIR spectroscopy after MB (Figure 10(b)), -OH, C=C, and Zr-O peaks shifted to different frequencies, confirming the interaction between these groups and our MB dye. Figure 10(c) shows the proposed model of MB and CV adsorption onto $Zr_3O/AC$, and similar observations were reported elsewhere [8]. From this study, the mechanism can be divided into these 4 fundamental steps:

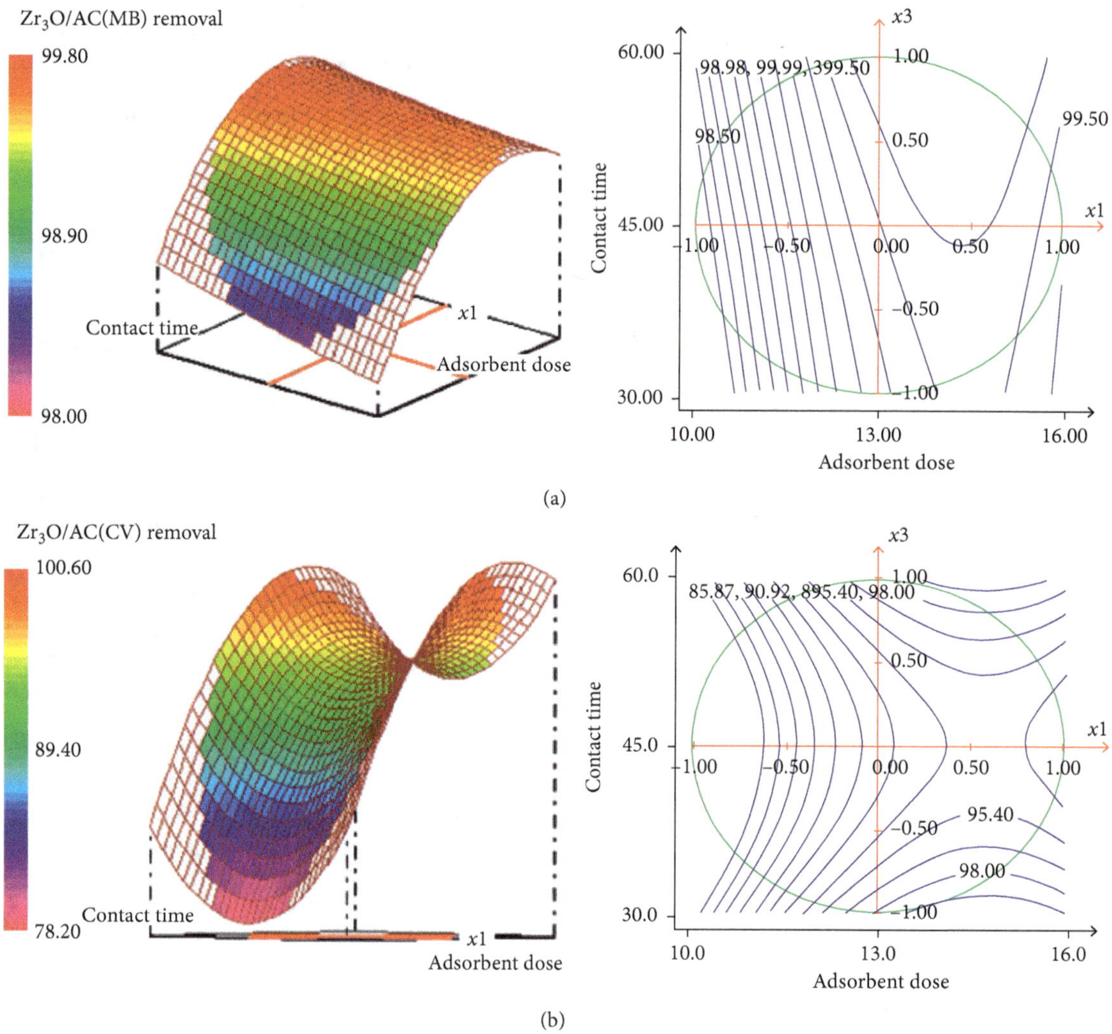


FIGURE 9: RSM and corresponding contour plot: (a) MB adsorption onto $Zr_3O$/AC as a function of adsorbent dose and contact time at the initial MB concentration constant; (b) CV adsorption onto $Zr_3O$/AC as a function of adsorbent dose and contact time at the initial MB concentration constant.

TABLE 8: Model validation.

| Conditions | Predicted | Experimental |
|---|---|---|
| Adsorbent dose = 12.66 mg | Adsorption of MB (%) onto $Zr_3O$/AC | |
| $V_m$ = 25 ml | Predicted | Experimental |
| pH of solution, | | |
| $C_0$ (MB) = 100 mg·L$^{-1}$ | | |
| $W$ = 30 rpm | 99.60 ± 0.07 | 99.25 |
| $T$°C = 24 ± 1°C | | |
| 58.08 min | | |
| Adsorbent dose = 14.92 mg | Adsorption of CV (%) onto $Zr_3O$/AC | |
| $V_m$ = 25 ml | Predicted | Experimental |
| pH of solution, | | |
| $C_0$ (MB) = 100 mg·L$^{-1}$ | | |
| $W$ = 30 rpm | 99.50 ± 0.22 | 99.01 |
| $T$°C = 24 ± 1°C | | |
| 31.35 min | | |

(i) The adsorption of MB and CV molecules inside the $Zr_3O$/AC pores and the interaction between Zr-O and nitrogen of CV and MB

(ii) The hydrogen bonding would be produced between hydroxyl functional groups of $Zr_3O$/AC and the nitrogen of CV and MB

(iii) The $\pi$-$\pi$ interaction between the double bonds of AC and molecule and that of pollutants

## 4. Conclusion

$Zr_3O$/AC was prepared by a facile chemical route and was characterized by different structural and textural techniques. The composite $Zr_3O$/AC was used as an efficient adsorbent for the MB and CV removal. $Zr_3O$/AC presented irregular particle size with a regular channel array and micro-mesopores. Also, numerous functional organic groups on the surface were observed by FTIR spectroscopy before and after MB and CV adsorption. Functional groups on the surface along with the porous structure, channel array, and physical interactions ($\pi$-$\pi$ and hydrogen) between the

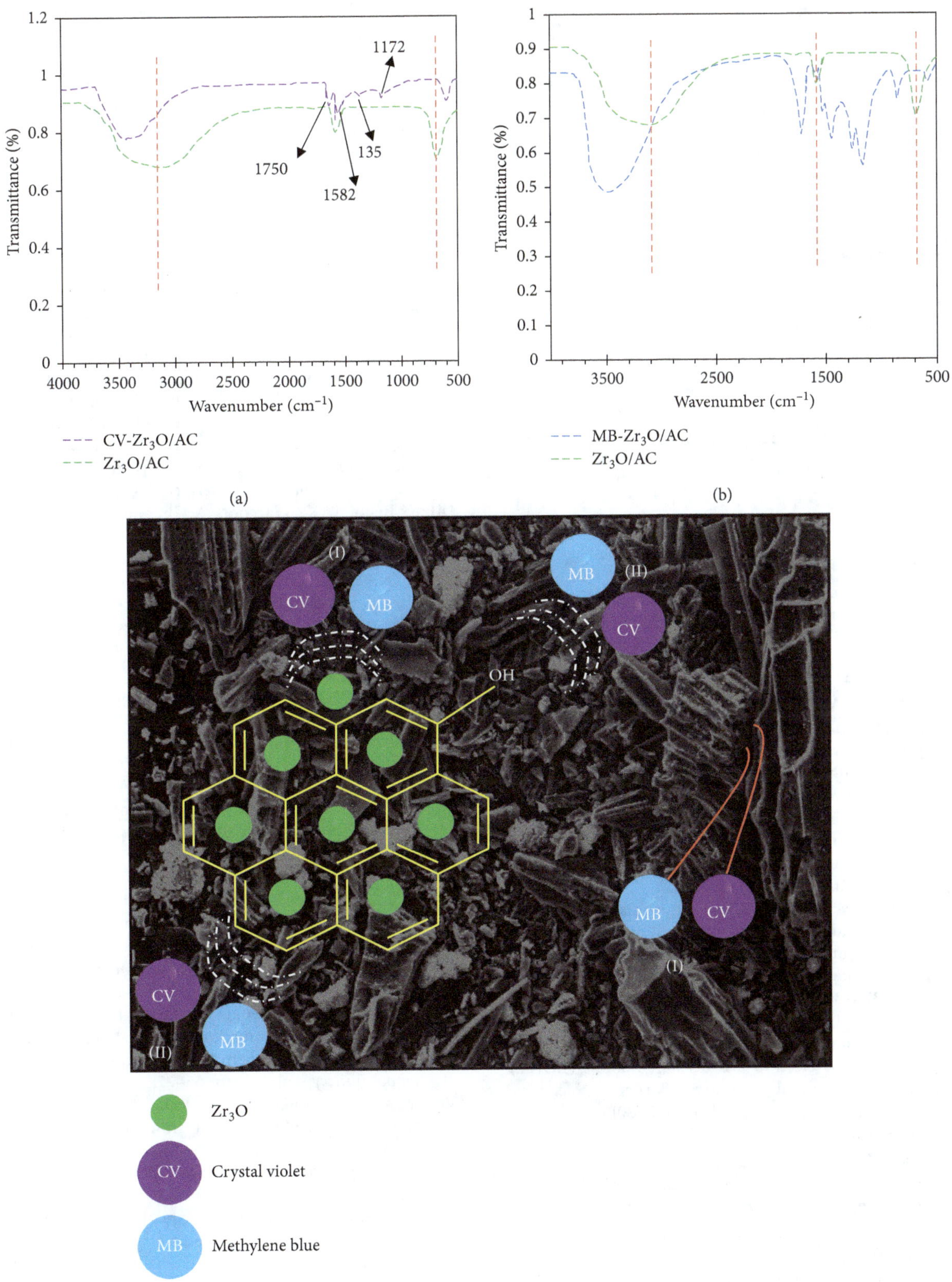


FIGURE 10: FTIR spectra of (a) $Zr_3O$/AC- and CV-loaded $Zr_3O$/AC, (b) $Zr_3O$/AC- and MB-loaded $Zr_3O$/AC, and (c) proposed adsorption mechanism.

$Zr_3O$/AC and the dyes are the main factors controlling the adsorption capacity. Langmuir and Freundlich models better described the adsorption process for MB and CV removal, respectively. The adsorption process followed the PSO kinetic model. Statistical response modeling together with the regression model helped us to optimize the experimental factors which provided a good fit with the experimental values.

## Conflicts of Interest

The authors declare that there no conflicts of interest regarding the publication of this paper.

## Acknowledgments

The authors acknowledge the support of Ibn Zohr University and MASCIR Foundation for the support and help during SEM analyses.

## References

[1] F. Ansari, M. Ghaedi, M. Taghdiri, and A. Asfaram, "Application of ZnO nanorods loaded on activated carbon for ultrasonic assisted dyes removal: experimental design and derivative spectrophotometry method," *Ultrasonics Sonochemistry*, vol. 33, pp. 197–209, 2016.

[2] A. M. M. Vargas, A. L. Cazetta, M. H. Kunita, T. L. Silva, and V. C. Almeida, "Adsorption of methylene blue on activated carbon produced from flamboyant pods (*Delonix regia*): study of adsorption isotherms and kinetic models," *Chemical Engineering Journal*, vol. 168, no. 2, pp. 722–730, 2011.

[3] Y. Li, Q. Du, T. Liu et al., "Comparative study of methylene blue dye adsorption onto activated carbon, graphene oxide, and carbon nanotubes," *Chemical Engineering Research and Design*, vol. 91, no. 2, pp. 361–368, 2013.

[4] M. N. Ashiq, M. Najam-Ul-Haq, T. Amanat, A. Saba, A. M. Qureshi, and M. Nadeem, "Removal of methylene blue from aqueous solution using acid/base treated rice husk as an adsorbent," *Desalination and Water Treatment*, vol. 49, no. 1–3, pp. 376–383, 2012.

[5] Y. Ju, J. Fang, X. Liu et al., "Photodegradation of crystal violet in $TiO_2$ suspensions using UV–vis irradiation from two microwave-powered electrodeless discharge lamps (EDL-2): products, mechanism and feasibility," *Journal of Hazardous Materials*, vol. 185, no. 2-3, pp. 1489–1498, 2011.

[6] Y. Miyah, A. Lahrichi, M. Idrissi, S. Boujraf, H. Taouda, and F. Zerrouq, "Assessment of adsorption kinetics for removal potential of crystal violet dye from aqueous solutions using Moroccan pyrophyllite," *Journal of the Association of Arab Universities for Basic and Applied Sciences*, vol. 23, no. 1, pp. 20–28, 2017.

[7] H. Ait Ahsaine, M. Ezahri, A. Benlhachemi et al., "Novel Lu-doped $Bi_2WO_6$ nanosheets: synthesis, growth mechanisms and enhanced photocatalytic activity under UV-light irradiation," *Ceramics International*, vol. 42, no. 7, pp. 8552–8558, 2016.

[8] H. A. Ahsaine, M. Zbair, Z. Anfar et al., "Cationic dyes adsorption onto high surface area "almond shell" activated carbon: kinetics, equilibrium isotherms and surface statistical modeling," *Materials Today Chemistry*, vol. 8, pp. 121–132, 2018.

[9] I. Ali and V. K. Gupta, "Advances in water treatment by adsorption technology," *Nature Protocols*, vol. 1, no. 6, pp. 2661–2667, 2007.

[10] H. A. Ahsaine, M. Zbair, and R. El Haouti, "Mesoporous treated sewage sludge as outstanding low-cost adsorbent for cadmium removal," *Desalination and Water Treatment*, vol. 85, pp. 330–338, 2017.

[11] C.-Y. Cao, J. Qu, F. Wei, H. Liu, and W.-G. Song, "Superb adsorption capacity and mechanism of flowerlike magnesium oxide nanostructures for lead and cadmium ions," *ACS Applied Materials & Interfaces*, vol. 4, no. 8, pp. 4283–4287, 2012.

[12] S.-W. Nam, D.-J. Choi, S.-K. Kim, N. Her, and K.-D. Zoh, "Adsorption characteristics of selected hydrophilic and hydrophobic micropollutants in water using activated carbon," *Journal of Hazardous Materials*, vol. 270, pp. 144–152, 2014.

[13] V. Ranjithkumar, S. Sangeetha, and S. Vairam, "Synthesis of magnetic activated carbon/α-$Fe_2O_3$ nanocomposite and its application in the removal of acid yellow 17 dye from water," *Journal of Hazardous Materials*, vol. 273, pp. 127–135, 2014.

[14] E. Repo, J. K. Warchoł, A. Bhatnagar, and M. Sillanpää, "Heavy metals adsorption by novel EDTA-modified chitosan–silica hybrid materials," *Journal of Colloid and Interface Science*, vol. 358, no. 1, pp. 261–267, 2011.

[15] T. Zhai, S. Xie, X. Lu et al., "Porous $Pr(OH)_3$ nanostructures as high-efficiency adsorbents for dye removal," *Langmuir*, vol. 28, no. 30, pp. 11078–11085, 2012.

[16] Y. Zhang, Y. Li, L. Yang, X. Ma, L. Wang, and Z.-F. Ye, "Characterization and adsorption mechanism of $Zn^{2+}$ removal by PVA/EDTA resin in polluted water," *Journal of Hazardous Materials*, vol. 178, no. 1–3, pp. 1046–1054, 2010.

[17] A. Hofmann, S. J. Clark, M. Oppel, and I. Hahndorf, "Hydrogen adsorption on the tetragonal $ZrO_2$(101) surface: a theoretical study of an important catalytic reactanty," *Physical Chemistry Chemical Physics*, vol. 4, no. 14, pp. 3500–3508, 2002.

[18] V. Hornebecq, C. Knöfel, P. Boulet, B. Kuchta, and P. L. Llewellyn, "Adsorption of carbon dioxide on mesoporous zirconia: microcalorimetric measurements, adsorption isotherm modeling, and density functional theory calculations," *Journal of Physical Chemistry C*, vol. 115, no. 20, pp. 10097–10103, 2011.

[19] A. Mullick and S. Neogi, "Acoustic cavitation induced synthesis of zirconium impregnated activated carbon for effective fluoride scavenging from water by adsorption," *Ultrasonics Sonochemistry*, vol. 45, pp. 65–77, 2018.

[20] W. Xiong, J. Tong, Z. Yang et al., "Adsorption of phosphate from aqueous solution using iron-zirconium modified activated carbon nanofiber: performance and mechanism," *Journal of Colloid and Interface Science*, vol. 493, pp. 17–23, 2017.

[21] S. Banerjee, S. R. Joshi, T. Mandal, and G. Halder, "Application of zirconium caged activated biochar alginate beads towards deionization of Cr(VI) laden water in a fixed bed column reactor," *Journal of Environmental Chemical Engineering*, vol. 6, no. 4, pp. 4018–4029, 2018.

[22] P. Suresh, J. J. Vijaya, and L. J. Kennedy, "Photocatalytic degradation of textile-dyeing wastewater by using a microwave combustion-synthesized zirconium oxide supported activated carbon," *Materials Science in Semiconductor Processing*, vol. 27, pp. 482–493, 2014.

[23] R. K. Prasad, "Degradation of biopolymeric pigments in distillery spentwash by electrocoagulation using copper anodes: statistical and canonical analysis," *Environmental Chemistry Letters*, vol. 8, no. 2, pp. 149–155, 2010.

[24] M. H. Kalavathy, I. Regupathi, M. Ganesa, and L. Rose, "Modelling, analysis and optimization of adsorption parameters for $H_3PO_4$ activated rubber wood sawdust using response surface methodology (RSM)," *Colloids and Surfaces B: Biointerfaces*, vol. 70, no. 1, pp. 35–45, 2009.

[25] J. N. Sahu, J. Acharya, and B. C. Meikap, "Response surface modeling and optimization of chromium(VI) removal from aqueous solution using Tamarind wood activated carbon in batch process," *Journal of Hazardous Materials*, vol. 172, no. 2-3, pp. 818–825, 2009.

[26] H. Yang, D. Zhou, Z. Chang, and L. Zhang, "Adsorption of crystal violet onto amino silica: optimization, equilibrium, and kinetic studies," *Desalination and Water Treatment*, vol. 52, no. 31–33, pp. 6113–6121, 2014.

[27] Z. Anfar, M. Zbair, H. A. Ahsaine, M. Ezahri, and N. El Alem, "Well-designed $WO_3$/activated carbon composite for Rhodamine B removal: synthesis, characterization, and modeling using response surface methodology," *Fullerenes, Nanotubes and Carbon Nanostructures*, vol. 26, no. 6, pp. 389–397, 2018.

[28] M. Zbair, H. Ait Ahsaine, and Z. Anfar, "Porous carbon by microwave assisted pyrolysis: an effective and low-cost adsorbent for sulfamethoxazole adsorption and optimization using response surface methodology," *Journal of Cleaner Production*, vol. 202, pp. 571–581, 2018.

[29] Z. Anfar, R. El Haouti, S. Lhanafi, M. Benafqir, Y. Azougarh, and N. El Alem, "Treated digested residue during anaerobic co-digestion of Agri-food organic waste: methylene blue adsorption, mechanism and CCD-RSM design," *Journal of Environmental Chemical Engineering*, vol. 5, no. 6, pp. 5857–5867, 2017.

[30] I. Langmuir, "The constitution and fundamental properties of solids and liquids. Part I. Solids," *Journal of the American Chemical Society*, vol. 38, no. 11, pp. 2221–2295, 1916.

[31] H. M. F. Freundlich, "Über die adsorption in lösungen," *Zeitschrift für Physikalische Chemie*, vol. 57A, pp. 385–470, 1906.

[32] M. Seredych, O. Mabayoje, M. M. Kolesnik, V. Krstic, and T. J. Bandosz, "Zinc (hydr)oxide/graphite based-phase composites: effect of the carbonaceous phase on surface properties and enhancement in electrical conductivity," *Journal of Materials Chemistry*, vol. 22, no. 16, pp. 7970–7978, 2012.

[33] Y. Sun, H. Li, G. Li, B. Gao, Q. Yue, and X. Li, "Characterization and ciprofloxacin adsorption properties of activated carbons prepared from biomass wastes by $H_3PO_4$ activation," *Bioresource Technology*, vol. 217, pp. 239–244, 2016.

[34] Y. Li, Q. Du, X. Wang et al., "Removal of lead from aqueous solution by activated carbon prepared from *Enteromorpha prolifera* by zinc chloride activation," *Journal of Hazardous Materials*, vol. 183, no. 1–3, pp. 583–589, 2010.

[35] S. Ghorai and K. K. Pant, "Equilibrium, kinetics and breakthrough studies for adsorption of fluoride on activated alumina," *Separation and Purification Technology*, vol. 42, no. 3, pp. 265–271, 2005.

[36] A. Sadezky, H. Muckenhuber, H. Grothe, R. Niessner, and U. Pöschl, "Raman microspectroscopy of soot and related carbonaceous materials: spectral analysis and structural information," *Carbon*, vol. 43, no. 8, pp. 1731–1742, 2005.

[37] W. Song, B. Gao, X. Xu et al., "Adsorption of nitrate from aqueous solution by magnetic amine-crosslinked biopolymer based corn stalk and its chemical regeneration property," *Journal of Hazardous Materials*, vol. 304, pp. 280–290, 2016.

[38] S. Vasudevan and J. Lakshmi, "The adsorption of phosphate by graphene from aqueous solution," *RSC Advances*, vol. 2, no. 12, pp. 5234–5242, 2012.

[39] S. Cheng, L. Zhang, H. Xia, and J. Peng, "Characterization and adsorption properties of La and Fe modified activated carbon for dye wastewater treatment," *Green Processing and Synthesis*, vol. 6, no. 5, p. 487, 2017.

[40] Z. Heidarinejad, O. Rahmanian, M. Fazlzadeh, and M. Heidari, "Enhancement of methylene blue adsorption onto activated carbon prepared from Date Press Cake by low frequency ultrasound," *Journal of Molecular Liquids*, vol. 264, pp. 591–599, 2018.

[41] M. E. Mahmoud, A. E. H. Abdou, and M. E. Sobhy, "Engineered nano-zirconium oxide-crosslinked-nanolayer of carboxymethyl cellulose for speciation and adsorptive removal of Cr(III) and Cr(VI)," *Powder Technology*, vol. 321, pp. 444–453, 2017.

[42] M. Zbair, Z. Anfar, H. Khallok, H. A. Ahsaine, M. Ezahri, and N. Elalem, "Adsorption kinetics and surface modeling of aqueous methylene blue onto activated carbonaceous wood sawdust," *Fullerenes, Nanotubes and Carbon Nanostructures*, vol. 26, no. 7, pp. 433–442, 2018.

[43] A. H. Jawad, R. A. Rashid, M. A. M. Ishak, and L. D. Wilson, "Adsorption of methylene blue onto activated carbon developed from biomass waste by $H_2SO_4$ activation: kinetic, equilibrium and thermodynamic studies," *Desalination and Water Treatment*, vol. 57, no. 52, pp. 25194–25206, 2016.

[44] G. Crini, H. N. Peindy, F. Gimbert, and C. Robert, "Removal of C.I. Basic Green 4 (Malachite Green) from aqueous solutions by adsorption using cyclodextrin-based adsorbent: kinetic and equilibrium studies," *Separation and Purification Technology*, vol. 53, no. 1, pp. 97–110, 2007.

[45] S. Lagergren, "Zur theorie der sogenannten adsorption geloester stoffe, kungliga svenska Vetenskapsakademiens," *Handlingar*, vol. 24, no. 4, pp. 1–39, 1898.

[46] Y. S. Ho and G. McKay, "Pseudo-second order model for sorption processes," *Process Biochemistry*, vol. 34, no. 5, pp. 451–465, 1999.

[47] W. J. Weber and J. C. Morris, "Intraparticule diffusion during the sorption of surfactants onto activated carbon," *Journal of Sanitary Engineering Division*, vol. 89, pp. 53–61, 1963.

[48] S. E. A. S. El-Deen and F.-S. Zhang, "Immobilisation of $TiO_2$-nanoparticles on sewage sludge and their adsorption for cadmium removal from aqueous solutions," *Journal of Experimental Nanoscience*, vol. 11, no. 4, pp. 239–258, 2016.

[49] Y. Chen, S. R. Zhai, N. Liu, Y. Song, Q. D. An, and X. W. Song, "Dye removal of activated carbons prepared from NaOH-pretreated rice husks by low-temperature solution-processed carbonization and $H_3PO_4$ activation," *Bioresource Technology*, vol. 144, pp. 401–409, 2013.

[50] M. A. Islam, S. Sabar, A. Benhouria, W. A. Khanday, M. Asif, and B. H. Hameed, "Nanoporous activated carbon prepared from karanj (*Pongamia pinnata*) fruit hulls for methylene blue adsorption," *Journal of the Taiwan Institute of Chemical Engineers*, vol. 74, pp. 96–104, 2017.

[51] O. Pezoti Junior, A. L. Cazetta, R. C. Gomes et al., "Synthesis of $ZnCl_2$-activated carbon from macadamia nut endocarp (*Macadamia integrifolia*) by microwave-assisted pyrolysis:

optimization using RSM and methylene blue adsorption," *Journal of Analytical and Applied Pyrolysis*, vol. 105, pp. 166–176, 2014.

[52] T. Wang, S. Tan, and C. Liang, "Preparation and characterization of activated carbon from wood via microwave-induced $ZnCl_2$ activation," *Carbon*, vol. 47, no. 7, pp. 1880–1883, 2009.

[53] D. Pathania, S. Sharma, and P. Singh, "Removal of methylene blue by adsorption onto activated carbon developed from Ficus carica bast," *Arabian Journal of Chemistry*, vol. 10, pp. S1445–S1451, 2017.

[54] S. K. Ghosh, A. K. Hajra, and A. Bandyopadhyay, "Air agitated tapered bubble column adsorber for hazardous dye (crystal violet) removal onto activated ($ZnCl_2$) carbon prepared from bamboo leaves," *Journal of Molecular Liquids*, vol. 240, pp. 313–321, 2017.

[55] S. Senthilkumaar, P. Kalaamani, and C. Subburaam, "Liquid phase adsorption of crystal violet onto activated carbons derived from male flowers of coconut tree," *Journal of Hazardous Materials*, vol. 136, no. 3, pp. 800–808, 2006.

[56] C. A. Başar, "Applicability of the various adsorption models of three dyes adsorption onto activated carbon prepared waste apricot," *Journal of Hazardous Materials*, vol. 135, no. 1-3, pp. 232–241, 2006.

[57] T. Kannan, C. Buvaneswari, and N. Palvannan, "Removal of plant poisoning dyes by adsorption on Tomato Plant Root and green carbon from aqueous solution and its recovery," *Desalination*, vol. 249, no. 3, pp. 1132–1138, 2009.

[58] M. Zbair, Z. Anfar, H. A. Ahsaine, N. El Alem, and M. Ezahri, "Acridine orange adsorption by zinc oxide/almond shell activated carbon composite: operational factors, mechanism and performance optimization using central composite design and surface modeling," *Journal of Environmental Management*, vol. 206, pp. 383–397, 2018.

[59] S. Manna, D. Roy, P. Saha, D. Gopakumar, and S. Thomas, "Rapid methylene blue adsorption using modified lignocellulosic materials," *Process Safety and Environmental Protection*, vol. 107, pp. 346–356, 2017.

[60] B. Zhang, X. Han, P. Gu, S. Fang, and J. Bai, "Response surface methodology approach for optimization of ciprofloxacin adsorption using activated carbon derived from the residue of desilicated rice husk," *Journal of Molecular Liquids*, vol. 238, pp. 316–325, 2017.

# 15

# Fabrication of Functionalized MOFs Incorporated Mixed Matrix Hollow Fiber Membrane for Gas Separation

**Haitao Zhu,[1,2] Xingming Jie,[1] and Yiming Cao[1]**

[1] *Dalian Nation Library for Clean Energy (DNL), Dalian Institute of Chemical Physics, Chinese Academy of Sciences, Dalian 116023, China*
[2] *University of Chinese Academy of Sciences, Beijing 100049, China*

Correspondence should be addressed to Xingming Jie; jiexm@dicp.ac.cn and Yiming Cao; ymcao@dicp.ac.cn

Academic Editor: Hossein Kazemian

The metal-organic framework (MOFs) of MIL-53 was functionalized by aminosilane grafting and then incorporated into Ultem®1000 polymer matrix to fabricate mixed matrix hollow fiber membrane (MMHFM) with high separation performance. SEM, XRD, and TGA were performed to characterize the functionalized MIL-53 and prepared MMHFM. The filler particles were embedded in membrane successfully and dispersed well in the polymer matrix. The incorporation of MOFs endowed MMHFM better thermal stability. Moreover, effects of solvent ratio in spinning dope, spinning condition, and testing temperature on gas separation performance of MMHFM were investigated. By optimizing dope composition, air gap distance, and bore fluid composition, MMHFM containing functionalized MIL-53 achieved excellent gas permeance and $CO_2/N_2$ selectivity. The $CO_2$ permeance increased from 12.2 GPU for pure Ultem HFM to 30.9 GPU and the ideal $CO_2/N_2$ selectivity was enhanced from 25.4 to 34.7 simultaneously. Additionally, gas permeance increased but the selectivity decreased with the temperature increase, which followed the solution-diffusion based transport mechanism.

## 1. Introduction

Membrane based gas separation and purification hold promise as energy-efficient technology if compared with traditional methods such as adsorption and cryogenic distillation [1–4]. However, the permeability-selectivity tradeoff described as Robeson upper bound restricted gas separation performance of polymeric membranes [5]. Several solutions have been proposed in past decades to further improve membrane performance. One of the most effective routes to enhance the separation performance is fabrication of mixed matrix membrane (MMM) that combines inorganic materials with polymer matrix [6, 7]. Regarding the filler materials, metal-organic frameworks (MOFs) are a satisfactory alternative to improve the permeation properties of membranes because of its large surface area, high porosity, adsorption capacity, and good compatibility [8–10].

Over the past few years, various MOFs have been incorporated into membranes to improve gas separation performance, such as Matrimid®5218/ZIF-8 [11], PBI/ZIF-7 [12], PI/$Mg_2$(dobdc) [13], and Ultem/ZIF-8 [14]. Appropriate structures of MOFs are efficient for gas separation application as previous references reported [15–19]. Couck et al. [20] revealed that the amino-functionalized MOFs $NH_2$-MIL-53 achieved excellent performance for $CO_2$ capture. Moreover, MMM based on $NH_2$-MIL-53 with improved structural feature and gas separation performance was synthesized by incorporating the filler particles in polyimide matrix. The membrane containing 25 wt% MOFs loading presented a 50% increment in $CO_2$ permeability, while retaining the separation selectivity [21]. The amino-functionalized MOFs filled MMMs have been also fabricated by Seoane et al. and the membrane performance was improved approaching the Robeson $CO_2/CH_4$ and $H_2/CH_4$ upper bound with 10 wt% MOFs loading [22]. In addition, other researchers have fabricated mixed matrix composite membrane by synthesizing MOFs continuous membrane on outer-side or inner-side of hollow fiber membrane to improve gas separation performance [23–26].

However, the filler particle aggregation and interfacial isolation are the major drawbacks that hinder the development of inorganic filler based MMMs [27, 28]. As a feasible method, the certain functions on MOFs could lead to enhanced interaction with functional groups of polymer chains [28]. For example, in NH2-MIL-53 incorporated MMM, the hydrogen bonding between the sulfone group of polysulfone and amine of MOFs enhanced compatibility at the polymer/filler interface [29]. Here, the present work prepared functionalized MOFs by grafting aminosilane to improve the interfacial affinity between filler particles and polymer matrix. The $-NH_2$ group on the functionalized MOFs could interact with carbonyl group of Ultem 1000 and the grafting alkyl chains also enhanced the adhesion with polymer matrix. Moreover, the dispersion of filler particles was uniform in the aid of silicane modification. In order to prepare asymmetric MMHFM with optimal membrane structure and remarkable separation performance, we investigated the important factors including solvent/nonsolvent ratio in spinning dope solution, spinning condition (air gap distance and bore fluid composition), and testing temperature.

## 2. Experimental

*2.1. Materials.* Ultem 1000 polymer was purchased from GE Plastics in pellet form and was dried 140°C for 12 h before use. Terephthalic acid ($H_2BDC$) was purchased from Sinopharm Chemical Reagent Co., Ltd. 3-Aminopropyltrimethoxysilane (APTMS) was purchased from J&K Scientific Ltd., Beijing, China. N,N-dimethylformamide (DMF), $Al(NO_3)_3 \cdot 9H_2O$, N-methyl-2-pyrrolidone (NMP), anhydrous ethanol (EtOH), n-hexane, petroleum ether, and aqueous ammonia (25 wt%) were purchased from Kemiou Chemical Reagent Co., Ltd., Tianjin, China. Polydimethylsiloxane (PDMS, Sylgard 184) was produced by Dow Chemicals.

*2.2. Synthesis of MIL-53.* MIL-53 was synthesized by the solvothermal method [30, 31]. First, $Al(NO_3)_3 \cdot 9H_2O$ (4.22 g) and $H_2BDC$ (1.245 g) were added into the mixed solvent of 20 mL of deionized water and 55 mL of DMF and magnetically stirred for 40 min. Then the solution was transferred into a Teflon-lined steel autoclave and heated in an air oven at 150°C for 48 h. The product was centrifuged and washed with deionized water three times, followed by immersing in DMF for 8 h. Afterward, the white powder was filtered from DMF and dried in an air oven for 12 h at 150°C. Finally, the fine powdered MIL-53 was further activated at 180°C under vacuum for 12 h.

*2.3. Functionalization of MIL-53.* MIL-53 particles (2.0 g) were dispersed in the mixed solvent of 2.6 g of deionized water, 60 mL of EtOH, and 1.36 g of aqueous ammonia. The suspension was stirred and sonicated in a round-bottomed flask at 60°C for 2 h. Then 1.44 g of APTMS was added dropwise into the mixture under stirring. The suspension reacted at room temperature for 12 h and then was centrifuged to obtain the modified product and washed with EtOH at least three times to remove residual APTMS. Finally, the aminosilane modified MIL-53 was dried under vacuum at 80°C for 24 h.

*2.4. Mixed Matrix Hollow Fiber Membrane Preparation*

*2.4.1. Preparation of Spinning Dope.* A certain amount of functionalized MIL-53 was added to a portion of the required NMP with stirred and sonicated alternatively for 45 min to obtain a well-dispersed suspension. Then MOFs filler was primed by addition of a dilute polymer solution into the suspension to ensure better adhesion. This mixture was stirred for 3 h and sonicated for 45 min until it displayed in uniform consistency. Then a required amount of EtOH was added with the remaining NMP into the suspension. Finally, the Ultem 1000 pellet was gradually added into the mixture to increase dope viscosity. After adding all the components, the suspension was mechanically stirred at 60°C for about 12 h to fully dissolve the polymer. For comparison purposes, the pure Ultem dope solution was prepared following an identical procedure. The polymer concentration in all dope solutions kept 33 wt% and the MOFs loading kept 15 wt% which was determined by

$$\text{MOFs loading (wt\%)} = \left(\frac{\text{wt. MOFs}}{\text{wt. MOFs} + \text{wt. polymer}}\right) \times 100\%. \quad (1)$$

*2.4.2. MMHFM Fabrication.* Asymmetric hollow fiber membranes in this study were fabricated via the dry-wet spinning method. The homogeneous dope and bore fluid were coextruded from a spinneret after dope solution was degassed; then the nascent fiber entered a water coagulant bath with a range of spinning parameters (Table 1) through the air gap distance and was wound up on a take-up drum. The spinning process was performed at 60°C and a filter was attached upstream of the spinneret to prevent impurity particles blocking the narrow spinneret orifice. Proper pressure was supplied by the nitrogen cylinder to maintain a constant spinning rate and a syringe pump was used to control the flow rate of bore fluid.

The as-fabricated fibers were rinsed in flowing tap water for about 24 h to remove the remaining solvent. Then the fibers were exchanged by immersing in three successive aliquots of EtOH for 30 min, followed by 30 min immersion in three successive aliquots of n-hexane. Afterward, the hollow fiber membranes were dried at 70°C for 3 h.

For sealing the microvoids or defects in the skin layer, each membrane module containing fibers was dip-coated by 3 wt% Sylgard 184 solution with a corresponding curing agent (10 wt% of Sylgard 184 PDMS) in petroleum ether. After coating for 30 min, the modules were placed into a 70°C oven and cured for 3 h after petroleum ether evaporation. The coated hollow fiber membranes were tested to evaluate gas permeation properties.

*2.5. Characterizations.* The morphologies of functionalized MOFs and membranes were examined by scanning electron

TABLE 1: Detailed spinning parameters of hollow fiber membrane fabrication.

| Spinning parameters | Value |
|---|---|
| Spinneret dimension (mm) | o.d./i.d. = 0.70 : 0.35 |
| Spinneret temperature (°C) | 60 ± 2 |
| Air gap distance (cm) | 0.8–5.0 |
| Bore fluid composition | NMP : $H_2O$ = 10 wt% : 90 wt%<br>NMP : $H_2O$ = 50 wt% : 50 wt%<br>NMP : $H_2O$ = 90 wt% : 10 wt% |
| Bore fluid flow rate (mL/min) | 0.27 ± 0.02 |
| External coagulant | Tap water |
| Temperature of external coagulant (°C) | 18 ± 2 |

microscopy (SEM, Zeiss Supra 55 Sapphire). MOFs particles were tested without any dispersion. The membrane samples were frozen and broken off in liquid nitrogen and then sputtered with gold before SEM testing. MOFs and hollow fiber membranes were characterized by powder X-ray diffraction (XRD) using a PANalytical Empyrean diffractometer with Cu K$\alpha$ radiation ($\lambda$ = 0.15406 nm). The basal spacing was calculated by the Bragg's equation:

$$n\lambda = 2d \sin\theta. \quad (2)$$

Thermal stability of functionalized MIL-53, pure Ultem HFM, and MMHFM were measured by thermo-gravimetric analysis (TGA) at a heating rate of 10°C/min from room temperature to 800°C under nitrogen atmosphere using a TGA instrument (Netzsch STA449F3).

*2.6. Pure Gas Permeation Measurement.* Gas permeation experiments were conducted at 35°C with a testing pressure of 0.5 MPa. The permeation flow rate was measured with soap bubble flow meter and permeance ($P/l$) was calculated by the following:

$$\frac{P}{l} = \frac{p_P V_P}{RT}\frac{V_m}{At\Delta p}, \quad (3)$$

where $p_p$ is the pressure of the permeate side (cmHg), $V_P$ is the volume of permeate gas during the test duration ($cm^3$), $V_m$ is the molar volume of ideal gas at standard condition [$cm^3$ (STP)/mol], $R$ is the universal gas constant, $T$ is the absolute temperature (K), $A$ is the permeation area ($cm^2$), $t$ is the test duration (s), and $\Delta p$ is the trans-membrane pressure difference (cmHg). The permeance ($P/l$) values are commonly reported in the gas permeation unit GPU, 1 GPU = $1 \times 10^{-6}$ $cm^3$ (STP)/($cm^2$·s·cmHg). The ideal selectivity of gases $i$ and $j$ is defined as follows:

$$\alpha_{(i/j)} = \frac{(P/l)_i}{(P/l)_j}. \quad (4)$$

For each sample, at least three membrane modules were prepared and tested as our previous work [3], and the average value of testing result (with standard deviation) was used to determine the gas separation performance.

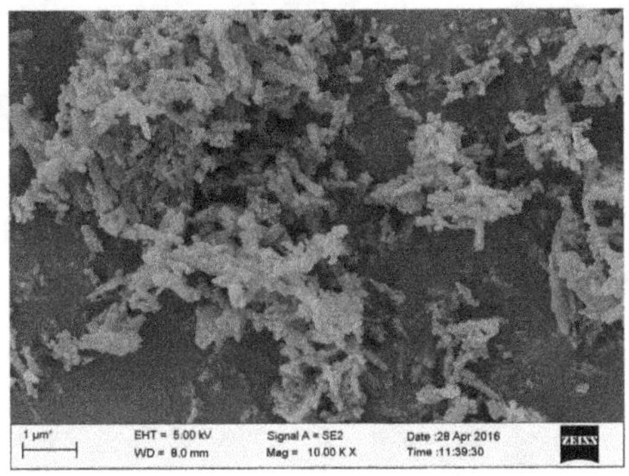

FIGURE 1: SEM image of functionalized MIL-53 powder.

## 3. Results and Discussion

*3.1. Characterizations.* Figure 1 depicts SEM image of the functionalized MIL-53 by grafting aminosilane and MOFs particles appeared in the clump form due to the particle agglomeration. Therefore, the functionalized MOFs used as filler in MMHFM should be sufficiently dispersed before adding into spinning dope solution. The outer and inner surface, cross section, and skin layer of prepared pure Ultem HFM were shown in Figure 2. Morphology of MMHFM with 15 wt% filler loading was presented in Figure 3. The distribution of MOFs particles in Ultem matrix was homogeneous and there was no obvious aggregation as indicated by EDX mapping. Moreover, no surface cracks or interfacial voids revealed the excellent compatibility between the MOFs filler and polymer matrix. Additionally, more finger-like pores were generated in the outer region of MMHFM, which should be caused by instantaneous demixing with addition of MOFs particles.

The X-ray diffraction (XRD) patterns of functionalized MIL-53 and hollow fiber membranes are shown in Figure 4. The Ultem/functionalized MIL-53 MMHFM exhibited new peaks compared with pure Ultem HFM, which indicated the presence of functionalized MIL-53 particles in the polymer matrix. The positions of extraneous peaks of MMHFM were not identical to the functionalized MOFs, suggesting the polymer chains intercalation into the clearance of functionalized MOFs particles. Additionally, the characteristic band of amorphous Ultem appeared around 16.1° that represented

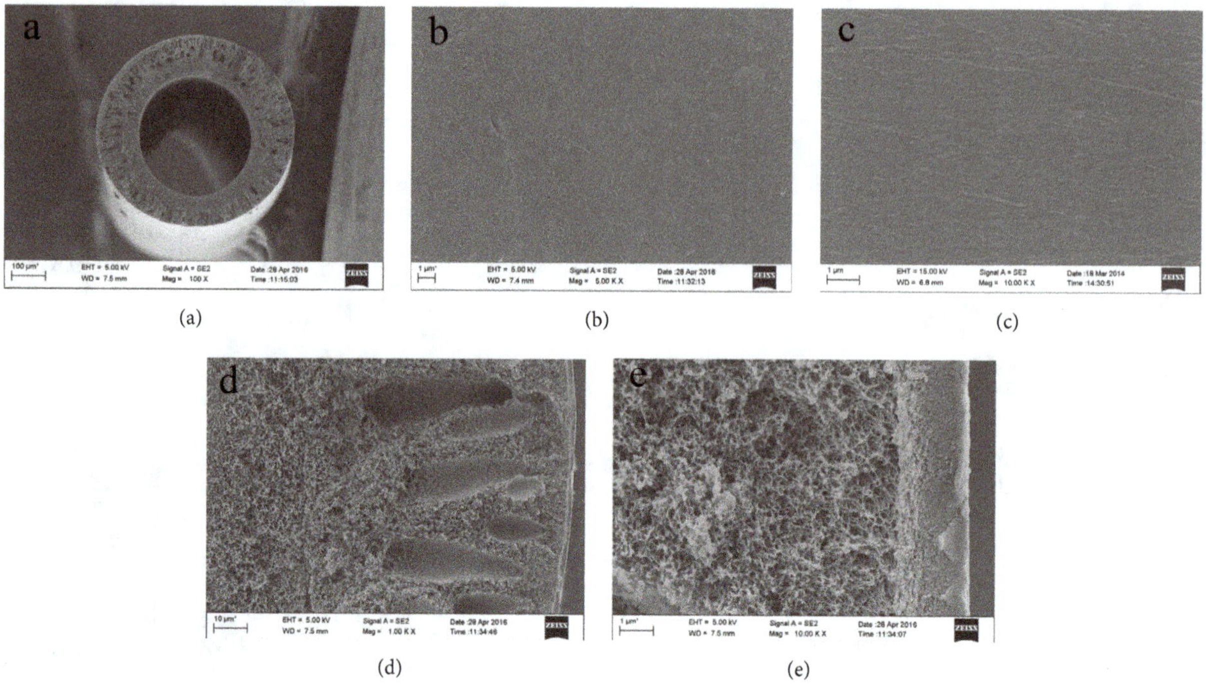


FIGURE 2: SEM images of cross section (a), outer surface (b), inner surface (c), and skin layer close-up (d, e) of pure Ultem HFM.

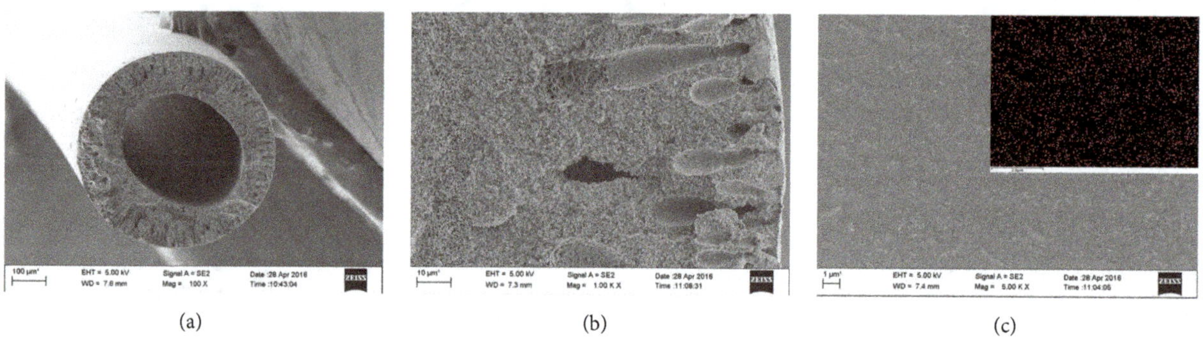


FIGURE 3: SEM images of cross section (a), skin layer close-up (b), and outer surface (c); the inset is EDX mapping of outer surface (Al: red).

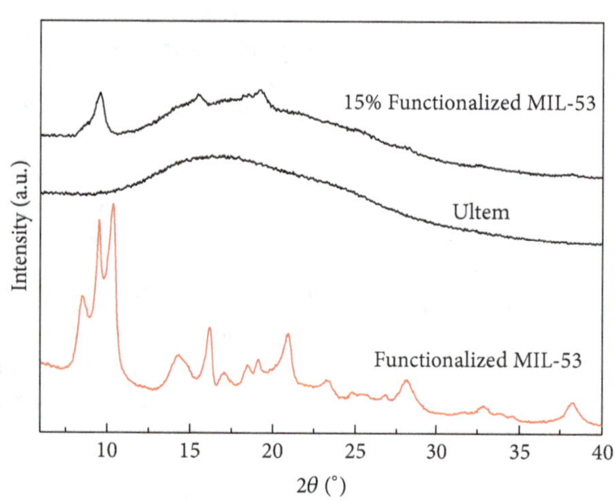


FIGURE 4: X-ray diffractions of functionalized MIL-53, pure Ultem HFM, and Ultem/functionalized MIL-53 MMHFM.

5.5 Å of the $d$-space value. The band of MMHFM was slightly shifted to lower region which corresponded to a larger $d$-space. This shift revealed the changed arrangement of polymer chains packing with MOFs addition.

As shown in Figure 5, the thermal gravimetric analysis of functionalized MIL-53, pure Ultem HFM, and MMHFM was performed to investigate the membrane thermal stability. A weight loss of functionalized MIL-53 before 140°C was caused by the adsorbed water evaporation. Another weight loss stage around 370–515°C was ascribed to decomposition of the grafted functional groups. At higher temperature, the organic linkers eliminated from the framework and functionalized MIL-53 transformed to $Al_2O_3$. With addition of functionalized MIL-53 particles, MMHFM achieved a higher decomposition temperature compared with pure Ultem HFM. The decomposition temperature increase for MOFs-containing MMHFM was attributed to higher energy

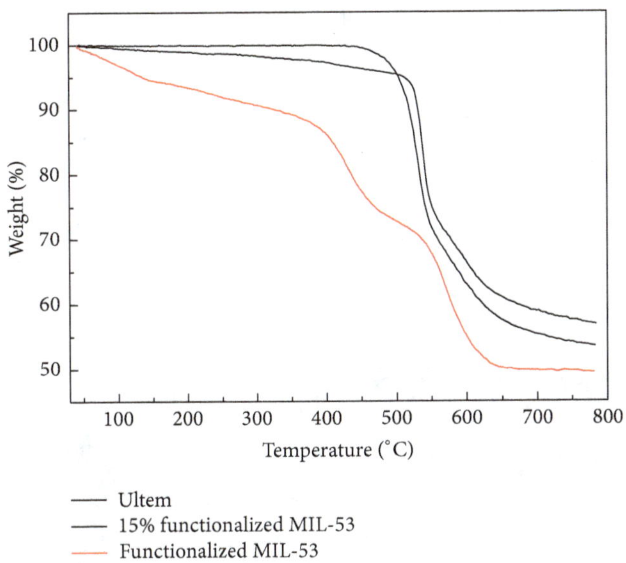


FIGURE 5: TGA curves of functionalized MIL-53, pure Ultem HFM, and Ultem/functionalized MIL-53 MMHFM.

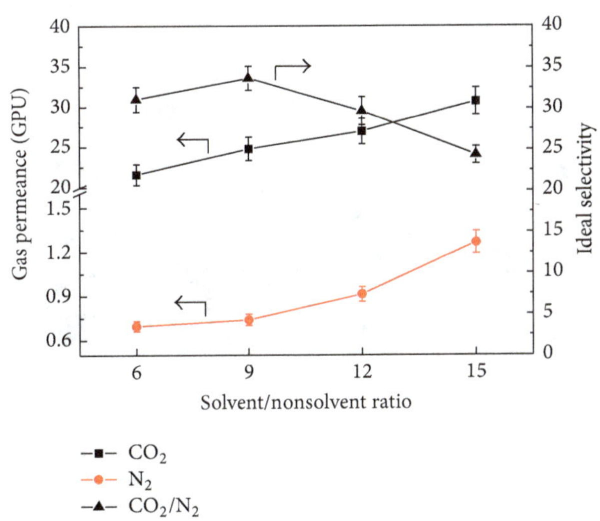


FIGURE 6: Gas permeance and ideal selectivity of MMHFM fabricated from spinning dope with different solvent/nonsolvent ratio.

required for polymer chains movement or segmentation via interaction between polymer matrix and functionalized MIL-53. Meanwhile, the shield effect of filler particles would also improve the thermal stability of MMHFM [32].

*3.2. Effect of Solvent/Nonsolvent Ratio in Spinning Dope.* The filler particles significantly affect the phase inversion as indicated by previous references [31, 33]. The addition of filler would cause the binodal line occurring at lower nonsolvent concentration which reduced the required nonsolvent for separation of dope solution. Figure 6 presents the effect of solvent/nonsolvent ratio in spinning dope on pure gas permeance and ideal selectivity of MMHFM containing the functionalized MIL-53. Gas permeance decreased with the decreased solvent/nonsolvent ratio which was ascribed to the

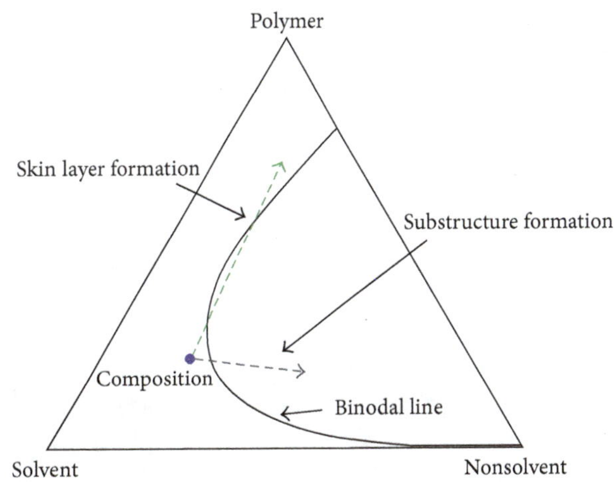


FIGURE 7: Phase diagram demonstrating phase inversion during dry-wet spinning process.

formation of thicker skin layer as Ismail and Lai reported [34]. The evaporation of more volatile nonsolvent could increase polymer concentration of the nascent fiber outer part, resulting in the increment of skin layer thickness. The ideal $CO_2/N_2$ selectivity reached the highest value of 33.7 at the solvent/nonsolvent ratio of 9. However, excessive nonsolvent in dope caused early phase separation and composition path went across binodal line as indicated by Figure 7 which would trigger defects in the fiber skin in spite of the relatively large thickness of skin layer, thus reducing the ideal selectivity. This was consistent with our previous work [31]. At the solvent/nonsolvent ratio above 9, on the contrary, the polymer concentration in outermost region of nascent MMHFM was not high enough to form defect-free skin layer. Consequently, the ideal selectivity dropped obviously accompanying with increasing permeance.

### 3.3. Effect of Spinning Condition

*3.3.1. Air Gap Distance.* Besides the dope composition, the spinning condition was also critical to membrane performance. The outer part cross section of MMHFM prepared through different air gap distances was not identical, especially for the skin layer, as shown in Figure 8. The MMHFM spun through 2 cm of air gap distance possessed fine asymmetric structure composed with thin dense skin layer and porous substructure. The functionalized MIL-53 incorporated MMHFM obtained 30.9 GPU of $CO_2$ permeance and also exhibited excellent $CO_2/N_2$ selectivity of 34.7, as presented in Table 2. In contrast, the MMHFM prepared through a larger air gap distance (5 cm) had a thicker skin layer (Figure 8(a)), which could be explained by the delayed demixing during phase inversion [34]. The thicker selective layer increased gas transport resistance, which resulted in reduced gas permeance. Additionally, the ideal selectivity also decreased obviously, which should be ascribed to the absorption of moisture through the relatively large air gap. When the air gap distance decreased to 0.8 cm, it was difficult to observe the dense skin layer as shown in Figure 8(c).

Table 2: Effect of spinning condition on gas permeance and ideal selectivity of MMHFM.

| Gap distance (cm) | NMP content (wt%) | Permeance (GPU) | | $CO_2/N_2$ selectivity |
|---|---|---|---|---|
| | | $CO_2$ | $N_2$ | |
| 5 | 50 | $20.4 \pm 1.0$ | $0.69 \pm 0.03$ | $29.7 \pm 0.7$ |
| 2 | 50 | $30.9 \pm 1.3$ | $0.89 \pm 0.03$ | $34.7 \pm 1.0$ |
| 0.8 | 50 | $39.8 \pm 1.4$ | $1.61 \pm 0.04$ | $24.8 \pm 1.0$ |
| 2 | 10 | $28.3 \pm 1.1$ | $1.01 \pm 0.04$ | $28.0 \pm 1.3$ |
| 2 | 90 | $24.8 \pm 1.4$ | $0.74 \pm 0.04$ | $33.6 \pm 1.5$ |

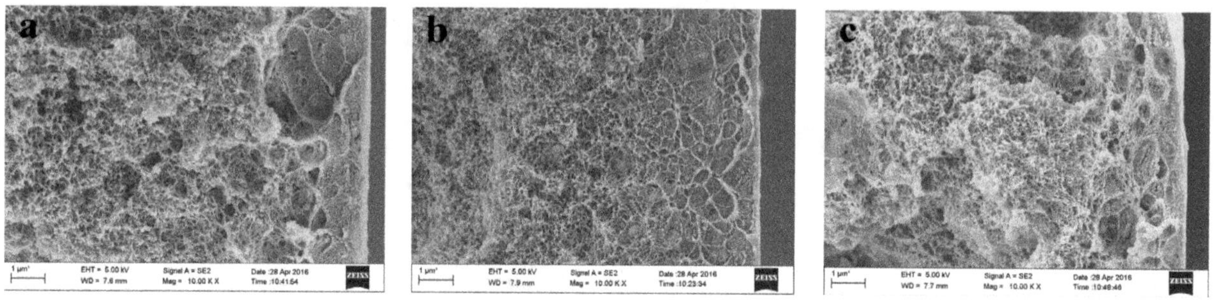

(a) (b) (c)

Figure 8: SEM images of outer part cross section of MMHFM prepared through different air gap distances ((a) 5 cm, (b) 2 cm, and (c) 0.8 cm).

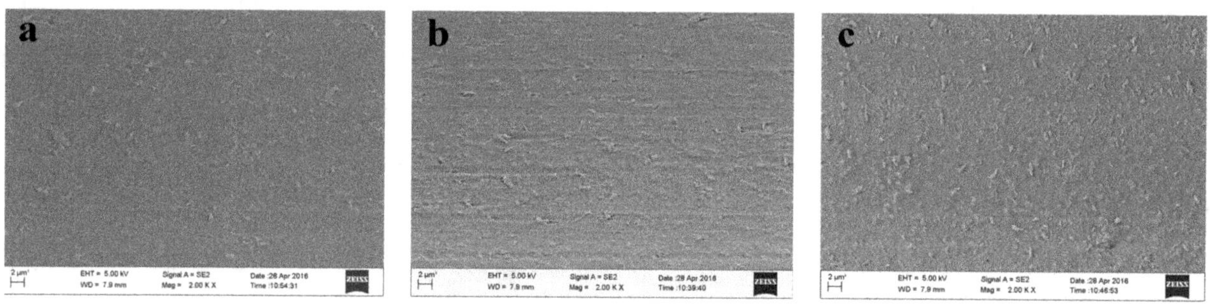

(a) (b) (c)

Figure 9: SEM images of inner surface of MMHFMs with different bore fluid compositions ((a) 10% NMP, (b) 50% NMP, and (c) 90% NMP).

Correspondingly, the nonselective pinhole defects would occur in the skin layer inevitably. Therefore, $CO_2$ permeance increased significantly to 39.8 GPU, but the ideal $CO_2/N_2$ selectivity was reduced to 24.8.

*3.3.2. Bore Fluid Composition.* The bore fluid composition was another important factor that influences the gas separation performance of MMHFM. Three different compositions of bore fluid were investigated in this work. The Ultem/functionalized MIL-53 MMHFM fabricated with the bore fluid containing 10 wt% NMP showed much lower $CO_2/N_2$ selectivity compared with that of MMHFM with 50 wt% NMP in the bore fluid. The high water content in the bore fluid would cause early phase separation through the air gap and quickly solidify the inner surface of MMHFM (Figure 9(a)), which would trigger the formation of defective skin layer. When the NMP content increased to 90 wt%, it could be observed that gas permeance decreased to 24.8 GPU for $CO_2$ and 0.74 GPU for $N_2$ (Table 2). The permeance decrease was ascribed to slow coagulation of polymer and formation of dense skin layer. Additionally, the hydrophobic functionalized MIL-53 particles tended to migrate to the inner surface presumably during the transfer between solvent and nonsolvent as indicated by Figure 9(c), which reduced the number of filler particles and fractional free volume in membrane. Therefore, the fast permeation channels decreased substantially, leading to relatively low gas permeance.

*3.4. Comparison of Gas Separation Performance with Pure Ultem HFM.* By varying the dope composition and spinning condition, the optimum permeation performance of MMHFM incorporated with functionalized MIL-53 was achieved. As shown in Table 3, $CO_2$ permeance increased significantly from 12.2 GPU for pure Ultem HFM to 30.9 GPU for MMHFM, which was attributed to the micropores of MOFs that offered extra permeation channels especially in the dense skin layer. The MOFs incorporation would disrupt polymer chain arrangement and thus expand interfacial

TABLE 3: Gas permeance and ideal selectivity of pure Ultem HFM and MMHFM.

| Fiber | Permeance (GPU) | | $CO_2/N_2$ selectivity |
|---|---|---|---|
| | $CO_2$ | $N_2$ | |
| Pure Ultem HFM | 12.2 ± 0.4 | 0.48 ± 0.02 | 25.4 ± 0.8 |
| MMHFM | 30.9 ± 1.3 | 0.89 ± 0.03 | 34.7 ± 1.0 |

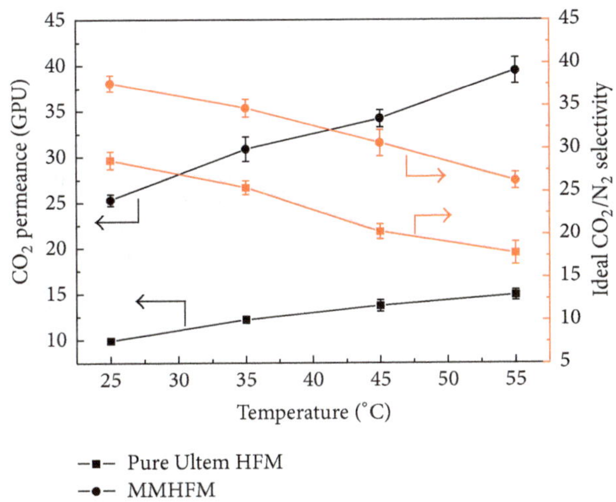


FIGURE 10: Gas permeation properties of pure Ultem HFM and MMHFM plotted as a function of testing temperature.

space, also reducing the permeation resistance. Moreover, the ideal $CO_2/N_2$ selectivity was enhanced to 34.7 by a 37% increase at the same time. The alkyl chains grafted on MOFs endowed good particle dispersion in polymer matrix and improved polymer/filler interfacial affinity, which avoided the nonselective void formation. Additionally, the amino-functionalized MIL-53 with $CO_2$ preferential adsorption also increased $CO_2$ solubility [28, 35] and then facilitated the improvement of $CO_2/N_2$ selectivity.

*3.5. Effect of Testing Temperature on Membrane Performance.* The effect of testing temperature on gas permeation properties was also investigated in this work. As shown in Figure 10, the trend of MMHFM was similar to that of pure Ultem HFM following the solution-diffusion based transport mechanism. When the measurement was conducted from 25°C to 55°C, the $CO_2$ permeance of MMHFM increased about 56% and about 50% for pure Ultem HFM. The increase in testing temperature not only accelerated the gas diffusion rate but also promoted polymer chain mobility, which lowered the required activation energy for gas permeation and thus increased gas permeance [36]. However, the ideal selectivity decreased with increasing temperature. In order to further investigate the temperature dependency of gas permeance, an Arrhenius type equation was used to represent this activated process, as shown by

$$P = P_0 \exp\left(\frac{-E_p}{RT}\right), \tag{5}$$

where $P$, $P_0$, $E_p$, $R$, and $T$ denote the gas permeation, preexponential factor, activation energy of permeation, universal gas constant, and the absolute temperature, respectively. The activation energy of permeation ($E_p$) is mainly affected by gas molecular size and solubility [37, 38]. The noninteracting gas with larger kinetic diameter like $N_2$ has greater $E_p$ than the gas with smaller kinetic diameter and better interaction with membrane like $CO_2$ [38, 39], so $N_2$ exhibits more temperature dependency of permeance. Hence, gas permeance increased but the selectivity decreased with temperature rising up. Additionally, MMHFM was not damaged by testing from lower temperature up to higher temperature and, returning to lower temperature several times, the gas separation performance was regained.

## 4. Conclusion

MMHFM containing 15 wt% functionalized MIL-53 was fabricated via the dry-wet phase inversion method. The functionalized MIL-53 particles dispersed homogeneously in the polymer matrix and good interfacial affinity was achieved without nonselective voids or cracks. MMHFM with incorporation of MOFs obtained a higher decomposition temperature compared with pure Ultem HFM because of the difficulty for polymer chain movement and the shield effect of filler particles. MMHFM spun from the dope with appropriate solvent ratio (solvent/nonsolvent ratio was equal to 9) achieved excellent separation performance. Moreover, the spinning condition was also optimized and was better to be chosen 2 cm of air gap distance and 50% NMP content in bore fluid composition. By optimizing dope composition, air gap distance, and bore fluid composition, $CO_2$ permeance and ideal $CO_2/N_2$ selectivity of MMHFM reached 30.9 GPU and 34.7, respectively. With respect to the testing temperature, the permeation trend of MMHFM was similar to that of pure Ultem HFM following the solution-diffusion mechanism.

## Competing Interests

The authors declare that there is no conflict of interests regarding the publication of this paper.

## Acknowledgments

This work was supported by the financial support from National Natural Science Foundation of China (no. 21436009).

## References

[1] R. W. Baker, "Future directions of membrane gas separation technology," *Industrial and Engineering Chemistry Research*, vol. 41, no. 6, pp. 1393–1411, 2002.

[2] A. Brunetti, F. Scura, G. Barbieri, and E. Drioli, "Membrane technologies for $CO_2$ separation," *Journal of Membrane Science*, vol. 359, no. 1-2, pp. 115–125, 2010.

[3] X. Jie, C. Duan, L. Wang et al., "Fabrication of an asymmetric 4,4′-oxydiphthalic anhydride-2,4,6- trimethyl-1,3-phenylenediamine/2,6-diaminotoluene copolyimide hollow fiber membrane and its performance for $CO_2$ separation," *Industrial and Engineering Chemistry Research*, vol. 53, no. 11, pp. 4442–4452, 2014.

[4] L. Xu, C. Zhang, M. Rungta, W. Qiu, J. Liu, and W. J. Koros, "Formation of defect-free 6FDA-DAM asymmetric hollow fiber membranes for gas separations," *Journal of Membrane Science*, vol. 459, pp. 223–232, 2014.

[5] L. M. Robeson, "The upper bound revisited," *Journal of Membrane Science*, vol. 320, no. 1-2, pp. 390–400, 2008.

[6] I. F. J. Vankelecom, E. Merckx, M. Luts, and J. B. Uytterhoeven, "Incorporation of zeolites in polyimide membranes," *Journal of Physical Chemistry*, vol. 99, no. 35, pp. 13187–13192, 1995.

[7] C. M. Zimmerman, A. Singh, and W. J. Koros, "Tailoring mixed matrix composite membranes for gas separations," *Journal of Membrane Science*, vol. 137, no. 1-2, pp. 145–154, 1997.

[8] H. B. Tanh Jeazet, C. Staudt, and C. Janiak, "Metal-organic frameworks in mixed-matrix membranes for gas separation," *Dalton Transactions*, vol. 41, no. 46, pp. 14003–14027, 2012.

[9] L. Cao, K. Tao, A. Huang, C. Kong, and L. Chen, "A highly permeable mixed matrix membrane containing CAU-1-NH2 for H2 and CO2 separation," *Chemical Communications*, vol. 49, no. 76, pp. 8513–8515, 2013.

[10] B. Li, H. Wang, and B. Chen, "Microporous metal-organic frameworks for gas separation," *Chemistry - An Asian Journal*, vol. 9, no. 6, pp. 1474–1498, 2014.

[11] M. J. C. Ordoñez, K. J. Balkus, J. P. Ferraris, and I. H. Musselman, "Molecular sieving realized with ZIF-8/Matrimid® mixed-matrix membranes," *Journal of Membrane Science*, vol. 361, no. 1-2, pp. 28–37, 2010.

[12] T. Yang, Y. Xiao, and T.-S. Chung, "Poly-/metal-benzimidazole nano-composite membranes for hydrogen purification," *Energy and Environmental Science*, vol. 4, no. 10, pp. 4171–4180, 2011.

[13] T.-H. Bae and J. R. Long, "$CO_2/N_2$ separations with mixed-matrix membranes containing $Mg_2$(dobdc) nanocrystals," *Energy & Environmental Science*, vol. 6, no. 12, pp. 3565–3569, 2013.

[14] Y. Dai, J. R. Johnson, O. Karvan, D. S. Sholl, and W. J. Koros, "Ultem ®/ZIF-8 mixed matrix hollow fiber membranes for $CO_2/N_2$ separations," *Journal of Membrane Science*, vol. 401-402, pp. 76–82, 2012.

[15] H. Hayashi, A. P. Côté, H. Furukawa, M. O'Keeffe, and O. M. Yaghi, "Zeolite A imidazolate frameworks," *Nature Materials*, vol. 6, no. 7, pp. 501–506, 2007.

[16] R. Banerjee, A. Phan, B. Wang et al., "High-throughput synthesis of zeolitic imidazolate frameworks and application to $CO_2$ capture," *Science*, vol. 319, no. 5865, pp. 939–943, 2008.

[17] J. Ploegmakers, S. Japip, and K. Nijmeijer, "Mixed matrix membranes containing MOFs for ethylene/ethane separation Part A: membrane preparation and characterization," *Journal of Membrane Science*, vol. 428, pp. 445–453, 2013.

[18] S. Japip, H. Wang, Y. Xiao, and T. S. Chung, "Highly permeable zeolitic imidazolate framework (ZIF)-71 nano-particles enhanced polyimide membranes for gas separation," *Journal of Membrane Science*, vol. 467, pp. 162–174, 2014.

[19] S. Japip, Y. Xiao, and T. Chung, "Particle-size effects on gas transport properties of 6FDA-Durene/ZIF-71 mixed matrix membranes," *Industrial & Engineering Chemistry Research*, vol. 55, no. 35, pp. 9507–9517, 2016.

[20] S. Couck, J. F. M. Denayer, G. V. Baron, T. Rémy, J. Gascon, and F. Kapteijn, "An amine-functionalized MIL-53 metal-organic framework with large separation power for $CO_2$ and $CH_4$," *Journal of the American Chemical Society*, vol. 131, no. 18, pp. 6326–6327, 2009.

[21] T. Rodenas, M. Van Dalen, E. García-Pérez et al., "Visualizing MOF mixed matrix membranes at the nanoscale: towards structure-performance relationships in $CO_2/CH_4$ separation over $NH_2$-MIL-53(Al)@PI," *Advanced Functional Materials*, vol. 24, no. 2, pp. 249–256, 2014.

[22] B. Seoane, C. Téllez, J. Coronas, and C. Staudt, "$NH_2$-MIL-53(Al) and $NH_2$-MIL-101(Al) in sulfur-containing copolyimide mixed matrix membranes for gas separation," *Separation and Purification Technology*, vol. 111, pp. 72–81, 2013.

[23] B. P. Biswal, A. Bhaskar, R. Banerjee, and U. K. Kharul, "Selective interfacial synthesis of metal-organic frameworks on a polybenzimidazole hollow fiber membrane for gas separation," *Nanoscale*, vol. 7, no. 16, pp. 7291–7298, 2015.

[24] A. J. Brown, N. A. Brunelli, K. Eum et al., "Interfacial microfluidic processing of metal-organic framework hollow fiber membranes," *Science*, vol. 345, no. 6192, pp. 72–75, 2014.

[25] F. Cacho-Bailo, S. Catalán-Aguirre, M. Etxeberría-Benavides et al., "Metal-organic framework membranes on the inner-side of a polymeric hollow fiber by microfluidic synthesis," *Journal of Membrane Science*, vol. 476, pp. 277–285, 2015.

[26] F. Cacho-Bailo, I. Matito-Martos, J. Perez-Carbajo et al., "On the molecular mechanisms for the $H_2/CO_2$ separation performance of zeolite imidazolate framework two-layered membranes," *Chemical Science*, vol. 8, no. 1, pp. 325–333, 2017.

[27] T.-S. Chung, L. Y. Jiang, Y. Li, and S. Kulprathipanja, "Mixed matrix membranes (MMMs) comprising organic polymers with dispersed inorganic fillers for gas separation," *Progress in Polymer Science*, vol. 32, no. 4, pp. 483–507, 2007.

[28] H. Zhu, L. Wang, X. Jie, D. Liu, and Y. Cao, "Improved interfacial affinity and $CO_2$ separation performance of asymmetric mixed matrix membranes by incorporating postmodified MIL-53(Al)," *ACS Applied Materials & Interfaces*, vol. 8, no. 34, pp. 22696–22704, 2016.

[29] B. Zornoza, A. Martinez-Joaristi, P. Serra-Crespo et al., "Functionalized flexible MOFs as fillers in mixed matrix membranes for highly selective separation of $CO_2$ from $CH_4$ at elevated pressures," *Chemical Communications*, vol. 47, no. 33, pp. 9522–9524, 2011.

[30] X.-D. Do, V.-T. Hoang, and S. Kaliaguine, "MIL-53(Al) mesostructured metal-organic frameworks," *Microporous and Mesoporous Materials*, vol. 141, no. 1–3, pp. 135–139, 2011.

[31] H. Zhu, X. Jie, L. Wang, G. Kang, D. Liu, and Y. Cao, "Effect of MIL-53 on phase inversion and gas separation performance of mixed matrix hollow fiber membranes," *RSC Advances*, vol. 6, no. 73, pp. 69124–69134, 2016.

[32] A. K. Zulhairun, B. C. Ng, A. F. Ismail, R. Surya Murali, and M. S. Abdullah, "Production of mixed matrix hollow fiber membrane for $CO_2/CH_4$ separation," *Separation and Purification Technology*, vol. 137, pp. 1–12, 2014.

[33] P. Anadão, L. F. Sato, R. R. Montes, and H. S. De Santis, "Polysulphone/montmorillonite nanocomposite membranes: effect of clay addition and polysulphone molecular weight on the membrane properties," *Journal of Membrane Science*, vol. 455, pp. 187–199, 2014.

[34] A. F. Ismail and P. Y. Lai, "Effects of phase inversion and rheological factors on formation of defect-free and ultrathin-skinned asymmetric polysulfone membranes for gas separation," *Separation and Purification Technology*, vol. 33, no. 2, pp. 127–143, 2003.

[35] S. Japip, K. Liao, and T. Chung, "Molecularly tuned free volume of vapor cross-linked 6FDA-durene/ZIF-71 MMMs for $H_2/CO_2$ separation at 150°C," *Advanced Materials*, vol. 29, no. 4, 2017.

[36] A. L. Khan, S. P. Sree, J. A. Martens, M. T. Raza, and I. F. J. Vankelecom, "Mixed matrix membranes comprising of matrimid and mesoporous COK-12: preparation and gas separation properties," *Journal of Membrane Science*, vol. 495, pp. 471–478, 2015.

[37] B. D. Freeman, "Basis of permeability/selectivity tradeoff relations in polymeric gas separation membranes," *Macromolecules*, vol. 32, no. 2, pp. 375–380, 1999.

[38] J. A. de Sales, P. S. O. Patrício, J. C. Machado, G. G. Silva, and D. Windmöller, "Systematic investigation of the effects of temperature and pressure on gas transport through polyurethane/poly(methylmethacrylate) phase-separated blends," *Journal of Membrane Science*, vol. 310, no. 1-2, pp. 129–140, 2008.

[39] W.-H. Lin and T.-S. Chung, "Gas permeability, diffusivity, solubility, and aging characteristics of 6FDA-durene polyimide membranes," *Journal of Membrane Science*, vol. 186, no. 2, pp. 183–193, 2001.

# A New Approach for Accurate Prediction of Liquid Loading of Directional Gas Wells in Transition Flow or Turbulent Flow

**Ruiqing Ming[1,2] and Huiqun He[2]**

[1]*Research Institute of Petroleum Exploration & Development, Beijing 100083, China*
[2]*CNPC Drilling Research Institute, Beijing 100195, China*

Correspondence should be addressed to Ruiqing Ming; 411478802@qq.com

Academic Editor: Franco Tassi

Current common models for calculating continuous liquid-carrying critical gas velocity are established based on vertical wells and laminar flow without considering the influence of deviation angle and Reynolds number on liquid-carrying. With the increase of the directional well in transition flow or turbulent flow, the current common models cannot accurately predict the critical gas velocity of these wells. So we built a new model to predict continuous liquid-carrying critical gas velocity for directional well in transition flow or turbulent flow. It is shown from sensitivity analysis that the correction coefficient is mainly influenced by Reynolds number and deviation angle. With the increase of Reynolds number, the critical liquid-carrying gas velocity increases first and then decreases. And with the increase of deviation angle, the critical liquid-carrying gas velocity gradually decreases. It is indicated from the case calculation analysis that the calculation error of this new model is less than 10%, where accuracy is much higher than those of current common models. It is demonstrated that the continuous liquid-carrying critical gas velocity of directional well in transition flow or turbulent flow can be predicted accurately by using this new model.

## 1. Introduction

Gas well is usually exploited by using depletion type. The ultimate recovery of pure gas reservoir is generally more than 90%, which is far more than the oil recovery [1]. Liquid loading is the most common phenomenon in gas wells. And the main cause of this phenomenon is that in the later period during the gas well production, the formation pressure, gas velocity, and liquid-carrying capacity will be reduced, and the part of formation water in the wellbore will stay in well bottom causing fluid accumulation [2, 3]. Liquid loading will create an increased backpressure on the formation and reduce production pressure differential, which decreases the gas rate and even kills the gas well [4, 5]. Based on the theory of hydrodynamics, the formation water can be drawn to the surface when current gas velocity is higher than critical gas velocity. Therefore, it is of great significance to accurately predict the critical gas velocity in order to avoid liquid loading and increase gas recovery rate.

Under the condition of vertical well and laminar flow ($10^4 \leq \text{Re} \leq 2 \times 10^5$), Turner et al. [6] assumed that droplets in gas wells are spherical and derived continuous liquid-carrying critical gas velocity formula. In order to make calculation convenient, the drag coefficient was taken as 0.44. In order to fit their experimental data, the calculated results were improved by 20% upward. Coleman et al. [7, 8] applied Turner model to their experimental data. They concluded that the Turner model is suitable for wellhead pressure less than 3.45 MPa and that the 20% upward adjustment is not necessary to calculate the critical gas velocity. Li et al. [9] assumed that droplets in gas wells are elliptical and derived a new formula for calculating the critical liquid-carrying gas velocity, whose result was only 1/3 of the result calculated by the Turner model. Later, many scholars have made various changes to the Turner model and the Li model [9], but they are only suitable for the vertical well and laminar flow. Based on the previous study [6–8], Nosseir et al. [10] considered the influence of flow regime on the calculation result. And they took drag coefficient as 0.2 and extended the application scope of the Turner model to transition flow and turbulence flow ($2 \times 10^5 \leq \mathrm{Re} \leq 10^6$). Based on the Turner model, Zhou and Yuan [11] believed that liquid concentration will also affect the continuous liquid-carrying capacity and thereby proposed a new model to calculate critical gas velocity. Based on the Li model [9], Luan and He [12] used dimensionless loss factor to evaluate gas energy loss and took the impact of the variation of gas lift efficiency into consideration and thereby derived a new formula to calculate critical gas velocity. At low pressure, the calculated results of the model are better than those calculated by the Turner model and the Li model.

Belfroid et al. [13] proposed that using the Turner model to predict critical gas velocity in directional well will cause large errors and adjusted the Turner model with the Fiedler shape function to fit well data for all inclination angles. Based on the force analysis of droplet in directional well, Yang et al. [14] put forward the calculation method of continuous liquid-carrying critical gas velocity in directional well and laminar flow condition. They believed that droplet is not impacted by the tubing wall and always rises up along the central line of the tubing. Based on the force analysis of liquid film in directional well, Chen et al. [15] derived a new formula for evaluating liquid-carrying capacity in directional well and laminar flow condition.

As known from the standard experimental drag curve, drag coefficient fluctuates heavily under the condition of transition flow and turbulence flow ($2 \times 10^5 \leq \mathrm{Re} \leq 10^6$). It is obviously unreasonable to take drag coefficient as a fixed value. Engelund and Hansen [16], Clift and Gauvin [17], and Yen [18] fitted the experimental data, respectively, and derived the formulas relating to drag coefficient and Reynolds number in laminar flow ($10^4 \leq \mathrm{Re} \leq 2 \times 10^5$). Later, Barati et al. [19] used a multigene Genetic Programming (GP) procedure to obtain a hyperbolic tangent function about the two parameters, whose result was more accurate than before in laminar flow.

At present, there are more and more directional gas wells under the condition of transition flow and turbulence flow, such as Sulige gas field in China. And the calculation model which meets the above conditions is not established. When the existing models are used to calculate the critical gas velocity, the computational error will be large. Therefore, in the directional gas well, it is necessary to consider the impact of Reynolds number on drag coefficient under the condition of transition flow and turbulence flow. So the calculation model needs to be modified.

This paper presents a new model to calculate critical gas velocity of directional well in transition flow and turbulence flow. The new model analyzes the force state of droplet in a new way. And gas field data in Chinese western block and northern block were employed to validate the new model.

## 2. New Model

Zhao et al. [20] did some experiments and concluded that the liquid phase in the directional well is dispersed into small droplets to be taken out of the wellbore by natural gas, which means that the droplet model is more reliable than the liquid film model in the process of liquid-carrying. We assume that the droplet is spherical and the collision between droplets is neglected. In critical state, the velocity of liquid phase and gas phase is basically the same, so the droplet is free from the friction of airstream. We conclude that the droplet will only be carried along the tubing out of wellbore; otherwise the horizontal component of $D$ (see Figure 1) cannot be balanced. The force balance is shown in Figure 1.

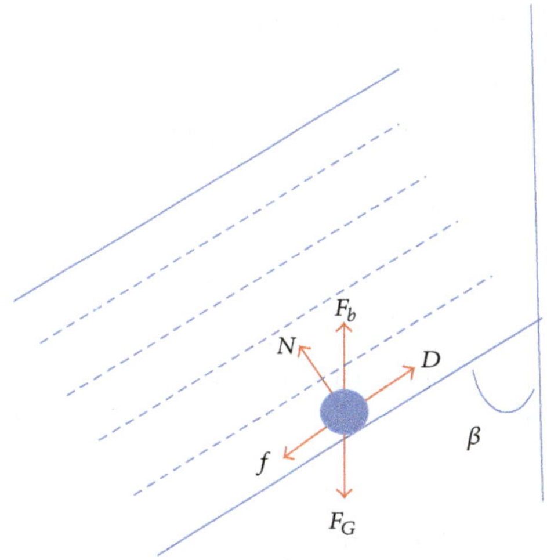


FIGURE 1: Force balance of droplet in directional well, where $\beta$ is the deviation angle, degrees, with $\beta = 90^\circ$ which represents the horizontal well; $D$ is the drag force of airstream on droplet, N; $F_G$ is the droplet gravity, N; $f$ is the wall friction to the droplet, N; $N$ is the bracing force, N; $F_b$ is the buoyancy force, N.

For the droplet (see Figure 1), we get the force balance:

$$F_b \cos\beta + D = F_G \cos\beta + f, \tag{1}$$

$$N + F_b \sin\beta = F_G \sin\beta. \tag{2}$$

It is assumed that the sphere surface is smooth and the equivalent diameter is $d$, m; $F_G$, $F_b$, and $D$ can be expressed as

$$\begin{gathered} F_G = \frac{\pi}{6} d^3 \rho_l g, \\ F_b = \frac{\pi}{6} d^3 \rho_g g, \\ D = \frac{\pi}{8} d^2 \rho_g C_d u^2, \end{gathered} \tag{3}$$

where $\rho_l$ is the liquid density, kg/m$^3$; $g$ is gravitational acceleration, m/s$^2$; $\rho_g$ is the gas density, kg/m$^3$; $C_d$ is the drag coefficient, dimensionless; $u$ is the critical gas velocity, m/s.

$f$ is the wall friction to the droplet, N, which can be expressed as

$$f = \lambda N \tag{4}$$

where $\lambda$ is the wall friction factor, dimensionless, which is related to tubing roughness and Reynolds number.

Combining (2), (3), and (4), we can express the wall friction as

$$f = \frac{\pi}{6} g\lambda d^3 \left(\rho_l - \rho_g\right) \sin\beta. \quad (5)$$

In addition, the droplet is affected by the surface tension that makes it complete and the inertia force that causes it to rupture. When the Weber number ranges from 20 to 30, the droplet will break. Turner et al. [6] concluded that if the maximum diameter droplet is taken out of the wellbore, liquid loading will not happen. Li et al. [1] took the Weber number as 30, and the maximum diameter of droplet can be expressed as

$$d_m = \frac{30\sigma}{\left(\rho_g u^2\right)}, \quad (6)$$

where $\sigma$ is the gas-liquid interfacial tension, N; $d_m$ is the maximum droplet diameter, m.

Substituting (3), (5), and (6) into (1), we can derive the general calculation model of continuous liquid-carrying critical gas velocity:

$$u = \sqrt[4]{\frac{40g\sigma\left[\left(\rho_l - \rho_g\right)\cos\beta + \lambda\left(\rho_l - \rho_g\right)\sin\beta\right]}{C_d {\rho_g}^2}}. \quad (7)$$

The wall friction factor is related to the wall roughness and Reynolds number, and the conventional friction factor ranges from 0.01 to 0.1. Li et al. [1] and Chen et al. [15] demonstrated that the wall friction factor has little impact on critical gas velocity. Taking $\lambda$ as 0.1, we can get

$$u = 4.5\sqrt[4]{\frac{0.1\sin\beta + \cos\beta}{C_d}}\left(\sqrt[4]{\frac{\left(\rho_l - \rho_g\right)\sigma}{{\rho_g}^2}}\right). \quad (8)$$

As known by the standard experimental drag curve, drag coefficient fluctuates heavily under the condition of transition flow and turbulence flow ($2 \times 10^5 \le \text{Re} \le 10^6$). It is obviously unreasonable to take drag coefficient as a fixed value. Thus, we use SPSS to conduct nonlinear fitting of the experimental data in transition flow and turbulence flow (see Table 1 and Figure 2).

As can be seen from Table 1 and Figure 2, $R$ square of cubic equation is 0.940, which shows that the fitting of this model is better than others. The $F$ statistic was 25.902, indicating that regression model fitting results are good. Therefore, we take the cubic model in transition flow and turbulence flow:

$$C_d = -3.316 \times 10^{-18}\text{Re}^3 + 7.3 \times 10^{-12}\text{Re}^2 - 4.918 \times 10^{-6}\text{Re} + 1.143. \quad (9)$$

It can be seen from Table 2 and (8) that, as Reynolds number increases, drag coefficient decreases and critical gas velocity increases in transition flow and turbulence flow. Nosseir et al. [10] proposed that drag coefficient can be taken as 0.2 in transient flow and turbulence flow ($2 \times 10^5 \le \text{Re} \le 10^6$). It can be clearly seen from Table 2 that when Reynolds number ranges from $3.2 \times 10^5$ to $10^6$, the result obtained by Nosseir model is less than the actual data and Nosseir model will cause considerable errors. Zhao et al. [21] used Nosseir model to predict critical gas velocity in Sulige gas field, but the calculation results have a large deviation.

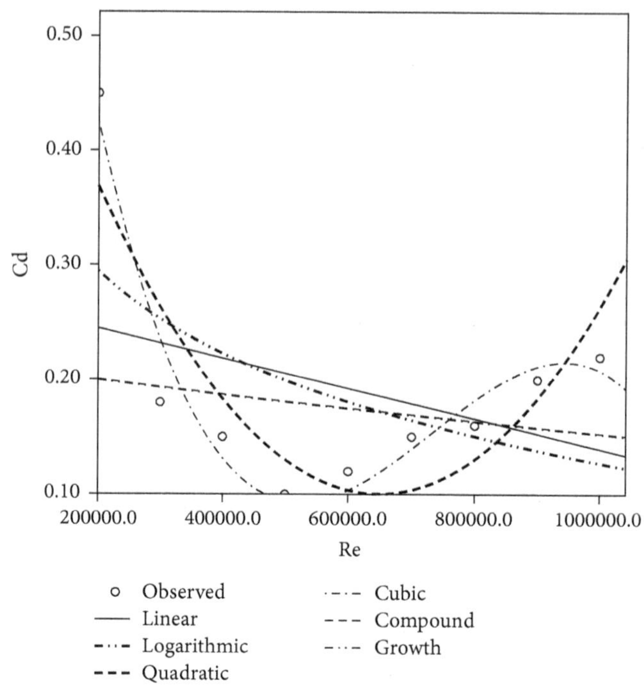


FIGURE 2: Nonlinear fitting graph.

In a word, the critical liquid-removal gas velocity of directional well in transient flow and turbulence flow can be written as follows:

$$u = B\left(\sqrt[4]{\frac{\left(\rho_l - \rho_g\right)\sigma}{{\rho_g}^2}}\right),$$
$$C_d = -3.316 \times 10^{-18}\text{Re}^3 + 7.3 \times 10^{-12}\text{Re}^2 - 4.918 \times 10^{-6}\text{Re} + 1.143. \quad (10)$$

$B$ number is a correction term, which can be introduced:

$$B = 4.5\sqrt[4]{\frac{0.1\sin\beta + \cos\beta}{C_d}}. \quad (11)$$

## 3. Correction Term

As can be seen from (10) and (11), the correction term is mainly dependent upon Reynolds number and deviation angle. As shown in Figure 3, the correction term is plotted as function of deviation angle at different Reynolds number.

It can be seen from Figure 3 and Table 3 that, with the increase of Reynolds number, the correction term increases significantly and then decreases slowly, but it shows a rising trend in general. Furthermore, with the increase of the

TABLE 1: Nonlinear fitting of Reynolds number and drag coefficient in transition flow and turbulence flow.

| Equation | Model summary | | | | | Parameter estimates | | | |
|---|---|---|---|---|---|---|---|---|---|
| | $R^2$ | $F$ | df1 | df2 | Sig. | Constant | $b1$ | $b2$ | $b3$ |
| Linear | 0.121 | 0.966 | 1 | 7 | 0.358 | 0.271 | $-1.317E-7$ | — | — |
| Logarithmic | 0.290 | 2.860 | 1 | 7 | 0.135 | 1.564 | −0.104 | — | — |
| Quadratic | 0.757 | 9.330 | 2 | 6 | 0.014 | 0.661 | $-1.728E-6$ | $1.330E-12$ | — |
| Cubic | 0.940 | 25.902 | 3 | 5 | 0.002 | 1.143 | $-4.918E-6$ | $7.300E-12$ | $-3.316E-18$ |
| Compound | 0.046 | 0.336 | 1 | 7 | 0.580 | 0.214 | 1.000 | — | — |
| Growth | 0.046 | 0.336 | 1 | 7 | 0.580 | −1.542 | $-3.353E-7$ | — | — |

TABLE 2: Fitting drag coefficient from (9).

| Reynolds number ($10^5$) | Fitting drag coefficient | Reynolds number ($10^5$) | Fitting drag coefficient |
|---|---|---|---|
| 2.1 | 0.40 | 3.9 | 0.14 |
| 2.3 | 0.36 | 4.1 | 0.13 |
| 2.4 | 0.34 | 4.2 | 0.12 |
| 2.6 | 0.30 | 4.4 | 0.11 |
| 2.9 | 0.25 | 4.7 | 0.10 |
| 3.0 | 0.24 | 4.8 | 0.10 |
| 3.1 | 0.22 | 4.9 | 0.10 |
| 3.2 | 0.21 | 5.0 | 0.09 |
| 3.4 | 0.18 | 5.5 | 0.09 |
| 3.7 | 0.15 | 7.0 | 0.14 |
| 3.8 | 0.14 | 8.0 | 0.18 |

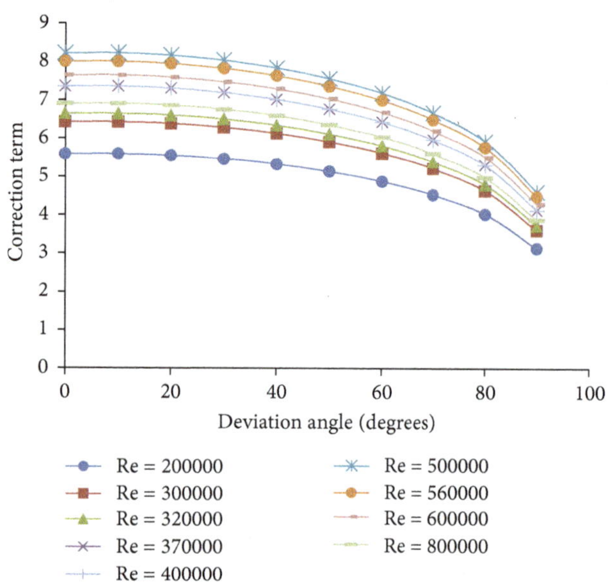


FIGURE 3: Relationship between deviation angle and correction coefficient under different Reynolds number.

correction term, the critical gas velocity increases and liquid-carrying capacity decreases. With the increase of deviation angle, the influence of Reynolds number on the correction term is becoming smaller. While at an identical Reynolds number, as the deviation angle increases, the correction term decreases, and this decreasing trend increases as the pipe deviates, which means that critical gas velocity decreases and liquid-carrying capacity is enhanced. This is because as the deviation angle increases, the gravitational force in the flow direction decreases which will reduce the critical gas velocity [22]. For the convenience of site application, the curve in Figure 3 is transformed into a correction term reference table (see Table 3). And the noticed point is that the table should be modified based on the actual gas field conditions.

## 4. Application of Field Data

In order to validate the applicability to directional gas well in transition flow and turbulence flow, a large amount of field data collected from Chinese gas reservoir including both northern block and western block is used to verify the prediction accuracy of the new model. In this paper, calculation results obtained by the new model will be compared with those from other models.

*4.1. Application of Field Data in Chinese Western Block.* Western block consists of 16 directional gas wells, which includes 4 unloaded wells and 12 loaded wells. Among all the directional gas wells in the block, the Reynolds number ranges from $2.3 \times 10^5$ to $7.7 \times 10^5$ and the deviation angle ranges from 24° to 50°. The detailed field data are listed in Table 4. And the gas rate is converted to the superficial velocity used to compare with the calculated critical gas velocity, as shown in column 7. Column 9 and column 11 represent the critical gas velocity calculated by Belfroid model and by new model, respectively. If the critical gas velocity is higher than current gas velocity, the well is considered to be loaded. On the contrary, if the critical gas velocity is lower than current gas velocity, water cannot accumulate in bottom hole and the well is considered to be unloaded. Column 10 and column 12 represent the predicting status from the two models. Column 13 represents the test status which is the actual status of 16 directional gas wells in Chinese western block. If the predicting status is consistent with the test status, then the predicted result is correct. On the other hand, if the predicting status is inconsistent with the test status, the predicted result is incorrect.

As can be seen in Table 4, when the Belfroid model is used to predict liquid loading status, 6 wells are predicted incorrectly, including well number 3, well number 4, well number 8, well number 9, well number 11, and well number 14, with 62.5% accuracy. The main reason is due to the fail to consider the influence of Reynolds number on drag coefficient in transition flow and turbulent flow. When the new model is used to predict liquid loading status, only 2 wells are predicted incorrectly, including well number 3 and well number 4, with up to 87.5% accuracy, which is in good agreement with the actual state. The prediction results show great improvement over the Belfroid model. Therefore, in the aspect of predicting liquid loading, the new model is better than the Belfroid model.

TABLE 3: Correction term table.

| Deviation angle (degrees) | Correction term | | | | | |
|---|---|---|---|---|---|---|
| | Re = 200000 | Re = 300000 | Re = 400000 | Re = 500000 | Re = 600000 | Re = 800000 |
| 0 | 5.589 | 6.429 | 7.357 | 8.216 | 7.646 | 6.908 |
| 10 | 5.593 | 6.433 | 7.361 | 8.221 | 7.649 | 6.912 |
| 20 | 5.553 | 6.387 | 7.308 | 8.162 | 7.595 | 6.863 |
| 30 | 5.469 | 6.289 | 7.197 | 8.038 | 7.479 | 6.758 |
| 40 | 5.336 | 6.137 | 7.023 | 7.842 | 7.298 | 6.595 |
| 50 | 5.148 | 5.921 | 6.775 | 7.566 | 7.041 | 6.362 |
| 60 | 4.892 | 5.627 | 6.438 | 7.191 | 6.691 | 6.046 |
| 70 | 4.542 | 5.224 | 5.977 | 6.676 | 6.213 | 5.613 |
| 80 | 4.037 | 4.643 | 5.313 | 5.934 | 5.522 | 4.989 |
| 90 | 3.143 | 3.615 | 4.137 | 4.621 | 4.299 | 3.885 |

Table 4: Field data in western block and liquid loading predicting results.

| Well number | Wellhead pressure (MPa) | Tubing diameter (in) | Gas rate ($10^4$ $m^3$/d) | Liquid rate ($m^3$/d) | Deviation angle (°) | Current gas velocity (m/s) | Reynolds number ($10^5$) | Belfroid model | | New model | | Test status |
|---|---|---|---|---|---|---|---|---|---|---|---|---|
| | | | | | | | | Critical gas velocity (m/s) | Predicting status | Critical gas velocity (m/s) | Predicting status | |
| 1 | 12.6 | 2.44 | 2.92 | 0.25 | 24 | 3.46 | 2.3 | 3.82 | Loaded | 4.23 | Loaded | Loaded |
| 2 | 8.9 | 2.44 | 2.36 | 0.2 | 30 | 2.93 | 3.5 | 3.27 | Loaded | 3.86 | Loaded | Loaded |
| 3 | 10.3 | 2.44 | 3 | 0.8 | 30 | 4.13 | 3.8 | 3.65 | Unloaded | 3.93 | Unloaded | Loaded |
| 4 | 15.3 | 2.44 | 3.44 | 0.35 | 50 | 5.44 | 2.7 | 4.79 | Unloaded | 5.21 | Unloaded | Loaded |
| 5 | 12.1 | 2.44 | 3.39 | 0.38 | 25 | 3.48 | 2.3 | 4.13 | Loaded | 4.27 | Loaded | Loaded |
| 6 | 4.6 | 2.44 | 1.48 | 0.65 | 45 | 1.54 | 2.6 | 2.49 | Unloaded | 2.87 | Unloaded | Unloaded |
| 7 | 21.9 | 2.44 | 4.51 | 1.25 | 26 | 3.94 | 3.1 | 4.88 | Loaded | 5.84 | Loaded | Loaded |
| 8 | 16.3 | 2.44 | 4.78 | 0.8 | 25 | 4.86 | 3.3 | 4.27 | Unloaded | 5.13 | Loaded | Loaded |
| 9 | 13.3 | 2.44 | 2.46 | 1.35 | 30 | 4.51 | 4.1 | 4.05 | Unloaded | 5.06 | Loaded | Loaded |
| 10 | 15.6 | 2.44 | 4.69 | 0.7 | 40 | 3.12 | 5.3 | 4.70 | Loaded | 5.17 | Loaded | Loaded |
| 11 | 23.2 | 2.44 | 4.57 | 1.2 | 50 | 6.25 | 6.5 | 5.71 | Unloaded | 6.90 | Loaded | Loaded |
| 12 | 2.5 | 2.44 | 2.6 | 0.4 | 30 | 5.22 | 7.7 | 3.92 | Unloaded | 4.53 | Unloaded | Unloaded |
| 13 | 10.1 | 2.44 | 2.18 | 1.4 | 28 | 4.63 | 6.3 | 3.84 | Unloaded | 4.42 | Unloaded | Unloaded |
| 14 | 19.4 | 2.44 | 5 | 2.3 | 45 | 4.97 | 3.2 | 4.26 | Unloaded | 5.27 | Loaded | Loaded |
| 15 | 17.5 | 2.44 | 4.01 | 1.25 | 32 | 2.63 | 2.7 | 3.98 | Loaded | 4.61 | Loaded | Loaded |
| 16 | 18 | 2.44 | 0.9 | 4.5 | 25 | 3.07 | 3.4 | 3.72 | Unloaded | 4.47 | Unloaded | Unloaded |

TABLE 5: Field data of 4 wells in northern block.

| Well number | A1 | A2 | B1 | B2 |
|---|---|---|---|---|
| Deviation angle/(°) | 24.4 | 30 | 33 | 30 |
| Reynolds number/$10^5$ | 3.5 | 2.7 | 3.9 | 5.3 |
| Tubing size/in | 2.44 | 2.44 | 2.99 | 2.99 |
| Wellhead pressure/MPa | 3 | 2.65 | 3 | 2.85 |
| Gas rate/$10^4$ $m^3 \cdot d^{-1}$ | 0.5 | 0.45 | 1.8 | 3 |
| Liquid rate/$m^3 \cdot d^{-1}$ | 0.14 | 0.27 | 0.97 | 1.80 |
| Test status | Loaded | Loaded | Unloaded | Unloaded |

TABLE 6: Calculated results of different models.

| Model | A1 | | A2 | | B1 | | B2 | |
|---|---|---|---|---|---|---|---|---|
| | Critical velocity/ (m/s) | Error | Critical velocity/ (m/s) | Error | Critical velocity/ (m/s) | Error | Critical velocity/ (m/s) | Error |
| Critical velocity | 2.91 | — | 0.9 | — | 2 | — | 1.83 | — |
| Turner model | 4.6 | 58% | 1.4 | 56% | 3.2 | 60% | 2.97 | 62% |
| Belfroid model | 4.0 | 37% | 1.33 | 48% | 2.8 | 40% | 2.67 | 46% |
| Chen model | 3.5 | 20% | 0.53 | 41% | 2.7 | 35% | 1.2 | 34% |
| New model | 2.83 | 3% | 0.84 | 7% | 1.95 | 3% | 1.74 | 5% |

*4.2. Application of Field Data in Chinese Northern Block.* There are 4 wells in Chinese northern block, including A1, A2, B1, and B2. A1 and A2 have been shut in due to bottom overstock liquid terrible. On the other hand, B1 and B2 are being produced. The detailed field data are listed in Table 5. In this paper, the new model and several common models are used to calculate critical liquid-carrying gas velocity. The results are shown in Table 6.

As can be seen in Tables 5 and 6, under the condition of transition flow and turbulent flow, the error of the calculated results by the new model is less than 10%, which is more accurate than the common methods. The Turner model, as a whole, overestimates the critical gas velocity for directional well in transition flow and turbulent flow, which is possibly due to the overlook of the impact of deviation angle and flow regime on critical gas velocity. The prediction results of Belfroid model are better than Turner model. But the prediction error is still large, which may be because the model fails to consider the impact of flow regime on critical gas velocity. In addition, the prediction error of Chen model [15] is also large. This is mainly because the model is obtained by force analysis of liquid film, but the droplet model is more reliable than the liquid film model in the process of liquid-carrying [20].

## 5. Conclusion

(1) The balance of forces acting on droplet in directional gas well is analyzed. By nonlinear simulation of the experiment data ($2 \times 10^5 \leq \text{Re} \leq 10^6$), we obtain a function relating to Reynolds number and drag coefficient. Eventually we derive a new model to predict critical gas velocity of directional well in transition flow and turbulent flow.

(2) A correction term formula is introduced. The impact of deviation angle and Reynolds number on the correction term is discussed. And a correction term reference table is given for the convenience of site application.

(3) A large quantity of field data collected from Chinese gas reservoir, including both northern block and western block, is used to verify the prediction accuracy of the new model. It can be consistently seen that the new model is superior to the several common models in predicting liquid loading of directional wells in transition flow and turbulent flow.

## Nomenclature

$\beta$: Deviation angle, degrees
$D$: Drag force of airstream on droplet, N
$F_G$: Droplet gravity, N
$f$: Wall friction to the droplet, N
$N$: Bracing force, N
$F_b$: Buoyancy force, N
$d$: Equivalent diameter, m
$d_m$: Maximum droplet diameter, m
$\rho_l$: Liquid density, kg/$m^3$
$g$: Gravity acceleration, m/$s^2$
$\rho_g$: Gas density, kg/$m^3$
$C_d$: Drag coefficient, dimensionless
$u$: Critical gas velocity, m/s
$\lambda$: Wall friction factor, dimensionless
$\sigma$: Gas-liquid interfacial tension, N
Re: Reynolds number, dimensionless
$B$: $B$ number, dimensionless.

## Conflicts of Interest

The authors declare that there are no conflicts of interest regarding the publication of this paper.

## Acknowledgments

This work was supported by project of CNPC Drilling Research Institute (Grant no. 2015F-2001-01).

## References

[1] L. Li, L. Zhang, B. Yang, Y. Yin, and D. Li, "Prediction method of critical liquid-carrying flow rate for directional gas wells," *Oil & Gas Geology*, vol. 33, no. 4, pp. 650–654, 2012.

[2] G. Chupin, B. Hu, T. Haugset, J. Sagen, and M. Claudel, "Integrated wellbore/reservoir model predicts flow transients in liquid-loaded gas wells," in *Proceedings of the SPE Annual Technical Conference and Exhibition*, Anaheim, Calif, USA, November 2007.

[3] W. Schiferli, S. P. C. Belfroid, S. Savenko, C. A. M. Veeken, and B. Hu, "Simulating liquid loading in gas wells," in *Proceedings of the 7th North American Conference on Multiphase Technology*, Banff, Canada, June , 2010.

[4] A. Fadairo, D. Femi-Oyewole, and O. Falode, "An improved predictive tool for liquid loading in a gas well," in *Proceedings of the SPE Nigeria Annual International Conference and Exhibition*, Lagos, Nigeria, August 5-7, 2013.

[5] J. F. Lea and H. V. Nickens, "Solving gas-well liquid-loading problems," *Journal of Petroleum Technology*, vol. 56, no. 4, pp. 30–36, 2004.

[6] R. G. Turner, M. G. Hubbard, and A. E. Dukler, "Analysis and prediction of minimum flow rate for the continuous removal of liquids from gas wells," *Journal of Petroleum Technology*, vol. 21, no. 11, pp. 1475–1482, 1969.

[7] S. B. Coleman, H. B. Clay, D. G. McCurdy, and L. H. Norris, "A new look at predicting gas-well load-up," *Journal of Petroleum Technology*, vol. 43, no. 3, pp. 329–333, 1991.

[8] S. B. Coleman, H. B. Clay, D. G. McCurdy, and L. H. Norris, "Understanding gas-well load-up behavior," *Journal of Petroleum Technology*, vol. 43, no. 3, pp. 334–338, 1991.

[9] M. Li, S. Li, and L. Sun, "New view on continuous-removal liquids from gas wells," *SPE Production & Facilities*, vol. 17, no. 1, pp. 42–46, 2002.

[10] M. A. Nosseir, T. A. Darwich, M. H. Sayyouh, and M. El Sallaly, "A new approach for accurate prediction of loading in gas wells under different flowing conditions," *SPE Production and Facilities*, vol. 15, no. 4, pp. 241–246, 2000.

[11] D. Zhou and H. Yuan, "A New model for predicting gas-well liquid loading," *SPE Production & Operations*, vol. 25, no. 2, pp. 172–181, 2010.

[12] G. Luan and S. He, "A new model for the accurate prediction of liquid loading in low-pressure gas wells," *Journal of Canadian Petroleum Technology*, vol. 51, no. 6, pp. 493–498, 2012.

[13] S. P. C. Belfroid, W. Schiferli, G. J. N. Alberts, C. A. M. Veeken, and E. Biezen, "Prediction onset and dynamic behaviour of liquid loading gas wells," in *Proceedings of the SPE Annual Technical Conference and Exhibition, Society of Petroleum Engineers*, pp. 21–24, Denver, Colo, USA, 2008.

[14] W. M. Yang, M. Wang, L. Chen, J. C. Wu, and P. X. Yu, "A prediction model on calculation of continuous liquid carrying critical production of directional gas wells," *Natural Gas Industry*, vol. 29, no. 5, pp. 82–84, 2009.

[15] D. C. Chen, Y. Yao, G. Fu, H. X. Meng, and S. X. Xie, "A new model for predicting liquid loading in deviated gas wells," *Journal of Natural Gas Science and Engineering*, vol. 34, pp. 178–184, 2016.

[16] F. Engelund and E. Hansen, *Monograph on Sediment Transport in Alluvial Streams*, Monograpsh Denmark Technical University, Hydraulic Lab, Denmark, 1967.

[17] R. Clift and W. H. Gauvin, "The motion of particles in turbulent gas streams," *Revista Argentina de Cardiología*, vol. 82, no. 3, pp. 14–28, 1970.

[18] B. C. Yen, *Sediment fall velocity in oscillating flow*, University of Virginia, USA, 1992.

[19] R. Barati, S. A. A. S. Neyshabouri, and G. Ahmadi, "Development of empirical models with high accuracy for estimation of drag coefficient of flow around a smooth sphere: an evolutionary approach," *Powder Technology*, vol. 257, no. 5, pp. 11–19, 2014.

[20] Z. J. Zhao, T. Liu, J. Xu, J. Zhu, and Y. Yang, "Stable fluid-carrying capacity of gas wells," *Natural Gas Industry*, vol. 35, no. 6, pp. 59–63, 2015.

[21] B. B. Zhao, X. H. Bai, D. J. Chen, X. R. Li, and W. B. Zhao, "Effect assessment of drainage gas recovery through velocity string and its new application area," *China Petroleum Machinery*, vol. 40, no. 11, pp. 62–65, 2012.

[22] S. Luo, M. Kelkar, E. Pereyra, and C. Sarica, "A new comprehensive model for predicting liquid loading in gas wells," *SPE Production and Operations*, vol. 29, no. 4, pp. 337–349, 2014.

# 17

# Vibrational Spectroscopies and Chemometry for Nondestructive Identification and Differentiation of Painting Binders

**Serena Carlesi,[1] Maurizio Becucci,[1,2] and Marilena Ricci[1]**

[1]*Department of Chemistry "Ugo Schiff", University of Florence, Via della Lastruccia 3-13, 50019 Sesto Fiorentino, Italy*
[2]*The European Laboratory for Non-Linear Spectroscopy (LENS), Via N. Carrara 1, 50019 Sesto Fiorentino (FI), Italy*

Correspondence should be addressed to Marilena Ricci; marilena.ricci@unifi.it

Academic Editor: Ewa Schab-Balcerzak

A comprehensive dataset of vibrational spectra of different natural organic binding media is presented and discussed. The binding media were applied on a glass substrate and analyzed after three months of natural ageing. The combination of Raman and FT-NIR spectroscopies allows for an improved identification of these materials as Raman technique is more informative about the skeletal vibrations, while FT-NIR spectroscopy is more sensitive to the substituents and polar groups. The experimental results are initially discussed in the framework of current spectral assignment. Then, multivariate analysis (PCA) is applied leading to differentiation among the samples. The two major principal components allow for a complete separation of the different classes of organic materials. Further differentiation within the same class is possible thanks to the secondary components. The loadings obtained from PCA are discussed on the basis of the spectral assignment leading to clear understanding of the physical basis of this differentiation process.

## 1. Introduction

The chemical nature of the materials used in painting as binding media is usually very diverse and complex, which makes their characterization a challenging task. A greater knowledge of the original binders used by the artist is of great interest to art historians because it allows them to distinguish between different painting techniques, which are often directly associated with the geographical, social, and chronological contexts in which the artist lived and worked. Information about the class of binders is also important for restorers, who often use solvents during the cleaning and restoration of artworks [1].

The identification and chemical characterization of various painting binders (protein-, polysaccharide-, and lipid-based) has been frequently addressed in published research, mainly using techniques such as gas chromatography/mass spectrometry (GC/MS) in combination with various offline or online derivatization procedures [2, 3]. Recent authors have proposed the use of spectroscopic techniques, as vibrational and electronic spectroscopic methods, coupled with multivariate analysis for the discrimination of different painting binders (animal glue, whole egg, and linseed oil) [2, 4–6] and also of organic compounds of the same class, like lipidic binders [7–10], subjected to natural ageing.

Separation methods (like gas chromatography, capillary electrophoresis, and liquid chromatography) and mass spectrometry present limitations for their application, such as large sample consumption, long experimental times, complex sample preparation, and multiparameter operating conditions. These are obvious problems for the analysis of samples of historical interest [4, 11, 12]. Spectroscopic techniques have the potential to overcome these limitations, as they allow nondestructive and contactless measurements, giving molecular information on complex macromolecules, and could be performed in situ by means of portable devices.

The work presented here explores the application of chemometric tools, such as principal component analysis (PCA) to combined FT-NIR and Raman data in order to enable us to distinguish among organic compounds of the same and different classes. With this aim in mind, we prepared 13 model samples containing traditional organic materials, commonly used by artists throughout history as binding media. These models samples were ageing naturally for three months at room temperature, without controlled environmental conditions, and then were further characterized by Raman and FT-NIR spectroscopy. The present model is not a realistic presentation of fully aged artworks, because the ageing processes of these materials last a much longer time. Nevertheless these models are useful for assessing the possibility of our data analysis.

Because these two vibrational spectroscopies obey different selection rules and provide often complementary information, combining the two could permit a greater amount of spectroscopic information to be extracted from each sample. Nevertheless, there are several different causes that can lead to random noise in Raman and FT-NIR spectra, thus making the identification of spectral bands more difficult, particularly in computational analyses, as in the present work. To identify small differences in the original spectra, first derivatives of the spectral data were used for PCs extraction [4, 6, 8, 9, 13]. As a result, an increase in sensitivity/specificity for differentiation capability and classification can be expected, due to the new dimensions added to the dataset [8, 9]. Principal Component Analysis (PCA) can be used to explain the variance-covariance structure of a set of data by the original variables' linear combination that form a new set of independent coordinates, called principal components (PC). The PCs are hierarchically ordered to describe the largest variance of the dataset, so that a smaller number of variables can account for most of the information from the original dataset [14]. The purpose of all multivariate data analysis is to decompose the data in order to detect and model the "hidden phenomena" or better to define the set of natural coordinates for these variables that show more clearly the similitudes and differences between the various samples [14, 15]. Therefore, the application of PCA to large spectral datasets can be exploited to extract meaningful information from spectral differences, allowing us to discriminate among different samples based on their composition [14, 15].

The results clearly show the potential of this methodology to extract the information needed to identify the painting materials that is contained in the FT-NIR and Raman spectra.

## 2. Materials and Methods

*2.1. Materials.* All neat binding media analyzed in this work, along with corresponding symbols, are listed in Table 1.

Each drying oil, as supplied, was applied as film on four microscope glass slides. Proteinaceous and polysaccharide binding media were prepared according to recipes from original treatises and artists' accounts [16–18]. Rabbit skin glue, strong glue, and fish glue were swollen and dissolved in warm demineralized water to give a solution of 5% (w/w). Egg white was beaten to form stiff peaks and left for 24 h; foam was skimmed and removed and then a solution of 50% (w/w) clear egg white in demineralized water was prepared. Egg yolk was extracted from egg by piercing the yolk and allowing the liquid to drip from the encasing film; the yolk was diluted in demineralized water to give a 50% (w/w) emulsion. Whole egg was prepared by mixing one part of egg white solution and two parts on egg yolk emulsion along with one part of vinegar. Casein was swollen in water for 24 h to give a 1.5% (w/w) solution; dilute ammonia solution was added until the swollen gel dissolved and excess ammonia was left to evaporate. Arabic gum was dissolved in warm demineralized water with a ratio of 1 to 2 (w/w), heating in a double boiler; the mixture was left to soak 48 h for a full absorption and finally, the obtained solution was strained through two layers of cheesecloth into a clean jar. Tragacanth gum was added to demineralized water in a ratio 1 : 30 (w/w) and left to swell in a mucilaginous suspension. Then, this suspension was strain in a similar way to that previously described for Arabic gum.

TABLE 1: The neat binding media analyzed with reporting the manufacturer and the corresponding identification symbol.

| Neat binder | Producer/manufacturer | Symbol |
|---|---|---|
| *Proteinaceous binders* | | |
| Rabbit skin glue | Zecchi | RG |
| Strong glue | Zecchi | SG |
| Fish glue | Zecchi | FG |
| Egg white | Local market | EW |
| Egg yolk | Local market | EY |
| Whole egg | Local market | WE |
| Casein | Zecchi | C |
| *Drying oils* | | |
| Linseed oil | Zecchi | LOZ |
| Linseed oil | Maimeri | LOM |
| Linseed oil | Ferrario | LOF |
| Poppy-seed oil | Zecchi | POZ |
| Poppy-seed oil | Talens | POT |
| Walnut oil | Zecchi | WOZ |
| Walnut oil | Ferrario | WOF |
| Stand-oil | Zecchi | SOLZ |
| *Polysaccharide media* | | |
| Arabic gum | Zecchi | AG |
| Tragacanth gum | Zecchi | TG |

All these organic binders and oils were individually applied as film on four microscope glass slides and left to dry on a shelf ($T$: ~20°C; RH: ~50–55%; light: solar light through a window). Hence, sixty-eight model samples for neat binders were prepared.

These materials were analyzed after an ageing of three months. As regards drying oils, they should be treated as "fresh" binding media because the drying process has not already been completed (according to previous studies, an oil paint can be regarded as aged only one year after its preparation [17, 19]).

*2.2. Fourier Transform-Near Infrared (FT-NIR) Measurements.* FT-NIR reflectance spectra were acquired by using a portable ALPHA FT-IR spectrometer (Bruker Optics) equipped with an external reflection module. Total reflection spectra (including both the specular and diffuse reflection contributions) were collected in situ, in the spectral range 7500–3900 cm$^{-1}$, with a resolution of 4 cm$^{-1}$, averaging over 128 scans. The spectral sampling distance used for each scan was 2 cm$^{-1}$. The background was acquired using a gold mirror as reference sample. In total, 272 spectra were recorded, at four different positions for each sample (16 spectra for each binder).

*2.3. Micro-Raman Spectroscopy.* Micro-Raman analysis was performed using a micro-Raman spectrometer (Renishaw 2000) equipped with a 785 nm diode laser, operating at 1.1 mW laser power on the sample. The laser was focused onto the sample using a 50x objective lens, so that an area of about 2 $\mu$m in diameter was analyzed. Spectra were recorded in the 2000–200 $cm^{-1}$ spectral range, with a spectral resolution of 4 $cm^{-1}$. The integration time and the spectral sampling distance used for each scan were 25 s and 1 $cm^{-1}$, respectively. Each spectrum was the average of three acquisitions. The spectrometer was frequency calibrated at the beginning of each measurement session by using the Raman signal from a silicon wafer at 520 $cm^{-1}$. No changes related to photochemical damage were observed on the samples after the Raman measurements. In total, 272 spectra were recorded, at four different positions for each sample (16 spectra for each binder).

*2.4. Data Analysis.* Micro-Raman and FT-NIR spectral datasets were preprocessed using the PCA method by means of The Unscrambler X version 10.3 software (CAMO).

Although micro-Raman spectra were registered between 2000 and 200 $cm^{-1}$, data analysis was performed in the 1800–750 $cm^{-1}$ range, since the most important bands of binding media are located in this spectral region. The FT-NIR spectra were analyzed in the 6000 and 3900 $cm^{-1}$ range, where the signal was well defined and not much affected by noise.

The FT-NIR and micro-Raman datasets were preprocessed separately to discern variability in the original spectra. In detail, the FT-NIR spectra were subjected to the Savitzky-Golay derivative algorithm [20] (2nd order polynomial, 15 data points) without any other preprocessing. The micro-Raman spectra were, instead, subjected to baseline subtraction, average reduction of variables (factor two), and first derivative calculation by applying the Savitzky-Golay algorithm (2nd order polynomial, 15 data points). Then, both the FT-NIR and micro-Raman spectra were vector normalized, as the use of absolute band heights could lead to erroneous sample classifications [7]. The combined spectra were obtained as follows: the first derivative Raman and FT-NIR spectra were joined, generating the new combined Raman-FT-NIR spectrum, 16 for each typology of sample.

Each new combined spectrum includes 1575 data points, with the first 1050 derived from the FT-NIR spectrum abscissa, related to the spectral region 6000–3900 $cm^{-1}$, and the remaining 525 points taken from the micro-Raman spectrum abscissa in the 1800–750 $cm^{-1}$ spectral region. Each combined spectrum is characterized by both fundamental and combination-overtone modes, as have been previously described in the FT-NIR and micro-Raman sections.

The Uncertainty Test was always performed to estimate the significance of variables. Then, PCA was performed by applying a mean centering onto data column-wise, and the Nonlinear Iterative Partial Least Squares (NIPALS) algorithm. Each model was always validated by Leave One Out (LOO) cross-validation and by a projection of evaluated sets of spectral data acquired by the same samples (four spectra for each type of sample).

## 3. Results and Discussion

This study involves the application of the Principal Component Analysis (PCA) on FT-NIR and micro-Raman spectroscopic data in order to characterize the spectral differences, in frequency and intensity, among various classes of binding media. The following discussion is divided into two main sections dealing with an initial characterization of FT-NIR and Raman spectra of each binder and a successive evaluation of results obtained from the statistical model.

*3.1. FT-NIR and Micro-Raman Spectra of Different Binders.* First, we discuss the proteinaceous binders such as casein, egg white, and glues; following, we analyze the egg yolk and whole egg, which differ from the other protein-based binding media for the lipid content [6, 18]. Finally we discuss the spectra of the other materials.

We report in Figure 1 the FT-NIR spectra (Figure 1(a)) and the Raman spectra (Figure 1(b)) of casein, egg white, and glues. The raw FT-NIR spectra of these proteinaceous binding media do not show significant spectral differences. The combination band of amide III (C-N stretching/N-H in plane bending), the first overtone of carbonyl stretching (~4600 $cm^{-1}$), and the combination band of the stretching and bending modes of NH (~4880 $cm^{-1}$) represent the characteristic signals of proteins. Furthermore, a weak absorption band, located at about 3988 $cm^{-1}$ and ascribed to the first C-N-C stretching overtone in amides, can be observed [6, 21]. For all these binders, the methylenic overtones and combination bands result to be broad and weak (Figure 1(a)). Nevertheless, some differences can be highlighted: in the spectra of the glues, the combination band of asymmetric methylenic C-H stretching and bending is located at 4376 $cm^{-1}$, while in the spectra related to casein and egg white, this absorption band is located at about 4340 $cm^{-1}$. Furthermore, this absorption band results to be more intense than the symmetric one, always for the case of glues.

The micro-Raman spectra of casein, egg white, and glues (Figure 1(b)) are characterized by the vibrations related to amide I (at about 1665 $cm^{-1}$), $CH_2$ scissoring (at about 1450 $cm^{-1}$), and amide III (at about 1245 $cm^{-1}$). A sharp band due to the aromatic ring breathing is present in all these spectra, which arises from phenylalanine amino acid residues (at about 1002 $cm^{-1}$) [6, 18]. This last peak is especially pronounced in egg white and casein, for which phenylalanine is present in higher concentrations than in collagen-based glues [18].

Furthermore, contributions due to tyrosine amino acid are present in all these protein-based binders (826–850, 1210, and 1605–1610 $cm^{-1}$) [18]. Moreover, in the spectra of casein and egg white, peaks ascribable to tryptophan amino acid (at approximately 1550 and 880 $cm^{-1}$) and to phosphoproteins (~954 $cm^{-1}$) can be found [18].

The FT-NIR spectra related to egg yolk and whole egg, shown in Figure 2(a), are characterized by the methylenic combination bands and their overtones, in the ranges 4250–4350 $cm^{-1}$ and 5600–5800 $cm^{-1}$, more intense with respect to the other protein-based binders, due to the lipid content of

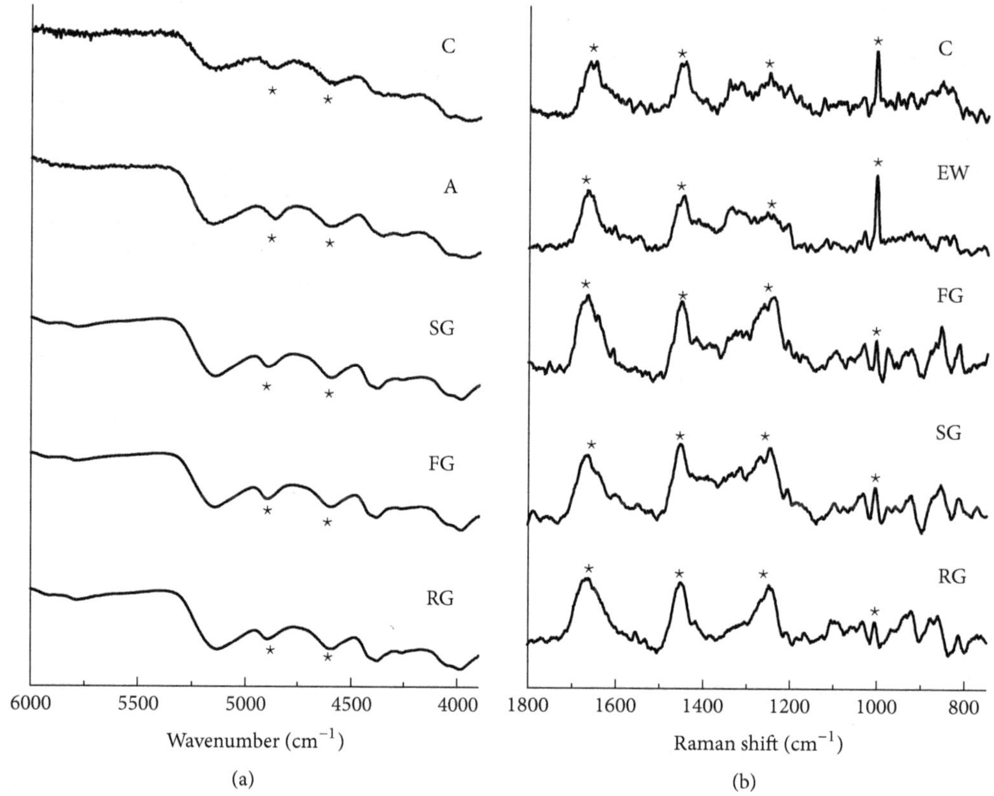


FIGURE 1: FT-NIR (a) and Raman (b) spectra of casein-C, egg white-EW, fish glue-FG, strong glue-SG, and rabbit glue-RG. Raman spectra are presented after background subtraction. Star symbols point to some specific spectral features discussed in the text. The assignments of the FT-NIR and Raman spectra are reported in Tables S1 and S2.

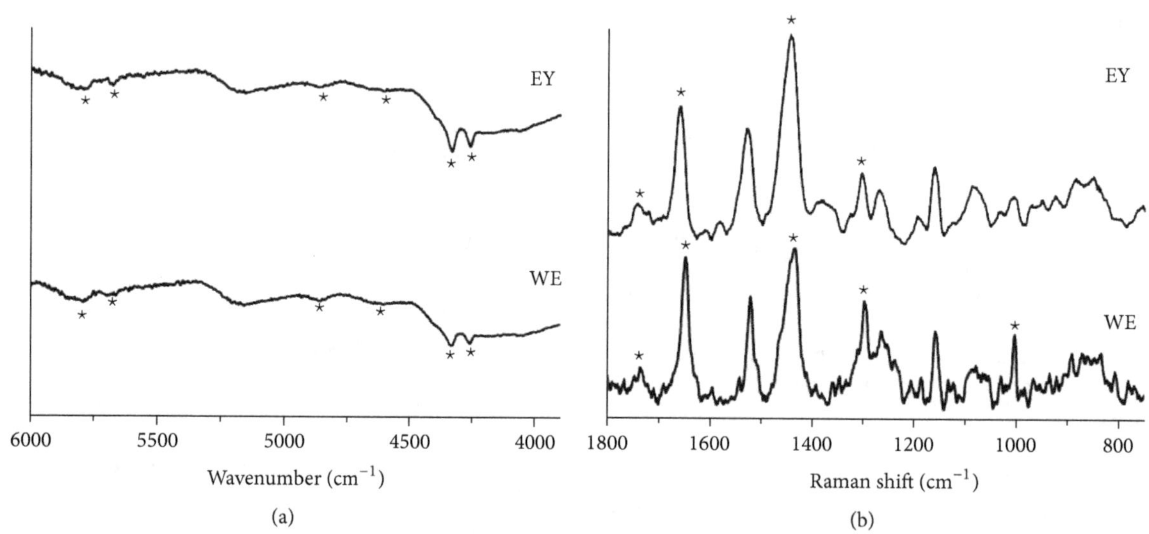


FIGURE 2: FT-NIR (a) and Raman (b) spectra of egg yolk-EY and whole egg-WE. Raman spectra are presented after background subtraction. Star symbols point to some specific spectral features discussed in the text. The assignments of the FT-NIR and Raman spectra are reported in Tables S3 and S4.

yolk [6, 21]. The shape and position of the methylenic combination bands from the lipid component in egg yolk/whole egg are different with respect to those from drying oils, thus providing distinctive features for their recognition [6, 21]. However, the main proteinaceous bands are all detectable, as can be seen in Figure 2(a).

Concerning the corresponding Raman spectra (Figure 2(b)), due to the presence of fatty acid esters, the characteristic

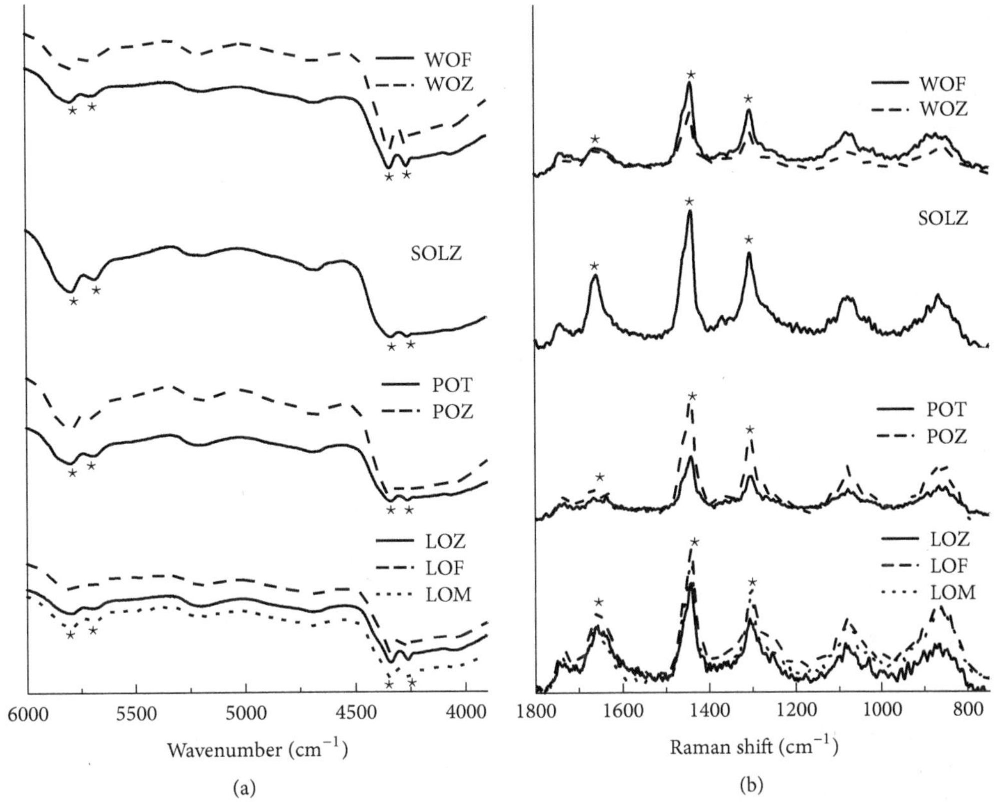


FIGURE 3: FT-NIR (a) and Raman (b) spectra of drying oils. Raman spectra are presented after background subtraction. Star symbols point to some specific spectral features discussed in the text. Symbols for each drying oil are taken from Table 1. The assignments of the FT-NIR and Raman spectra are reported in Tables S5 and S6.

carbonyl vibration at 1740 cm$^{-1}$ is present. Furthermore, bands due to the following vibrations are predominant: 1302 cm$^{-1}$ (in-phase methylene twist) [18], 1442 cm$^{-1}$ (scissoring mode of methylene) [18], and 1655 cm$^{-1}$, (*cis*-double bond stretching) [22]. This last band has a frequency close to the usual value at which the amide I band is observed (1665 cm$^{-1}$) [18, 22]. Egg yolk and whole egg have the lowest values for polyunsaturated fatty acids in comparison with drying oils [2, 8]. Nevertheless, characteristic proteinaceous bands are present, as can be seen in Figure 2(b). The phenylalanine ring breathing (~1002 cm$^{-1}$) [18, 22] results to be more intense in whole egg spectrum, because it presents a major proteinaceous content for the preparation (it was prepared mixing one part of egg white with two parts of egg yolk and one part of vinegar).

FT-NIR spectra of drying oils are shown in Figure 3(a). As regards drying oils, the main near infrared absorption bands are represented by the combination of methylenic C-H stretching and bending (between 4250 and 4350 cm$^{-1}$) and first $CH_2$ stretching overtone bands (in the 5700–5800 cm$^{-1}$ spectral range). The second overtone of ester carbonyl stretching mode is observed at 5176 cm$^{-1}$, whereas the combination of ester C=O and methylenic C-H stretching modes occurs at about 4690 cm$^{-1}$ [6, 8]. Raw FT-NIR spectra of drying oils do not show significant spectral differences between each other. In addition, the FT-NIR spectrum related to stand-oil does not show any information that could be ascribed to the effect of the heating process used during its preparation.

On the other hand, the major Raman bands (shown in Figure 3(b)) are represented by symmetric rock in *cis*-double bond (~1265 cm$^{-1}$), in-phase methylene twist mode (~1302 cm$^{-1}$), scissoring mode of methylene (~1442 cm$^{-1}$), *cis*-double bound stretching mode (~1655 cm$^{-1}$), and ester stretching (~1745 cm$^{-1}$), as discussed by different authors [2, 8]. Therefore, the drying oils Raman spectra predominantly contain bands arising from vibrations of the hydrocarbon chains. As can be seen in Figure 3(b), the major difference between the different drying oils is located at the band 1655 cm$^{-1}$ that is more intense for linseed oil and stand-oil. Linseed oil, indeed, contains around 52–55% of linolenic acid in its chemical composition, whereas poppy-seed oil contains 0.4–0.6% and walnut oil contains 10–15% [2, 22]. Concerning stand-oil, the heating process generates a complex set of reactions leading to different products, including systems with conjugated double bonds [23].

Broad bands characterize the FT-NIR spectra of gums, as shown in Figure 4(a). The characteristic signals of polysaccharides are represented by the combination bands of O-H stretching and bending (~5140 cm$^{-1}$), of O-H bending and C-O stretching (~4800 cm$^{-1}$) and of C-H and C-C/C-O-C stretching modes (~4010 cm$^{-1}$) [21]. In addition, a very broad

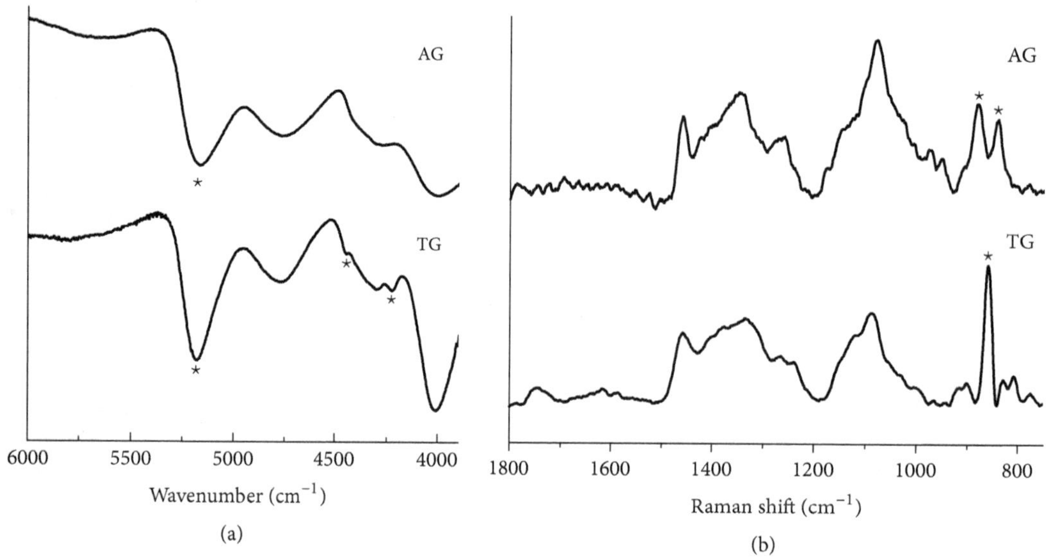


FIGURE 4: FT-NIR (a) and Raman (b) spectra of tragacanth gum-TG and Arabic gum-AG. Raman spectra are presented after background subtraction. Star symbols point to some specific spectral features discussed in the text. The assignments of the FT-NIR and Raman spectra are reported in Tables S7 and S8.

band ascribed to combination band of C-H stretching and bending, at about 4270 cm$^{-1}$, can be observed [21].

As regards the near infrared region, the combination band of O-H stretching and bending represents the main difference between two gums. In the spectrum of tragacanth gum, this band is shifted of around 20 cm$^{-1}$ towards longer wavenumbers in comparison with Arabic gum.

The Raman spectra, shown in Figure 4(b), do not present intense peaks in the 1800–1500 cm$^{-1}$ spectral range. This feature already differentiates the gums from the other binding media previously described. These materials, indeed, do not contain aromatic or aliphatic C=C double bonds. In addition, no amide I and amide III bands are present, in contrast to proteinaceous binders [22, 24]. The rather broad peaks in the 1500–1200 cm$^{-1}$ region are ascribed to C-H deformations, whereas the peaks at lower wavenumbers are associated with C-C and C-O symmetrical stretching modes (1200–950 cm$^{-1}$) and C-O-C sugar ring vibrations (950–800 cm$^{-1}$) [22, 24]. The Raman spectrum of Arabic gum shows a doublet at around 850 cm$^{-1}$ (882 cm$^{-1}$ and 841 cm$^{-1}$), whereas tragacanth gum has a singlet (857 cm$^{-1}$). This feature allows distinguishing between the two gums.

*3.2. PCA Applied on Combined First Derivative FT-NIR and Micro-Raman Spectra.* In this section, the potentialities of PCA to improve the description of each class of binders are evaluated and discussed.

The application of a multivariate analytical tool, such as PCA, to the combined first derivative FT-NIR and Raman spectra allows the concomitant extraction of important information. The obtained statistical model explains a significant percentage (83%) of the cumulative variability by means of the first four PCs. In detail, PC1 explains 49% of the total variance contained in the data, whereas PC2, PC3, and PC4 explain 19, 11, and 4% of the total variance, respectively.

Focusing on the two-dimensional score plot in the space defined by PC1 and PC2 (Figure 5) four different data groups can be individuated: (1) gums; (2) drying oils; (3) egg yolk and whole egg; (4) glues, casein, and egg white.

In Figures 6(a), 6(b), and 6(c), the loading plots of PC1 and PC2 are reported, highlighting the contributions that can be ascribed to *polysaccharide-*, *lipid-*, and *protein-*based binding media, respectively. In all loading plots, the "plus" symbol is associated with a first derivative contribution concordant with the starting combined spectra, while a "minus" sign is used when its behavior is reversed. Gums dataset displays a negative score value for PC1, whereas it has a positive score value for PC2 (Figure 5). As can be seen in Figure 6(a), the greatest contributions to PC1 related to polysaccharide-based binding media are represented by the combination band of O-H stretching and bending (~ 5140 cm$^{-1}$) [21], the combination band of C-H and C-C/C-O-C stretching modes (~4010 cm$^{-1}$) [21], and the C-O-C sugar ring vibrations (950–800 cm$^{-1}$) [22, 24]. As regards PC2, the main highlighted polysaccharide vibrations are the combination band of O-H stretching and bending and the C-O-C sugar ring vibrations (Figure 6(a)). So, the first two PCs allow describing polysaccharide materials, immediately focusing on their principal spectral properties in NIR and Raman regions but do not allow any further differentiation.

Focusing, instead, on drying oils, these binders datasets contribute with a positive score value to PC1, whereas they result to be very close to a zero value for PC2 (Figure 5). PC1, shown in Figure 6(b), is characterized by bands associated with drying oils, such as the $CH_2$ stretching first overtone band (in the 5700–5800 cm$^{-1}$ spectral range) [6, 8], combination band of methylenic C-H stretching and bending (between 4250 and 4350 cm$^{-1}$) [6, 8], scissoring mode of methylene (1442 cm$^{-1}$) [2, 8], and in-phase methylene twist mode (~1302 cm$^{-1}$) [2, 8]. In PC2, their main contributions

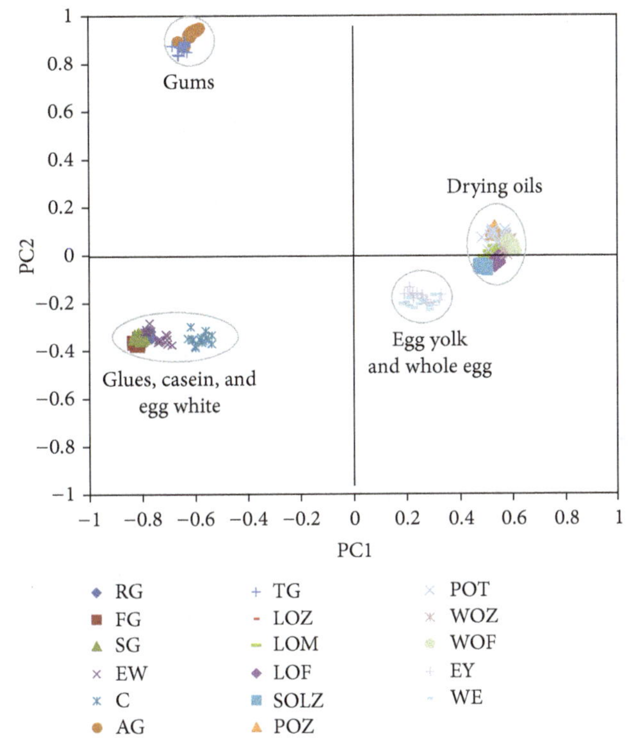


Figure 5: PCA applied to combined FT-NIR and Raman first derivative spectra, taking into account all neat binding media: score plot of PC1-PC2.

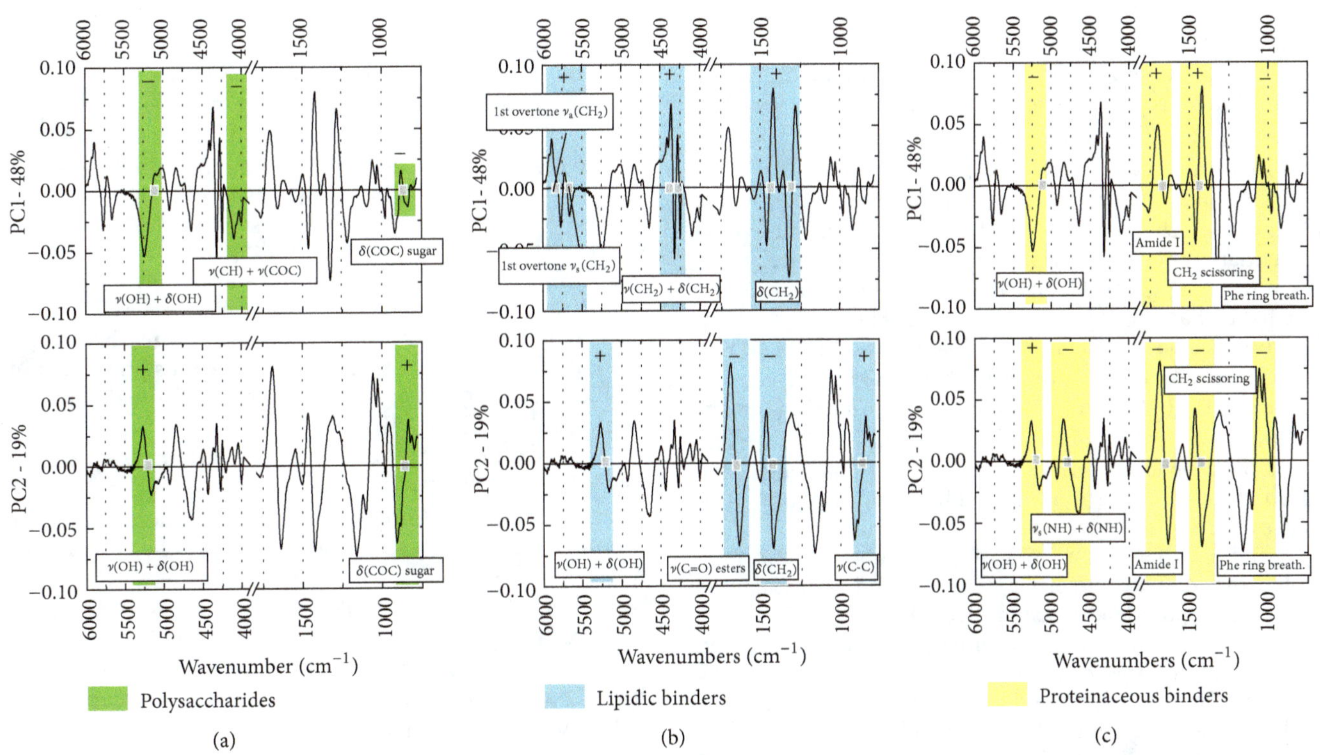


Figure 6: PCA applied to combined FT-NIR and Raman first derivative spectra, taking into account all neat binding media: score plot of PC1-PC2 (a); loading plots of PC1 and PC2, highlighting in the different panels the bands associated with different materials: polysaccharides (a), lipids (b), and proteinaceous materials (c).

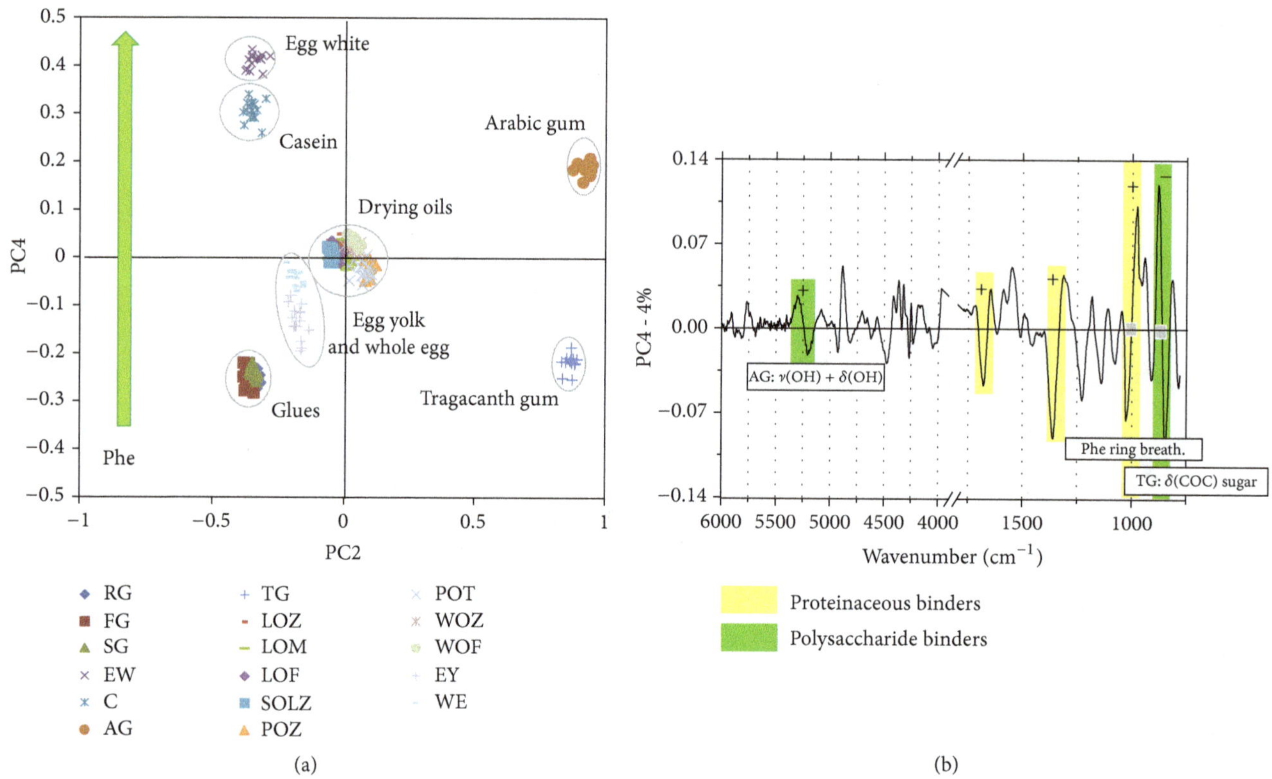


FIGURE 7: PCA applied to combined FT-NIR and Raman spectra, taking into account all neat binding media: score plot of PC2-PC4 (a) and loading plot of PC4 (b), highlighting proteinaceous and polysaccharide contributions.

are related to combination band of methylenic C-H stretching and bending, in NIR region, and scissoring mode of methylene, in Raman region. These contributions cause a very low score value for PC2, because they are opposed: the former is positive, while the latter is negative (Figure 6(b)). Also, for drying oils this statistical model immediately allows to highlight their characteristic spectral bands. Nevertheless, no differentiation between various drying oils is obtained.

Concerning glues, C, and EW display a negative score value for PC1 and a negative one for PC2 (Figure 5). As regards proteinaceous binders, in the loading plot of PC1 (Figure 6(c)), specific contributions due to amide I (~1665 $cm^{-1}$), $CH_2$ scissoring (~1450 $cm^{-1}$), and phenylalanine ring breathing (1002 $cm^{-1}$) vibrations can be observed [18, 22]. Nevertheless, PC2 better describes protein-based binding media. Indeed, in the corresponding loading plot (Figure 6(c)) intense contributions related to the combination band of the stretching and bending modes of NH at about 4880 $cm^{-1}$, amide I mode, $CH_2$ scissoring vibration, and phenylalanine ring breathing mode can be highlighted [6, 21]. Instead, EY and WE are located at an intermediate position between drying oils and the other protein-based binding media (Figure 5). These binders are characterized by a lipid content that is completely absent in C, EW, and glues.

As previously discussed, the purpose of all multivariate data analysis is to decompose the data in order to detect and model the "hidden phenomena." Taking into mind this principle/precept, the PC4 results to be notably interesting. As we can observe in the space defined by the score plot of PC2 and PC4 (Figure 7(a)) a clear separation between AG and TG can be observed as well as between EW, C, and glues. Indeed, the loading plot of PC4, reported in Figure 7(b), is characterized by two main contributions: phenylalanine ring breathing and C-O-C ring vibration. Concerning phenylalanine ring breathing vibration, PC4 defines a trend that describes the different content of this amino acid in proteinaceous binding media. The peak at 1002 $cm^{-1}$, as previously described, is especially pronounced in EW and C Raman spectra, for which phenylalanine is present in higher concentrations than in collagen-based glues (a total of approximately 12% of tyrosine plus phenylalanine is detected in egg and casein, compared to approximately 4% total in animal glue) [18].

On the other hand, C-O-C ring vibration, also described by PC4, shows the specific frequency related to TG: the characteristic singlet at 857 $cm^{-1}$ [22, 24]. Additionally, another minor contribution, ascribable to gums, is the combination band of OH stretching and bending. In this case, the corresponding frequency is 5180 $cm^{-1}$, which is close to that observed for AG near infrared spectrum (Figure 4(a)).

In Table 2 a summary of the main spectral properties highlighted by this statistical model for the different classes of binding media is reported.

## 4. Conclusions

In this work, we have analyzed model films of protein-, polysaccharide-, and lipid-based binding media after a natural

Table 2: Main near infrared and Raman bands, characteristic for different classes of binding media, highlighted by PCA applied on combined spectra referred to all neat binders.

| Class of binders | FT-NIR | Micro-Raman |
|---|---|---|
| Polysaccharide-based | $\nu$(OH) + $\delta$(OH)<br>$\nu$(CH) + $\nu$(COC) | $\delta$(COC) sugar ring vibrations |
| Lipid-based | 1st overtone $CH_2$<br>$\nu(CH_2) + \delta(CH_2)$<br>[$\nu$(OH) + $\nu$(OH)] | $\delta(CH_2)$<br>$\nu$(C-C) |
| Protein-based | $\nu$(NH) + $\delta$(NH)<br>$\nu$(OH) + $\nu$(OH) | Amide I<br>$\delta(CH_2)$<br>Phe ring breathing |

ageing of three months, highlighting the applicability of FT-NIR and Raman spectroscopies for their characterization. Assignment of main near infrared and Raman bands revealed significant differences between spectra related to various classes of binders. These differences are emphasized by PCA applied on the corresponding combined first derivative spectra. The application of PCA on combined first derivative spectra has allowed immediately visualizing the main spectral properties, in both regions investigated, of these materials. The proposed method represents a novel noninvasive and nondestructive approach for their identification and differentiation.

Additionally, the results described highlighted the capability of this methodology to extract information related to hidden variance. Therefore, we differentiated egg white and casein between each other and from glues, based on different content in the amino acid phenylalanine. It has been also possible to distinguish between Arabic gum and tragacanth gum. Nevertheless, we did not observe any differentiation within drying oils or glues groups. This result indicates and suggests the need for further PCA modelling of each such group in order to fine-tune the analysis and to understand the characteristic features of the different classes of binders.

This work represents a contribution to the development of nondestructive, optical methods that could provide knowledge of the binder media used by artists. Their identification is often quite difficult due to their complex molecular nature. The development of these methods, which could act without detriment of the sample, helps not only in complementing historical research, but also in the process of conservation and restoration of artworks.

## Conflicts of Interest

The authors declare that there are no conflicts of interest regarding the publication of this article.

## Acknowledgments

Authors wish to thank Dr. M. Picollo, IFAC-CNR, Italy, for making available the ALPHA-FTIR spectrometer for this research. This work was supported by Italian MIUR (PRIN, Grant 2010329WPF_007) and by University of Florence (ex60%). Support from Ente Cassa di Risparmio di Firenze (Grant no. 2014.0405A2202.8044) is also kindly acknowledged.

## References

[1] B. H. Stuart, *Analytical Techniques in Materials Conservation*, John Wiley & Sons, New York, NY, USA, 2007.

[2] E. Manzano, J. García-Atero, A. Dominguez-Vidal, M. J. Ayora-Cañada, L. F. Capitán-Vallvey, and N. Navas, "Discrimination of aged mixtures of lipidic paint binders by Raman spectroscopy and chemometrics," *Journal of Raman Spectroscopy*, vol. 43, no. 6, pp. 781–786, 2012.

[3] M. P. Colombini, A. Andreotti, I. Bonaduce, F. Modugno, and E. Ribechini, "Analytical strategies for characterizing organic paint media using gas chromatography/mass spectrometry," *Accounts of Chemical Research*, vol. 43, no. 6, pp. 715–727, 2010.

[4] N. Navas, J. Romero-Pastor, E. Manzano, and C. Cardell, "Raman spectroscopic discrimination of pigments and tempera paint modelsamples by principal component analysis on first-derivative spectra," *Journal of Raman Spectroscopy*, vol. 41, no. 11, pp. 1486–1493, 2010.

[5] N. Navas, J. Romero-Pastor, E. Manzano, and C. Cardell, "Benefits of applying combined diffuse reflectance FTIR spectroscopy and principal component analysis for the study of blue tempera historical painting," *Analytica Chimica Acta*, vol. 630, no. 2, pp. 141–149, 2008.

[6] M. Vagnini, C. Miliani, L. Cartechini, P. Rocchi, B. G. Brunetti, and A. Sgamellotti, "FT-NIR spectroscopy for non-invasive identification of natural polymers and resins in easel paintings," *Analytical and Bioanalytical Chemistry*, vol. 395, no. 7, pp. 2107–2118, 2009.

[7] A. Pallipurath, J. Skelton, P. Ricciardi, S. Bucklow, and S. Elliott, "Multivariate analysis of combined Raman and fibre-optic reflectance spectra for the identification of binder materials in simulated medieval paints," *Journal of Raman Spectroscopy*, vol. 44, no. 6, pp. 866–874, 2013.

[8] S. Carlesi, M. Ricci, C. Cucci et al., "Multivariate analysis of combined fourier transform near-infrared spectrometry (FT-NIR) and Raman datasets for improved discrimination of drying oils," *Applied Spectroscopy*, vol. 69, no. 7, pp. 865–876, 2015.

[9] S. Carlesi, M. Ricci, C. Cucci, C. Lofrumento, M. Picollo, and M. Becucci, "Multivariate analysis of combined reflectance FT-NIR and micro-Raman spectra on oil-paint models," *Microchemical Journal*, vol. 124, pp. 703–711, 2016.

[10] J. Bueno and I. K. Lednev, "Advanced statistical analysis and discrimination of gunshot residue implementing combined Raman and FT-IR data," *Analytical Methods*, vol. 5, no. 22, pp. 6292–6296, 2013.

[11] A. Zoppi, C. Lofrumento, M. Ricci, E. Cantisani, T. Fratini, and E. M. Castellucci, "A novel piece of Minoan art in Italy: the first spectroscopic study of the wall paintings from Phaistos," *Journal of Raman Spectroscopy*, vol. 43, no. 11, pp. 1663–1670, 2012.

[12] J. Romero-Pastor, N. Navas, S. Kuckova, A. Rodríguez-Navarro, and C. Cardell, "Collagen-based proteinaceous binder-pigment interaction study under UV ageing conditions by MALDI-TOF-MS and principal component analysis," *Journal of Mass Spectrometry*, vol. 47, no. 3, pp. 322–330, 2012.

[13] C. Pasquini, "Near infrared spectroscopy: fundamentals, practical aspects and analytical applications," *Journal of the Brazilian Chemical Society*, vol. 14, no. 2, pp. 198–219, 2003.

[14] M. Otto, *Chemometrics: Statistics and Computer Application in Analytical Chemistry*, John Wiley & Sons, 2007.

[15] K. H. Esbensen, "Multivariate Data Analysis—In Practice," CAMO Software AS., 2009.

[16] C. Cennini and F. Frezzato, "Il Libro Dell'arte," N. Pozza, 2003.

[17] L. Masschelein-Kleiner, *Ancient Binding Media, Varnishes and Adhesives*, ICCROM, Rome, Italy, 2nd edition, 1995.

[18] A. Nevin, I. Osticioli, D. Anglos, A. Burnstock, S. Cather, and E. Castellucci, "Raman spectra of proteinaceous materials used in paintings: a multivariate analytical approach for classification and identification," *Analytical Chemistry*, vol. 79, no. 16, pp. 6143–6151, 2007.

[19] I. Bonaduce, L. Carlyle, M. P. Colombini et al., "New insights into the ageing of linseed oil paint binder: a qualitative and quantitative analytical study," *PLoS ONE*, vol. 7, no. 11, Article ID e49333, 2012.

[20] A. Savitzky and M. J. E. Golay, "Smoothing and differentiation of data by simplified least squares procedures," *Analytical Chemistry*, vol. 36, no. 8, pp. 1627–1639, 1964.

[21] D. A. Burns and E. W. Ciurczak, *Handbook of Near-infrared Analysis*, CRC Press, Boca Raton, Fla, USA, 2008.

[22] P. Vandenabeele, B. Wehling, L. Moens, H. Edwards, M. De Reu, and G. Van Hooydonk, "Analysis with micro-Raman spectroscopy of natural organic binding media and varnishes used in art," *Analytica Chimica Acta*, vol. 407, no. 1-2, pp. 261–274, 2000.

[23] O. Zovi, L. Lecamp, C. Loutelier-Bourhis, C. M. Lange, and C. Bunel, "Stand reaction of linseed oil," *European Journal of Lipid Science and Technology*, vol. 113, no. 5, pp. 616–626, 2011.

[24] L. Brambilla, C. Riedo, C. Baraldi et al., "Characterization of fresh and aged natural ingredients used in historical ointments by molecular spectroscopic techniques: IR, Raman and fluorescence," *Analytical and Bioanalytical Chemistry*, vol. 401, no. 6, pp. 1827–1837, 2011.

# 18

# 3-Mercaptopropyltrimethoxysilane Modified Diatomite: Preparation and Application for Voltammetric Determination of Lead (II) and Cadmium (II)

**Dinh Quang Khieu,[1] Bui Hai Dang Son,[1] Vo Thi Thanh Chau,[2] Pham Dinh Du,[3] Nguyen Hai Phong,[1] and Nguyen Thi Diem Chau[4]**

[1] *College of Science, Hue University, Hue City 530000, Vietnam*
[2] *Faculty of Technology, Industrial University of Ho Chi Minh City, Quang Ngai Campus, Quang Ngai City 570000, Vietnam*
[3] *Faculty of Natural Science, Thu Dau Mot University, Thu Dau Mot City 590000, Vietnam*
[4] *Department of Agriculture and Rural Development, Tam Ky City, Quang Nam Province 560000, Vietnam*

Correspondence should be addressed to Dinh Quang Khieu; dqkhieu@hueuni.edu.vn

Academic Editor: Pedro Ávila-Pérez

In this study, functionalized diatomite was prepared by grafting of 3-mercaptopropyltrimethoxysilane (MPTMS) to diatomite (MPTMS-diatomite). The diatomite with thermal treatment from 100 to 700°C was functionalized by MPTMS under dry and humid conditions. The obtained MPTMS-diatomite was characterized by X-ray diffraction (XRD), thermal gravity-differential scanning calorimeter (TG-DSC), and Fourier transformation infrared (FT-IR). The results showed that an increase in treatment temperature seems to reduce the loading of MPTMS onto diatomite. The humidity of diatomite was favorable for the grafting of functional groups on the surface. The possible mechanisms of MPTMS loading to diatomite (MPTMS-diatomite) were also proposed. The performance of a carbon paste electrode (CPE) modified with MPTMS-diatomite in the simultaneous determination of Cd(II) and Pb(II) ions was addressed.

## 1. Introduction

Heavy metals (Hg(II), Pb(II), Cd(II), Cu(II), Co(II), Ni(II), Zn(II), etc.), which are commonly found in wastewaters discharged from chemical manufacturing, mining, extractive metallurgy, nuclear, and other industries, cause danger to humans and aquatic animals and plants and therefore must be controlled urgently [1]. The detection of heavy metal ions in water samples has received much attention over the past few years. The commonly used methods for metal analysis are atomic absorption spectrometry (AAS) and electrothermal atomic absorption spectrometry (ETAAS), inductively coupled plasma-mass spectrometry (ICP-MS), inductively coupled plasma atomic emission spectrometry (ICP-AES), and inductively coupled plasma optical emission spectrometry (ICP-OES) [2–4]. These techniques can assist quantitative analyses of the total amount of toxic metals in water samples. However, disadvantages such as instrument costs and complex operation procedures may limit the use of these techniques.

Anodic stripping voltammetry (ASV) is a useful electrochemical method for detecting trace metals because of its wide linear dynamic range and low detection limit which results from a preconcentration step performed directly in the voltammetric cell. Several porous materials have been utilized as electrode modifiers with the aim of preconcentrating the analyte at the electrode surface via a selective reaction due to special function groups grafted to electrode surface [5–10].

Diatomite, which consists of a wide variety of sized diatoms in a structure containing up to 80–90% pores, is a pale-coloured, soft, lightweight sedimentary rock composed principally of silica fossils of aquatic unicellular alga [11]. Diatomite has received considerable attention for its unique combination of physical and chemical properties including

TABLE 1: Chemical composition of diatomite (wt.%).

| Elements | K | Al | Si | Ti | Fe | Others |
|---|---|---|---|---|---|---|
| Raw diatomite | 0.34 | 13.07 | 76.02 | 1.31 | 7.68 | 1.58 |

large porosity, fine particle size, and chemical inertness [12]. Silanol groups often occur on the diatomite surface especially on the natural diatomite surface. The density of the silanol groups can be controlled by hydrating diatomite. When these silanol groups combine with organosilane through a series of chemical reaction, the functional groups at the terminal end of the organosilanes can be anchored to the surface of the diatomite to form surface activated products. The diatomite in combination with surface coated functional groups offers a potential materials for adsorption [13, 14] and modifying electrodes to determine the dye [13]. Recently, the usage of silica-based materials (especially organically modified silicates) for modifying glassy carbon electrodes (GCEs) applied to the voltammetric detection of metal ions such as Hg(II), Cd(II), and Pb(II) has been extensively reported [6, 7]. However the use of organosilane modified diatomite as an electrode modifier less pronounced so far [15]. To the best of our knowledge, the use of a 3-mercaptopropyltrimethoxysilane (MPTMS) which grafted diatomite for modifying electrode in connection with the electrochemical analysis of Pb (II) and Cd (II) has not yet been reported.

In this paper, the preparation of MPTMS modified diatomite was demonstrated. The effects of thermal treatment of diatomite and the humidity of diatomite on MPTMS functionalization and the possible proposed mechanisms of MPTMS grafting diatomite surface were discussed. The MPTMS-diatomite which was used to modify electrode for simultaneous determination of Pb(II) and Cd(II) in aqueous solution by differential pulse anodic stripping voltammetry method (DP-ASV) was also studied.

## 2. Experimental

*2.1. Materials.* Diatomite sample was collected from Phu Yen province, Vietnam, after removing organic stuff by repeated sedimentation and desiccation at 100°C. The 3-mercaptopropyltrimethoxysilane (MPTMS, >99%, Merck) and chloroform (Merck) were used as functional reagent and solvent, respectively. The chemical composition of diatomite analyzed by EDS was listed in Table 1.

All working solutions of reagents were prepared from chemicals for analysis (Merck, Germany). Acetate buffer solutions (ABS) of pH = 4,5 were prepared from $CH_3COOH$ and $CH_3COONa$ (Merck). pH of the ABS was adjusted by adding 1.0 M nitric acid or 1.0 M sodium hydroxide solutions into the ABS and then checking by pH meter. Working solution of Cd(II) and Pb(II) was prepared daily from stock solution (1000 ppm, Merck, Germany). Nafion ($C_7HF_{13}O_5S{\cdot}C_2F_4$, Merck) was used as an adhesive and electric conductivity material.

*2.2. The Functionalization of Diatomite with MPTMS (MPTMS-Diatomite).* The diatomite was calcinated at 100, 300, 500, and 700°C for 3 hours. The incorporation of MPTMS into calcinated diatomite was conducted under dried and humid conditions. For the dry condition, 2 grams of calcinated diatomite was mixed with 2 g of MPTMS and 36 mL of chloroform in a 250 mL flask covered with watch glass. After mixing the slurry for 4 days at ambient temperature by a magnetic stirrer, it was washed with 10 mL of chloroform and filtered through a Buchner funnel. The separated solid was dried at 70°C for 24 hours. The functionalized diatomite is denoted as 100D, 300D, 500D, and 700D. The number implies temperature at which diatomite was heat-treated prior to functionalization. For the humid condition, the calcinated diatomite was hydrated for 2 hours in a desiccator containing saturated NaCl solution at the bottom before functionalization. Two grams of hydrated diatomite was mixed with 2 g of MPTMS for two hours. The solid was filtered and the extra MPTMS was extracted with chloroform in a Soxhlet until the solution in flask was clear. The resulting samples are denoted as 100H, 300H, 500H, and 700H. With the aim of studying the applicability of the material MPTMS-diatomite, the use of MPTSE-diatomite as an electrode modifier was performed.

### 2.3. Voltammetric Procedure

*2.3.1. Preparation of Working Electrode-Modified Electrode.* The preparation procedure was carried out as [16]. In brief, glassy carbon electrode (GCE) was first polished with 0.05 $\mu$m $Al_2O_3$ slurries on a polishing cloth and then rinsed ultrasonically with 2 M $HNO_3$, absolute ethanol, and double-distilled water. Subsequently, the GCE was electrochemically cleaned by cyclic potential scan between −1.5 V and +1.5 V in 0.5 M ABS pH 4.5 with the scan rate of 0.10 V $s^{-1}$. After that, the electrode was rinsed with double-distilled water. Nafion-MPTMS-diatomite composite was prepared by dispersing 5.0 mg MPTMS-diatomite powder into 1.0 mL Nafion (0.1 wt%) and sonicated. The GCE was coated with 5 $\mu$L of the composite and dried at room temperature (~25°C) (denoted as MPTMS-diatomite/GCE).

*2.3.2. Voltammetric Procedure.* Solution under study (final volume of 10 mL) containing 0.1 M ABS (pH 4.5) and 100 ppb Cd(II) and 200 ppb Pb(II) was transferred into the electrochemical cell with the 3 electrodes. Then, voltammetric measurement was conducted as follows:

(i) *For surveying voltammetric characteristics of Cd(II) and Pb(II) on the MPTMS-diatomite/GCE*: Cd(II) and Pb(II) was accumulated on the surface of the modified electrode at a potential of −1.2 V ($E_{acc}$) for an accumulated time of 60 s ($t_{acc}$). During this stage, the electrode was rotated at a constant rate of 1000 rpm. After that, the electrode rotation was off. Then, cyclic voltammograms were recorded from −1.2 V to +0.2 V (forward potential scan) and subsequently from +0.2 V to −1.2 V (reverse potential scan) at a scan rate of 0.1 V $s^{-1}$.

(ii) *For differential pulse anodic stripping voltammetric (DP-ASV) determination of Cd(II) and Pb(II) on electrode MPTMS-diatomite/GCE*: during the DP-ASV procedure, the Cd(II) and Pb(II) were accumulated on the surface of the rotating modified electrode as done for the above procedure. After that, the electrode rotation was off for 10 s and then DP-ASV voltammograms were recorded from −1.2 V to +0.2 V at a scan rate of 0.02 V $s^{-1}$ and suitable differential voltammetry technique parameters. During the potential anodic scan, Cd(0) and Pb(0) on the electrode surface were oxidized to Cd(II) and Pb(II). DP-ASV voltammograms of blank solution (without Cd(II) and Pb(II) and prepared from double-distilled water) were similarly recorded before each measurement. The Cd(II) and Pb(II) quantitation was done by standard addition method.

*2.4. Apparatuses.* The phases were monitored by powder X-ray diffraction (XRD), recorded on 8D Advance Bucker, Germany with Cu$K_\alpha$ radiation in the range of from 20° to 60° with a scan step size of 0.01° and a scan step time of 0.04 s. The amounts of MPTMS incorporated into diatomite were evaluated by TG-DSC SETARAM under ambient atmosphere. Morphology of diatomite was observed by scanning electron microscopy (SEM.Jeole-3432) and transmission electron microscope (TEM) (JEOL JEM-2100F). Fourier transform infrared (FT-IR) spectra of the typical samples (1 wt% sample + 99 wt% KBr) were collected on Shimadzu FT-IR 8010M; the wavenumber range was 500–4000 $cm^{-1}$. A CPA-HH5 Computerized Polarographic Analyzer (produced by Laboratory of Computer Application to Chemistry, Institute of Chemistry, Vietnam Academy of Science and Technology) was used for voltammetry experiments. All measurements were done in the cell with three electrodes: a GCE with a diameter of 2.8 ± 0.1 mm used for formatting the modified electrode as working electrode, Ag/AgCl/3 M KCl as reference electrode, and a platinum wire as auxiliary electrode. All measurements were carried out at room temperature.

## 3. Results and Discussion

*3.1. Preparation of MPTMS Modified Diatomite (MPTMS-Diatomite).* Figure 1 shows XRD patterns of diatomite calcinated at 100°C, 300°C, 500°C, and 700°C. The raw diatomite (dried at 100°C) consisted of mainly amorphous structure. At 300°C, the characterized peak of quartz was observed indicating that the silica in amorphous form was crystallized to form quartz crystallites. The diffraction intensity of quartz phase increases with an increase in calcinated temperature. Figure 2 shows SEM and TEM observation. Morphology of raw diatomite shows a cylindrical structure with a *ca.* length of 10–20 $\mu$m and an internal pore diameter of about 2-3 $\mu$m. There are nearly regular arrays of submicron pore in a diameter of *ca.* 100 nm in the wall.

In order to investigate the surface characteristics of raw and calcined diatomite, FT-IR analyses were carried out in the range 400–4000 $cm^{-1}$. The main absorption bands for raw diatomite, as shown in Figure 3, were observed at 466, 528, 690, 798, 914, 1022, 1153, 3622, and 3695 $cm^{-1}$. The absorption peaks around 466 and 528 $cm^{-1}$ represent to the

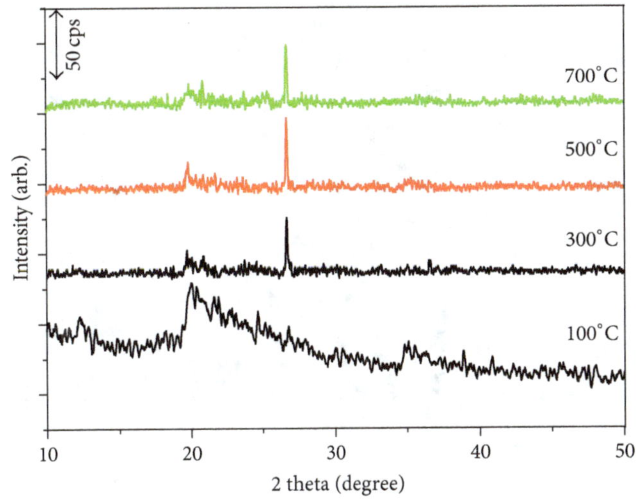


FIGURE 1: XRD patterns of diatomite calcined at different temperatures.

Si–O–Si bending vibration [17]. The 690 and 798 $cm^{-1}$ bands reflect SiO–H vibration while the 910 $cm^{-1}$ band corresponds to the Si–O stretching of silanol group. The bands at 1153 and 1022 $cm^{-1}$ are strong due to stretching vibration of the siloxane (–Si–O–Si–) groups [18]. The band at 1639 $cm^{-1}$ corresponds to H–O–H bending vibration of water. The bands at 3695 and 3622 are attributed to the internal silanol due to strongly interacting vicinal OH [19]. The surface is also predominantly covered by physically adsorbed water in the pore and water bounded to the surface hydroxyl groups *via* H-bonds [19, 20]. By removing adsorbed water, upon heating the diatomite at 700°C, the OH groups on diatomite surface may be removed. The disappearance of the OH groups may have a significant effect on MPTMS grafting to diatomite, which is dependent on not only the pore structure of diatomite but also the surface functional groups.

Figure 4 shows DSC and TG diagrams of diatomite functionalized by MPTMS at dried conditions. The DSC and TG curves were similar for all samples. The exothermic peak, together with no any ignition loss at around 300°C, should be attributed to the crystallization of amorphous silica to quartz. The result also agrees with the fact that the crystallization of quartz was detected at 300°C by XRD analysis. However, even for sample 300D–700D which was calcinated at 300°C for 3 hours before functionalization, this peak is also observed. This can be explained by a reconstruction to form amorphous silica during silane treatment. The exothermic peaks and ignition losses at around 520°C should be assigned to the decomposition and oxidation of MPTMS incorporated into diatomite. The hydrolysis of MPTMS into silanols occurs in water. The silane incorporated into diatomite is due to the condensation of silanols of diatomite and silane. In this work, it is suggested that the silane is strongly bonded to diatomite surface as silane is not removed by washing chloroform, dispersed in water, or heated at 100°C. Then the amount of MPTMS incorporated into diatomite is proportional to the amount of ignition loss in range 100°C to 700°C calculated by

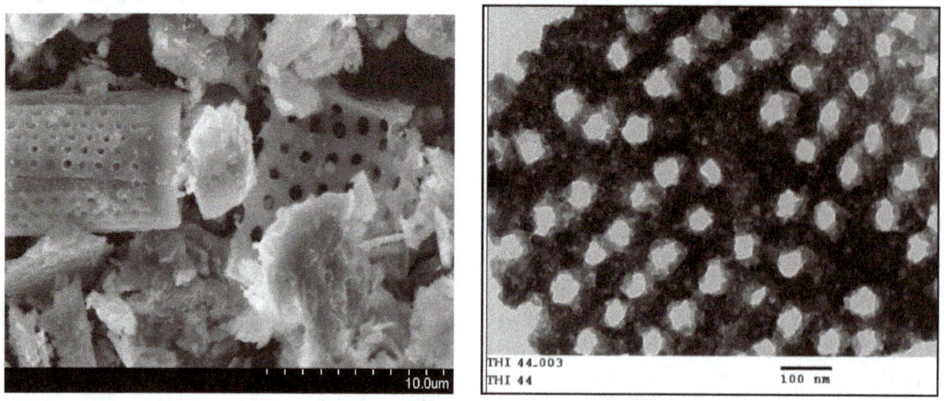


(a) (b)

FIGURE 2: (a) SEM and (b) TEM observation of raw diatomite.

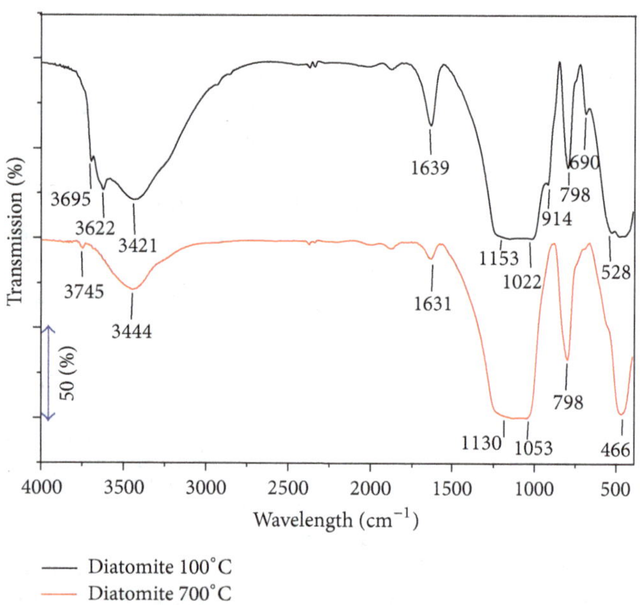


FIGURE 3: Infrared spectra of (a) raw and (b) calcined diatomite at 700°C.

TG. Therefore, the ignition loss (100–700°C) was assumed as MPTMS loading to diatomite. The amount of MPTMS seems to increase slightly from 5.6% to 6.3% for D100 and D300 and decrease remarkably with a further increase in thermal treatment temperature (see Table 2).

Figure 5 shows TG-DSC diagrams of functionalized diatomite under humid conditions. The curves were similar to all samples. The thermal behaviors were also similar to the case of samples under dried conditions except for ignition loss together exothermic peaks at around 300°C observed for most cases. From Table 2, the amount of MPTMS incorporated into diatomite seems to be higher in the samples in dried conditions and the diatomite thermal treated at 100°C–300°C is most favorable for MPTMS functionalization.

There are two types of silanols, isolated and H-bonded silanols on diatomite surface [20]. At room temperature, both types of silanols are H-bonded with water. With an increase in temperature, dehydration occurs. At first, the desorption of water and the exposure of more and more isolated silanols were favorable for the silanols of silane adsorbing onto diatomite surfaces. This accounts for the fact that the high amount of MPTMS was incorporated into diatomite calcinated in the range of 100–300°C. The calcinated diatomite exposed to water-saturated medium could create more silanols, which consequently adsorb MPTMS more strongly than the diatomite under dried conditions. As the temperature increases to more than 300°C, the silanols begin to condense to form siloxane bridges that are not favorable for coupling reactions. In the case of hydrated diatomite, two decompositions of silane incorporated into diatomite are observed at around 320°C and 520°C in TG-DSC curves instead of only 520°C in the case of the dried diatomite. Based on the results reported by Johansson et al. [21] about adsorption of silane coupling agents onto kaolin surfaces, we suggest two possible mechanisms by which MPTMS react with diatomite surface. The first proposed mechanism involves four steps:

(1) MPTMS was converted to the reactive silanol formed by hydrolysis

$$\text{SH-}\left(\text{CH}_2\right)_3\text{Si}\left(\text{OCH}_3\right)_3 + 3\text{H}_2\text{O} \Longleftrightarrow \text{SH-}\left(\text{CH}_2\right)_3\text{Si}\left(\text{OH}\right)_3 + 3\text{HOCH}_3 \quad (1)$$

(2) Condensation of the organosilane to oligomers is presented:

$$(n+1)\ \text{SH-}(\text{CH}_2)_3\text{Si}(\text{OH})_3 \longleftrightarrow \text{HO}-\underset{\text{OH}}{\overset{\text{SH}}{\text{Si}}}-\left[\text{O}-\underset{\text{OH}}{\overset{\text{SH}}{\text{Si}}}-\text{OH}\right]_n + n\,\text{H}_2\text{O} \quad (2)$$

TABLE 2: The amount of MPTES incorporated into diatomite estimated by TG.

| | Dried condition | | | | Humid condition | | | |
|---|---|---|---|---|---|---|---|---|
| | 100D | 300D | 500D | 700D | 100H | 300H | 500H | 700H |
| Ignition loss at *ca.* 100°C (%) | 5.67 | 0.00 | 0.00 | 0.00 | 3.71 | 2.45 | 0 | 0.00 |
| Ignition loss at *ca.* 300°C (%) | 0.00 | 0.00 | 0.00 | 0.00 | 0.00 | 2.02 | 1.31 | 0.80 |
| Ignition loss at *ca.* 520°C (%) | 5.44 | 3.76 | 2.55 | 1.82 | 6.99 | 5.69 | 1.70 | 0.97 |
| Total ignition loss > 100°C (%) | 5.44 | 3.76 | 2.55 | 1.82 | 6.99 | 7.71 | 3.01 | 1.77 |

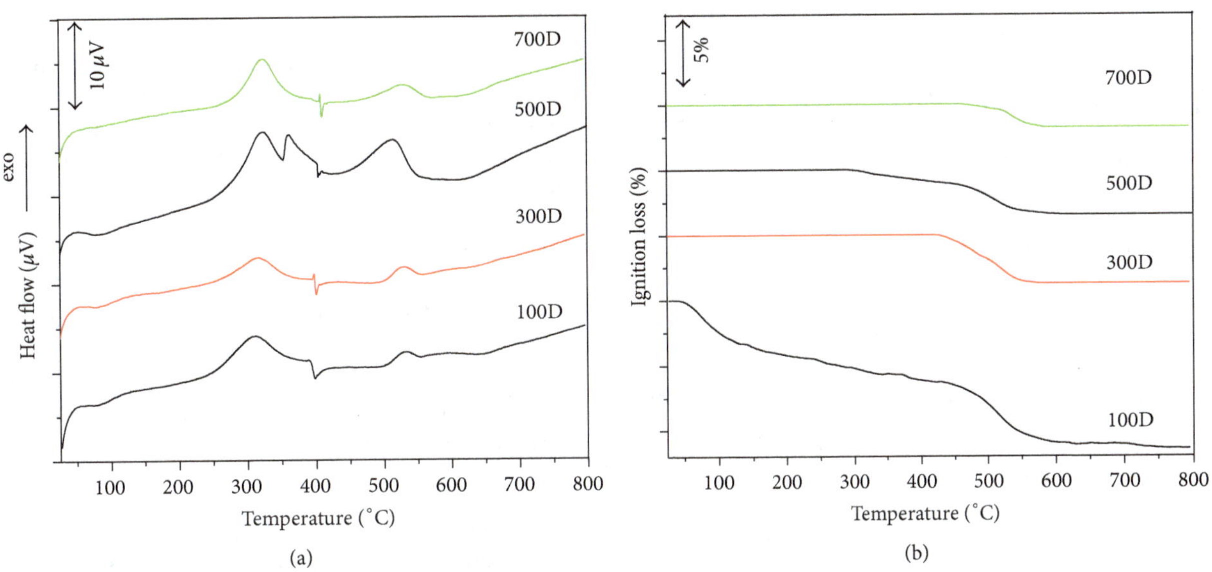


FIGURE 4: (a) DSC diagrams; (b) TG diagrams of diatomite functionalized by MPTMS at dried condition.

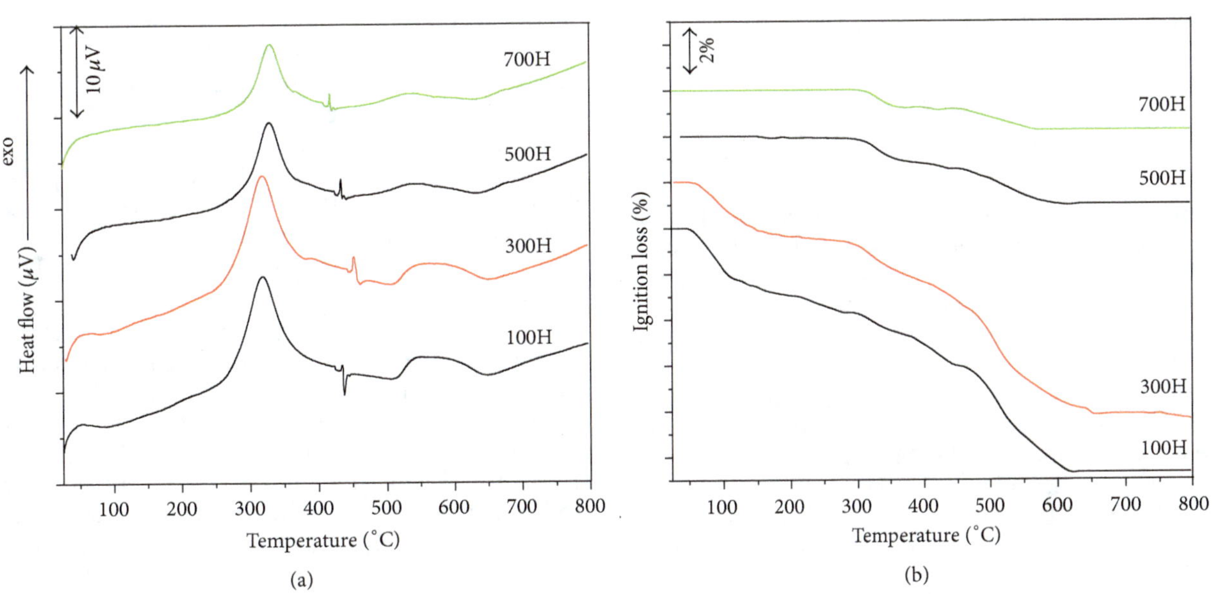


FIGURE 5: (a) DSC diagrams; (b) TG diagrams of functionalized diatomite under humid conditions.

(3) Formation of hydrogen bonds between the oligomers and the OH groups on the diatomite surface is presented:

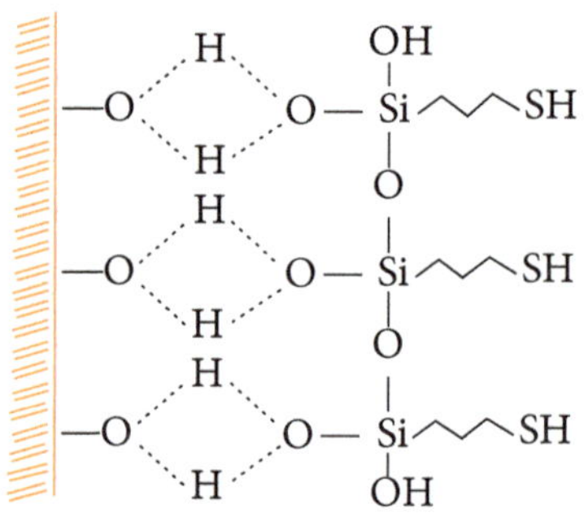

(3)

(4) Finally, a covalent linkage is formed under drying

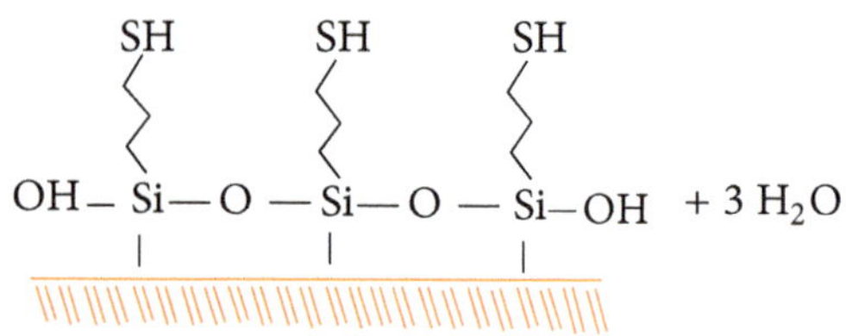

(4)

The second proposed mechanism involves only two steps:

(1) MPTMS is converted to the reactive silanol form by hydrolysis (R: $CH_2$-$CH_2$-$CH_2$-SH; X: $OCH_3$)

$$\text{SH-}(CH_2)_3\,Si(OCH_3)_3 + 3H_2O \Longleftrightarrow \text{SH-}(CH_2)_3\,Si(OH)_3 + 3HOCH_3 \quad (5)$$

(2) The silanol groups react directly with hydroxyl groups on the diatomite surface

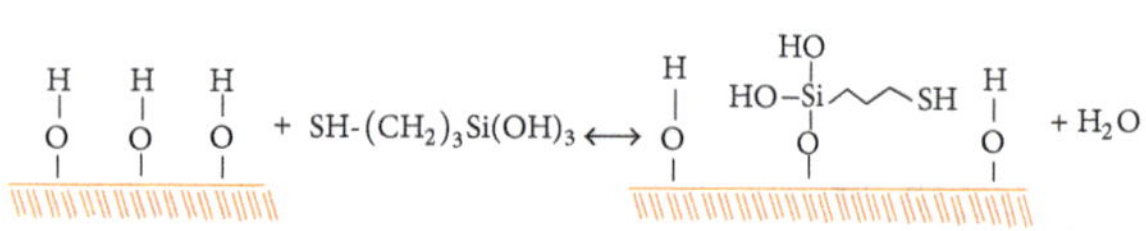

(6)

It is assumed that silane incorporated in diatomite by the first mechanism is more stable than one by the second mechanism. It means that it will be decomposed at higher temperature with the silane incorporated by the latter. For silane treatment with dried diatomite, the only ignition loss and exothermic peak at around 520°C could be assigned to the decomposition of silane grafting through the first mechanism. When diatomite is hydrated before silane treatment, the incorporation of MPTMS might occur through both mechanisms. Then, ignition loss and exothermic peak at around 300°C in TG-DSC for 100–700H might be assigned to decomposition of silanes in which MPTMS is grafted through the second mechanism.

The effect of hydrated times on the loading of MPTMS was also investigated. The diatomite calcined at 100°C for 3 hours was hydrated in a desiccators containing saturated NaCl solution for 0, 1, 2, 3, 6, 8, and 10 h. The functionalization was conducted as in the case of humid conditions. The amount of MPTMS incorporated into diatomite estimated from TG analysis was listed in Table 3.

Table 3: The total amount of ignition loss of MPTMS-diatomite in the range 100°C–700°C.

| Hydrated time | 0 h | 1 h | 2 h | 3 h | 6 h | 8 h | 10 h |
|---|---|---|---|---|---|---|---|
| Ignition loss (%) | 6.99 | 7.21 | 7.91 | 9.88 | 8.04 | 8.90 | 6.59 |

As can be seen, the amount of MPTMS peaked around 3 hours and then decreased as the time of hydrate was prolonged. Those results provide further evidence that water moisture favors the coupling reaction of silanols of silane and diatomite. However, the hydrated time prolonged means that there was too much water in the channel of diatomite, and the silane molecules will be significantly hydrated and condense each other instead of being bound to the wall and be removed under chloroform reflux condition; that is, the loading amount will decrease. Figure 7 shows TEM and SEM observation of MPTMS modified diatomite. The inclusions were highly dispersed on the diatomite surface (Figure 6(a)). However, pores in diatomite surface around 40–50 nm were also observed by TEM (Figure 6(b)). This means that the MPTMS-diatomite is combining highly porous structure with the binding properties of the thiol groups, leading to a high number of reactive sites that are readily accessible for the target analyte (Pb(II) and Cd(II)).

*3.2. Voltammetric Characteristics of Cd(II) and Pb(II) on MPTMS-Diatomite/GCE and Limit of Detection (LOD).* Figure 7 shows anodic voltammetric curves using electrodes modified by MPTMS-diatomite containing different MPTMS amounts. The percentage number in parenthesis presents the ignition loss which is assigned to as MPTMS amount. Cd(II) and Pb(II) ion showed anodic peaks at approximate −0.806 and −0.584 V in 0.1 M ABS (pH 4.5), respectively. For bare GCE and Nafion/GCE the broad oxidation peaks were observed. The intensity of anodic peak increases with an increase in MPTMS amount from 6.59% to 9.88%. A well-defined oxidation peak and high current response were observed using the MPTMS (9.88%)-diatomite electrode with highest MPTMS amount. Then, the MPTMS-diatomite with the highest amount of incorporated MPTMS was selected for modifying electrodes.

$K_3[Fe(CN)_6]/K_4[Fe(CN)_6]$ was selected as a probe to evaluate the performance of the prepared electrodes by cyclic voltammetry. Figure 8 presents the electrochemical responses of bare GCE and MPTMS-diatomite/GCE in 1.0 mM $K_3Fe(CN)_6/K_4Fe(CN)_6$ ($10^{-3}$ M and 0.1 M KCl) solution. It is clear that the cathodic-anodic curves using electrode modified by MPTMS-diatomite were broad compared to that of bare GCE, indicating that the electron-transfer kinetics of $Fe(CN)_6^{3-}/Fe(CN)_6^{4-}$ are gradually prevented, which is probably related to the higher ohmic drop value in the electrode process [22].

The redox behavior of Cd(II) and Pb(II) was investigated by cyclic voltammetry in a potential range from −1.2 to +0.2 V as shown in Figure 9. The Cd(II) and Pb(II)

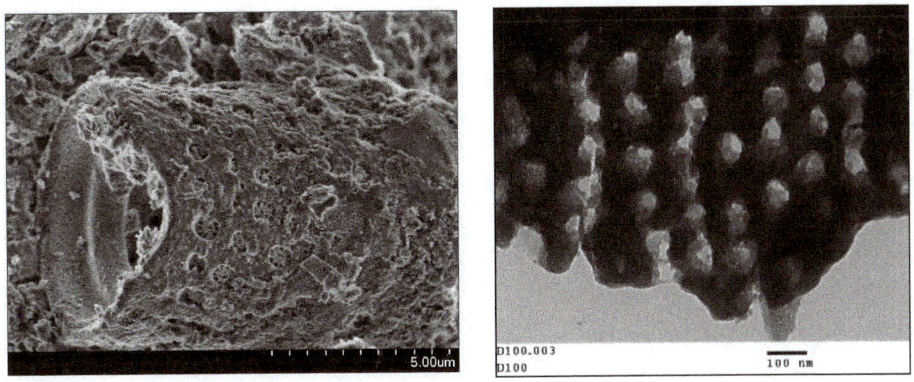


(a) (b)

FIGURE 6: SEM (a) and (b) TEM observation of diatomite modify MPTMS.

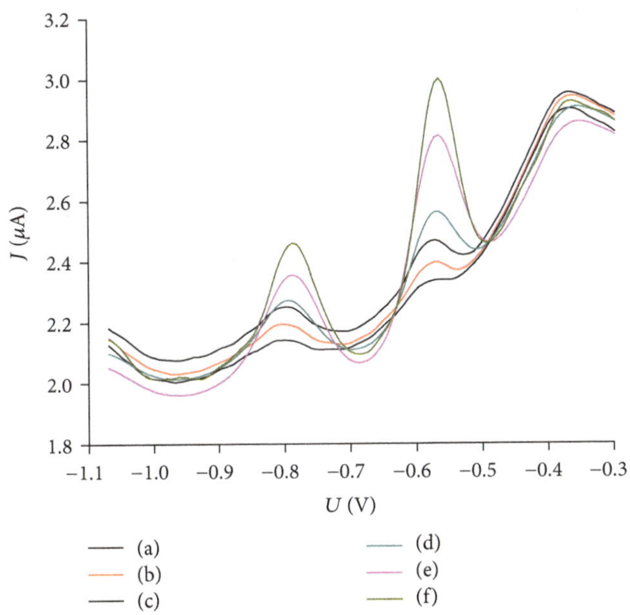


FIGURE 7: DP-ASV voltammograms for Cd(II) and Pb(II) using GCE modified by MPTMS-diatomite with differential amounts. (a) Bare GCE; (b) Nafion/GCE; (c) MPTMS (6.99%)-diatomite/GCE; (d) MPTMS (7.21%)-diatomite/GCE; (e) MPTMS (8.04%)-diatomite/GCE; (f) MPTMS (9.88%)-diatomite/GCE recorded in 0.1 M ABS (pH 4.5) containing 100 ppb Cd(II) and 200 pbb Pb(II); scan rate 0.10 V $s^{-1}$.

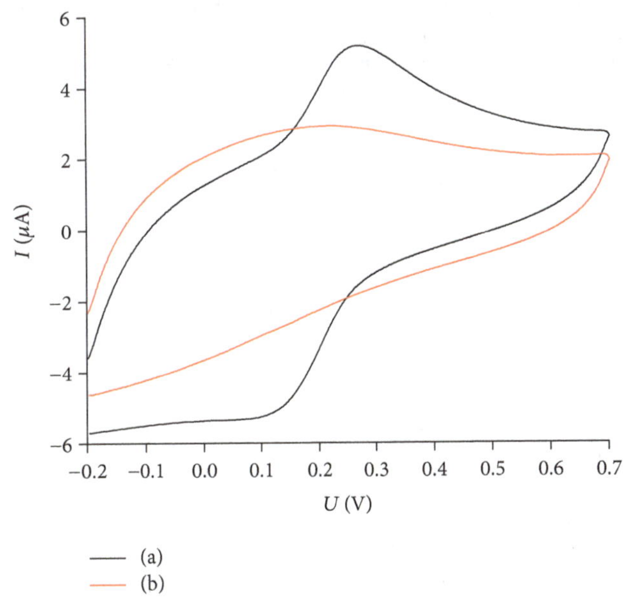


FIGURE 8: Cyclic voltammetric curves recorded in $10^{-3}$ M $K_3[Fe(CN)_6]/K_4[Fe(CN)_6]$ solutions containing 0.1 M KCl; scan rate 0.10 V $s^{-1}$. (a) GCE; (b) diatomite-MPTMS/GCE.

anodic and cathodic peaks were observed for the diatomite-MPTMS/GCE; the peak separation of anodic and cathodic potentials for Cd(II) and Pb(II) were 0.226 V and 0.139 V, respectively. Therefore, the electrochemical process of Cd(II) and Pb(II) on the electrode surfaces was *quasi*-reversible. The analytical signals for electrode surface were stable for at least 5 measuring cycles with the relative standard deviation (RSD) from 2.75 to 3.5%. These results indicated very good reproducibility for Cd(II) and Pb(II) detection with the developed electrode. The anodic peak current of Cd(II) and Pb(II) ($I_P$) was higher than the cathodic peak current, *ca.* 1.5-fold and 16.5-fold, respectively. It means that the anodic curves can be used for analytical signals.

The thiol group binds Cd(II) and Pb(II) to surface complexes because of its high affinity to metal ions. The metals were accumulated in electrode due to reduction reaction and then dissolved in solution through oxidation reaction. The electrochemical reaction occurred as follows [23, 24]:

$$
\begin{aligned}
&(\equiv \mathrm{SH})_2 + \mathrm{M}\,(\mathrm{Cd\,(II)\ or\ Pb\,(II)}) \longleftrightarrow \equiv \mathrm{SM} + 2\mathrm{H}^+ \quad \text{Complex step}\\
&\equiv \mathrm{SM} + 2e \longleftrightarrow \equiv \mathrm{SM}^0 \quad \text{Accumulation step}\\
&\equiv \mathrm{SM}^0 - 2e + 2\mathrm{H}^+ \longleftrightarrow (\equiv \mathrm{SH})_2 + 2\mathrm{M\,(II)} \quad \text{Stripping step}
\end{aligned}
\tag{7}
$$

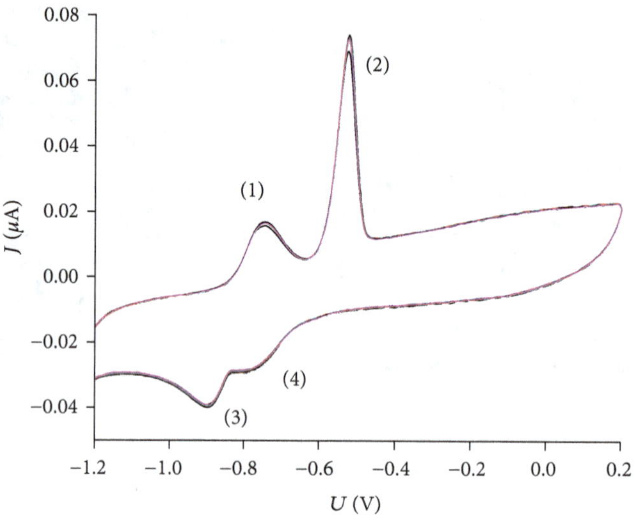


FIGURE 9: Cyclic voltammetric curves recorded in 0.1 M ABS (pH 4.5) containing 100 ppb Cd(II) and 200 pbb Pb(II); scan rate 0.10 V $s^{-1}$. (1): $I_{a,Cd}$ = 0.0152 mA, RSD = 2.75% ($n$ = 5), and $E_{a,Cd}$ = −0.751 mV. (2): $I_{a,Pb}$ = 0.0627 mA, RSD = 2.83% ($n$ = 5), and $E_{a,Pb}$ = −0.525 mV. (3): $I_{c,Cd}$ = 0.0102 mA, RSD = 3.50% ($n$ = 5), and $E_{c,Cd}$ = −0.897 mV. (4): $I_{c,Pb}$ = 0.0038 mA, RSD = 2.96% ($n$ = 5), and $E_{c,Pb}$ = −0.758 mV.

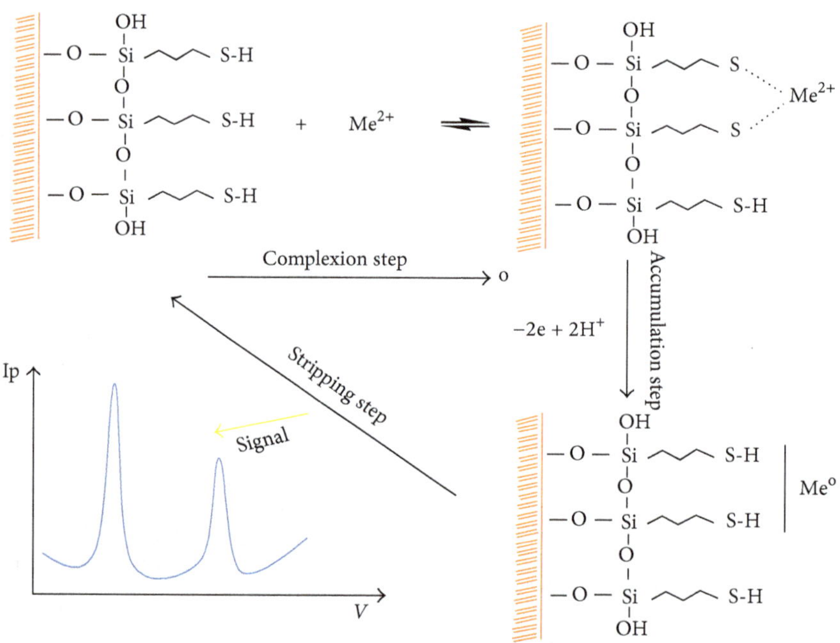


FIGURE 10: Proposed representation of preconcentration and stripping mechanism of Cd(II) and Pb(II) on MPTMS-diatomite/GCE.

Electrochemical process of Pb(II) and Cd(II) on electrodes modified by MPTMS-diatomite was illustrated by Figure 10.

Figure 11 shows the DP-ASV of Cd(II) and Pb(II) with different concentrations. An obvious anodic peaks were observed at around −0.82 and −0.61 V. These observations can be attributed to the oxidation species of Cd(II) and Pb(II), respectively, which have been deposited into the surface of MPTMS-diatomite/GCE during reduction processes. The intensity of anodic peak increased with an increase in Cd(II) and Pb(II) concentration from 20 to 300 ppb. A calibration plot of the anodic current response versus Cd(II) concentration and Pb(II) is presented in the insets of Figure 11. The results show that the current peak response was linear to the Cd(II) concentration with a $R^2$ of 0.999 in the range of 20–300 ppb and the Pb(II) concentration with $R^2$ of 0.994 in the range of 20–150 ppb. The limit of detection (LOD) was calculated on the basis of (3.3 $S_a/b$) criteria ($S_a$ represents the standard deviation of the intercept while $b$ represents the slope of the calibration curve defined for the LOD concentration range (20–300 ppb)). The LOD for Cd(II) and Pb(II) calculated was 15.9 ppb and 6.9 ppb, respectively.

Table 4 shows a comparison of MPTMS-diatomite-GCE developed with other GCEs modified with other nanoparticles for simultaneous determination of Pb(II) and Cd(II). As

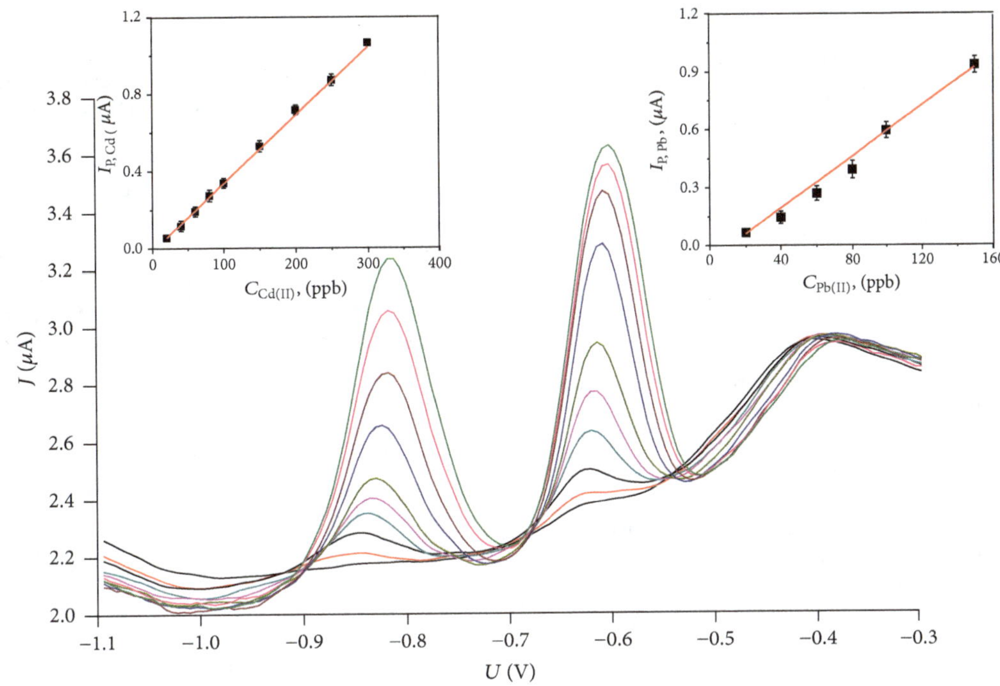


FIGURE 11: The DP-ASV voltammograms of Cd(II) and Pb(II). Conditions: the concentrations in the range of 20 to 300 ppb of the Cd(II) and Pb(II); other conditions: 0.1 M ABS (pH 4.5); $E_{acc} = -1.2$ V; $t_{acc} = 60$ s; pulse amplitude ($\Delta E$) = 50 mV; pulse time = 40 ms; potential step = 6 mV; $v = 20$ mV $s^{-1}$; $\omega = 2000$ rpm (inset (left): a plot of stripping peak current with concentration of Cd(II); inset (right): a plot of stripping peak current with concentration of Pb(II)).

TABLE 4: Comparison of GCE modified with MPTMS-diatomite with other GCEs modified with other nanoparticles.

| Electrode | Method | Linear range (ppb) | LOD (ppb) | References |
|---|---|---|---|---|
| MgO–Nafion/GCE | DP-ASV, ABS pH = 5.0 | Cd(II): 1.1–23.6<br>Pb(II): 2.1–414 | — | [22] |
| BiNPs/GCE | DP-ASV, ABS pH = 4.5 | 5.0–60.0 | Cd(II): 0.4; Pb(II): 0.8 | [25] |
| AuNPs/GCE | DP-ASV, ABS pH = 5.0 | — | Cd(II): 20.7; Pb(II): 11.2 | [26] |
| Padpa/Ru/GCE | DP-ASV, HCl 0,1 M | — | Cd(II): 21.7; Pb(II): 9.8 | [27] |
| Diatomite-MPTMS/GCE | DP-ASV; ABS; pH = 4.5 | Cd(II): 20–300<br>Pb(II): 20–150 | Cd(II): 15.9; Pb(II): 6.9 | This work |

Glassy carbon electrode (GCE); bismuth nanoparticles (BiNPs); gold nanoparticles (AuNPs); tris(2,2′-bipyridyl)Ru(II) (Padpa/Ru).

seen in Table 4, the GCE modified with MPTMS-diatomite developed in the present work shows good LOD compared with other similar materials/GCE reported previously in the literature [21–24]. This indicates that MPTMS-diatomite is a potential material for electrode modifiers.

## 4. Conclusions

The effects of functionalization conditions on the loading of MPTMS in diatomite were investigated. Diatomite from Phu Yen consists of mainly amorphous structure. The crystallization to form quartz occurs at more than 300°C. The diatomite with thermal treatment in the range of 100–300°C is favorable for functionalization. The humidity of diatomite also affects the functionalization level significantly. The MPTMS loading around 9.8% peaked as diatomite was hydrated for 3 hours. We have demonstrated that MPTMS-diatomite is useful to prepare chemically modified glassy carbon electrodes. The electrode modified by MPTMS-diatomite exhibited potential for the use of the simultaneous determination of cadmium and lead by DP-ASV. The stripping peak currents of the two metal ions had linear relationships with the concentrations in the range of 20 to 300 ppb ($i$ ($\mu$A) = −0.0202 + 0.0036 $C$ (ppb), $R^2 = 0.9997$) for Cd(II) and 20 to 150 ppb ($i$ ($\mu$A) = −0.1179 + 0.0069 $C$ (ppb), $R^2 = 0.9943$) for Pb(II). The LOD for Cd(II) and Pb(II) calculated was 15.9 ppb and 6.9 ppb.

## Conflicts of Interest

The authors declare that there are no conflicts of interest regarding the publication of this paper.

## References


[1] L. Cui, Y. Wang, L. Gao et al., "EDTA functionalized magnetic graphene oxide for removal of Pb(II), Hg(II) and Cu(II) in water treatment: adsorption mechanism and separation property," *Chemical Engineering Journal*, vol. 281, pp. 1–10, 2015.

[2] F. A. Aydin and M. Soylak, "Separation, preconcentration and inductively coupled plasma-mass spectrometric (ICP-MS) determination of thorium(IV), titanium(IV), iron(III), lead(II) and chromium(III) on 2-nitroso-1-naphthol impregnated MCI GEL CHP20P resin," *Journal of Hazardous Materials*, vol. 173, no. 1-3, pp. 669–674, 2010.

[3] J. Otero-Romaní, A. Moreda-Piñeiro, P. Bermejo-Barrera, and A. Martin-Esteban, "Inductively coupled plasma-optical emission spectrometry/mass spectrometry for the determination of Cu, Ni, Pb and Zn in seawater after ionic imprinted polymer based solid phase extraction," *Talanta*, vol. 79, no. 3, pp. 723–729, 2009.

[4] K. Ndung'u, S. Hibdon, and A. R. Flegal, "Determination of lead in vinegar by ICP-MS and GFAAS: evaluation of different sample preparation procedures," *Talanta*, vol. 64, no. 1, pp. 258–263, 2004.

[5] A. Maghear, M. Etienne, M. Tertiş, R. Sandulescu, and A. Walcarius, "Clay-mesoporous silica composite films generated by electro-assisted self-assembly," *Electrochimica Acta*, vol. 112, pp. 333–341, 2013.

[6] M. M. Abdel-Galeil, M. M. Ghoneim, H. S. El-Desoky, T. Hattori, and A. Matsuda, "Anodic stripping voltammetry determination of lead ions using highly sensitive modified electrodes based on multi-walled carbon nanotube," *Journal of Chemistry and Biochemistry*, vol. 2, no. 2, 2014.

[7] P. Zhang, S. Dong, G. Gu, and T. Huang, "Simultaneous determination of $Cd^{2+}$, $Pb^{2+}$, $Cu^{2+}$ and $Hg^{2+}$ at a carbon paste electrode modified with ionic liquid-functionalized ordered mesoporous silica," *Bulletin of the Korean Chemical Society*, vol. 31, no. 10, pp. 2949–2954, 2010.

[8] A. Sánchez, S. Morante-Zarcero, D. Pérez-Quintanilla, I. Sierra, and I. Del Hierro, "Determination of Hg(II) in natural waters using a carbon paste electrode modified with hybrid mesostructured silica nanoparticles," *Sensors and Actuators, B: Chemical*, vol. 163, no. 1, pp. 38–43, 2012.

[9] P. Kula, Z. Navrátilová, P. Kulová, and M. Kotoucek, "Sorption and determination of Hg(II) on clay modified carbon paste electrodes," *Analytica Chimica Acta*, vol. 385, no. 1-3, pp. 91–101, 1999.

[10] Z. Navrátilová and P. Kula, "Clay modified electrodes: present applications and prospects," *Electroanalysis*, vol. 15, no. 10, pp. 837–846, 2003.

[11] Y. Al-Degs, M. A. M. Khraisheh, and M. F. Tutunji, "Sorption of lead ions on diatomite and manganese oxides modified diatomite," *Water Research*, vol. 35, no. 15, pp. 3724–3728, 2001.

[12] M. A. Al-Ghouti, M. A. M. Khraisheh, S. J. Allen, and M. N. Ahmad, "The removal of dyes from textile wastewater: a study of the physical characteristics and adsorption mechanisms of diatomaceous earth," *Journal of Environmental Management*, vol. 69, no. 3, pp. 229–238, 2003.

[13] J. Zhang, T. Ding, Z. Zhang, L. Xu, and C. Zhang, "Enhanced adsorption of trivalent arsenic from water by functionalized diatom silica shells," *PLoS ONE*, vol. 10, no. 4, Article ID e0123395, 2015.

[14] Y. Yu, J. Addai-Mensah, and D. Losic, "Functionalized diatom silica microparticles for removal of mercury ions," *Science and Technology of Advanced Materials*, vol. 13, no. 1, Article ID 015008, 2012.

[15] I. K. Tonlé, E. Ngameni, H. L. Tcheumi, V. Tchiéda, C. Carteret, and A. Walcarius, "Sorption of methylene blue on an organoclay bearing thiol groups and application to electrochemical sensing of the dye," *Talanta*, vol. 74, no. 4, pp. 489–497, 2008.

[16] T. Zhang, L. Zeng, L. Han et al., "Ultrasensitive electrochemical sensor for p-nitrophenyl organophosphates based on ordered mesoporous carbons at low potential without deoxygenization," *Analytica Chimica Acta*, vol. 822, pp. 23–29, 2014.

[17] G. Rytwo, D. Tropp, and C. Serban, "Adsorption of diquat, paraquat and methyl green on sepiolite: experimental results and model calculations," *Applied Clay Science*, vol. 20, no. 6, pp. 273–282, 2002.

[18] R. A. Shawabkeh and M. F. Tutunji, "Experimental study and modeling of basic dye sorption by diatomaceous clay," *Applied Clay Science*, vol. 24, no. 1-2, pp. 111–120, 2003.

[19] M. A. M. Khraisheh, M. A. Al-Ghouti, S. J. Allen, and M. N. Ahmad, "Effect of OH and silanol groups in the removal of dyes from aqueous solution using diatomite," *Water Research*, vol. 39, no. 5, pp. 922–932, 2005.

[20] P. Yuan, D. Q. Wu, H. P. He, and Z. Y. Lin, "The hydroxyl species and acid sites on diatomite surface: a combined IR and Raman study," *Applied Surface Science*, vol. 227, no. 1–4, pp. 30–39, 2004.

[21] U. Johansson, A. Holmgren, W. Forsling, and R. L. Frost, "Adsorption of silane coupling agents onto kaolinite surfaces," *Clay Minerals*, vol. 34, no. 2, pp. 239–246, 1999.

[22] Z. Wu, C. Xu, H. Chen et al., "Mesoporous MgO nanosheets: 1,6-hexanediamin-assisted synthesis and their applications on electrochemical detection of toxic metal ions," *Journal of Physics and Chemistry of Solids*, vol. 74, no. 7, pp. 1032–1038, 2013.

[23] C. Hu, K. Wu, X. Dai, and S. Hu, "Simultaneous determination of lead(II) and cadmium(II) at a diacetyldioxime modified carbon paste electrode by differential pulse stripping voltammetry," *Talanta*, vol. 60, no. 1, pp. 17–24, 2003.

[24] S. Ramezani, M. Ghobadi, and B. N. Bideh, "Voltammetric monitoring of Cd (II) by nano-$TiO_2$ modified carbon paste electrode sensitized using 1,2-bis-[o-aminophenyl thio] ethane as a new ion receptor," *Sensors and Actuators, B: Chemical*, vol. 192, pp. 648–657, 2014.

[25] D. Yang, L. Wang, Z. Chen, M. Megharaj, and R. Naidu, "Anodic stripping voltammetric determination of traces of Pb(II) and Cd(II) using a glassy carbon electrode modified with bismuth nanoparticles," *Microchimica Acta*, vol. 181, no. 11-12, pp. 1199–1206, 2014.

[26] X. Xu, G. Duan, Y. Li et al., "Fabrication of gold nanoparticles by laser ablation in liquid and their application for simultaneous electrochemical detection of $Cd^{2+}$, $Pb^{2+}$, $Cu^{2+}$, $Hg^{2+}$," *ACS Applied Materials and Interfaces*, vol. 6, no. 1, pp. 65–71, 2014.

[27] E. A. Khudaish, M. M. Al-Hinaai, and S. H. Al-Harthi, "A solid-state sensor based on tris(2,2ı-bipyridyl)ruthenium(II)/poly(4-aminodiphenylamine) modified electrode: characterization and applications," *Sensors and Actuators, B: Chemical*, vol. 185, pp. 478–487, 2013.

# 19

# Application of Central Composite Design in the Adsorption of Ca(II) on Metakaolin Zeolite

**Upenyu Guyo, Lycenter Yard Phiri, and Fidelis Chigondo**

*Department of Chemical Technology, Midlands State University, Private Bag 9055, Senga, Gweru, Zimbabwe*

Correspondence should be addressed to Fidelis Chigondo; chigondofj@gmail.com

Academic Editor: Ana Moldes

Metakaolin zeolite-A was synthesized from thermally activated kaolin clay and characterized by Fourier Transform Infrared Spectroscopy and X-Ray Diffraction Spectroscopy. The effects of pH (2–10), contact time (10–180 min), initial concentration (5–120 $mgL^{-1}$), and dosage (0.1–2 g) and their interactions were investigated using response surface methodology following a central composite design. Optimum removal (87.70%) was obtained at pH 6, contact time 180 min, initial concentration 40.0 $mgL^{-1}$, and adsorbent dosage 1.0 g by Excel Solver using the GRG solving method. The adsorption data fitted best to the Langmuir model with correlation coefficient $R^2 = 0.993$ and Chi-square value $\chi^2 = 4.76$. The Freundlich isotherm gave a correlation coefficient $R^2 = 0.933$ and $\chi^2 = 37.91$. The adsorption process followed the pseudo-second-order model. The calculated thermodynamic parameters showed that the adsorption process was endothermic and not thermodynamically spontaneous. The studied zeolite-A can therefore be used as a promising adsorbent for the removal of Ca(II) ions from aqueous solutions.

## 1. Introduction

Zeolites are crystalline, microporous, hydrated aluminosilicates formed by the sharing of oxygen atoms in the framework of aluminum and silicon tetrahedrals [1]. Synthetic zeolites are used commercially more often than natural zeolites due to the purity of their crystalline products, the uniformity of particles, the greater thermal stability, and the fact that they can be engineered with a wide variety of properties and pore sizes [2]. Zeolite synthesis involves the hydrothermal crystallization of aluminosilicate gels in the presence of alkali hydroxides [3]. The type of zeolite is affected by composition of the reaction mixture, the nature of the reactants and their pretreatments, temperature of the process, reaction time, and pH of the reaction mixture. The main problems in zeolite application are availability and cost of raw materials, specifically the silica source [4]. The use of commercial silica in zeolite synthesis is affected by variations in reactivity and selectivity apart from being expensive [5]. Cheaper raw materials such as clay minerals could be an alternative starting material for zeolite synthesis. Kaolin is a cheap clay mineral whose deposits are found vastly around mining areas in Zimbabwe. Many researchers have reported the synthesis of zeolites from kaolin under different conditions [6–11]. Kaolin is not stable under severe alkaline conditions and as such it is desirable to convert it into a more reactive form (metakaolin) by calcination at elevated temperatures. Because of their unique porous properties, zeolites could be applied in the removal of hardness in domestic water which is mainly due to calcium and magnesium ions. Water supplies with total hardness of more than 500 mg/L are not suitable for domestic consumption [12]. Hard water is responsible for the formation of deposits in boilers, cooling towers, water pipes, and household facilities, as well as diverse influences on the cleaning performance of detergents. Traditionally, water purification plants used lime and soda ash for the removal of hardness. One of the main drawbacks of this process is the generation of large amounts of liquid sludge as well as the need for recarbonation of the treated water [13]. The use of polyphosphates in removing water hardness is also practiced but results in massive discharge of phosphates in water bodies which lead to eutrophication [14].

Various other methods including membrane precipitation, phytoextraction, flocculation, solvent extraction, ultrafiltration, reverse osmosis, electrodialysis, and adsorption have been studied for the removal of a wide variety of cations from water streams including magnesium and calcium [15–17]. Most of these are either inefficient or expensive and result in the generation of large amounts of sludge. Activated carbon is considered to be a particularly competitive and effective process for the removal of cations but has been hampered by the high costs associated with production and regeneration of spent carbon. The use of low-cost ion exchangers such as zeolites is an attractive method for the removal of metals from aqueous solutions. In this work, we report on the synthesis of zeolite-A from Zimbabwean kaolin and its application in the sorption of calcium ions from aqueous solutions. All the process independent variables have been optimized collectively by applying response surface methodology (RSM), a mathematical statistical technique which optimizes the process and evaluates the relationship between a set of independent variables [18–20]. Moreover, it evaluates the relative significance of the various independent variables even in the presence of complex factor-factor interactions [21]. The adsorbent was characterized using Fourier Transform Infrared (FTIR) Spectroscopy and XRD (X-Ray Diffraction) Spectroscopy.

## 2. Materials and Methods

*2.1. Chemicals and Solutions.* All chemicals used in this work were obtained from Skylabs (Zimbabwe) and were used as received without further purification. A pH meter (Az-8601, China) was employed for pH measurement and 0.1 M HCl and 0.1 M NaOH were used for pH adjustment. $CaCl_2$ was used for the preparation of stock solution using double deionized water. All glassware used was soaked in 10% (v/v) nitric acid solution for 24 h and cleaned repeatedly with double deionized water.

*2.2. Sample Collection and Pretreatment.* The kaolin clay used in this study was collected from Chegutu, Zimbabwe. The kaolin clay sample was pulverized and sieved through a 90 $\mu$m sieve. The pulverized kaolin clay material was digested in 6 M HCl to remove iron and washed with distilled water followed by drying overnight at 120°C. The treated kaolin clay was exposed to 900°C for 1 h where its structure was destroyed [22]. This removed all incorporated hydrocarbons and was dehydroxylated to form an activated amorphous material called metakaolin, a material which accepts and readily exchanges sodium as a guest in its structure.

*2.3. Zeolitization Using the Hydrothermal Transformation Method.* The zeolite gel was prepared by dissolving metakaolin in sodium hydroxide solutions [23]. Metakaolin (20 g) was refluxed and stirred in 100 mL of sodium hydroxide (1, 3, and 5 M) at a temperature of 100°C at specified reaction times (1, 2, and 4 h) [5]. The metakaolin-sodium hydroxide slurries were filtered by suction filtration and washed several times with deionized water to remove excess unreacted sodium hydroxide. The reaction products were dried at a temperature of 120°C in the oven for 24 h. Triplicate samples were prepared for each NaOH concentration and reaction time.

*2.4. Characterization of the Zeolites.* The Fourier Transform Infrared (FTIR) Spectroscopy spectra of the metakaolin zeolite (MZ) were obtained using a Thermo Fisher Scientific spectrometer (Nicolet 6700, USA) in the scanning range 4000–400 $cm^{-1}$. A Bruker D8 Advance X-ray diffractometer (Germany) using nickel filtered Cu K$\alpha$ radiation (40 kV, 40 mA) was used for identification of the structural phases.

*2.5. Adsorption Batch Studies.* All experiments were carried out in batch mode in 250 mL conical flasks. Several experimental parameters (solution pH: 2–10; contact time: 10–180 min; adsorbent dosage: 0.1–2.0 g; and initial metal ion concentration: 5–120 mg/L) were investigated following a central composite design (CCD). A predetermined adsorbent mass was added to 100 mL solution in a conical flask containing $Ca^{2+}$ cations at a concentration according to CCD. The mixture was agitated on a shaker at 150 rpm for a specified time at room temperature. After agitation, the solid matrix was separated from the rest of the solution by filtration using Whatman number 42 filter paper followed by centrifugation to get the supernatant solution. The determination of $Ca^{2+}$ ions in aqueous solution was done by complexometric titration using standardized 0.1 M EDTA (ethylenediaminetetraacetic acid) in the presence of Eriochrome Black T as an indicator. Typically, 100 mL of each supernatant solution was pipetted into a 250 mL conical flask, 4 mL of 1.0 M $NH_3$ solution and 2-3 drops of Eriochrome Black T were added, and the mixture was titrated with 0.1 M EDTA solution until the color changed from pink to blue [12].

The percentage removal ($R$) of Ca(II) ions from aqueous solutions was determined using the following equation:

$$R = \frac{(C_o - C_e)}{C_o} \times 100\%. \tag{1}$$

The equilibrium adsorption amount ($q_e$) was calculated according to the mass balance equation:

$$q_e = \frac{(C_o - C_e)\,V}{M}, \tag{2}$$

where $q_e$ is the equilibrium amount (mg $g^{-1}$) of Ca(II) ions adsorbed per gram of metakaolin zeolite adsorbent, $C_o$ and $C_e$ are the Ca(II) concentrations (mg$L^{-1}$) in the solution initially and after adsorption, respectively, $V$ is the volume ($L$) of the solution, and $M$ is the mass (g) of the adsorbent used in the experiment.

The experimental domain used is presented in Table 1. Preliminary experiments were done to obtain the extreme levels of the independent input variables. Design Expert Version 5 (Stat-Ease, USA) was used to generate the CCD experiments, optimize the levels of the independent variables, and evaluate the interactions of the process parameters. In generating the CCD, the experimental domain for independent factors for high and low was entered in terms of alpha value of 2 to give 30 experiments with 6 center points. The percentage removal of Ca(II) ions was taken as the dependent variable. The Design Expert software was also

TABLE 1: Experimental domain.

| | pH | Contact time (min) | Initial concentration (mg $L^{-1}$) | Dosage (g) |
|---|---|---|---|---|
| Min. point | 2 | 10 | 5 | 0.1 |
| Max. point | 10 | 180 | 120 | 2 |

used for statistical, regression, and graphical analysis of the obtained data. The general Logit $R$ transformation quadratic model was proposed to be

$$\begin{aligned}\text{Logit } R = & b_0 + b_1 A + b_2 B + b_3 C + b_4 D + b_5 A^2 + b_6 B^2 \\ & + b_7 C^2 + b_8 D^2 + b_9 AB + b_{10} AC + b_{11} AD \\ & + b_{12} BC + b_{13} BD + b_{14} CD,\end{aligned} \tag{3}$$

where $R$ is the percentage removal of Ca(II) ions, $A$ is the pH, $B$ is the contact time in min, $C$ is the initial concentration in mg/L, and $D$ is the dosage in g. The optimum values of the independent input variables were obtained by Excel Solver software applying the GRG function.

*2.6. Desorption Studies.* 0.2 g of the adsorbent was left in contact with 100 mL of 30 mg/L of Ca(II) for 180 min at 150 rpm. The spent adsorbent was filtered, washed, and dried at 55°C for 24 h. The dried spent adsorbent was left in contact with different concentrations of NaOH (0.1, 0.5, 1.0, and 2.0 M) for 500 min and then filtered and washed several times with deionized water. The desorbed adsorbent was used for adsorption of the Ca(II) ions and the regeneration percentage was determined.

## 3. Results and Discussion

### 3.1. Characterization of the Adsorbent

*3.1.1. FTIR.* Zeolites were synthesized at different reaction times and different NaOH concentrations and the zeolite that was similar to zeolite-A was selected for use. Figure 1 shows the FTIR spectra of the metakaolin zeolites synthesized at an optimum time of 4 h at different NaOH concentrations. The spectrum of the selected zeolite (C2, Figure 1) had a band at 3458.53 $cm^{-1}$ corresponding to O-H vibration. There was structural T-O stretching and deformation at 1642.33 $cm^{-1}$ and a peak at 1021.10 $cm^{-1}$ that corresponded to internal tetrahedron asymmetrical vibration. A peak at 1421.37 $cm^{-1}$ might be due to Si-O asymmetric stretching whilst bands at 723.79 and 1072.79 $cm^{-1}$ were due to internal symmetrical stretching vibration of (Al-Si)-O. The 473.46 $cm^{-1}$ band was due to internal vibration of (Al-Si)-O bending and the 558.95 $cm^{-1}$ band corresponded to the formation of double four-membered rings by external linkages [5]. The zeolite-A framework possesses good exchange capacity, making it a good material as an ion exchange agent. It may thus be concluded that IR spectra of the C2 spectra are closely complementary to zeolite-A; therefore, the C2 zeolites were used for batch adsorption studies.

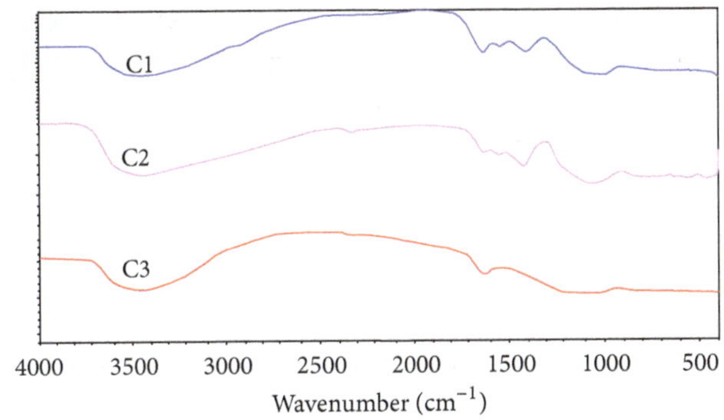


FIGURE 1: FTIR spectra of metakaolin zeolites (Cl, 4 h, 3 M NaOH zeolite; C2, 4 h, 5 M NaOH zeolite; and C3, 4 h, 1 M NaOH zeolite).

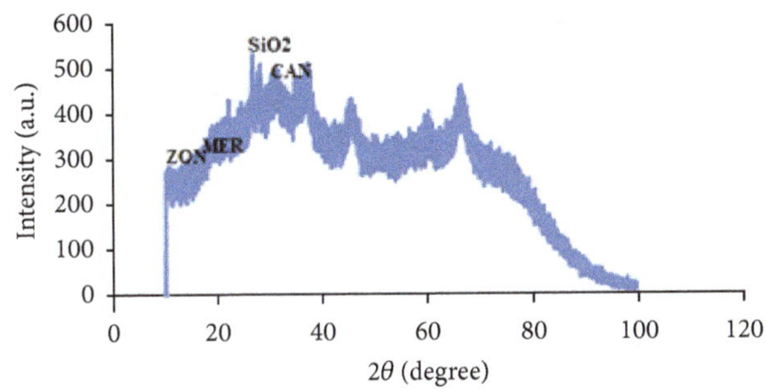


FIGURE 2: XRD pattern for metakaolin zeolite-A.

*3.1.2. XRD Analysis.* The XRD data (Figure 2) showed four crystallographic phases at $2\theta$ degree values of 10.09 (ZON), and 12.69 (MER), 26.69 (Quartz), and 30.28 (CAN) are quite consistent with the two theta values of zeolite-A (LTA) showing that the pure phase of crystalline zeolite-A was produced.

*3.2. Adsorption Batch Studies.* Adsorption studies using MZ were done on an aqueous Ca(II) solution applying a central composite design [24] developed from the experimental domain presented in Table 1. The investigation was carried out following a full central composite design consisting of 30 experiments, which included several replicate experiments to check for consistency of the adsorption procedure. The average Ca(II) removal from aqueous solutions using the prepared adsorbent over six replicate runs (midpoint experimental conditions) was 39.73 ± 0.41% (at 95% confidence). The results obtained were consistent as evidenced by the small relative standard deviation (1.04%). The CCD batch method was therefore capable of providing consistent results and hence was suitable for use in the investigation of the adsorption of Ca(II) ions onto MZ.

*3.2.1. Response Surface Methodology Modeling.* The experimental variable settings used for this investigation together with the actual experimental results are presented in Table 2.

The proposed full Logit $R$ transformation model (see (3)) was fitted on the experimental data obtained in Table 2. The least squares regression results for the adsorption of Ca(II)

TABLE 2: Experimental variables and results.

| Run | $A$: pH | $B$: contact time (min) | $C$: initial concentration ($mgL^{-1}$) | $D$: dosage (g) | Ca(II) removal (%) |
|---|---|---|---|---|---|
| 1 | 2 | 95 | 62.5 | 1.05 | 32.0 |
| 2 | 4 | 52.5 | 33.75 | 0.575 | 57.8 |
| 3 | 4 | 52.5 | 33.75 | 1.525 | 62.2 |
| 4 | 4 | 52.5 | 91.25 | 1.525 | 26.8 |
| 5 | 4 | 52.5 | 91.25 | 0.575 | 24.7 |
| 7 | 4 | 137.5 | 33.75 | 0.575 | 74.1 |
| 8 | 4 | 137.5 | 33.75 | 1.525 | 78.5 |
| 9 | 4 | 137.5 | 91.25 | 1.525 | 28.5 |
| 10 | 4 | 137.5 | 91.25 | 0.575 | 25.8 |
| 11 | 6 | 10 | 62.5 | 1.05 | 36.0 |
| 12 | 6 | 95 | 62.5 | 1.05 | 39.2 |
| 13 | 6 | 95 | 62.5 | 1.05 | 39.2 |
| 14 | 6 | 95 | 62.5 | 1.05 | 40.0 |
| 15 | 6 | 95 | 62.5 | 2 | 42.4 |
| 16 | 6 | 95 | 62.5 | 1.05 | 40.0 |
| 17 | 6 | 95 | 62.5 | 0.1 | 36.0 |
| 18 | 6 | 95 | 62.5 | 1.05 | 40.0 |
| 19 | 6 | 95 | 62.5 | 1.05 | 40.0 |
| 20 | 6 | 95 | 120 | 1.05 | 22.9 |
| 21 | 6 | 180 | 62.5 | 1.05 | 43.2 |
| 22 | 8 | 52.5 | 33.75 | 0.575 | 66.7 |
| 23 | 8 | 52.5 | 33.75 | 1.525 | 72.6 |
| 24 | 8 | 52.5 | 91.25 | 1.525 | 30.1 |
| 25 | 8 | 52.5 | 91.25 | 0.575 | 26.8 |
| 26 | 8 | 137.5 | 33.75 | 1.525 | 75.6 |
| 27 | 8 | 137.5 | 33.75 | 0.575 | 59.3 |
| 28 | 8 | 137.5 | 91.25 | 1.525 | 34.5 |
| 29 | 8 | 137.5 | 91.25 | 0.575 | 32.3 |
| 30 | 10 | 95 | 62.5 | 1.05 | 47.2 |

TABLE 3: Least squares regression data for Ca(II) removal.

| Factor | Coefficient estimate | DF | Standard error | $t$ for $H_0$ Coeff. = 0 | Prob > $\lvert t \rvert$ | VIF |
|---|---|---|---|---|---|---|
| Intercept | 2.16972 | 1 | 0.2366 | | | |
| $A$: pH | 0.062119 | 1 | 0.0110 | 5.64648 | <0.0001 | 1.01478 |
| $B$: contact time | 0.008357 | 1 | 0.0015 | 5.48405 | <0.0001 | 8.79209 |
| $C$: concentration | −0.0863 | 1 | 0.0065 | −13.279 | <0.0001 | 47.681 |
| $D$: dosage | 0.149761 | 1 | 0.0463 | 3.23304 | 0.0042 | 1.01478 |
| $C^2$ | 0.0005 | 1 | 0.0000 | 10.147 | <0.0001 | 44.4523 |
| $BC$ | −8.10$E$ − 05 | 1 | 0.0000 | −3.6024 | 0.0018 | 14.1953 |
| $R^2$ | 0.984 | | | | | |

after elimination of the insignificant terms ($p > 0.05$) during stepwise multiple regression are summarized in Table 3.

All the $p$ values associated with the coefficients of pH ($A$), contact time ($B$), initial concentration ($C$), dosage concentration ($D$), initial concentration squared term ($C^2$), and the interaction between contact time and initial concentration term ($BC$) were smaller than 0.05 (Table 3), indicating a significant effect of these factors on the percentage removal of calcium by metakaolin zeolite. The interaction between contact time and initial concentration ($BC$) has an antagonistic

TABLE 4: Results for ANOVA for the model.

| Source | Sum of squares | DF | Mean square | $F$-value | Prob > $F$ |
|---|---|---|---|---|---|
| Model | 13.4078 | 6 | 2.23463 | 203.981 | <0.0001 |
| Residual | 0.219102 | 20 | 0.010955 | | |
| Lack of fit | 0.21761 | 15 | 0.014507 | 48.6273 | 0.0002 |
| Pure error | 0.001492 | 5 | 0.000298 | | |
| Cor. total | 13.6269 | 26 | | | |

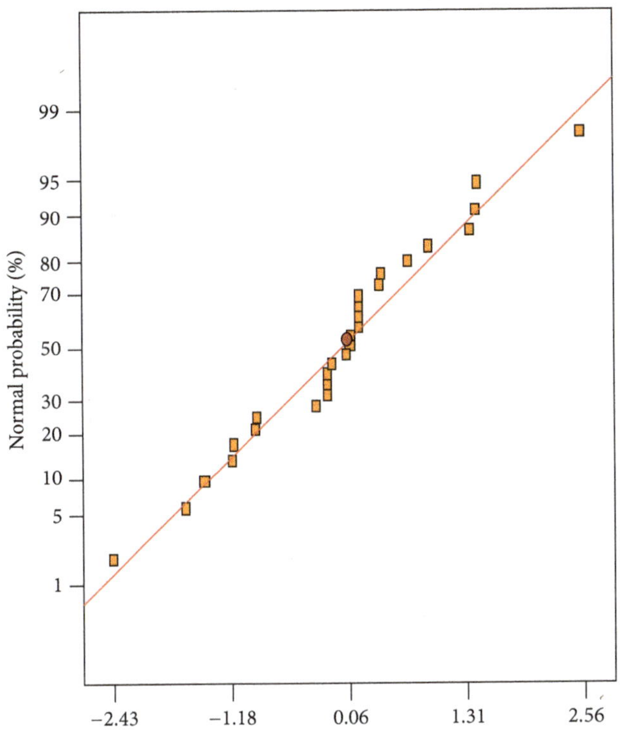


FIGURE 3: Normal plot of residuals.

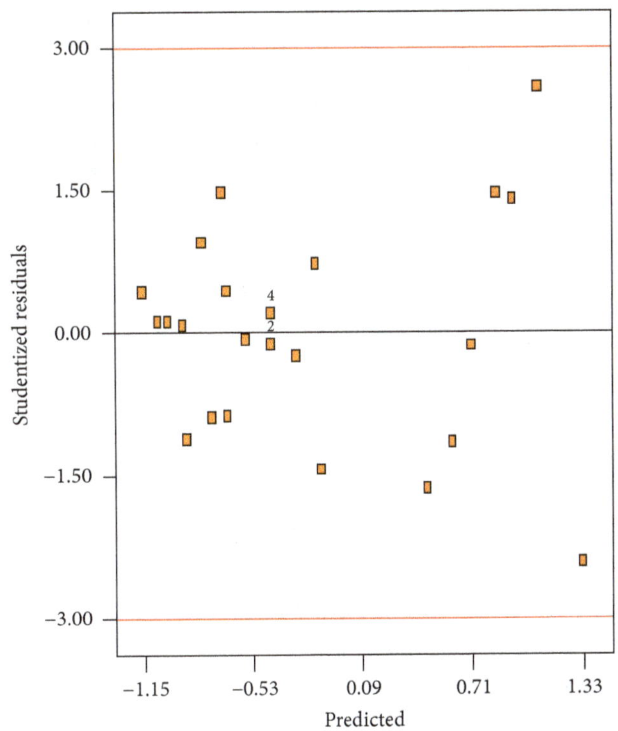


FIGURE 4: Plot of standard residuals versus predicted values.

effect on the removal of calcium as shown by the negative value of the associated coefficient. The final model obtained after elimination of the insignificant term was

$$\text{Logit}\,R = b_0 + b_1A + b_2B + b_3C + b_4D + b_5C^2 + b_6BC. \quad (4)$$

The results of the analysis of variance for the regression model are presented in Table 4.

The $p$ value for the model was less than 0.05 ($p < 0.0001$), indicating that the model was statistically significant. Moreover, a higher regression coefficient ($R^2 = 0.984$) supported the significance of the model [25]. Model validity was evaluated by normal probability plot which was produced by the Design Expert software. Figure 3 shows that the residuals were close to the straight line and were scattered randomly around it with no particular pattern, showing no abnormal behavior [26].

The plot of standard residuals versus predicted values showed a random distribution of residuals (difference between the experimental and the predicted values), with no particular pattern showing well behaved residuals (Figure 4) [27]. Moreover, it showed no outliers supporting the statistical validity of the model.

*3.2.2. Model Interpretation.* A 3-dimensional surface plot gives vital information on the behavior of the experimental design system. This aids the evaluation of the effects of the independent variables on the response factor. Figure 5 shows the effect of pH and contact time on % removal of Ca(II) ions at constant dosage amount and initial Ca(II) concentration. An increase in both pH and contact time increased the percentage removal of Ca(II). Since the sorbent had abundant binding sites, the gradual occupancy of these sites over different times resulted in an increase in the uptake of Ca(II). The lower percentage removal observed in acidic pH medium is due to protonation of adsorbent functional groups or competition of $H^+$ with Ca(II) ions to bind and occupy the active sites of the adsorbent. On the other hand, higher pH medium promotes formation of insoluble metal hydroxides such as $Ca(OH)_2$ which are filtered off.

The linear increase in adsorption efficiency of MZ with increasing adsorbent dosage indicated the accessibility of a

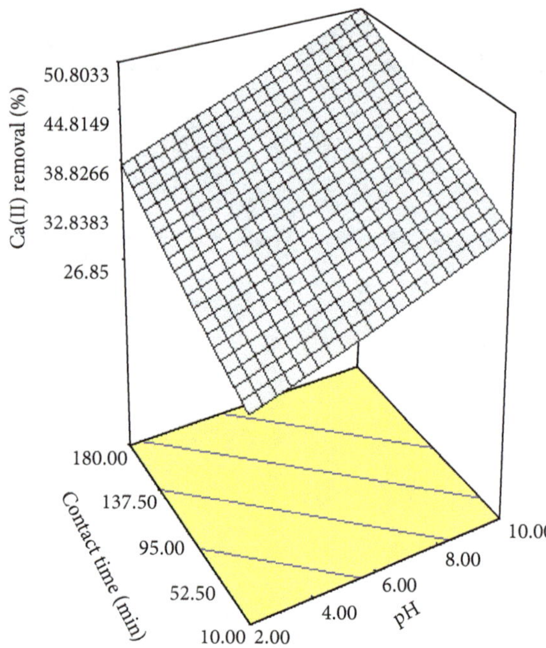


FIGURE 5: 3D surface plot for total hardness removal.

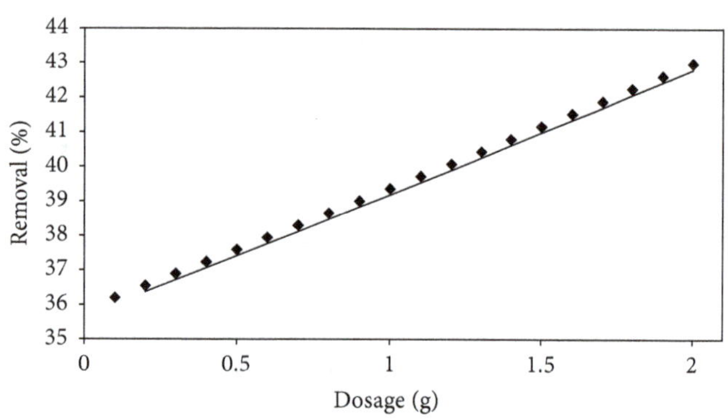


FIGURE 6: Effect of adsorbent dosage.

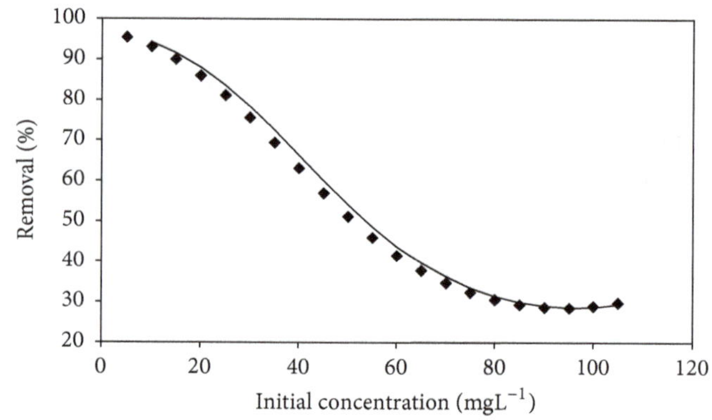


FIGURE 7: Effect of initial concentration.

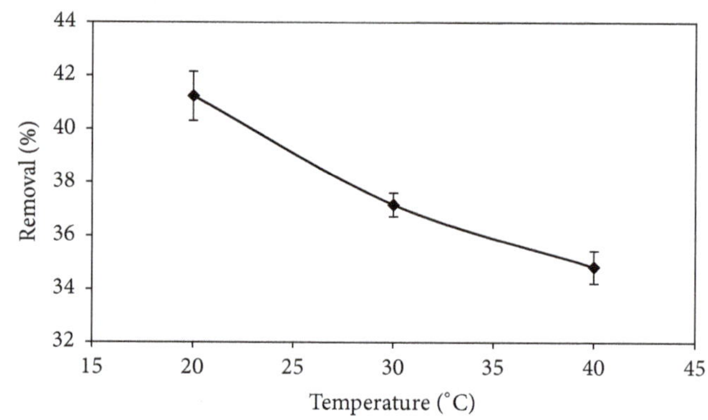


FIGURE 8: Effect of temperature.

larger number of sorption sites at higher dosage to adsorb Ca(II) ions (Figure 6).

The effect of initial metal concentration on the removal efficiency of the zeolite adsorbent is illustrated in Figure 7. The initial concentration-profile plot shows a remarkable decrease in removal efficiency as initial concentration is increased. At low concentration, most of the metal ions in the solution interact with active sites on the adsorbent, resulting in high percentage removal [28].

*3.2.3. Effect of Temperature.* The amount of Ca(II) adsorbed on MZ as a function of solution temperature is shown in Figure 8. The removal efficiency decreased as temperature was increased from 20 to 40°C, showing that low temperature favors Ca(II) removal from aqueous solutions. The decrease of Ca(II) removal may be attributed to the Ca(II) ions escaping from the solid phase with the rise in temperature of solutions [29]. The optimum temperature of 20°C was used for all further experiments.

*3.2.4. Adsorption Thermodynamics.* Thermodynamic parameters such as standard Gibb's free energy change ($\Delta G^o$), the entropy change ($\Delta S^o$), and the enthalpy change ($\Delta H^o$) were calculated from the following equations:

$$\begin{aligned} \Delta G^o &= -RT \ln K_d, \\ K_d &= \frac{q_e}{C_e}, \\ \Delta G^o &= \Delta H^o - T\Delta S^o, \\ \ln K_d &= \frac{\Delta S^o}{R} + \frac{\Delta H^o}{RT}, \end{aligned} \quad (5)$$

where $R$ is the ideal gas constant, $T$ is the temperature in K, and $K_d$ is the thermodynamic equilibrium constant. The parameter $q_e$ is the amount of Ca(II) adsorbed at equilibrium in mg/L and $C_e$ is the equilibrium concentration of Ca(II) ions in the solution. The intercept and the slope of the Van't Hoff plots of $\ln K_d$ versus $1/T$ were used to determine the $\Delta S^o$ and $\Delta H^o$ values, respectively. Table 5 shows the thermodynamic parameters. The calculated $\Delta H^o$ value was positive, showing that the adsorption process was endothermic. A slightly positive $\Delta S^o$ indicated some randomness at the solid/liquid interface during the sorption process. The $\Delta G^o$ values were all positive, showing that the sorption of metals was not thermodynamically spontaneous.

TABLE 5: Thermodynamic parameters.

| Temperature (K) | $\Delta G^o$ (kJmol$^{-1}$) | $\Delta H^o$ (kJmol$^{-1}$) | $\Delta S^o$ (kJmol$^{-1}$) |
|---|---|---|---|
| 293.15 | +1.284 | +3.953 | +0.0091 |
| 303.15 | +1.193 | | |
| 313.15 | +1.102 | | |

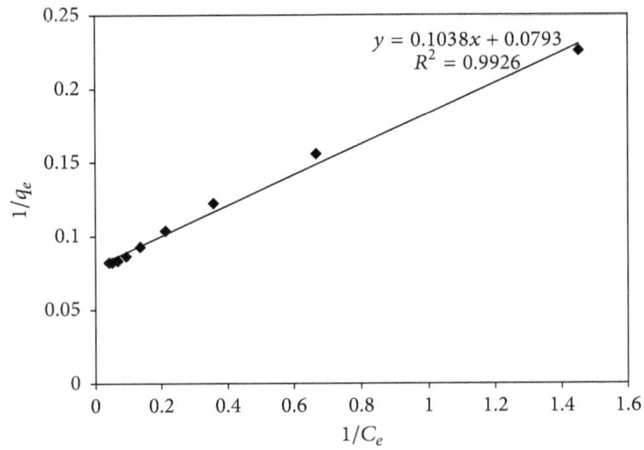


FIGURE 9: Langmuir isotherm model.

3.2.5. *Adsorption Isotherms.* Adsorption isotherms are useful in determining the nature of the interaction between the adsorbate and the adsorbent. In this study, the Freundlich [30] and Langmuir [31] adsorption isotherms were used to interpret the experimental data.

The linearized form of the Langmuir isotherm model can be represented by

$$\frac{1}{q_e} = \frac{1}{q_{\max}} + \left(\frac{1}{bq_{\max}}\right)\frac{1}{C_e}, \quad (6)$$

where $q_e$ is the amount of metal ion adsorbed at equilibrium, $q_{\max}$ is the maximum adsorption capacity (mg g$^{-1}$), $C_e$ (mgL$^{-1}$) is the equilibrium concentration, and $b$ is the equilibrium Langmuir constant. A plot of $1/q_e$ versus $1/C_e$ (Figure 9) gave a straight line with $1/q_{\max}$ as intercept and $1/bq_{\max}$ as slope, and hence $q_{\max}$ and $b$ could be estimated (Table 6).

Monolayer sorption with a heterogeneous energetic distribution of active sites, coupled by interaction between the adsorbed molecules, is assumed by the Freundlich isotherm. The linearized Freundlich isotherm equation is given by

$$\ln q_e = \ln K_F + \frac{1}{n}\ln C_e, \quad (7)$$

where $q_e$ (mg g$^{-1}$) is the amount of metal ion adsorbed at equilibrium by the adsorbent, $K_F$ (mg g$^{-1}$) and $n$ are constants representing the adsorption capacity and intensity of adsorption, respectively, and $C_e$ (mgL$^{-1}$) is the equilibrium concentration. A plot of $\ln q_e$ versus $\ln C_e$ (Figure 10) gave a straight line, and $K_F$ and $n$ were estimated from the intercept and slope, respectively (Table 6). The calculated values of $n$

TABLE 6: Equilibrium studies parameters.

| Isotherm | Parameter | Value |
|---|---|---|
| Langmuir | $q_{\max}$ (mg g$^{-1}$) | 12.610 |
| | $b$ (L mg$^{-1}$) | 0.764 |
| | $\chi^2$ | 4.76 |
| | $R^2$ | 0.993 |
| Freundlich | $KF$ (mg g$^{-1}$) | 5.608 |
| | $n$ | 3.578 |
| | $\chi^2$ | 37.91 |
| | $R^2$ | 0.933 |

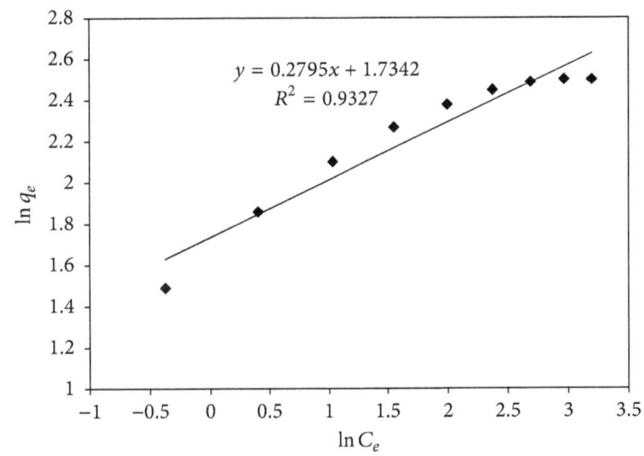


FIGURE 10: Freundlich isotherm model.

(3.578) were between 1 and 10 indicating favorability of the adsorption of Ca(II) ions on the adsorbent.

The Langmuir isotherm model correlation coefficient ($R^2 = 0.993$) was significantly closer to unity relative to that of the Freundlich isotherm model ($R^2 = 0.933$). Moreover, the Chi-square value for the Langmuir model ($\chi^2 = 4.76$) was lower than that of Freundlich model ($\chi^2 = 37.91$), indicating that the Freundlich model could not fit all the experimental data well.

3.2.6. *Adsorption Kinetics.* In the current study, the pseudo-first-order and pseudo-second-order models were used for determining the mechanism of the adsorption process. The pseudo-first-order model which assumes that the rate of adsorption site occupation is proportional to the number of unoccupied sites is expressed by the Lagergren equation [32] (Lagergren, 1898) in the linear form as follows:

$$\log(q_e - q_t) = \log q_e - \frac{t \cdot k_1}{2.3}, \quad (8)$$

where $q_e$ and $q_t$ are the amount of Ca(II) adsorbed (mg g$^{-1}$) onto MZ at equilibrium and at any time $t$, respectively, $k_1$ (min$^{-1}$) is the rate constant of pseudo-first-order model, and $t$ is the time (min). A plot of $\log(q_e - q_t)$ versus $t$ (Figure 11) was used to estimate the values of $k_1$ and $q_e$ from the slope and the intercept, respectively.

Table 7: Kinetic parameters.

| Experimental value | Pseudo-first-order kinetic model | | | Pseudo-second-order kinetic model | | |
|---|---|---|---|---|---|---|
| $q_e$ (mg g$^{-1}$) | $q_e$ (mg g$^{-1}$) | $k_1$ (min$^{-1}$) | $R^2$ | $q_e$ (mg g$^{-1}$) | $k_1$ (g mg$^{-1}$ min$^{-1}$) | $R^2$ |
| 2.77 | 1.01 | 0.01 | 0.964 | 2.81 | 0.0318 | 0.991 |

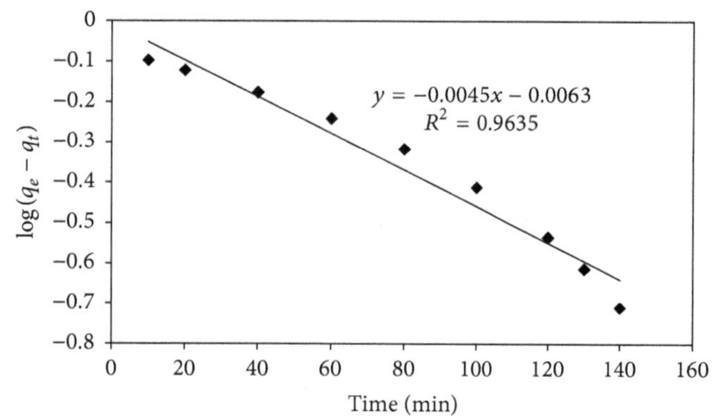


Figure 11: Pseudo-first-order model.

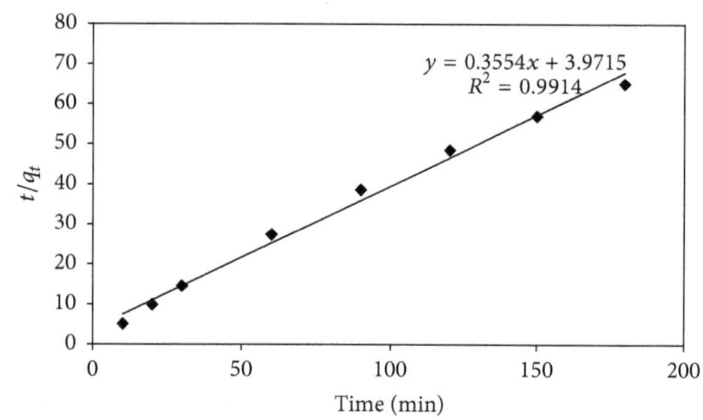


Figure 12: Pseudo-second-order kinetic model.

The second-order kinetic model can be expressed in the linear form as follows [33]:

$$\frac{t}{q_t} = \frac{t}{q_e} + \frac{1}{k_2 (q_e)^2}, \quad (9)$$

where $k_2$ is the equilibrium rate constant of pseudo-second-order model (g mg$^{-1}$ min$^{-1}$). The values of $q_e$ and $k_2$ (Table 7) were determined from the slope and intercept of the plot of $(t/q_t)$ versus $t$, respectively (Figure 12). The pseudo-second-order kinetic model correlation coefficient was $R^2 = 0.991$, suggesting the applicability of the model to the adsorption of Ca(II) by MZ. Furthermore, the $q_{e,\text{cal}}$ value (2.81 mg g$^{-1}$) for the pseudo-second-order kinetic reaction was found to be in better agreement with the $q_{e,\text{exp}}$ value (2.77 mg g$^{-1}$).

*3.2.7. Desorption Studies.* The reusability of the spent adsorbent was studied by desorption/regeneration experiments, to evaluate the economical feasibility of the method. Regeneration percentage of the spent adsorbent versus NaOH is

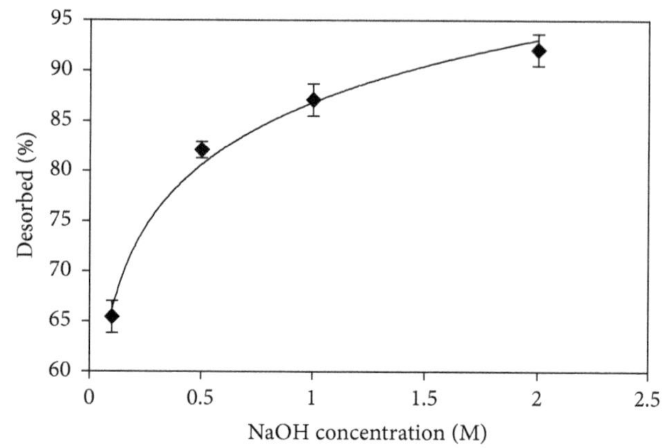


Figure 13: Desorption of Ca(II) ions from the spent adsorbent.

illustrated in Figure 13. Maximum recovery of the adsorbed metal ions was achieved at 2 M NaOH concentration leading to 90% desorption of the adsorbed Ca(II).

## 4. Conclusions

The present study shows that metakaolin zeolite has much potential as an adsorbent for the removal of Ca(II) from aqueous solutions. The influences of pH, adsorbent dosage, contact time, and initial concentration on adsorption of the metal ion and their interactions were investigated by a response surface methodology and ANOVA. The Excel Solver showed that optimum removal (87.70%) was obtained at a pH, contact time, adsorbent dosage, and initial Ca(II) concentration of 6, 180 min, 1.0 g, and 40.0 mg L$^{-1}$, respectively. Confirmatory experimental analysis under the optimized conditions gave average Ca(II) removal of 87.26 ± 0.31%. The adsorption process was found to be endothermic and not thermodynamically spontaneous. The Langmuir isotherm model best described the adsorption mechanism with maximum adsorption capacity of 12.61 mg g$^{-1}$. The kinetic study was performed based on pseudo-first-order and pseudo-second-order equations and the adsorption followed pseudo-second-order kinetic model. The results obtained showed that metakaolin zeolite can be used as an effective adsorbent for removal of Ca(II) ions from aqueous solutions.

## Conflicts of Interest

The authors declare that there are no conflicts of interest regarding the publication of this paper.

## Acknowledgments

The authors are grateful to the Department of Chemical Technology, Midlands State University, Gweru, Zimbabwe, for providing the research facilities.

## References

[1] H. Liu, S. Peng, L. Shu, T. Chen, T. Bao, and R. L. Frost, "Magnetic zeolite NaA: synthesis, characterization based on metakaolin and its application for the removal of $Cu^{2+}$, $Pb^{2+}$," *Chemosphere*, vol. 91, no. 11, pp. 1539–1546, 2013.

[2] M. Gougazeh and J.-C. Buhl, "Synthesis and characterization of zeolite A by hydrothermal transformation of natural Jordanian kaolin," *Journal of the Association of Arab Universities for Basic and Applied Sciences*, vol. 15, no. 1, pp. 35–42, 2014.

[3] H. S. Ibrahim, T. S. Jamil, and E. Z. Hegazy, "Application of zeolite prepared from Egyptian kaolin for the removal of heavy metals: II. Isotherm models," *Journal of Hazardous Materials*, vol. 182, no. 1-3, pp. 842–847, 2010.

[4] T. T. Walek, F. Saito, and Q. Zhang, "The effect of low solid/liquid ratio on hydrothermal synthesis of zeolites from fly ash," *Fuel*, vol. 87, no. 15-16, pp. 3194–3199, 2008.

[5] C.-F. Wang, J.-S. Li, L.-J. Wang, and X.-Y. Sun, "Influence of NaOH concentrations on synthesis of pure-form zeolite A from fly ash using two-stage method," *Journal of Hazardous Materials*, vol. 155, no. 1-2, pp. 58–64, 2008.

[6] T. S. Jamil, H. S. Ibrahim, I. H. Abd El-Maksoud, and S. T. El-Wakeel, "Application of zeolite prepared from Egyptian kaolin for removal of heavy metals: I. Optimum conditions," *Desalination*, vol. 258, no. 1-3, pp. 34–40, 2010.

[7] A. A. Ismail, R. M. Mohamed, I. A. Ibrahim, G. Kini, and B. Koopman, "Synthesis, optimization and characterization of zeolite A and its ion-exchange properties," *Colloids and Surfaces A: Physicochemical and Engineering Aspects*, vol. 366, no. 1-3, pp. 80–87, 2010.

[8] A. R. Loiola, J. C. R. A. Andrade, J. M. Sasaki, and L. R. D. da Silva, "Structural analysis of zeolite NaA synthesized by a cost-effective hydrothermal method using kaolin and its use as water softener," *Journal of Colloid and Interface Science*, vol. 367, no. 1, pp. 502–508, 2012.

[9] M. L. Mignoni, D. I. Petkowicz, N. R. C. Fernandes Machado, and S. B. C. Pergher, "Synthesis of mordenite using kaolin as Si and Al source," *Applied Clay Science*, vol. 41, no. 1-2, pp. 99–104, 2008.

[10] D. M. El-Mekkawi and M. M. Selim, "Removal of $Pb^{2+}$ from water by using Na-Y zeolites prepared from Egyptian kaolins collected from different sources," *Journal of Environmental Chemical Engineering*, vol. 2, no. 1, pp. 723–730, 2014.

[11] S. Chandrasekhar and P. N. Pramada, "Microwave assisted synthesis of zeolite A from metakaolin," *Microporous and Mesoporous Materials*, vol. 108, no. 1-3, pp. 152–161, 2008.

[12] M. N. Sepehr, M. Zarrabi, H. Kazemian, A. Amrane, K. Yaghmaian, and H. R. Ghaffari, "Removal of hardness agents, calcium and magnesium, by natural and alkaline modified pumice stones in single and binary systems," *Applied Surface Science*, vol. 274, pp. 295–305, 2013.

[13] A. Dimirkou and M. K. Doula, "Use of clinoptilolite and an Fe-overexchanged clinoptilolite in $Zn^{2+}$ and $Mn^{2+}$ removal from drinking water," *Desalination*, vol. 224, no. 1-3, pp. 280–292, 2008.

[14] Z. Xue, Z. Li, J. Ma et al., "Effective removal of $Mg^{2+}$ and $Ca^{2+}$ ions by mesoporous LTA zeolite," *Desalination*, vol. 341, no. 1, pp. 10–18, 2014.

[15] L. Fu, J. Wang, and Y. Su, "Removal of low concentrations of hardness ions from aqueous solutions using electrodeionization process," *Separation and Purification Technology*, vol. 68, no. 3, pp. 390–396, 2009.

[16] J. N. Apell and T. H. Boyer, "Combined ion exchange treatment for removal of dissolved organic matter and hardness," *Water Research*, vol. 44, no. 8, pp. 2419–2430, 2010.

[17] L. Seifi, A. Torabian, H. Kazemian et al., "Adsorption of BTEX on surfactant modified granulated natural zeolite nanoparticles: parameters optimizing by applying taguchi experimental design method," *Clean - Soil, Air, Water*, vol. 39, no. 10, pp. 939–948, 2011.

[18] M. Roosta, M. Ghaedi, A. Daneshfar, R. Sahraei, and A. Asghari, "Optimization of the ultrasonic assisted removal of methylene blue by gold nanoparticles loaded on activated carbon using experimental design methodology," *Ultrasonics Sonochemistry*, vol. 21, no. 1, pp. 242–252, 2014.

[19] M. Roosta, M. Ghaedi, A. Daneshfar, and R. Sahraei, "Experimental design based response surface methodology optimization of ultrasonic assisted adsorption of safaranin O by tin sulfide nanoparticle loaded on activated carbon," *Spectrochimica Acta - Part A: Molecular and Biomolecular Spectroscopy*, vol. 122, pp. 223–231, 2014.

[20] M. K. Satapathy and P. Das, "Optimization of crystal violet dye removal using novel soil-silver nanocomposite as nanoadsorbent using response surface methodology," *Journal of Environmental Chemical Engineering*, vol. 2, no. 1, pp. 708–714, 2014.

[21] M. Savasari, M. Emadi, M. A. Bahmanyar, and P. Biparva, "Optimization of Cd (II) removal from aqueous solution by ascorbic acid-stabilized zero valent iron nanoparticles using response surface methodology," *Journal of Industrial and Engineering Chemistry*, vol. 21, pp. 1403–1409, 2015.

[22] C. A. Ríos, C. D. Williams, and M. A. Fullen, "Nucleation and growth history of zeolite LTA synthesized from kaolinite by two different methods," *Applied Clay Science*, vol. 42, no. 3-4, pp. 446–454, 2009.

[23] P. Thuadaij and N. Nuntiya, "Synthesis of Na-x hydrate zeolite from fly ash and amorphous silica from rice husk ash by fusion with caustic soda prior to incubation," in *Proceedings of the International Conference on Chemistry and Chemical Process IPCBEE*, vol. 10, pp. 69–74, 2011.

[24] S. Sugashini and K. M. M. S. Begum, "Optimization using central composite design (CCD) for the biosorption of Cr(VI) ions by cross linked chitosan carbonized rice husk (CCACR)," *Clean Technologies and Environmental Policy*, vol. 15, no. 2, pp. 293–302, 2013.

[25] A. Fakhri, "Application of response surface methodology to optimize the process variables for fluoride ion removal using maghemite nanoparticles," *Journal of Saudi Chemical Society*, vol. 18, no. 4, pp. 340–347, 2014.

[26] A. C. Cameron and P. K. Trivedi, *Regression analysis of count data*, vol. 53 of *Econometric Society Monographs*, Cambridge University Press, Cambridge, England, 2nd edition, 2013.

[27] D. C. Montgomery and E. A. Peck, *Introduction to Linear Regression Analysis*, John Wiley & Sons, New York, NY, USA, 1982.

[28] M. Moyo, U. Guyo, G. Mawenyiyo, N. P. Zinyama, and B. C. Nyamunda, "Marula seed husk (Sclerocarya birrea) biomass

as a low cost biosorbent for removal of Pb(II) and Cu(II) from aqueous solution," *Journal of Industrial and Engineering Chemistry*, vol. 27, pp. 126–132, 2015.

[29] U. Guyo, J. Mhonyera, and M. Moyo, "Pb(II) adsorption from aqueous solutions by raw and treated biomass of maize stover—a comparative study," *Process Safety and Environmental Protection*, vol. 93, pp. 192–200, 2015.

[30] H. M. Freundlich, "Over the adsorption in solution," *Journal of Phyical Chemisttry*, vol. 57, pp. 384–470, 1906.

[31] I. Langmuir, "The adsorption of gases on plane surfaces of glass, mica and platinum," *The Journal of the American Chemical Society*, vol. 40, no. 9, pp. 1361–1403, 1918.

[32] S. Lagergren, "About the theory of so-called adsorption of soluble substances," *Kungliga Svenska Vetenskapsakademiens Handlingar*, vol. 24, pp. 1–39, 1898.

[33] Y. S. Ho and G. McKay, "Pseudo-second order model for sorption processes," *Process Biochemistry*, vol. 34, no. 5, pp. 451–465, 1999.

20

# Determination of 2-Propenal Using Headspace Solid-Phase Microextraction Coupled to Gas Chromatography–Time-of-Flight Mass Spectrometry as a Marker for Authentication of Unrefined Sesame Oil

**Ahmad Rois Mansur,[1,2] Tae Gyu Nam,[1] Hae Won Jang,[1] Yong-Sun Cho,[1] Miyoung Yoo,[1] Dongwon Seo,[1] and Jaeho Ha[1]**

[1]*Korea Food Research Institute, Seongnam, Gyeonggi 13539, Republic of Korea*
[2]*Department of Food Biotechnology, Korea University of Science and Technology, Daejeon 34113, Republic of Korea*

Correspondence should be addressed to Jaeho Ha; jaeho.ha@wikim.re.kr

Academic Editor: Maria B. P. P. Oliveira

Ascertaining the authenticity of the unrefined sesame oil presents an ongoing challenge. Here, the determination of 2-propenal was performed by headspace solid-phase microextraction (HS-SPME) under mild temperature coupled to gas chromatography with time-of-flight mass spectrometry, enabling the detection of adulteration of unrefined sesame oil with refined corn or soybean oil. Employing this coupled technique, 2-propenal was detected in all tested refined corn and soybean oils but not in any of the tested unrefined sesame oil samples. Using response surface methodology, the optimum extraction temperature, equilibrium time, and extraction time for the HS-SPME analysis of 2-propenal using carboxen/polydimethylsiloxane fiber were determined to be 55°C, 15 min, and 15 min, respectively, for refined corn oil and 55°C, 25 min, and 15 min, respectively, for refined soybean oil. Under these optimized conditions, the adulteration of unrefined sesame oil with refined corn or soybean oils (1–5%) was successfully detected. The detection and quantification limits of 2-propenal were found to be in the range of 0.008–0.010 and 0.023–0.031 $\mu$g mL$^{-1}$, respectively. The overall results demonstrate the potential of this novel method for the authentication of unrefined sesame oil.

## 1. Introduction

Unrefined sesame oil has long been considered as a popular health food due to its considerable content of bioactive compounds such as phytosterols and lignans. It is most commonly consumed as a nutritive flavor enhancer in many Asian countries [1, 2]. For instance, approximately 73,000 tonnes of sesame seeds and 600 tonnes of unrefined sesame oils were imported annually in Korea during 2010–2013 to compensate for the limited domestic production of sesame seeds [3]. The retail price of unrefined sesame oil in Korea is 5 to 30 times higher than that of other refined edible oils. Therefore, the economically motivated adulteration of sesame oil, with potential impact on the oil quality, nutritional value, and safety to consumers, is a critical problem. Adulterated sesame oils are most commonly manufactured by blending pure unrefined sesame oil with cheaper and lower quality edible oils such as refined corn and soybean oils [4, 5].

In the past two decades, various analytical techniques combined or not with chemometric methods have been established for the verification of sesame oil authenticity. For instance, several nonseparation/nondestructive techniques (e.g., isotope ratio mass spectrometry, infrared spectroscopy, and nuclear magnetic resonance spectroscopy) have been applied for the detection of adulterated sesame oils [1, 4, 6]. Previous studies have also employed techniques such as gas chromatography (GC) coupled with a flame ionization detector [7] or mass spectrometer [8, 9], high performance liquid chromatography with a refractive index detector [10], an evaporative light scattering detector [5, 11] or a fluorescence detector [12], an electronic nose [13], and real-time PCR [14]. Despite the recent advances in the analytical methods available for the detection of sesame oil adulteration, the minimum adulteration detection levels remain relatively high. Currently, it is possible to detect the presence of refined corn oil in unrefined sesame oil at ≥10% [1, 4] and that of refined soybean oil at ≥3% [1]. This limitation arises mainly from the fact that sesame oil and oil adulterants (e.g., corn and soybean oils) have very similar physical, chemical, and biological properties. In addition, chemometric methods using unreliable marker compounds obtained from both sesame oil and oil adulterants might not be able to adequately detect the presence of oil adulterants in sesame oil. Additionally, the techniques described above are often time-consuming, expensive, and require the use of harmful solvents, complex technology, and expertise. Further research is therefore highly necessary in order to establish a new approach and identify specific compounds as reliable markers for the authentication of unrefined sesame oil.

Headspace solid-phase microextraction (HS-SPME) is a simple, fast, sensitive, solvent-free, and cost-efficient method for the extraction of volatile compounds in complex matrices such as those present in edible oils [15]. In addition, GC equipped with time-of-flight mass spectrometry (GC-TOF/MS) allows the acquisition of full mass spectra even for volatile compounds at very low concentrations [16]. Recently, headspace coupled to GC-TOF/MS has been employed to analyze volatile compounds for the purpose of sesame oil authentication [16]. In this line, Ma et al. [15] have reported that the volatile aldehydes present in edible oils can be extracted by HS-SPME under mild temperature conditions. 2-Propenal is one of the highly reactive aldehydes detected in thermally processed edible oils, formed from the constituents of glycerides such as the glycerol moieties of triacyl and diacyl glycerides. It can also be generated by degradation of unsaturated fatty acid backbones [17]. Despite extensive studies on the pathway and mechanism of 2-propenal formation from oil constituents [18, 19], the determination of 2-propenal under mild temperature and its application as a detection marker of sesame oil adulteration remain vastly unexplored. The present study, therefore, aims to develop a combined approach of HS-SPME analysis under mild temperature and GC-TOF/MS for the determination of 2-propenal as a volatile marker to detect unrefined sesame oil adulteration with refined corn or soybean oil.

## 2. Materials and Methods

*2.1. Materials.* Twelve brands of unrefined sesame oil, six brands of refined corn oil, and five brands of refined soybean oil used in this study were manufactured by food companies in Korea. All samples were purchased from a local supermarket (Seongnam, Korea); they were all pure and qualified products. The shelf life of commercial oil samples is at least twelve months. The samples were stored in a dark and dry environment prior to analysis. To prevent any changes in chemical constituents, samples were analyzed immediately after unpacking in the laboratory. 2-Propenal was purchased from Sigma-Aldrich (St. Louis, MO, USA).

*2.2. Chromatographic Conditions and Data Processing.* The extracted volatiles were analyzed using a GC-TOF/MS instrument (LECO Corp., St. Joseph, MI, USA) equipped with Agilent 6890N (Agilent Technologies, Santa Clara, CA, USA). The dimensions of the primary (DB-WAX) and the secondary (DB-1701) columns were 60 m × 0.25 mm internal diameter (ID) × 0.25 $\mu$m film thickness and 20 m × 0.10 mm ID × 0.40 $\mu$m film thickness (Agilent J&W GC Columns, USA), respectively. The temperatures of the GC inlet and transfer line were set at 220°C. High purity helium (99.99%) was used as the carrier gas at a constant flow rate of 1.20 mL $min^{-1}$ and a split ratio of 20 : 1. The primary oven temperature was maintained at 40°C for 2 min and then ramped to 230°C at a rate of 20°C $min^{-1}$, where it was held for 10 min. The secondary oven was always operated at a temperature 10°C higher than that of the primary oven throughout the analysis. The detector voltage was set to 1600 V with an electron energy of 70 eV. An acquisition delay of 400 s was employed and the mass spectra were collected in the mass range of $m/z$ 33–650, at an acquisition rate of 10 spectra $s^{-1}$. The temperature of the ion source was set at 200°C.

The raw chromatograms were first processed using ChromaTOF® software (LECO Corp, version 4.22) for peak finding, baseline correction, deconvolution, library matching, and area calculation. The peaks were extracted with a signal-to-noise ratio (S/N) of 20. The peaks were identified tentatively by comparison of the obtained mass spectra with those available in mass spectral databases (NIST and Wiley), with a minimum similarity match of 800. The area of each peak was calculated using the unique mass of each detected volatile compound. After the initial processing, all the peak information from different chromatograms was aligned using the Statistical Compare (SC) option in the ChromaTOF software in order to obtain an accurate comparison of the volatile profiles of sesame oil and oil adulterants (corn and soybean oils). Following the alignment process, the basic statistical information from different peak calculations was compared in order to identify the volatile compounds with potentially significant class differences between volatiles from sesame oil and the oil adulterants. This comparison was necessary in order to identify compounds that could be used as potential markers for the detection of sesame oil adulteration. The results of the SC process were exported as .csv files and used for further analysis.

### 2.3. Preliminary Experiments

*2.3.1. Selection of Volatile Markers for Authentication of Sesame Oil.* The initial step of the preliminary experiment involved the selection of volatile marker compounds for the detection of sesame oil adulteration. Commercial manual sampling SPME devices and 85 $\mu$m carboxen/polydimethylsiloxane (CAR/PDMS) fiber (Supelco, PA, USA) were selected for the extraction of volatile compounds in oil samples. Oil samples (each 2 g) were heated at 55°C for 15 min (equilibrium time) in a 20 mL glass vial equipped with a PTFE/silicone septum (Supelco, PA, USA) and extracted using a CAR/PDMS fiber for 15 min. After extraction, the SPME fiber was immediately thermally desorbed in the injection port of the GC-TOF/MS. Next, the volatile compounds present in refined corn and

TABLE 1: Experimental factors and their coded and actual values used for HS-SPME optimization using response surface methodology.

| Factor | Symbol | Coded levels | | |
|---|---|---|---|---|
| | | −1 | 0 | +1 |
| Extraction temperature (°C) | $X_1$ | 25 | 40 | 55 |
| Equilibrium time (min) | $X_2$ | 15 | 30 | 45 |
| Extraction time (min) | $X_3$ | 5 | 10 | 15 |

soybean oils but not in unrefined sesame oils were selected as adulteration detection markers. Among these compounds, 2-propenal was selected and used for the optimization of the HS-SPME conditions.

*2.3.2. Fiber Selection for the Extraction of 2-Propenal.* Four SPME fibers, (i) 85 $\mu$m CAR/PDMS, (ii) 50/30 $\mu$m divinylbenzene/CAR/PDMS (DVB/CAR/PDMS), (iii) 65 $\mu$m PDMS/DVB, and (iv) 100 $\mu$m PDMS, were evaluated to select the most suitable fiber for the extraction of 2-propenal in oil samples. A sample of corn or soybean oil (2 g) was heated at 55°C for 15 min (equilibrium time) in a glass vial, extracted using the four selected SPME fibers for 15 min, and then thermally desorbed in the GC-TOF/MS. The adsorption efficiencies on the SPME fibers were evaluated by comparison of the peak areas of 2-propenal.

*2.3.3. Temperature Selection for the Extraction of 2-Propenal.* The final step in the preliminary set of experiments was the determination of the most appropriate temperature for the extraction of 2-propenal. The extraction temperature was varied systematically from 25 to 100°C, while the equilibrium and extraction times were both held constant at 15 min. In the present study, CAR/PDMS and DVB/CAR/PDMS fibers were chosen because of their higher extraction efficiencies compared to those of the PDMS/DVB and PDMS fibers. Under different extraction temperatures, the 2-propenal in corn or soybean oils (each 2 g) was extracted using CAR/PDMS or DVB/CAR/PDMS fiber. Based on the obtained results, the CAR/PDMS fiber and extraction temperatures of 25–55°C were selected for the optimization of the HS-SPME analysis using response surface methodology (RSM).

*2.4. Optimization of HS-SPME Conditions Using RSM.* The effect of three different SPME factors including the extraction temperature ($X_1$, 25–55°C), equilibrium time ($X_2$, 15–45 min), and extraction time ($X_3$, 5–15 min) was evaluated by RSM (Table 1). The analytical design consisted of 34 experimental points in a randomized order, and six replicates were performed for the center point and used for the estimation of the pure error sum of squares (Supp. Table 1 in Supplementary Material available online at https://doi.org/10.1155/2017/9106409). The experimental points were divided into two blocks, each consisting of 17 experiments. 2-Propenal in corn or soybean oils (each 2 g) was extracted using CAR/PDMS fiber under the extraction conditions based on the design described in Supp. Table 1 in Supplementary Material. The obtained experimental data were then fitted to the following nonlinear polynomial equation:

$$Y = \beta_0 + \Sigma\beta_i X_i + \Sigma\beta_{ii} X_i^2 + \Sigma\beta_{ij} X_i X_j, \quad (1)$$

where $Y$ is the response value predicted by the model; $\beta_0$ is a constant; and $\beta_i$, $\beta_{ii}$, and $\beta_{ij}$ are the linear, quadratic, and interaction coefficients, respectively. The adequacy of the predicted model was evaluated using modeling analysis, coefficients of determination ($R^2$), and lack-of-fit tests.

*2.5. Determination of 2-Propenal in Sesame Oils Adulterated with Corn or Soybean Oils.* Samples of adulterated sesame oil were prepared shortly before analysis by randomly blending unrefined sesame oils with refined corn or soybean oils. The concentrations of refined corn or soybean oils in the adulterated sesame oil were 1, 5, and 10% (w/w). Three corn oils and three soybean oils were used as refined adulterants and 18 admixtures were prepared. Each oil sample was analyzed quantitatively in triplicate. 2-Propenal in the pure and adulterated oils was extracted using HS-SPME under the optimal conditions determined by RSM.

A stock solution of 2-propenal (750 mg $L^{-1}$) was prepared in absolute ethanol (Burdick & Jackson, MI, USA) [20] and stored at 4°C prior to analysis. The stock solution was subsequently diluted with absolute ethanol to produce intermediate standard solutions at concentrations ranging from 0.02 to 0.76 $\mu$g $mL^{-1}$. Each standard solution was prepared shortly before exposure to the CAR/PDMS fiber. Each standard solution (2 g) was extracted and injected in the GC-TOF/MS (six replicates) in order to obtain calibration curves. The average peak areas determined from the analysis of each of these standard solutions were used to quantify the concentration of 2-propenal in the oil samples and determine the limits of detection (LOD) and limits of quantification (LOQ).

*2.6. Statistical Analysis.* The response surface modeling was performed using the Minitab statistical package (Minitab Inc., USA, version 17). The differences between the treatments employed in the preliminary experiments were determined using one-way analysis of variance (ANOVA) followed by Tukey's test ($p < 0.05$). A two-sample $t$-test was performed in order to establish the differences between the predicted and observed peak areas of 2-propenal in corn and soybean oils ($p < 0.05$). These two statistical tests were conducted using the IBM SPSS Statistics software (SPSS Inc., USA, version 20).

## 3. Results and Discussion

*3.1. Preliminary Experiments.* The volatile compounds in the tested sesame, corn, and soybean oils were extracted using CAR/PDMS fiber and analyzed by GC-TOF/MS. The volatile compounds found only in corn and soybean oils were then selected as markers for the detection of sesame oil adulteration. Among the several marker compounds, 2-propenal was selected as an adulteration marker owing to its high peak area and reproducibility (data not shown). 2-Propenal was not found in all the tested unrefined sesame

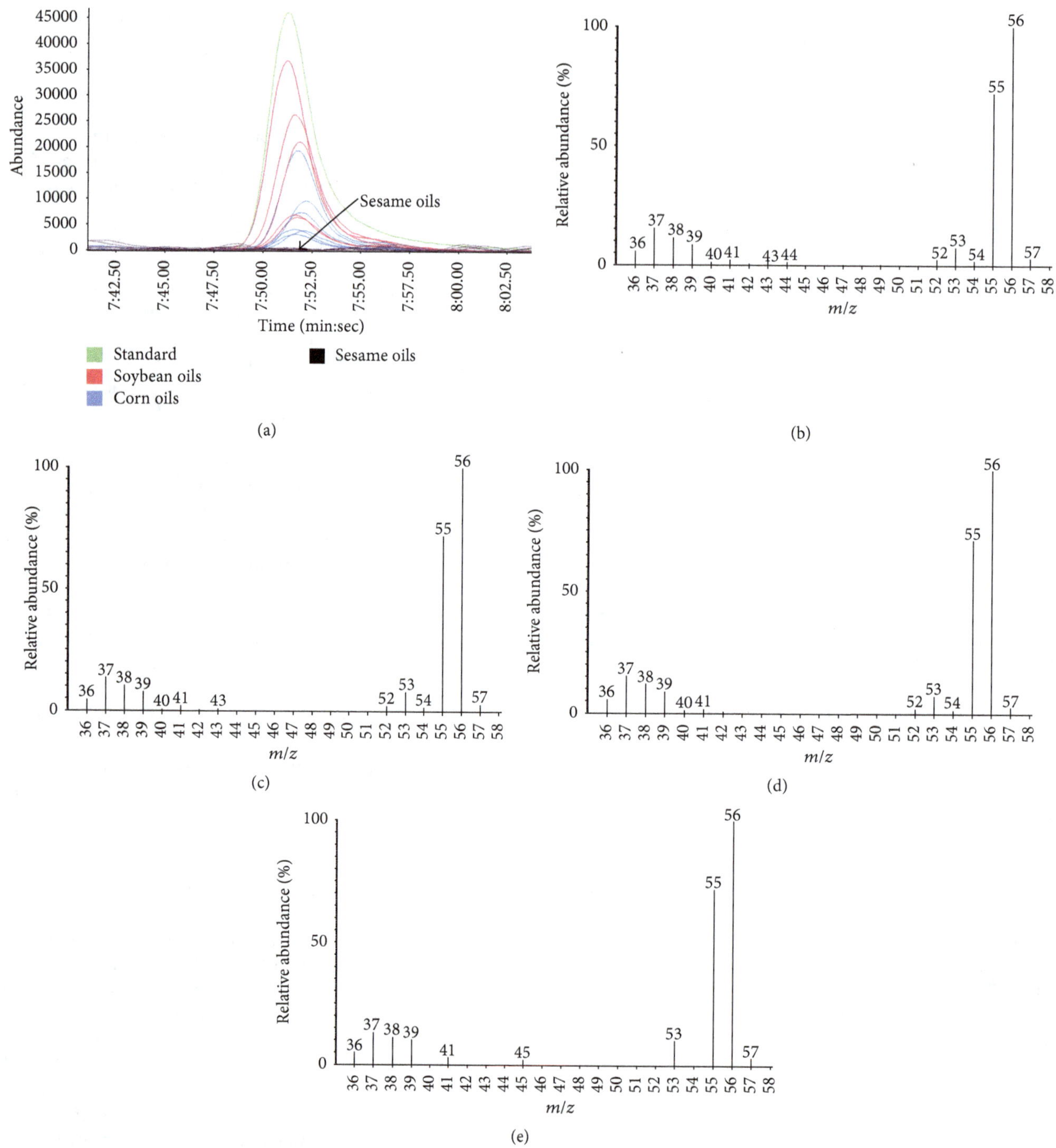

FIGURE 1: Chromatograms of 2-propenal in six corn oils (blue line), five soybean oils (red line), and twelve sesame oils (black line) analyzed using GC-TOF/MS (a). Similar MS spectra of 2-propenal in corn oil (b), soybean oil (c), authentic standard (d), and Wiley library (e) were detected at 70 eV of electron energy.

oils, while it was found in all the tested refined corn and soybean oils (Figure 1(a)). The presence of 2-propenal in corn and soybean oils (Figures 1(a), 1(b), and 1(c)) was confirmed by comparing the results to the retention time and mass spectra of standard solution containing 2-propenal at known concentrations (Figure 1(d)), as well as by comparison with a spectral library (Figure 1(e)). The mass spectrum (molecular ion at $m/z$ 56 and fragment ions at $m/z$ 37, 38, 39, 53, and 55) of 2-propenal was obtained by the electron ionization mass spectrometry.

In order to determine the most suitable fiber to extract 2-propenal from corn and soybean oils, four SPME fibers (CAR/PDMS, DVB/CAR/PDMS, PDMS/DVB, and PDMS) were evaluated. Selection of the appropriate fiber coating is

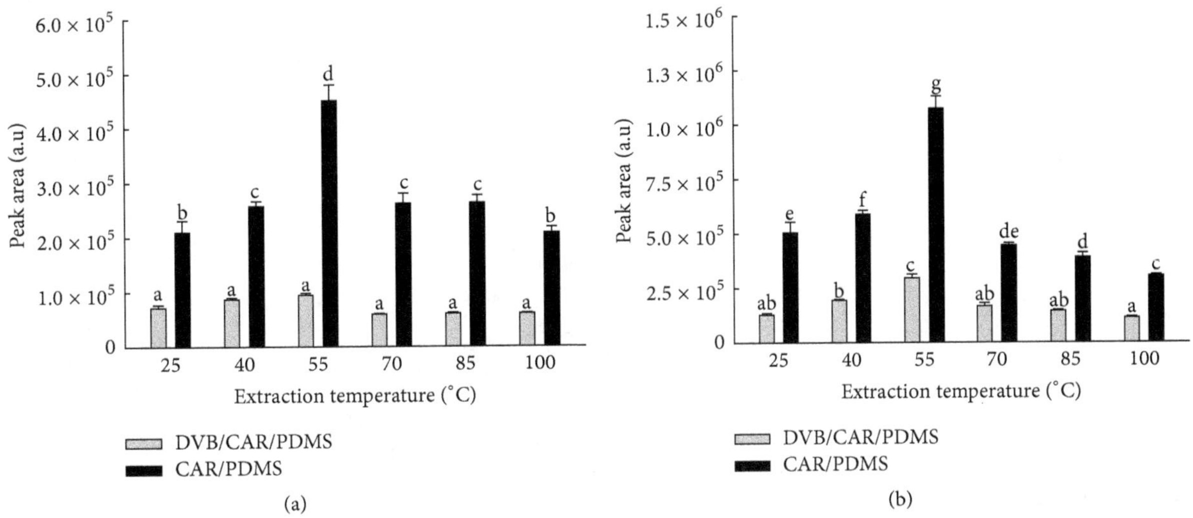


FIGURE 2: Comparison of DVB/CAR/PDMS and CAR/PDMS fibers on the extraction efficiencies of 2-propenal from corn (a) and soybean (b) oils at different temperatures (25–100°C). Different letters indicate the significant differences at $p < 0.05$.

one of the most critical steps in the optimization of SPME method. The suitability of a particular fiber coating for the analyte of interest is related directly to the properties of the extraction phase, which in turn determine the overall selectivity and reliability of the method [21]. The extraction efficiency of the SPME fibers was determined by comparing the peak areas of 2-propenal. PDMS and PDMS/DVB fibers were excluded from further analysis after a preliminary set of experiments. These two fibers had much lower extraction efficiencies compared to those of the CAR/PDMS and DVB/CAR/PDMS fibers (data not shown), indicating that the polarity of these two fibers makes them unsuitable for the extraction of 2-propenal in oil samples.

The performance of CAR/PDMS and DVB/CAR/PDMS fibers in the extraction of 2-propenal from oil samples was then further investigated and compared under different extraction temperatures (25–100°C). Results in Figure 2 show that the use of CAR/PDMS fiber resulted in a higher 2-propenal peak area ($p < 0.05$) for both corn and soybean oils compared to that determined using DVB/CAR/PDMS fiber at all extraction temperatures. The peak areas of 2-propenal in both corn and soybean oils extracted with CAR/PDMS fiber at 55°C were significantly higher ($p < 0.05$) than those obtained using the same fiber at other extraction temperatures. The peak areas of 2-propenal in corn and soybean oils extracted using CAR/PDMS fiber were found to decrease markedly at temperatures above 70°C. Generally, heating increases the vapor pressure and enhances mass transfer processes, thereby facilitating the release of analytes into the headspace. Nonetheless, excessive heating can have an adverse effect on the adsorption of analytes by the fiber coating as a result of reduction of the distribution constant [22]. Accordingly, CAR/PDMS fiber and extraction temperatures of 25–55°C were chosen as the starting point for further optimization of the HS-SPME conditions for the extraction of 2-propenal from refined corn and soybean oils using RSM.

*3.2. Optimization of HS-SPME Using RSM.* The mechanism of HS-SPME is based on the equilibrium distribution of analytes between three phases (fiber coating, headspace, and sample) and, therefore, the vapor pressure of the volatile analytes in the vial can affect significantly the extraction of volatile analytes present in the headspace by SPME. The main SPME variables/factors affecting the vapor pressure and equilibrium state of volatile compounds in the headspace are the extraction temperature, equilibrium time, and extraction time [22]. The drawback of the preliminary experiments described above, in which the effect of one SPME variable/factor (the extraction temperature) was evaluated while the remaining parameters (e.g., equilibrium and extraction time) remained constant, is that the optimization of a single variable is unlikely to produce the most optimal conditions for the extraction of 2-propenal. In contrast, RSM enables the adequate evaluation of the relative influence of all the experimental variables during the optimization process, even in the presence of complex interaction relationships [23]. RSM employs an initial modeling step, followed by the determination of the optimal parameter ranges. The modeling process is performed by adjusting first- or second-order polynomial equations to the data obtained experimentally, followed by the ANOVA.

In this study, the experimental response ($Y$, peak area of the 2-propenal) was modeled as a function of linear, quadratic, and interaction effects of the extraction temperature ($X_1$), equilibrium time ($X_2$), and extraction time ($X_3$). The results show that the predicted quadratic polynomial models provide a good fit to the experimental data, exhibiting relatively high values for the coefficient of determination ($R^2$) of 0.83 and 0.98 for corn and soybean oils, respectively (Table 2). In general, the closer the $R^2$ value is to unity, the better the fit between the theoretical model and experimental data is. The results of the ANOVA show that both the linear and quadratic parameters were not significant ($p > 0.05$) in corn oil (Table 3). Similarly, interactions between the parameters, except for the interaction between the extraction temperature and equilibrium time, also did not produce a significant effect ($p > 0.05$). Thus, the interplay between the

TABLE 2: Predicted quadratic polynomial models for the responses (peak area of 2-propenal) in refined oil samples analyzed in this study.

| Samples | Quadratic polynomial equation | $R^2$ |
|---|---|---|
| Corn oil | $Y = -11810 + 4138X_1 + 1023X_2 + 5927X_3 + 14.6X_1^2 + 42.1X_2^2 + 22X_3^2 - 86.0X_1X_2 + 131X_1X_3 - 55X_2X_3$ | 0.83 |
| Soybean oil | $Y = 614083 + 212110X_1 - 13998X_2 + 154996X_3 + 59306X_1^2 - 52701X_2^2 + 20003X_3^2 - 15609X_1X_2 + 47040X_1X_3 - 7266X_2X_3$ | 0.98 |

*Y*: peak area of 2-propenal (a.u).

TABLE 3: The significance probability ($p$ value and $F$-value) of regression coefficients and lack of fit in the final reduced models.

| Samples | | Linear effects | | | Quadratic effects | | | Interaction effects | | | Blocks | Regression | Lack of fit |
|---|---|---|---|---|---|---|---|---|---|---|---|---|---|
| | | $X_1$ | $X_2$ | $X_3$ | $(X_1)^2$ | $(X_2)^2$ | $(X_3)^2$ | $X_1X_2$ | $X_1X_3$ | $X_2X_3$ | | | |
| Corn oil | $p$ value[a] | 0.486 | 0.828 | 0.676 | 0.836 | 0.551 | 0.972 | 0.043 | 0.285 | 0.650 | 0.953 | 0.000 | 0.381 |
| | $F$-value | 0.50 | 0.05 | 0.18 | 0.04 | 0.37 | 0.00 | 4.65 | 1.20 | 0.21 | 0.00 | 10.29 | 2.03 |
| Soybean oil | $p$ value[a] | 0.000 | 0.080 | 0.000 | 0.001 | 0.002 | 0.188 | 0.080 | 0.000 | 0.403 | 0.069 | 0.000 | 0.713 |
| | $F$-value | 772.33 | 3.36 | 412.40 | 16.18 | 12.77 | 1.84 | 3.35 | 30.39 | 0.73 | 3.64 | 125.17 | 0.74 |

[a]Significance level was determined at $p < 0.05$.

TABLE 4: Optimum levels for the experimental factors and predicted versus observed 2-propenal peak area.

| Samples | Optimum conditions | | | D-optimal | Peak area of 2-propenal (a.u)[a] | |
|---|---|---|---|---|---|---|
| | $X_1$ (°C) | $X_2$ (min) | $X_3$ (min) | | Predicted | Observed |
| Corn oil | 55 | 15 | 15 | 0.9814 | 403478 | 416882 |
| Soybean oil | 55 | 25 | 15 | 1.000 | 1113987 | 1008313 |

[a]No significant difference between predicted and observed 2-propenal peak area verified by two-sample $t$-test, $p < 0.05$ ($n = 3$).

TABLE 5: Concentrations of 2-propenal in corn, soybean, sesame, and adulterated sesame oils.

| Oil adulterants[a] | 2-Propenal ($\mu g\,g^{-1}$ oil) | | | | |
|---|---|---|---|---|---|
| | 100 : 0[b] | 10 : 90 | 5 : 95 | 1 : 99 | 0 : 100 |
| CO 1 : SSO | 1.302 ± 0.085[c] | 0.115 ± 0.004 | 0.084 ± 0.005 | 0.062 ± 0.003 | ND[d] |
| CO 2 : SSO | 0.304 ± 0.021 | 0.037 ± 0.002 | 0.025 ± 0.001 | ND | ND |
| CO 3 : SSO | 0.578 ± 0.024 | 0.075 ± 0.008 | 0.056 ± 0.007 | 0.031 ± 0.002 | ND |
| SBO 1 : SSO | 3.557 ± 0.040 | 0.085 ± 0.007 | 0.057 ± 0.003 | 0.018 ± 0.004 | ND |
| SBO 2 : SSO | 2.630 ± 0.030 | 0.073 ± 0.006 | 0.035 ± 0.007 | ND | ND |
| SBO 3 : SSO | 0.609 ± 0.030 | 0.025 ± 0.003 | 0.008 ± 0.001 | ND | ND |

[a]Oil adulterants consisted of three different brands of corn or soybean oils; each oil adulterant was analyzed quantitatively in triplicate. [b]Corn or soybean oil/sesame oil (w/w). [c]All data are presented as mean ± standard deviation ($n = 3$); [d]not detected; CO: corn oil; SBO: soybean oil; SSO: sesame oil.

extraction temperature and equilibrium time was established as the primary determining term that may cause significant ($p < 0.05$) changes in the observed response (peak area of 2-propenal). For soybean oil, on the other hand, both the linear (extraction temperature and extraction time) and quadratic (extraction temperature and equilibrium time) parameters were found to play a significant role ($p < 0.05$) influencing the total peak area of the 2-propenal. Further, the interaction between the extraction temperature and extraction time was also found to be significant ($p < 0.05$). The predicted models provided an excellent fit to the experimental data for both oil samples examined in this study, with a regression of $p < 0.05$ and an insignificant ($p > 0.05$) lack of fit. In addition, no significant differences ($p > 0.05$) were found between the two experimental blocks, demonstrating that there were no significant differences between measurements obtained at different experimental periods.

Table 4 shows the optimum values for the independent HS-SPME variables/factors obtained using the "response optimizer" option available in the Minitab statistical package, which afforded composite desirability values of 0.98 and 1.00 for corn and soybean oils, respectively. In order to validate the proposed model, triplicate experiments were conducted under the optimum experimental conditions and the results were compared to those obtained by predictions. The validation results revealed that the experimental values of the 2-propenal peak area were not significantly different ($p < 0.05$) from the predicted values, thereby confirming the validity and adequacy of the predictive models. The optimum HS-SPME extraction temperature, equilibrium time, and extraction time for corn and soybean oils were determined to be 55°C, 15 min, 15 min, and 55°C, 25 min, 15 min, respectively.

*3.3. Application of 2-Propenal as a Detection Marker of Sesame Oil Adulteration.* To date, there is no consensus in the literature as to the origin of 2-propenal in edible oils. In this study, we confirmed that 2-propenal was not found in the tested unrefined sesame oil samples equilibrated and extracted at 55°C (Figure 1(a)), while it was detected in all the tested refined corn and soybean oils even at room temperature (Figure 2). In a previous study, 2-propenal was produced via thermal oxidation of vegetable oils at temperatures above 100°C [18]. This finding may be attributed to the thermal oxidation of refined corn and soybean oils during industrial refining stages because of the high temperatures (≥200°C) used. As shown in Supp. Figure 1 in Supplementary Material, thermal oxidation of unrefined sesame oil at 150°C resulted in the formation of 2-propenal. However, pressing sesame seeds without refining is a common process for the production of sesame oil in Asia. A possible reason is that the process for unrefined sesame oil production does not involve a refining stage using the high temperatures; therefore, 2-propenal may not be formed during sesame oil production.

The formation of lipid oxidation products including 2-propenal in refined corn and soybean oils may also be due to the lack of considerable amounts of antioxidants. The refining process for edible oil results in a huge loss of antioxidants such as $\alpha$-tocopherol due to the high operating temperatures [24]. Suja et al. [25] have reported that the addition of antioxidants in edible oils such as soybean and sunflower oils is effective in the inhibition of the lipid oxidation. Unrefined sesame oil, on the other hand, possesses high contents of endogenous antioxidants such as sesamolin, sesamin, sesamol, and phenolic compounds [26, 27]. In addition, sesame seed roasting prior to oil extraction can improve significantly the oxidative stability of the obtained oils as a result of the antioxidants produced via Maillard reactions. The oxidative stability is known to be positively correlated with the increasing roasting temperature of sesame seeds [28]. The natural antioxidants in unrefined sesame oil are likely able to inhibit the lipid oxidation, preventing the formation of 2-propenal.

In this study, unrefined sesame oil samples were randomly adulterated with corn or soybean oil samples at concentrations of 1, 5, and 10% (w/w). The authentic and adulterated sesame oils were extracted using HS-SPME under the optimum conditions described in Table 4 and analyzed by GC-TOF/MS. The concentration of 2-propenal in the adulterated sesame oils was then determined in order to investigate the applicability of 2-propenal as a marker for the detection of sesame oil adulteration. The results show that the concentration of 2-propenal was found to increase gradually with the increasing content of added corn or soybean oils (Table 5). 2-Propenal in sesame oil samples adulterated with refined corn or soybean oils was detected even at the lowest concentration of 1%. According to the results of the admixtures presented in Table 5, 2-propenal can be used as a marker for the discrimination of unrefined sesame oil adulteration with the refined oils tested in this work.

The analytical parameters for the evaluation of the proposed method are shown in Table 6. The linear range

Table 6: Linear regression equation, correlation coefficient, linear range, limits of detection, and quantification of the developed methods for 2-propenal extraction.

| Extraction conditions[a] | Linear regression equation | $R^2$ | Linear range ($\mu$g mL$^{-1}$) | LOD ($\mu$g mL$^{-1}$) | LOQ ($\mu$g mL$^{-1}$) |
|---|---|---|---|---|---|
| Corn oil ($X_1$: 55°C, $X_2$: 15 min, $X_3$: 15 min) | $Y = 220777x - 1468.90$ | 0.9999 | 0.02–0.76 | 0.010 | 0.031 |
| Soybean oil ($X_1$: 55°C, $X_2$: 25 min, $X_3$: 15 min) | $Y = 225128x - 566.02$ | 0.9999 | 0.02–0.76 | 0.008 | 0.023 |

[a]2-Propenal standard was extracted using a CAR/PDMS fiber under optimal conditions for corn or soybean oil samples; LOD: limits of detection; LOQ: limits of quantification.

covers from 0.02 to 0.76 $\mu$g mL$^{-1}$ for the 2-propenal authentic standard extracted using HS-SPME under optimal conditions for corn or soybean oil samples, and all the correlation coefficients were >0.99. The LOD and LOQ were found in the range of 0.008–0.010 and 0.023–0.031 $\mu$g mL$^{-1}$, respectively. Overall, the results confirm that the developed method can adequately detect the presence of refined corn and soybean oils at concentrations between 1 and 5% (w/w) in adulterated sesame oil.

## 4. Conclusion

The present work demonstrates an optimized HS-SPME method coupled to GC-TOF/MS for the effective extraction and detection of 2-propenal in refined corn and soybean oils. The application of 2-propenal as a marker allowed the detection of unrefined sesame oil adulteration with refined corn or soybean oil even at very low adulteration levels of 1–5% (w/w). In future work, this developed method will be further tested for the detection of unrefined sesame oil adulteration with other potential oil adulterants such as refined canola and rapeseed oils.

## Disclosure

Jaeho Ha's present address is World Institute of Kimchi, Gwangju 61755, Republic of Korea.

## Conflicts of Interest

The authors declare that there are no conflicts of interest associated with this publication.

## Authors' Contributions

Ahmad Rois Mansur and Tae Gyu Nam contributed equally to this work.

## Acknowledgments

This work was supported by Main Research Program (E0152202-02, 2016) of Korea Food Research Institute funded by the Ministry of Science, ICT & Future Planning, Republic of Korea. The authors are grateful to Mr. Byeong Ho Kim (LECO Korea CO., LTD) for his expert technical assistance.

## References

[1] J. Kim, G. Jin, Y. Lee, H. S. Chun, S. Ahn, and B. H. Kim, "Combined analysis of stable isotope, 1H NMR, and fatty acid to verify sesame oil authenticity," *Journal of Agricultural and Food Chemistry*, vol. 63, no. 40, pp. 8955–8965, 2015.

[2] W.-H. Wu, "The contents of lignans in commercial sesame oils of Taiwan and their changes during heating," *Food Chemistry*, vol. 104, no. 1, pp. 341–344, 2007.

[3] Food and Agriculture Organization of the United Nations, "Crops and livestock products," http://faostat.fao.org/beta/en/#data/TP, 2016.

[4] H.-Y. Seo, J. Ha, D.-B. Shin et al., "Detection of corn oil in adulterated sesame oil by chromatography and carbon isotope analysis," *Journal of the American Oil Chemists' Society*, vol. 87, no. 6, pp. 621–626, 2010.

[5] Y. W. Park, P.-S. Chang, and J. Lee, "Application of triacylglycerol and fatty acid analyses to discriminate blended sesame oil with soybean oil," *Food Chemistry*, vol. 123, no. 2, pp. 377–383, 2010.

[6] L. Xu, Z.-H. Ye, S.-M. Yan et al., "Combining local wavelength information and ensemble learning to enhance the specificity of class modeling techniques: identification of food geographical origins and adulteration," *Analytica Chimica Acta*, vol. 754, pp. 31–38, 2012.

[7] D. Peng, Y. Bi, X. Ren, G. Yang, S. Sun, and X. Wang, "Detection and quantification of adulteration of sesame oils with vegetable oils using gas chromatography and multivariate data analysis," *Food Chemistry*, vol. 188, pp. 415–421, 2015.

[8] F.-F. Ai, J. Bin, Z.-M. Zhang et al., "Application of random forests to select premium quality vegetable oils by their fatty acid composition," *Food Chemistry*, vol. 143, pp. 472–478, 2014.

[9] L. Zhang, P. Li, X. Sun et al., "Classification and adulteration detection of vegetable oils based on fatty acid profiles," *Journal of Agricultural and Food Chemistry*, vol. 62, no. 34, pp. 8745–8751, 2014.

[10] D.-S. Lee, E.-S. Lee, H.-J. Kim, S.-O. Kim, and K. Kim, "Reversed phase liquid chromatographic determination of triacylglycerol composition in sesame oils and the chemometric detection of adulteration," *Analytica Chimica Acta*, vol. 429, no. 2, pp. 321–330, 2001.

[11] W.-J. Lee, N.-W. Su, M.-H. Lee, and J.-T. Lin, "Assessment of authenticity of sesame oil by modified Villavecchia Test and HPLC-ELSD analysis of triacylglycerol profile," *Food Research International*, vol. 53, no. 1, pp. 195–202, 2013.

[12] S. Aued-Pimentel, E. Takemoto, R. Antoniassi, and E. S. G. Badolato, "Composition of tocopherols in sesame seed oil: An indicative of adulteration," *Grasas y Aceites*, vol. 57, no. 2, pp. 205–210, 2006.

[13] Z. Hai and J. Wang, "Detection of adulteration in camellia seed oil and sesame oil using an electronic nose," *European Journal of Lipid Science and Technology*, vol. 108, no. 2, pp. 116–124, 2006.

[14] J. He, W. Xu, Y. Shang et al., "Development and optimization of an efficient method to detect the authenticity of edible oils," *Food Control*, vol. 31, no. 1, pp. 71–79, 2013.

[15] C. Ma, J. Ji, C. Tan et al., "Headspace solid-phase microextraction coupled to gas chromatography for the analysis of aldehydes in edible oils," *Talanta*, vol. 120, pp. 94–99, 2014.

[16] F. Zhao, J. Liu, X. Wang, P. Li, W. Zhang, and Q. Zhang, "Detection of adulteration of sesame and peanut oils via volatiles by GC×GC-TOF/MS coupled with principal components analysis and cluster analysis," *European Journal of Lipid Science and Technology*, vol. 115, no. 3, pp. 337–347, 2013.

[17] J. F. Stevens and C. S. Maier, "Acrolein: Sources, metabolism, and biomolecular interactions relevant to human health and disease," *Molecular Nutrition & Food Research*, vol. 52, no. 1, pp. 7–25, 2008.

[18] A. Ewert, M. Granvogl, and P. Schieberle, "Isotope-labeling studies on the formation pathway of acrolein during heat processing of oils," *Journal of Agricultural and Food Chemistry*, vol. 62, no. 33, pp. 8524–8529, 2014.

[19] A. Shibata, M. Uemura, M. Hosokawa, and K. Miyashita, "Formation of acrolein in the autoxidation of triacylglycerols with different fatty acid compositions," *Journal of the American Oil Chemists' Society*, vol. 92, no. 11-12, pp. 1661–1670, 2015.

[20] M. Kächele, Y. B. Monakhova, T. Kuballa, and D. W. Lachenmeier, "NMR investigation of acrolein stability in hydroalcoholic solution as a foundation for the valid HS-SPME/GC-MS quantification of the unsaturated aldehyde in beverages," *Analytica Chimica Acta*, vol. 820, pp. 112–118, 2014.

[21] G. Ouyang, D. Vuckovic, and J. Pawliszyn, "Nondestructive sampling of living systems using in vivo solid-phase microextraction," *Chemical Reviews*, vol. 111, no. 4, pp. 2784–2814, 2011.

[22] A. Spietelun, A. Kloskowski, W. Chrzanowski, and J. Namieśnik, "Understanding solid-phase microextraction: Key factors influencing the extraction process and trends in improving the technique," *Chemical Reviews*, vol. 113, no. 3, pp. 1667–1685, 2013.

[23] M. A. Bezerra, R. E. Santelli, E. P. Oliveira, L. S. Villar, and L. A. Escaleira, "Response surface methodology (RSM) as a tool for optimization in analytical chemistry," *Talanta*, vol. 76, no. 5, pp. 965–977, 2008.

[24] J. Wei, L. Chen, X. Qiu et al., "Optimizing refining temperatures to reduce the loss of essential fatty acids and bioactive compounds in tea seed oil," *Food and Bioproducts Processing*, vol. 94, pp. 136–146, 2015.

[25] K. P. Suja, J. T. Abraham, S. N. Thamizh, A. Jayalekshmy, and C. Arumughan, "Antioxidant efficacy of sesame cake extract in vegetable oil protection," *Food Chemistry*, vol. 84, no. 3, pp. 393–400, 2004.

[26] Z. Konsoula and M. Liakopoulou-Kyriakides, "Effect of endogenous antioxidants of sesame seeds and sesame oil to the thermal stability of edible vegetable oils," *LWT- Food Science and Technology*, vol. 43, no. 9, pp. 1379–1386, 2010.

[27] S. W. Lee, M. K. Jeung, M. H. Park, S. Y. Lee, and J. Lee, "Effects of roasting conditions of sesame seeds on the oxidative stability of pressed oil during thermal oxidation," *Food Chemistry*, vol. 118, no. 3, pp. 681–685, 2010.

[28] E. Choe and D. B. Min, "Mechanisms and factors for edible oil oxidation," *Comprehensive Reviews in Food Science and Food Safety*, vol. 5, no. 4, pp. 169–186, 2006.

21

# Hetarylazopyrazolone Dyes Based on Benzothiazole and Benzimidazole Ring Systems: Synthesis, Spectroscopic Investigation, and Computational Study

**Ebru Aktan and Tahsin Uyar**

*Faculty of Science, Department of Chemistry, Gazi University, Teknikokullar, 06500 Ankara, Turkey*

Correspondence should be addressed to Ebru Aktan; ebruaktan@gazi.edu.tr

Academic Editor: James W. Gauld

In this study, the synthesized coupling component 1-(2-benzothiazolyl)-3-methylpyrazol-5-one reacted with diazotised heterocyclic amines to afford six novel hetarylazopyrazolone dyes. These azo dyes based on benzothiazole and benzimidazole ring systems were characterized by spectral methods and elemental analyses. The solvatochromic behaviors of these dyes in various solvents were evaluated. The ground state geometries of the dyes were optimized using density functional theory (DFT). Solvent, acid-base, and substituent influences on the wavelength of the maximum absorption were examined in detail. Time-dependent density functional theory (TD-DFT) calculations were performed to obtain the absorption spectra of the dyes in various solvents and the results compared with experimental values. Besides, frontier molecular orbitals (FMO) analysis for the dyes is also described from the computational process.

## 1. Introduction

The ease of availability of raw materials and ease of preparation [1], wide range of uses as dyestuff [2, 3], advanced applications such as liquid crystal displays (LCDs) [4], nonlinear optical (NLO) devices [5], dye-sensitized solar cells (DSSCs) [6], optoelectronic systems [7, 8], and wide application areas in biological and medical studies [9] are all desirable characteristics of azo dyes.

Various experimental and theoretical studies on the structure, tautomeric behavior, dyeing properties, and vibrational analysis of azo dyes [10–19] and their metal complexes [20–22] have been done. Due to their chemical significance and spectroscopic properties, hetarylazo dyes were studied extensively by spectroscopic and theoretical methods. Their higher tinctorial strength and brighter dyeing make azo dyes based on heterocyclic amines superior than aniline based azo dyes [23–28]. Some hetarylazopyrazolone dyes have not only potential use as dye stuff, but also other potential nontextile applications such as optical materials [29].

Many studies via theoretical calculations have been performed for determination of the structural, spectroscopic properties, and tautomeric behaviors of azo dyes [30, 31]. Density functional theory (DFT) and time-dependent density functional theory (TD-DFT) have been one of the most widely used methods in computational chemistry [32–37].

We have previously reported the synthesis, characterization, tautomeric behavior, and solvent effects on their absorption spectra of some hetarylazopyrazolone dyes with thiazole moiety together with their theoretical studies [38]. In continuation of this study, we report here the synthesis and characterization of a series of new monoazo dyes based on the benzothiazole and benzimidazole ring systems. Optimization calculations of the dyes in the ground state were done at the DFT B3LYP/6-31G(d,p) level. Absorption spectra based on the DFT-optimized structures were obtained by the time-dependent density functional theory (TD-DFT/B3LYP) level with the 6-31G(d,p) basis set.

## 2. Experimental

*2.1. General.* The reagents of analytical grade were purchased from commercial sources and used without any further purification. $^1$H NMR spectra were measured with a Bruker

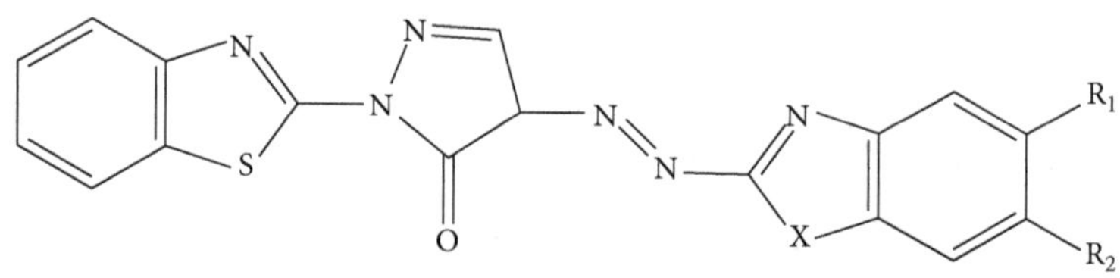

3a–3f

| Dye | Substituents | |
|---|---|---|
| 3a | X = S | $R_1 = R_2 = H$ |
| 3b | X = S | $R_1 = H$ $R_2 = Cl$ |
| 3c | X = S | $R_1 = H$ $R_2 = OCH_3$ |
| 3d | X = S | $R_1 = R_2 = CH_3$ |
| 3e | X = NH | $R_1 = R_2 = H$ |
| 3f | X = NH | $R_1 = R_2 = CH_3$ |

FIGURE 1: The general route for the synthesis of dyes 3a–3f.

TABLE 1: Characterization data and elemental analysis of dyes 3a–3f.

| Dye number | Molecular formula (mol.wt.) | Color[a] | Cryst. solvent | Yield[b] (%) | Melting point (°C) | C% Calc. | C% Found | H% Calc. | H% Found | N% Calc. | N% Found | S% Calc. | S% Found |
|---|---|---|---|---|---|---|---|---|---|---|---|---|---|
| 3a | $C_{18}H_{12}N_6OS_2$ (392.5) | Orange | Acetic acid | 87 | 255–257 | 55.09 | 44.90 | 3.08 | 3.11 | 21.41 | 20.94 | 16.34 | 16.09 |
| 3b | $C_{18}H_{11}ClN_6OS_2$ (426.9) | Orange | Acetic acid | 95 | 265–267 | 50.64 | 50.27 | 2.60 | 2.77 | 19.69 | 19.92 | 15.02 | 14.97 |
| 3c | $C_{19}H_{14}N_6O_2S_2$ (422.5) | Dark red | Ethanol[c] | 56 | 248–250 | 54.01 | 53.96 | 3.34 | 3.34 | 19.89 | 19.53 | 15.18 | 15.03 |
| 3d | $C_{20}H_{16}N_6OS_2$ (420.5) | Orange | Ethanol | 94 | 270-271 | 57.12 | 56.73 | 3.84 | 3.56 | 19.99 | 19.42 | 15.25 | 14.82 |
| 3e | $C_{18}H_{13}N_7OS$ (375.4) | Orange | Acetic acid/water | 76 | 294-295 | 57.59 | 57.20 | 3.49 | 3.52 | 26.12 | 25.88 | 8.54 | 8.11 |
| 3f | $C_{20}H_{17}N_7OS$ (403.5) | Orange | DMF/water | 81 | 303-304 | 59.54 | 58.93 | 4.25 | 4.33 | 24.3 | 23.77 | 7.95 | 7.64 |

[a]Powder.
[b]Yield of crude product.
[c]Washed with ethanol.

DPX-400 MHz spectrometer at room temperature in DMSO-$d_6$ with tetramethylsilane as the internal reference. UV-visible absorption spectra were recorded on Analytik Jena Specord 200 double-beam spectrophotometer, using dimethyl sulfoxide (DMSO), dimethylformamide (DMF), acetonitrile, methanol, acetic acid, and chloroform as solvent. Infrared spectra were taken in KBr on a Mattson 1000 spectrophotometer. Mass spectra were measured with a Micromass UK Platform II LC-MS spectrometer. Elemental analyses were performed with a LECO-CHNS-9320 instrument. Melting points were obtained with a Gallenkamp capillary melting apparatus and they are uncorrected.

*2.2. Synthesis.* The methods used for the syntheses of 2-hydrazinobenzothiazole (1) and coupling component 1-(2-benzothiazolyl)-3-methylpyrazol-5-one (2) were described in the previous part of our study [38].

*2.2.1. Synthesis of Hetarylazopyrazolone Dyes (3a–3f).* Diazotisation of various amines was affected with nitrosylsulfuric acid. General procedure was described below for 1-(2-benzothiazolyl)-3-methyl-4-(2-benzothiazolylazo)pyrazol-5-one and the other dyes were synthesized in a similar way. The characteristic properties and elemental analysis of the dyes are shown in Table 1. The synthesized dyes (3a–3f) are shown in Figure 1.

*2.2.2. Synthesis of 1-(2-Benzothiazolyl)-3-methyl-4-(2-benzothiazolylazo)pyrazol-5-one (1).* 2-Aminobenzothiazole (0.002 mol) was dissolved in icy acetic acid-propionic acid mixture (3 mL/2 mL) and was cooled to −5°C in ice-salt bath. Nitrosylsulfuric acid, prepared with dissolving sodium nitrite (0.002 mol, 0.14 g) in sulfuric acid (4 mL), was added to the heterocyclic amine dropwise in 30 min at −5°C. The mixture was stirred in cold for additional 2 hours. Urea was added to the mixture in order to decompose the unreacted nitrous acid. 1-(2-Benzothiazolyl)-3-methylpyrazol-5-one (0.002 mol, 0.46 g) was dissolved in KOH solution (10 mL 0.2 M) and cooled. The prepared coupling compound solution was added to the diazonium solution in half an hour. The mixture was stirred for additional 2 hours at 0–5°C and $Na_2CO_3$ solution was added till the mixture had a value of pH 5-6. The mixture was stirred for 30 min. Water was added to precipitate the product. The product was filtered, washed with water, and air-dried. Recrystallization was performed in ethanol to obtain the pure orange compound (Yield: 87% (m.p. 255–257°C)).

*2.3. Computational Details.* All calculations were carried out with Gaussian09 program [39]. The geometries of the dyes were fully optimized and calculated by B3LYP method and 6-31+G(d) basis set without any symmetry constraints. Vibrational frequency analysis confirms that all the optimized geometries correspond to minima on the potential energy surface. Time-dependent density functional theory/integral equation formalism polarisable continuum model (TD-DFT/IEFPCM) method has been used to obtain electronic absorption spectra of the dyes [40]. The (default) nonequilibrium procedure was selected for the TD-DFT calculations [41].

## 3. Results and Discussion

*3.1. Molecular Structure.* Azo dyes can exist in four possible tautomeric forms, namely, the keto-azo forms A and D, keto-hydrazo form B, and enol-azo form C as shown in Figure 2.

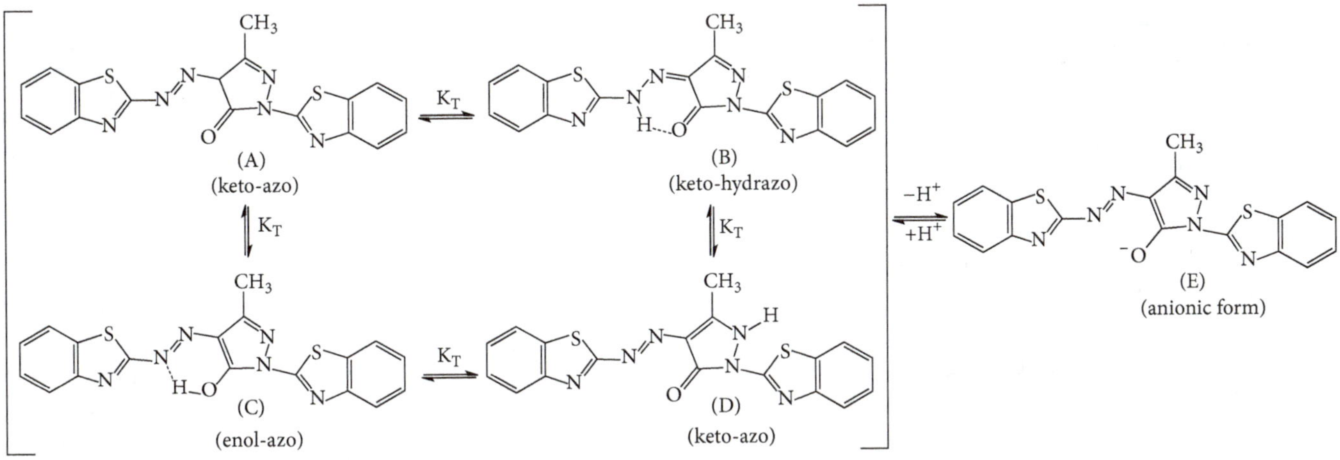


FIGURE 2: Tautomeric equilibrium of dye 3a.

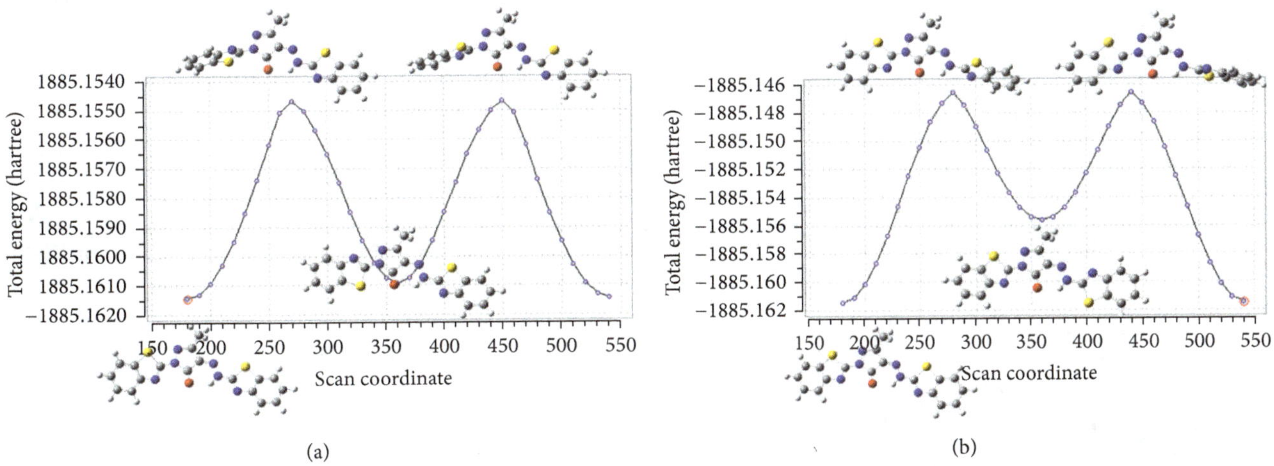


FIGURE 3: (a) Scanning graph of the dye 3a on dihedral angle (S24–C7–N22–C13). (b) Scanning graph of the dye 3a on dihedral angle (N26–C16–N19–N27).

The deprotonation of these four tautomers lead to common anion form E.

$^1$H NMR spectrum of dye 3a exhibits a broad peak at 14.94 ppm, which was attributed to NH of the hydrazo tautomer. This peak was observed at 15.00 ppm for dye 3b. These results suggest that dyes 3a and 3b are present as keto-hydrazo form in DMSO-$d_6$. The absence of hydroxyl (OH), imino (NH), and hydrazo (NH) peaks in the $^1$H NMR spectra suggests that dyes 3c, 3d, 3e, and 3f exist in keto-azo (A) form or in anionic form (E). The $^1$H NMR spectra are measured in DMSO-$d_6$ for the dyes except dye 3c, which is recorded in $CDCl_3$. Dye 3b showed a singlet peak at 2.30 ppm for methyl group at pyrazolone ring. The $^1$H NMR spectra for the dyes 3a–3f showed a singlet peak from 2.30 to 2.42 ppm for pyrazolone $CH_3$. Aromatic protons of the dyes 3a–3f appear at 7.00–8.14 ppm. Dye 3c showed a singlet peak at 3.90 ppm for $OCH_3$ proton at benzothiazole ring and a peak at 2.40 ppm for methyl group at pyrazolone ring. Dye 3d showed a singlet peak at 2.31 ppm for methyl groups at benzothiazole ring and two doubled peak at 7.79 ppm and 7.99 ppm for aromatic protons at benzothiazole ring. $^1$H NMR spectra of dyes 3e and 3f exhibit a broad peak at 13.30 and 13.24 ppm, respectively, which were due to NH benzimidazole proton.

FT-IR spectra of dyes 3a–3f showed a band at 1638–1684 $cm^{-1}$ for carbonyl group and a band at 3390–3445 $cm^{-1}$ for NH group. These results suggest that these dyes are predominantly in keto-hydrazo (B) form in solid state. The imidazole NH peaks of the dyes 3e and 3f were recorded at 3194 $cm^{-1}$ and 3147 $cm^{-1}$. The other $\nu_{max}$ values of 3021–3070 $cm^{-1}$ (aromatic C–H) and 2915–2994 $cm^{-1}$ (aliphatic C–H) were also recorded.

The molecular ion $[M]^+$ peaks of the dyes are observed at the expected $m/z$ values as the main peak. Dyes give $[C_{11}H_8N_5OS]^+$ ion peak at $m/z = 258$ except dye 3c with relative intensities 10–33%. The ion peak $[C_4N_2H_3]^+$ is observed at 79 (10%) for dye 3c. $^1$H NMR, FT-IR, and mass spectra results of the dyes are shown in Table 2.

A detailed potential energy scan was performed on dihedral angles (S24–C7–N22–C13) (Figure 3(a)) and (N26–C16–N19–N27) (Figure 3(b)) in the range of 0°–360° at HF/6-31G(d) level by 10° intervals to reveal all possible conformations of the dye 3a.

TABLE 2: Spectral data for dyes 3a–3f.

| Dye number | $^1$H-NMR[a] (ppm, DMSO-$d_6$) | FT-IR ($cm^{-1}$, KBr) | MS (m/e) |
|---|---|---|---|
| 3a | 2.30 (s, 3H, pyr $CH_3$)<br>7.35 (m, 2H, bth)<br>7.48 (m, 2H, bth)<br>7.59 (d, 1H, bth)<br>7.87 (t, 1H, bth)<br>7.94 (d, 1H, bth)<br>8.05 (t, 1H, bth)<br>14.94 (b, NH) | 3245 (N–H), 3070 (Ar C–H),<br>2915 (Al C–H), 1682 (C=O)<br>1595, 1523 (C=C, C=N) | 393 $[M]^+$ (%100)<br>258 (%33) |
| 3b | 2.30 (s, 3H, pyr $CH_3$)<br>7.33–8.14 (m, 7H, Ar CH)<br>15.00 (b, NH) | 3212 (N–H), 3062 (Ar C–H),<br>2994, 2978 (Al C–H)<br>1684 (C=O), 1597, 1523 (C=C, C=N) | 427 $[M]^+$ (%100)<br>258 (%31) |
| 3c* | 2.40 (s, 3H, pyr $CH_3$)<br>3.90 (s, 3H, bth $OCH_3$)<br>7.00–8.10 (m, 7H, Ar CH) | 3453 (N–H), 3070 (Ar C–H), 2940 (Al C–H)<br>1670 (C=O), 1606, 1529 (C=C, C=N) | 423 $[M]^+$ (%100)<br>79 (%10) |
| 3d | 2.31 (s, 6H, dimetbth 2x$CH_3$)<br>2.47 (s, 3H, pyr $CH_3$)<br>7.42 (t, 1H, bth)<br>7.52 (s, 1H, dimetbth)<br>7.61 (s, 1H, dimetbth)<br>7.79 (d, 1H, bth) | 3420 (N–H), 3060 (Ar C–H),<br>2968, 2919 (Al C–H), 1638 (C=O)<br>1557, 1519 (C=C, C=N) | 421 $[M]^+$ (%100)<br>258 (%16) |
| 3e | 2.41 (s, 3H, pyr $CH_3$)<br>7.21–7.97 (m, 6H, Ar CH)<br>13.30 (b, bimi NH) | 3416 (N–H), 3194 (benzimidazole N–H),<br>3057 (Ar C–H), 2935 (Al C–H),<br>1658 (C=O), 1621, 1570 (C=C, C=N) | 376 $[M]^+$ (%100)<br>258 (%43) |
| 3f | 2.33 (s, 6H, dimetbimi-2x$CH_3$)<br>2.42 (s, 3H, pyr $CH_3$)<br>7.34–8.14 (m, 6H, Ar CH)<br>13.23 (b, dimethylbenzimidazole NH) | 3390 (N–H), 3147 (benzimidazole N–H),<br>3021 (Ar C–H)<br>2967 (Al C–H), 1669 (C=O),<br>1616, 1529 (C=C, C=N) | 404 $[M]^+$ (%100)<br>258 (%10) |

[a]s, singlet; d, doublet; t, triplet; m, multiplet; b, broad.
pyr: pyrazolone, bth: benzothiazole, and dimetbth: dimethylbenzothiazole.
bimi: benzimidazole; dimetbimi: dimethylbenzimidazole.
*$^1$H-NMR in $CHCl_3$.

Then, final geometry optimizations were performed within the framework of DFT/B3LYP at the level of 6-31+G(d) without any geometrical restriction. The most stable conformation of the keto-hydrazo tautomer is shown in Figure 4 with atomic numbering.

*3.2. Effect of Solvent.* The UV-Vis absorption spectra of the dyes 3a–3f were recorded over the range of $\lambda$ between 300 and 700 nm and the results are given in Table 3.

$\lambda_{max}$ values of dye 3a, dye 3b, and dye 3c exhibit bathochromic shift in methanol, DMF, and DMSO. These dyes show only one maximum and also only one tautomeric form in all solvents. The isosbestic points are around 442 nm, 435 nm, and 465 nm, respectively. Absorption spectra of dye 3b in various solvents are given in Figure 5. It was determined that the absorption maxima of the dye 3d in methanol, DMF, and DMSO show no significant change, despite blue shift in chloroform and acetic acid. The difference between the experimental absorption maxima of dyes in the most apolar solvent chloroform and in the most polar solvent DMSO we used lies between 5 and 50 nm while the calculated absorption maxima of each dye do not change significantly with solvent polarity (max. 4 nm). The difference between the experimental and calculated absorption maxima lies between 0 and 36 nm. The absorption maxima of dye 3e and dye 3f vary less with the solvent. All of the absorption bands of the dye 3e and dye 3f show shoulders in a short wavelength, and their symmetries are distorted. The shoulders can be attributed either to the interference of a tautomer over the dominant one or to the transitions $\pi$-$\pi^*$, n-$\pi^*$, or charge transfer transitions.

Maximum absorption values and oscillator strengths of the dyes were calculated by using the TD-DFT/IEFPCM method with 6-31G(d,p) basis set in different solvents and the results are shown in Table 4.

*3.3. Effect of Acid and Base.* Acid-base effects on the absorption spectra of the dyes were investigated. Changes in the maximum absorption wavelength of the dyes with different solvents obtained by addition of the acid-base solution are also provided in Table 5.

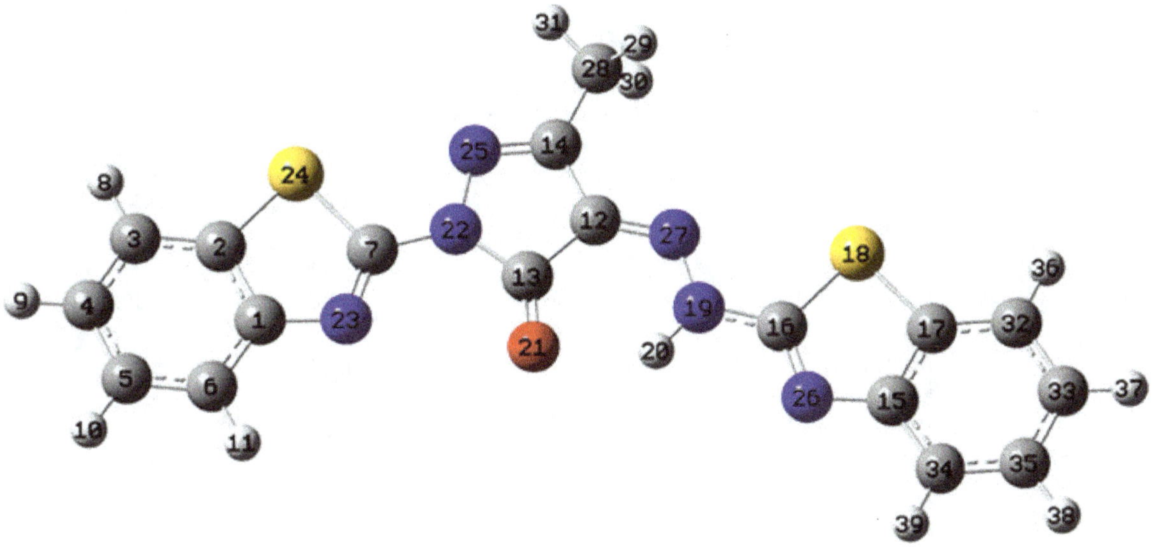

FIGURE 4: Computed optimized ground state geometry with atomic numbering of dye 3a using DFT.

TABLE 3: Influence of solvent on $\lambda_{max}$ (nm) of dyes 3a–3f.

| Dye number | | Chloroform | Acetic acid | Methanol | Acetonitrile | DMF (log $\varepsilon$) | DMSO | $\Delta\lambda_{max}$ [a] |
|---|---|---|---|---|---|---|---|---|
| 3a | Exp. | 419 | 408 | 457 | 411 | 467 (4.443) | 469 | 50 |
| | Calc.[b] | 433 (0.9087) | 432 (0.8890) | 434 (0.8672) | 432 (0.8779) | 434 (0.9028) | 433 (0.8990) | |
| 3b | Exp. | 421 | 409 | 459 | 412 | 469 (4.460) | 469 | 48 |
| | Calc. | 439 (0.9685) | 438 (0.9548) | 438 (0.9502) | 437 (0.9543) | 439 (0.9742) | 439 (0.9715) | |
| 3c | Exp. | 450 | 434 | 467 | 469 | 469 (4.569) | 469 | 19 |
| | Calc. | 488 (0.8218) | 487 (0.8212) | 486 (0.8268) | 488 (0.8240) | 490 (0.8392) | 490 (0.8372) | |
| 3d | Exp. | 450 | 425 | 465 | 462 | 469 (4.774) | 473 | 23 |
| | Calc. | 462 (0.8391) | 461 (0.8168) | 462 (0.7946) | 462 (0.7981) | 464 (0.8060) | 463 (0.8026) | |
| 3e | Exp. | 452 | 397[s], 449 | 401[s], 442 | 400[s], 449 | 453 (4.576) | 461 | 9 |
| | Calc. | 447 (0.9126) | 449 (0.9206) | 446 (0.9126) | 447 (0.9158) | 449 (0.9307) | 449 (0.9286) | |
| 3f | Exp. | 464 | 461 | 460 | 462 | 459 (4.649) | 469 | 5 |
| | Calc. | 480 (0.8510) | 482 (0.8528) | 480 (0.8510) | 481 (0.8544) | 483 (0.8710) | 483 (0.8688) | |

[a] $\lambda_{max}$(DMSO) − $\lambda_{max}$($CHCl_3$); [b] oscillator strengths are given in the parentheses; [s] shoulder.

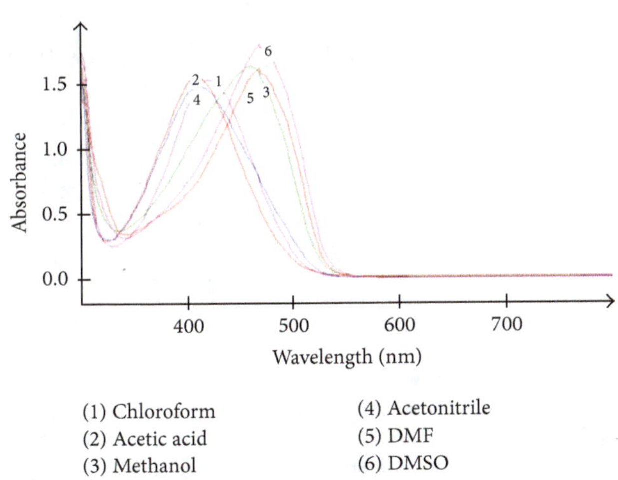


FIGURE 5: Absorption spectra of dye 3b in various solvents.

When acid was added to the methanol solutions of the dyes 3c, 3d, and 3e, the absorption maxima shift to the bathochromic area while the absorption maxima of the dyes 3a and 3b shift to the hypsochromic area. By the addition of KOH to methanol solutions of the dyes hypsochromic shifts were observed for dye 3f and the other dyes did not represent a significant change by their absorption maxima.

The absorption spectra of the dyes in chloroform are sensitive to the addition of piperidine while piperidine addition to DMF and DMSO solutions led to no notable change in the absorption spectra. When piperidine was added to chloroform solution the absorption peaks obtained are similar to the ones measured in DMF and DMSO solutions. There is no significant change in the absorption spectra of the dyes by the addition of trifluoroacetic acid to the chloroform solutions. Absorption spectra of dye 3a in acid and base are given in Figure 6.

TABLE 4: Calculated structural parameters (bond lengths in Å, angles in degree) of the dyes 3a–3f.

| Parameter | Dyes | | | | | |
|---|---|---|---|---|---|---|
| | 3a | 3b | 3c | 3d | 3e | 3f |
| *Bond lengths* | | | | | | |
| C2-S24 | 1.759 | 1.759 | 1.759 | 1.759 | 1.759 | 1.759 |
| S24-C7 | 1.781 | 1.780 | 1.781 | 1.781 | 1.781 | 1.781 |
| C7-N23 | 1.290 | 1.290 | 1.290 | 1.290 | 1.290 | 1.290 |
| C1-N23 | 1.388 | 1.388 | 1.388 | 1.388 | 1.388 | 1.388 |
| C7-N22 | 1.389 | 1.390 | 1.389 | 1.389 | 1.389 | 1.389 |
| N22-N25 | 1.405 | 1.405 | 1.405 | 1.405 | 1.404 | 1.404 |
| C13-O21 | 1.229 | 1.229 | 1.229 | 1.229 | 1.229 | 1.229 |
| C12-N27 | 1.307 | 1.306 | 1.309 | 1.308 | 1.310 | 1.311 |
| N27-N19 | 1.319 | 1.321 | 1.317 | 1.318 | 1.319 | 1.317 |
| N19-C16 | 1.381 | 1.380 | 1.382 | 1.382 | 1.382 | 1.382 |
| C16-S18(N18) | 1.771 | 1.773 | 1.774 | 1.772 | 1.371 | 1.371 |
| C16-N26 | 1.294 | 1.294 | 1.294 | 1.295 | 1.310 | 1.310 |
| S18(N18)-C17 | 1.763 | 1.761 | 1.762 | 1.763 | 1.391 | 1.391 |
| N26-C15 | 1.386 | 1.385 | 1.385 | 1.386 | 1.389 | 1.389 |
| *Bond angles* | | | | | | |
| S24-C7-N23 | 117.5 | 117.5 | 117.5 | 117.5 | 117.5 | 117.5 |
| N23-C7-N22 | 124.4 | 124.3 | 118.1 | 124.4 | 118.1 | 118.1 |
| C7-N22-N25 | 118.0 | 118.0 | 118.0 | 118.0 | 118.0 | 118.0 |
| N25-C14-C12 | 110.4 | 110.3 | 110.4 | 110.4 | 110.4 | 110.5 |
| N22-C13-O21 | 129.8 | 129.8 | 129.7 | 129.7 | 129.7 | 127.6 |
| C13-C12-N27 | 127.2 | 127.1 | 127.1 | 127.2 | 127.0 | 127.0 |
| N27-N19-C16 | 120.3 | 120.2 | 120.4 | 120.4 | 119.9 | 120.0 |
| N19-C16-S18 | 120.8 | 120.6 | 120.8 | 120.8 | 121.9 | 121.9 |
| C16-N26-C15 | 110.1 | 110.1 | 110.1 | 110.0 | 104.1 | 104.0 |
| *Torsion angles* | | | | | | |
| S24-C7-N22-N25 | 0.0 | 0.0 | 0.0 | 0.0 | 0.0 | 0.0 |
| N22-N25-C14-C28 | 180.0 | 180.0 | 180.0 | 180.0 | 180.0 | 180.0 |
| O21-C13-C12-N27 | 0.0 | 0.0 | 0.0 | 0.0 | 0.0 | 0.0 |
| N27-N19-C16-N26 | 180.0 | 180.0 | 180.0 | 180.0 | 180.0 | 180.0 |
| N19-C16-N26-C15 | 180.0 | 180.0 | 180.0 | 180.0 | 180.0 | 180.0 |

TABLE 5: Absorption maxima of dyes in acidic and basic solutions.

| Dye number | Methanol | Methanol + HCl | Methanol+ KOH | $CHCl_3$ | $CHCl_3$ + TFAA | $CHCl_3$ + piperidin | DMF | DMF + piperidin | DMSO | DMSO + piperidin |
|---|---|---|---|---|---|---|---|---|---|---|
| 3a | 457 | 440 | 459 | 418 | 418 | 469 | 469 | 469 | 469 | 469 |
| 3b | 462 | 440 | 463 | 422 | 420 | 469 | 469 | 469 | 469 | 469 |
| 3c | 467 | 479 | 466 | 450 | 452 | 469 | 469 | 469 | 472 | 472 |
| 3d | 465 | 477 | 465 | 450 | 449 | 469 | 469 | 469 | 469 | 469 |
| 3e | 400[s], 442 | 453 | 400[s], 442 | 452 | 451 | 457 | 457 | 457 | 461 | 461 |
| 3f | 460 | 462 | 449 | 464 | 464 | 464 | 462 | 462 | 469 | 469 |

TFAA: trifluoroacetic acid.
[s]shoulder.

Absorption spectra of the dyes 3a–3f were taken in pure chloroform and chloroform/DMF mixtures of 80/20, 60/40, 40/60, 20/80 (v/v), and pure DMF to understand whether the bathochromic shift results from tautomerization or ionization in basic chloroform solution (Table 6). Even a slight increase in the percentage of DMF in chloroform solution shifted the absorption maxima of the dyes 3a–3d to a sudden longer wavelength. These results suggest that

TABLE 6: The $\lambda_{max}$ values of dyes in different percentages of chloroform/DMF mixture.

| Dye number | Chloroform | 20% DMF | 40% DMF | 60% DMF | 80% DMF | DMF |
|---|---|---|---|---|---|---|
| 3a | 419 | 445 | 466 | 466 | 467 | 467 |
| 3b | 422 | 460 | 461 | 469 | 469 | 469 |
| 3c | 451 | 469 | 469 | 469 | 469 | 469 |
| 3d | 443 | 453 | 469 | 469 | 469 | 469 |
| 3e | 453 | 459 | 459 | 457 | 453 | 452 |
| 3f | 465 | 468 | 469 | 467 | 467 | 457 |

TABLE 7: The maximum absorption wavelengths (nm) of dyes at different temperature and concentration in chloroform and DMF.

| Dye number | Chloroform | | | DMF | | |
|---|---|---|---|---|---|---|
| | Concentrated (25°C) | Concentrated (40°C) | Diluted (25°C) | Concentrated (25°C) | Concentrated (40°C) | Diluted (25°C) |
| 3a | 419 | 417 | 418 | 470 | 468 | 467 |
| 3b | 421 | 420 | 422 | 469 | 469 | 469 |
| 3c | 450 | 450 | 450 | 469 | 469 | 469 |
| 3d | 450 | 449 | 450 | 469 | 469 | 469 |
| 3e | 452 | 452 | 452 | 452 | 452 | 453 |
| 3f | 464 | 464 | 463 | 457 | 457 | 459 |

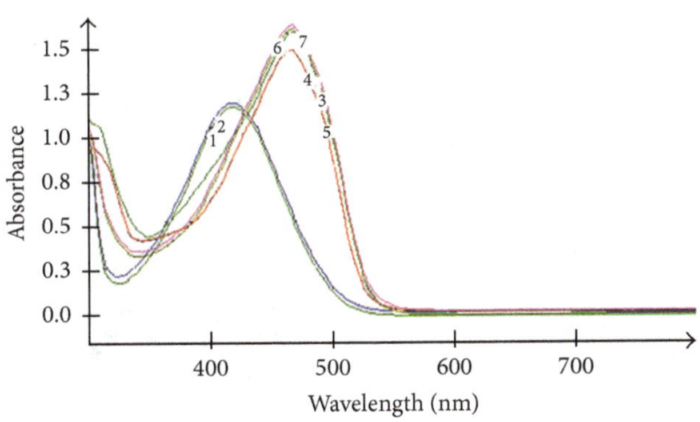


FIGURE 6: Absorption spectra of dye 3a in acid and base.

these dyes absorbed at long wavelength exist in anionic form. But such a shift was not observed for the dyes 3e and 3f.

*3.4. Effect of Concentration and Temperature.* The effect of molarity and temperature on the absorption spectra of the dyes 3a–3f was studied and the results are listed in Table 7. Therefore, chloroform and DMF solutions were diluted 50% and the absorption spectra were recorded at 25°C versus 40°C. The difference between the $\lambda_{max}$ values of the neutral molecule and the anion is in the range 1–48 nm. The dyes can easily ionize even in neutral polar solvents and the equilibrium resulting from the solvent change is an ionic equilibrium rather than a tautomeric equilibrium. As seen in Table 7, the absorption spectra are not affected by the concentration and temperature. The reason for this can be ionic equilibria.

*3.5. Effect of Substituent.* As seen in Table 4, the substituent effect is mostly observed in the most apolar solvent chloroform among all the solvents. The substituent affected $\lambda_{max}$ values less in DMF and DMSO when compared to other solvents. The substituent effect on $\lambda_{max}$ values in chloroform is evaluated and the results can be summarized as follows:

(i) The introduction of chlorine group into the sixth position of the benzothiazole ring did not have a notable effect on the absorption maxima. On the other hand, substitution of the methoxy group on the sixth position in dye 3c and methyl groups on the fifth and sixth position in dye 3d had the equal effect on $\lambda_{max}$ values resulting in 31 nm bathochromic shift.

(ii) The substituted methyl groups on the fifth and sixth positions on the benzimidazole ring in dye 3f caused 12 nm increase in bathochromic shift when it is compared with dye 3e.

(iii) As the electron donating ability of the substituents on the benzothiazole and benzimidazole in diazo compound increases, more electron donation occurs and in parallel to this bathochromic shift also increases.

*3.6. Frontier Molecular Orbitals (FMOs).* DFT computational results given in Table 7 reveal that the energy gaps ($\Delta E = E_{LUMO} - E_{HOMO}$) for benzothiazole moiety bearing azo dyes 3a–3d are 3.08, 3.04, 2.85, and 3.01 eV, respectively, while those for the benzimidazole moiety bearing azo dyes 3e and 3f are 3.04 and 2.87 eV, respectively. The frontier orbitals LUMO+1, LUMO, HOMO, and HOMO+1 are represented in Figure 7. HOMOs and LUMOs of the dyes 3a–3d

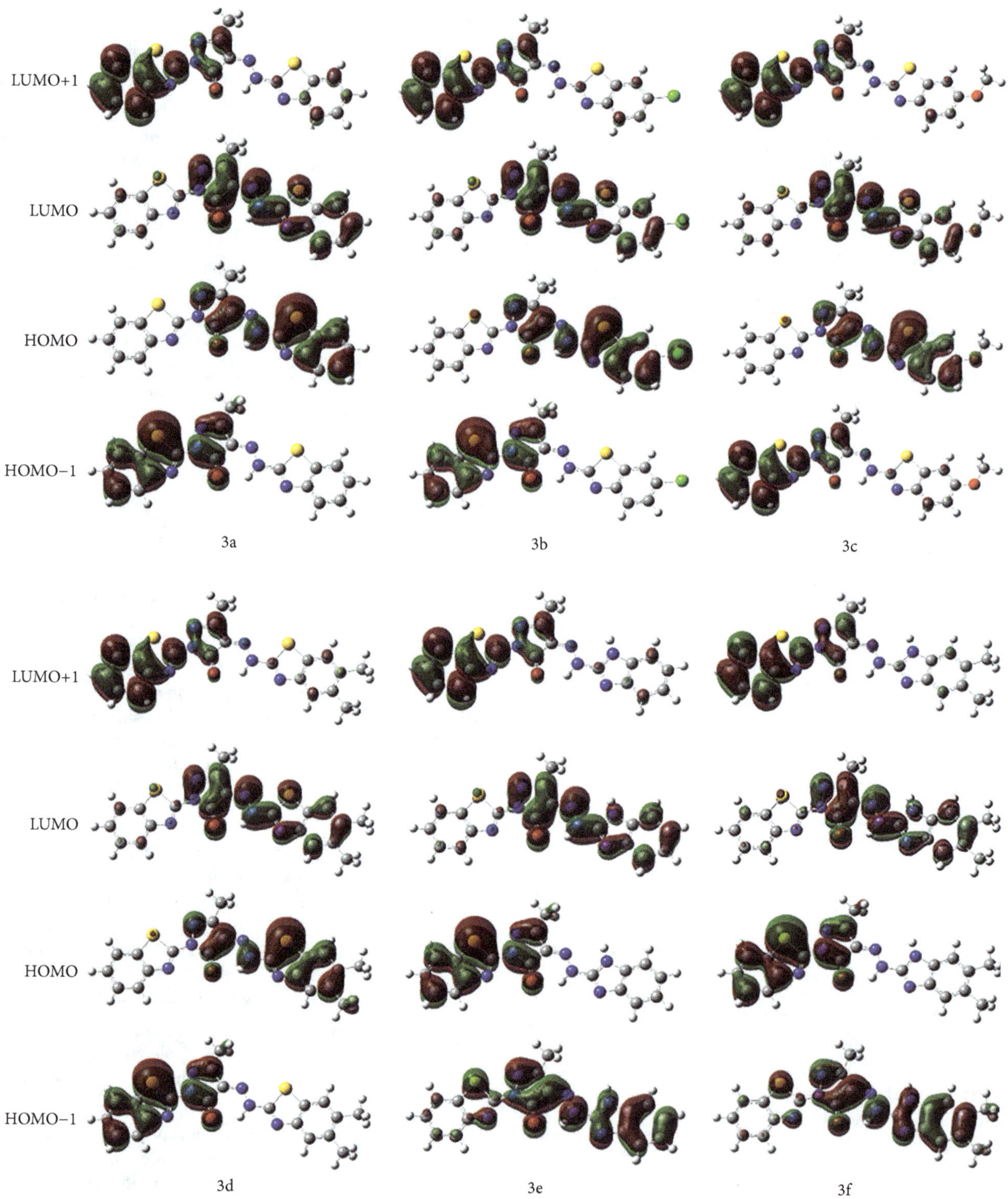


FIGURE 7: The frontier molecular orbitals of the dyes 3a–3f.

are mainly localized on pyrazole and benzothiazolyl diazo moiety. The main peaks in UV-Vis spectra are attributed to the HOMO-LUMO electronic transitions. As seen from Table 8, the increase in the wavelength values is in agreement with bathochromic shifts and $\Delta E$ decreasing. The electron densities in the HOMOs of the dyes 3e and 3f are mainly localized on pyrazole and benzothiazole moiety while electron densities in their LUMOs are localized on pyrazole and benzimidazolyl diazo moiety.

## 4. Conclusions

A series of azo dyes based on benzothiazolyl pyrazole and simultaneously benzothiazolyl and benzimidazolyl diazo

Table 8: HOMO and LUMO energies (eV), energy gaps ($\Delta E$), and $\lambda_{max}$ values (nm) of azo dyes 3a–3f calculated in chloroform.

| Dye | $\lambda_{max(calc.)}$ [a] | $E_{LUMO}$ | $E_{HOMO}$ | $Egap_{(calc)}$ |
|---|---|---|---|---|
| 3a | 433 (0.9087) | −3.26 | −6.35 | 3.08 |
| 3b | 439 (0.9685) | −3.33 | −6.37 | 3.04 |
| 3c | 488 (0.8218) | −3.17 | −6.02 | 2.85 |
| 3d | 462 (0.8391) | −3.21 | −6.22 | 3.01 |
| 3e | 447 (0.9126) | −3.18 | −6.22 | 3.04 |
| 3f | 480 (0.8510) | −3.13 | −5.99 | 2.87 |

[a]Oscillator strength values are given in parentheses.

moieties were synthesized and characterized by elemental analyses, FT-IR, $^1$H NMR spectra, and DFT calculations. Absorption spectra of the dyes were reported both experimentally and theoretically by TD-DFT calculations. The absorption maxima of the hetarylazopyrazolone dyes bearing benzothiazolyl diazo moiety showed bathochromic shifts because these dyes give easily acidic protons to the solvent like DMF and DMSO. The maximum absorption values obtained with TD-DFT/B3LYP method using nonequilibrium procedure are in a good agreement with the experimental results.

## Conflicts of Interest

The authors declare that there are no conflicts of interest regarding the publication of this paper.

## Acknowledgments

This work has been supported by the Gazi University Research Fund through Grant no. 05/2002-08.

## References


[1] H. Zollinger, *Color Chemistry: Synthesis, Properties and Applications of Organic Dyes and Pigments*, Wiley VCH, 3rd edition, 2003.

[2] J. Koh, H. Kim, J. Lee, and M. Eom, "Dyeing and fastness properties of phthalimide-based alkali-clearable azo disperse dyes on poly(ethylene terephthalate)," *Coloration Technology*, vol. 125, no. 6, pp. 322–327, 2009.

[3] G. Hussain, M. Ather, M. U. A. Khan et al., "Synthesis and characterization of chromium (III), iron (II), copper (II) complexes of 4-amino-1-(p-sulphophenyl)-3-methyl-5-pyrazolone based acid dyes and their applications on leather," *Dyes and Pigments*, vol. 130, pp. 90–98, 2016.

[4] X. Tong and Y. Zhao, "Multiple photochemical processes in liquid crystalline azo dye-doped liquid crystals," *Chemistry of Materials*, vol. 21, no. 17, pp. 4047–4054, 2009.

[5] A. V. Uklein, E. V. Ouskova, and V. Y. Gayvoronsky, "Nonlinear optical response of the nematic liquid crystal doped with polymer-azo-dye complex under picosecond laser pulses excitation," *Functional Materials*, vol. 22, no. 1, pp. 20–26, 2015.

[6] S. B. Novir and S. M. Hashemianzadeh, "Density functional theory study of new azo dyes with different $\pi$-spacers for dye-sensitized solar cells," *Spectrochimica Acta Part A: Molecular and Biomolecular Spectroscopy*, vol. 143, pp. 20–34, 2015.

[7] R. V. Solomon, R. Jagadeesan, S. A. Vedha, and P. Venuvanalingam, "A DFT/TDDFT modelling of bithiophene azo chromophores for optoelectronic applications," *Dyes and Pigments*, vol. 100, no. 1, pp. 261–268, 2014.

[8] M. Matsuoka, *Infrared Absorbing Dyes*, Plenum Press, New York, NY, USA, 1990.

[9] S. Erişkin, N. Şener, S. Yavuz, and I. Şener, "Synthesis, characterization, and biological activities of 4-imino-3-arylazo-4H-pyrimido[2,1-b][1,3]benzothiazole-2-oles," *Medicinal Chemistry Research*, vol. 23, no. 8, pp. 3733–3743, 2014.

[10] B. Minisini, G. Messager, I. Piyanzina, N. Delorme, and J.-F. Bardeau, "Vibrational analysis of [4-[(*E*)-phenylazo] phenyl] ethanol based on the comparison between the experimental and DFT calculated raman spectra," *Journal of Structural Chemistry*, vol. 55, no. 5, pp. 843–851, 2014.

[11] M. S. Deshmukh and N. Sekar, "A combined experimental and TD-DFT investigation of three disperse azo dyes having the nitroterephthalate skeleton," *Dyes and Pigments*, vol. 103, pp. 25–33, 2014.

[12] S. Harisha, J. Keshavayya, B. E. K. Swamy, and C. C. Viswanath, "Dyes and pigments synthesis, characterization and electrochemical studies of azo dyes derived from barbituric acid," *Dyes and Pigments*, vol. 136, pp. 742–753, 2017.

[13] S. A. Basaif, M. A. Hassan, and A. A. Gobouri, "$AlCl_3$-Catalyzed diazocoupling of 1-(aryl/hetaryl)-3-phenyl-1*H*-pyrazol-2-in-5-ones in aqueous medium. Synthesis of hetaryl-azopyrazolones and their application as disperse dyes," *Dyes and Pigments*, vol. 72, no. 3, pp. 387–391, 2007.

[14] B. Babür, N. Seferoğlu, E. Aktan, T. Hökelek, E. Şahin, and Z. Seferoğlu, "Phenylazoindole dyes 3: determination of azo-hydrazone tautomers of new phenylazoindole dyes in solution and solid state," *Journal of Molecular Structure*, vol. 1081, pp. 175–181, 2015.

[15] T. Aksungur, Ö. Arslan, N. Seferolu, and Z. Seferolu, "Photophysical and theoretical studies on newly synthesized N,N-diphenylamine based azo dye," *Journal of Molecular Structure*, vol. 1099, pp. 543–550, 2015.

[16] Y. Li, B. O. Patrick, and D. Dolphin, "Near-infrared absorbing azo dyes: synthesis and X-ray crystallographic and spectral characterization of monoazopyrroles, bisazopyrroles, and a boron-azopyrrole complex," *The Journal of Organic Chemistry*, vol. 74, no. 15, pp. 5237–5243, 2009.

[17] B. Babür, N. Seferoğlu, E. Aktan, T. Hökelek, E. Şahin, and Z. Seferoğlu, "Phenylazoindole dyes 2: the molecular structure characterizations of new phenylazo indoles derived from 1,2-dimethylindole," *Dyes and Pigments*, vol. 103, pp. 62–70, 2014.

[18] S. Kinali, S. Demirci, Z. Çalişir, M. Kurt, and A. Ataç, "DFT, FT-IR, FT-Raman and NMR studies of 4-(substituted phenylazo)-3,5- diacetamido-1H-pyrazoles," *Journal of Molecular Structure*, vol. 993, no. 1-3, pp. 254–258, 2011.

[19] B. Minisini, G. Messager, I. Piyanzina, N. Delorme, and J.-F. Bardeau, "Vibrational analysis of [4-[(E)-phenylazo] phenyl] ethanol based on the comparison between the experimental and DFT calculated raman spectra," *Journal of Structural Chemistry*, vol. 55, no. 5, pp. 843–851, 2014.

[20] A. Pramanik and G. Das, "Molecular to supramolecular structure: influence of coordination environment in azo-dye complexes," *Crystal Growth and Design*, vol. 8, no. 8, pp. 3107–3113, 2008.

[21] K. J. Al-Adilee, A. K. Abass, and A. M. Taher, "Synthesis of some transition metal complexes with new heterocyclic thiazolyl azo

dye and their uses as sensitizers in photo reactions," *Journal of Molecular Structure*, vol. 1108, pp. 378–397, 2016.

[22] N. Turan, R. Adigüzel, K. Buldurun, and E. Bursal, "Spectroscopic, thermal and antioxidant properties of novel mixed ligand-metal complexes obtained from saccharinate complexes and azo dye ligand (Mnppa)," *International Journal of Pharmacology*, vol. 12, no. 2, pp. 92–100, 2016.

[23] Ç. K. Atay, Y. Kara, M. Gökalp, I. Kara, T. Tilki, and F. Karci, "Disazo dyes containing pyrazole and indole moieties: Synthesis, characterization, absorption characteristics, theoretical calculations, structural and electronic properties," *Journal of Molecular Liquids*, vol. 215, pp. 647–655, 2016.

[24] F. Karci and F. Karci, "Synthesis of some novel pyrazolo[5,1-c][1,2,4]triazine derivatives and investigation of their absorption spectra," *Dyes and Pigments*, vol. 76, no. 1, pp. 97–103, 2008.

[25] Ç. Karabacak and O. Dilek, "Synthesis, solvatochromic properties and theoretical calculation of some novel disazo indole dyes," *Journal of Molecular Liquids*, vol. 199, pp. 227–236, 2014.

[26] F. Karci and E. Bakan, "New disazo pyrazole disperse dyes: synthesis, spectroscopic studies and tautomeric structures," *Journal of Molecular Liquids*, vol. 206, pp. 309–315, 2015.

[27] N. Sener, I. Sener, S. Yavuz, and F. Karci, "Synthesis, absorption properties and biological evaluation of some novel disazo dyes derived from pyrazole derivatives," *Asian Journal of Chemistry*, vol. 27, no. 8, pp. 3003–3012, 2015.

[28] F. Karci, N. Şener, M. Yamaç, I. Şener, and A. Demirçali, "The synthesis, antimicrobial activity and absorption characteristics of some novel heterocyclic disazo dyes," *Dyes and Pigments*, vol. 80, no. 1, pp. 47–52, 2009.

[29] A. Emandi, I. Serban, and R. Bandula, "Synthesis of some new solvatochromic 1(4)-substituted Pyrazol-5-one Azo derivatives," *Dyes and Pigments*, vol. 41, no. 1-2, pp. 63–77, 1999.

[30] B. Babür, N. Seferoğlu, and Z. Seferoğlu, "A ratiometric fluorescence chemosensor based on a coumarin-pyrazolone hybrid: the synthesis and an investigation of the photophysical, tautomeric and anion binding properties by spectroscopic techniques and DFT calculations," *Tetrahedron Letters*, vol. 56, no. 17, pp. 2149–2154, 2015.

[31] B. Çatikkaş, E. Aktan, and E. Yalçin, "Vibrational and electronic investigations, NLO, FMO analysis on a hetarylazoindole disperse dye by density functional theory," *Journal of Molecular Structure*, vol. 1117, pp. 218–226, 2016.

[32] J. Jayabharathi, V. Thanikachalam, M. Venkatesh Perumal, and N. Srinivasan, "A physiochemical study of azo dyes: DFT based ESIPT process," *Spectrochimica Acta—Part A: Molecular and Biomolecular Spectroscopy*, vol. 83, no. 1, pp. 200–206, 2011.

[33] Ü. Ö. Özdemir, E. Aktan, F. Ilbiz et al., "Characterization, antibacterial, anticarbonic anhydrase II isoenzyme, anticancer, electrochemical and computational studies of sulfonic acid hydrazide derivative and its Cu(II) complex," *Inorganica Chimica Acta*, vol. 423, pp. 194–203, 2014.

[34] E. Yalçin, S. Achelle, Y. Bayrak, N. Seferolu, A. Barsella, and Z. Seferolu, "Styryl-based NLO chromophores: synthesis, spectroscopic properties, and theoretical calculations," *Tetrahedron Letters*, vol. 56, no. 20, pp. 2586–2589, 2015.

[35] G. Elmaci, E. Aktan, N. Seferoğlu, T. Hökelek, and Z. Seferoğlu, "Synthesis, molecular structure and computational study of (Z)-2-((*E*)-4-nitrobenzylidene)hydrazone)-1,2-diphenylethan-1-one," *Journal of Molecular Structure*, vol. 1099, pp. 83–91, 2015.

[36] S. Alyar, H. Alyar, U. O. Ozdemir et al., "Synthesis, characterization, antibacterial activity and quantum chemical studies of N′-acetyl propane sulfonic acid hydrazide," *Journal of Molecular Structure*, vol. 1094, pp. 237–245, 2015.

[37] Ü. Özmen Özdemir, A. Altuntaş, A. B. Gündüzalp, F. Arslan, and F. Hamurcu, "New aromatic/heteroaromatic propanesulfonylhydrazone compounds: synthesis, physical properties and inhibition studies against carbonic anhydrase II (CAII) enzyme," *Spectrochimica Acta. Part A: Molecular and Biomolecular Spectroscopy*, vol. 128, pp. 452–460, 2014.

[38] E. Aktan, N. Ertan, and T. Uyar, "Synthesis, characterization and theoretical study of new hetarylazopyrazolone dyes and investigation of their absorption spectra," *Journal of Molecular Structure*, vol. 1060, pp. 215–222, 2014.

[39] M. J. Frisch, G. W. Trucks, H. B. Schlegel et al., *Gaussian 09, Revision A.02*, Gaussian Inc, Wallingford, Conn, USA, 2009.

[40] J. Tomasi, B. Mennucci, and E. Cancès, "The IEF version of the PCM solvation method: an overview of a new method addressed to study molecular solutes at the QM ab initio level," *Journal of Molecular Structure: THEOCHEM*, vol. 464, no. 1–3, pp. 211–226, 1999.

[41] M. Cossi and V. Barone, "Time-dependent density functional theory for molecules in liquid solutions," *Journal of Chemical Physics*, vol. 115, no. 10, pp. 4708–4717, 2001.

# 22

# LIBS Analysis of Geomaterials: Comparative Study of Basalt Plasma Induced by TEA $CO_2$ and Nd:YAG Laser in Air at Atmospheric Pressure

**Jelena Savovic,[1] Milos Momcilovic,[1] Sanja Zivkovic,[1] Andrei Stancalie,[2] Milan Trtica,[1] and Miroslav Kuzmanovic[3]**

[1] *Vinča Institute of Nuclear Sciences, University of Belgrade, Mike Alasa 12-14, 11001 Belgrade, Serbia*
[2] *National Institute for Laser, Plasma and Radiation Physics, Center for Advanced Laser Technologies (CETAL), Atomistilor 409, 077125 Magurele, Romania*
[3] *Faculty of Physical Chemistry, University of Belgrade, Studentski trg 12-16, 11000 Belgrade, Serbia*

Correspondence should be addressed to Jelena Savovic; lelas@vin.bg.ac.rs

Academic Editor: J. O. Caceres

We present a study of the plasma generated by transversely excited atmospheric (TEA) $CO_2$ laser irradiation of a basalt sample. The plasma was induced in air at atmospheric pressure. The same sample was also analyzed using a commercial LIBS system based on Nd:YAG laser and time-gated detection. The main plasma parameters, temperature, and electron number density were determined and analytical capabilities of the two systems compared. Despite differences in laser wavelength, pulse duration, applied fluence, and signal detection scheme, the two systems are comparable in terms of element detectability and limits of detection. In both cases, all elements usually present in geological samples were identified. The estimated limits of detection for most elements were below 100 ppm, while for Cu, Cr, and Sr they were around or below 10 ppm. The obtained results led to the conclusion that simple, cost-effective TEA $CO_2$ LIBS system can find applications for geological explorations.

## 1. Introduction

The data on the chemical composition of geomaterials are of fundamental importance for exploitation and use of georesources, environmental studies, and space explorations. Because of that, development of rapid and accurate instrumental analytical methods for chemical analysis of geological materials is of prime importance. At the same time, the analysis of geomaterials is often a challenging analytical task, as it requires determination of a large number of elements, over a wide range of concentrations, in a variety of complex matrices.

The well-established spectrochemical techniques that are mainly used in this field are inductively coupled plasma optical emission spectroscopy (ICP-OES), inductively coupled plasma mass spectrometry (ICP-MS), and X-ray fluorescence (XRF) spectroscopy. In recent years, a considerable progress has been made in the development of another spectrochemical technique, laser-induced breakdown spectroscopy (LIBS) [1]. LIBS combines capability of providing fast multielemental analysis with no sample pretreatment, potential for in situ and remote analysis, the ability to perform surface and depth profiling, and the ability to provide isotopic ratio information additionally to elemental composition [2]. Regarding the analysis of geomaterials, LIBS has certain advantages compared to ICP and XRF. ICP-OES and ICP-MS have better analytical figures of merit such as accuracy and precision and limits of detection compared to LIBS. However, many geological samples are silicate rocks, and the main benefit of LIBS compared to ICP is that there is no need for solid sample digestion, that is, no need for complex and time-consuming dissolution procedures. On the other hand, XRF spectrometry is a rapid method for in-field elemental analysis of geological samples and requires simple and minimal sample preparation (e.g., fusion or preparation of pressed pellets) [3]. These characteristics are also common to LIBS. At the same time, compared to XRF, LIBS has higher sensitivity for elements like Mg, Si, Al, Ca, and K and especially for light elements like Li, Be, B, C, F, and Na. Also, unlike ICP and XRF, LIBS may be easily applied for depth profiling by repeated ablation of the same location. In summary, each of the aforementioned analytical techniques is characterized by its own strengths and weaknesses, and the choice of the "best" technique is dictated by the analytical problem that has to be solved. In that sense, LIBS may be considered as spectrochemical technique complemental to ICP and XRF.

Due to its unique characteristics, LIBS can be used for elemental analysis of almost any type of material. However, the vast majority of LIBS applications deal with compact solid samples and, among them, perhaps the most analyzed sample types were geological materials. LIBS has been successfully applied, both in laboratory setting and in the field, for chemical analysis of rocks, soils, sediments, and other natural materials [2, 5–10]. One of the main problems in quantitative LIBS analysis of complex matrices (such as geological samples) comes from the matrix effects which can reduce the accuracy and precision of the analysis [1, 2, 11]. A number of calibration strategies were proposed for compensation of matrix effects in LIBS, for example, different normalization procedures as well as calibration-free approach [2, 9].

In this work, a recently developed LIBS system, based on transversely excited atmospheric (TEA) $CO_2$ laser and time-integrated signal detection [12], was applied to the qualitative analysis of basalt samples in air at atmospheric pressure. The basaltic rock was chosen as a representative of geological samples. Basalt is a very important rock as it is the most common rock in Earth's crust and is also an abundant rock on other satellites and planets, for example, Moon and Mars [13]. TEA $CO_2$ laser-based LIBS has been previously applied for the analysis of basalt under simulated Martian atmospheric conditions ($CO_2$ gas; pressure: 9 mbar) [4]. It was shown that this LIBS system provides good analytical capabilities for geological studies under low-pressure $CO_2$ gas. Although LIBS under different atmospheric conditions offers the possibility to improve the resolution, signal intensity, and signal-to-noise ratio or enable specific applications, it also introduces a degree of complexity in the experimental setup. Thus, the present study was undertaken to examine the possibility of using TEA $CO_2$ laser-based LIBS for the analysis of geologic samples in air at atmospheric pressure. The same sample was also analyzed using a commercial LIBS system based on Nd:YAG laser and time-gated detection.

Currently, the most common LIBS system used for the analysis of geological samples comprises Nd:YAG laser and time-gated detection [5–10]. Compared to it, the main advantage of TEA $CO_2$ laser-based LIBS is simplification of the instrumentation. The characteristics of TEA $CO_2$ laser (wavelength and pulse duration) make this laser especially suitable for time-integrated measurements, which precludes the use of delay generators and expensive gated detectors required for time-resolved measurements. Differences in laser wavelength and pulse duration for the two laser sources (10.6 $\mu$m, 100 ns for TEA $CO_2$ compared to 1.064 nm, 7 ns for Nd:YAG) inevitably influence the ablation process, the plasma parameters, and plasma expansion dynamics. Excitation wavelength has a strong influence on the laser-target and laser-plasma interactions. Mass ablation rate per pulse ($m$) is higher for shorter wavelengths ($m \sim \lambda^{-4/9}$) [14]. However, when operating with infrared wavelengths, dominant mechanism for laser-plasma interaction is *Inverse Bremsstrahlung* (IB) absorption. The rate of IB absorption ($\alpha$IB) increases as the laser wavelength increases ($\alpha\text{IB} \sim \lambda^3$) [14]. Thus, for Nd:YAG plasma, a higher ablation efficiency may be expected, while laser-plasma interaction is expected to be much stronger in the case of $CO_2$ laser. Efficient IB absorption reheats the plasma, causing an additional plasma excitation and expansion. In LIBS, only a fraction of the analyte mass is excited and is capable of producing detectable optical emission. Higher excitation efficiency in $CO_2$ plasma may compensate the lower ablation efficiency. In addition, the dimensions and lifetime of plasma increase with pulse duration. The plasma takes longer to decay and, hence, the emission lasts longer. With the long-pulse irradiation, the excitation of the plasma is rather gradual, and, hence, it is suitable to obtain highly excited plasma.

The analytical capabilities of the two LIBS systems were compared in terms of detectability and limits of detection (LODs). A simplified single-sample method was applied for estimation of LOD [15]. Elemental composition of the basalt sample was determined by ICP analysis. Also, plasma parameters, temperature, and electron number density for plasma created by $CO_2$ and Nd:YAG laser pulses were evaluated and compared.

## 2. Experimental

*2.1. The Sample.* The rock sample used in the present work was the same as the one used in our previous study [4], that is, a Tertiary basaltic rock from the Balkan Peninsula, classified according to the TAS scheme as basaltic trachyandesite [16]. A sample was prepared by cutting the original natural basaltic rock to a piece about 35 mm long and ~5 mm thick. The sample was used as is, without any preparation. Elemental composition of the sample was determined by ICP analysis. Samples for ICP analysis were prepared by alkaline fusion procedure [17]. For the major elements, 40 mg of powdered material was fused with LiBO flux and then dissolved in dilute $HNO_3$. For the trace elements, $HF/HClO_4/HNO_3$ dissolution was undertaken using 100 mg of sample. Five analyses were completed to determine measurement precision. Elemental composition of the sample is given in Table 1.

*2.2. TEA $CO_2$ LIBS System.* The experimental setup is described in prior publication [4, 18] and is briefly summarized here. A commercial TEA $CO_2$ laser system, developed at Vinča Institute, was used as the energy source for plasma generation on the basalt samples [19]. Pulsed TEA $CO_2$ laser emits radiation at 10.6 $\mu$m, with an output peak power typically in the order of megawatts. The laser-optical pulse has a gain-switched spike, followed by a slowly decaying tail. The full width at half maximum (FWHM) of the spike

TABLE 1: Elemental composition of basaltic rock obtained using ICP-OES analysis [4].

| Concentration* (wt%) | | | | | | | | | |
|---|---|---|---|---|---|---|---|---|---|
| $SiO_2$ | $TiO_2$ | $Al_2O_3$ | $Fe_2O_3$ | MnO | MgO | CaO | $Na_2O$ | $K_2O$ | $P_2O_5$ |
| 53 | 1.1 | 12.2 | 5.7 | 0.10 | 4.8 | 4.8 | 2.4 | 5.7 | 1.2 |
| Concentration ($\mu$g/g) | | | | | | | | | |
| Cr | Ni | Co | V | Cu | Zn | Ba | Sr | Zr | B |
| 300 | 100 | 60 | 120 | 70 | 130 | 2000 | 740 | 160 | 127 |

*The relative expanded standard uncertainty ranged from 1–3%, in 100 ppm concentration range, to 7–15% for the highest measured concentrations.

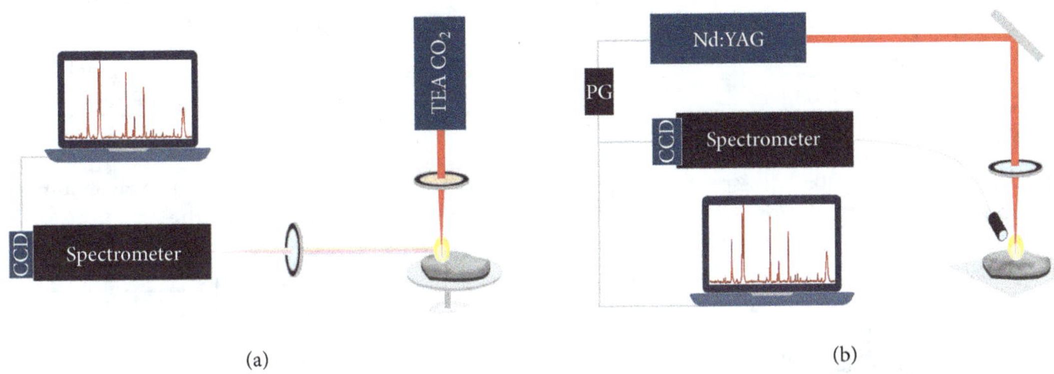


FIGURE 1: Schematic diagram of the experimental setup used for LIBS analysis. (a) TEA $CO_2$ laser, nongated signal detection; (b) Nd:YAG laser, gated signal detection.

is about 100 ns, and the tail is about 2 $\mu$s. About 35% of the total irradiated laser energy is consisted in the initial spike. Typical output pulse energy was 160 mJ. The sample surface was irradiated by laser light focused using a ZnSe lens (focal length: 130 mm). The angle of incidence of the laser beam with respect to the surface was 90°. The sample was mounted on a rotating sample holder. Single-shot craters well separated in space were achieved by rotation of the sample, while repeated ablation of the same circular path was avoided by lateral movement of the sample.

The optical emission from the plasma was viewed in the direction parallel to the target surface (Figure 1(a)). A time-integrated space-resolved laser-induced plasma spectroscopy (TISR-LIPS) was applied [12, 20–23]. Spatial resolution was achieved by changing the position of the plasma along the direction of the incoming laser beam. The focusing lens and the target holder were placed on positioning stages (three-axis translation stage; travel: 25 mm; resolution: 10 $\mu$m) to maintain focus during experiments. For spatially resolved measurements, it was essential to find plasma region, where spectral continuum background intensity is low, while the discrete line signal intensity is high. The optimization was performed by monitoring the signal-to-background ratio (SBR) of several elements (Mg: 285.2 nm, Ca: 393.4 nm, Mn: 403.08 nm, and Fe: 358.12 nm lines) as a function of the viewing position. Different spatial regions of the plasma were sequentially examined by moving the target holder in steps of 100 $\mu$m. Highest SBR values were obtained at distances 2 mm away from the sample surface. Thus, in all reported experiments, the observing position was set to 2 mm from the target surface. The horizontal part of the plume was projected by an achromatic lens on the entrance slit of the monochromator (entrance slit width: 30 $\mu$m; height: 1 mm; magnification: 1 : 1). The monochromator was equipped with diffraction grating with 600 lines/mm (dispersion: 0.7 nm/mm; blaze at 330 nm in the first order). For the time-integrated measurements, Apogee Alta F1007 CCD camera was used. The CCD consists of 1024 × 122 pixels, each of 12 × 12 microns. The total active area is 12.3 × 1.46 mm.

The laser was operated at 1.3 Hz repetition rate, while the shutter of the nongated CCD detector was opened for 30 s. Measurement time was much longer than plasma lifetime; however, the main contribution to the time-integrated spectrum comes from a very limited temporal window. Averaging of many acquired spectra was a procedure used to attenuate the variation of spectral emission, that is, to compensate pulse-to-pulse variations. Each recorded LIBS spectrum is an average spectrum, obtained by accumulation of consecutive spectra from 40 different locations on the sample. This procedure was as a rule repeated in triplicate, and the resulting spectra were averaged. Thus, all reported spectra correspond to an average of 120 laser pulses. A relative standard deviation of measured LIBS signal intensity was 10–15%.

*2.3. Nd:YAG LIBS System.* Basalt sample was also analyzed by using commercial LIBS system based on pulsed Nd:YAG laser and time-gated detection (Figure 1(b)). This was a LIBSCAN-100 system which was purchased from Applied Photonics Ltd. (UK). As laser source, it is equipped with a Q-switched Nd:YAG laser emitting at 1064 nm (model Quantel Big Sky CFR Ultra GRM), pulse duration of 7 ns, frequency of 10 Hz, and energy up to 100 mJ. The system possesses 8

TABLE 2: Spectroscopic constants of the Fe II lines (Figure 3) used in Boltzmann plot temperature determination.

| Line number | Transition | Wavelength (nm) | $g_kA_{ki}$ ($s^{-1}$) | $E_k$ (eV) | Acc |
|---|---|---|---|---|---|
| 1 | $3d^6(^5D)4s$ - $3d^6(^5D)4p$ | 256.253 | $1.07E+09$ | 5.8232247 | C+ |
| 2 | | 256.347 | $6.04E+08$ | 5.8755874 | B+ |
| 4 | | 258.258 | $3.50E+08$ | 5.8755874 | B+ |
| 5 | | 258.587 | $7.15E+08$ | 4.79323573 | B+ |
| 7 | | 259.154 | $3.40E+08$ | 5.8232247 | C+ |
| 9 | | 261.382 | $4.24E+08$ | 4.84893717 | B+ |
| 10 | | 261.762 | $2.93E+08$ | 4.81789816 | B |
| 11 | | 262.829 | $3.50E+08$ | 4.83702227 | B |
| 3 | $3d^6(^3P2)4s$ - $3d^6(^3P2)4p$ | 257.436 | $9.60E+08$ | 7.3973237 | C+ |
| 6 | $3d^6(^1G2)4s$ - $3d^6(^1G2)4p$ | 258.795 | $1.69E+09$ | 8.9430267 | B+ |
| 8 | $3d^6(^1I)4s$ - $3d^6(^1I)4p$ | 259.278 | $4.38E+09$ | 8.8565247 | B+ |

spectrometer channels covering 182–1057 nm spectral range. The acquisition is made on an integration time of 1.1 ms (the smallest) with 1.27 $\mu$s delay time from the laser pulse. The signals were averaged on 50 laser pulses. The statistical variability in the LIBS signal was approximately 10%.

A diameter of a focused laser spot was ~1 mm for TEA $CO_2$ laser and ~1.4 mm for Nd:YAG laser, that is, in both cases well above the grain sizes of the rock sample (<1 mm) [4]. Thus, the influence of grain size on accuracy and precision of the LIBS analyses was not expected to be large [24]. The main contribution to uncertainty comes from fluctuations due to laser energy and high degree of surface roughness.

## 3. Results and Discussion

LIBS spectra of basalt were recorded in air at atmospheric pressure. All reported spectra are spectra averaged over many laser shots, and the measurements were taken at a number of locations on the sample surface (a fresh surface was exposed to each incident laser pulse) to avoid the problem with sample heterogeneity. Spectra obtained on the basaltic rock sample with a single TEA $CO_2$ laser pulse using a nongated detection, covering the spectral region from 250 to 520 nm, are shown in Figure 2. The spectra were taken 2 mm from the target surface. For comparison, spectra obtained using LIBS system equipped with pulsed Nd:YAG laser and time-gated detection are also shown. Most of the elements present in the sample were identified (Table 1), including those present in concentrations around or below 100 ppm, like Co, Cr, Ni, and V. As it may be seen from Figure 2, there is no difference with regard to element detectability for the two LIBS systems used, despite the fact that the applied laser intensities differ by one order of magnitude (56 MW/cm$^2$ and 460/740 MW/cm$^2$ for TEA $CO_2$ laser and Nd:YAG laser, resp.). Thus, it may be concluded that excitation conditions in $CO_2$ plasma are suitable for detection of all elements usually present in geological samples.

The dissociation, ionization, and excitation processes taking place in plasma are related to characteristic parameters of plasma such as temperatures ($T$) and electron number densities ($n_e$). Thus, it was of interest to determine these parameters for plasma created by $CO_2$ and Nd:YAG laser pulses. The excitation temperature was evaluated by Boltzmann's plot method [25] using the equation

$$\ln\left[\frac{\lambda_{ki}I_{ki}}{g_kA_{ki}}\right] = \ln\left[\frac{hcN}{4\pi Z(T)}\right] - \frac{E_k}{k_BT}, \quad (1)$$

where $\lambda_{ki}$ is the wavelength of the emitted light, $I_{ki}$ is the integrated line intensity of the transition involving an upper level ($k$) and a lower level ($i$), $g_k$ is statistical weight of upper level, $A_{ki}$ is the transition probability, $h$ is Planck's constant, $c$ is the velocity of light in vacuum, $N$ is the total number density, $Z(T)$ is the partition function, $E_k$ is excitation energy, and $k_B$ is the Boltzmann constant. For the determination of excitation temperature of the plasma, eleven iron ionic (Fe II) lines were selected as shown in Figure 3. Considering a relatively narrow spectral interval used, the chromatic sensitivity was not taken into account.

The line identifications and different spectroscopic parameters such as wavelength, transition strength ($g_k A_{ki}$), energy $E_k$, and accuracy of transition probability (B+ $\leq$ 7%; B $\leq$ 10%; C+ $\leq$ 18%) are listed in Table 2 based on reference data [26]. In case of TEA $CO_2$ LIBS, the whole set of spectral lines listed in Table 2 was used to obtain Boltzmann's plot, whereas in case of Nd:YAG LIBS, lines numbers 3 and 8 were excluded as they were not resolved.

The estimated temperature of $CO_2$ laser-induced plasma was 14300 K, while that of Nd:YAG plasma was 19900 K and 21900 K for 50 mJ and 80 mJ pulses, respectively (Figure 4).

For both LIBS systems, under applied experimental conditions, high temperature plasma was created. Higher temperatures were obtained in Nd:YAG laser-induced plasma. However, it should be noted that estimated plasma temperatures are not directly comparable, because time-integrated measurements for $CO_2$ laser plasma provide average values for the spatially selected plasma zone, while for the Nd:YAG LIBS the values correspond to a specific time window.

The electron number density was determined from the measured *Stark* width of Al I 396.15 nm line (Figure 5).

Deconvolution of Stark width from the experimental profile was done by applying the following formula [27, 28]:

$$w_m = w_V = 0.5346 * w_L + \left(0.2169 * w_L^2 + w_G^2\right)^{1/2}, \quad (2)$$

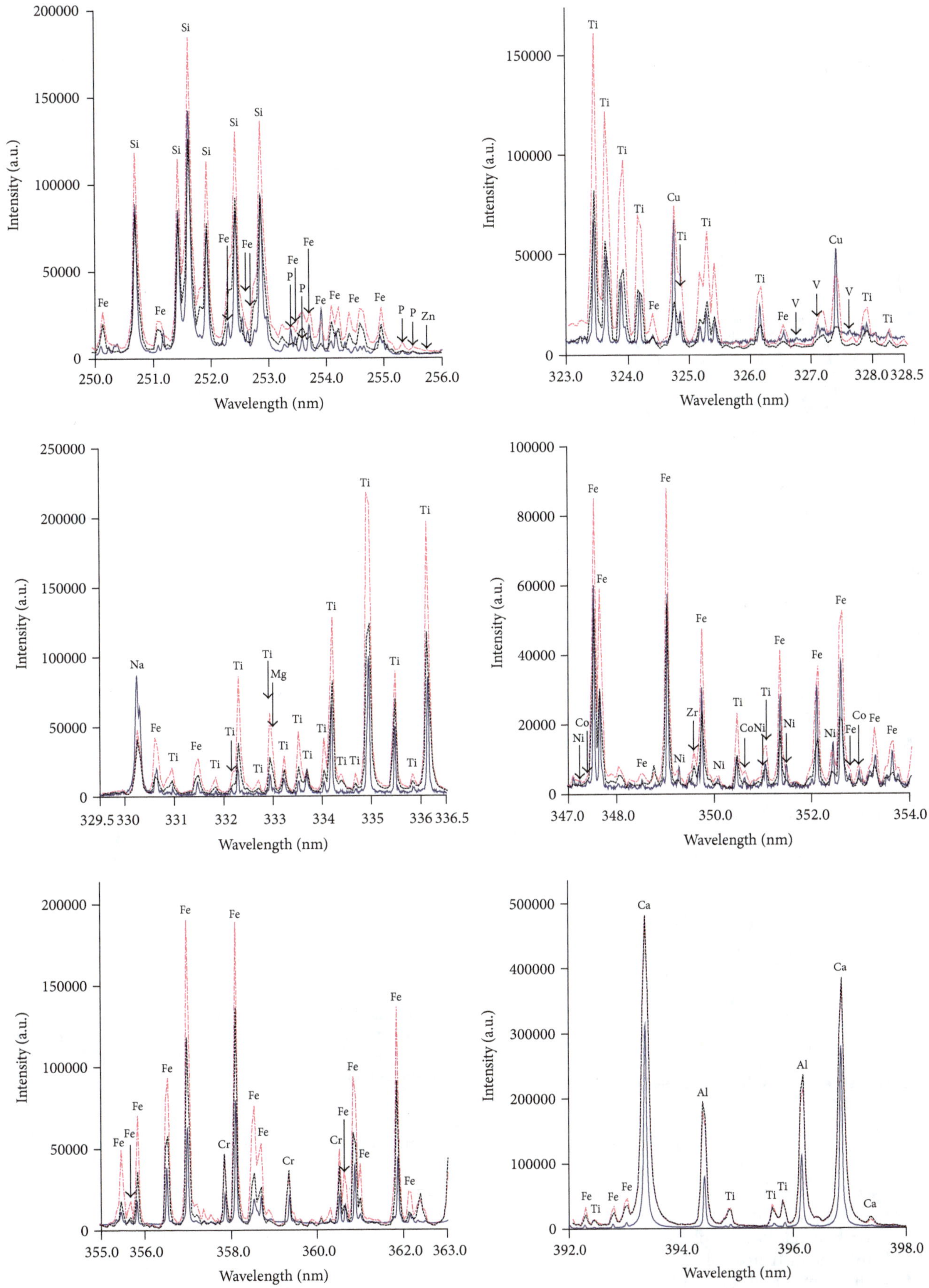


Figure 2: Continued.

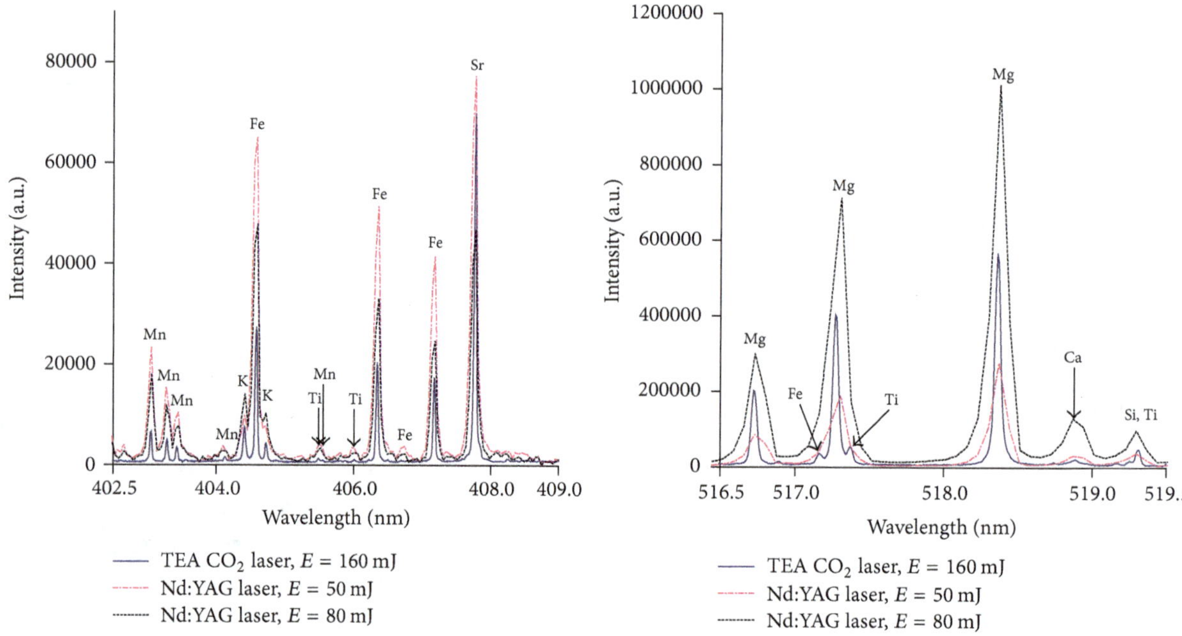


FIGURE 2: LIBS spectra obtained on the basalt sample with a single TEA $CO_2$ laser pulse using a nongated detection and with Nd:YAG laser and time-gated detection (delay: 1.27 $\mu$s; integration time gate: 1.1 ms). In both cases, the plasma was induced in air at atmospheric pressure.

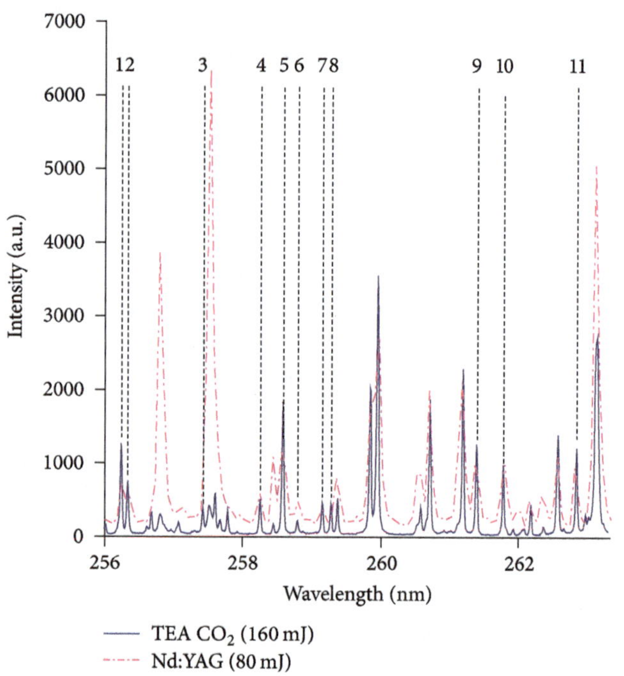


FIGURE 3: Fe II spectral lines (listed in Table 1) used for calculation of plasma temperature.

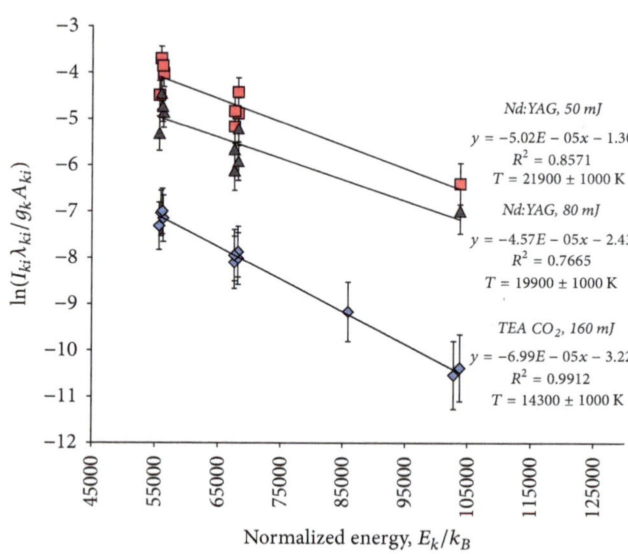


FIGURE 4: Boltzmann's plot for laser-induced plasma of basalt at atmospheric pressure utilizing Fe II transition lines listed in Table 1. $T_{exc} = -1/\text{Slope}$. The error bars account for combined uncertainties of line intensity measurements and of transition probability accuracy (Table 2).

where $w_L = w_S$, $w_G = (w_D^2 + w_i^2)^{0.5} \approx w_i$, and $w_m$, $w_V$, $w_L$, $w_S$, $w_G$, $w_i$, and $w_D$ are measured, Voigt, Lorentzian, Stark, Gaussian, instrumental, and Doppler width, respectively. For TEA $CO_2$ LIBS, the instrumental width was measured to be FWHM = 0.025 nm, and, for Nd:YAG LIBS, FWHM = 0.04 nm. An approximate formula was used to determine $n_e$ from deconvoluted Stark widths with Stark broadening parameter taken from [29]:

$$n_e\left[\text{cm}^{-3}\right] = \frac{10^{16}\Delta\lambda_{\text{Stark}}\ [\text{nm}]}{2\omega\ [\text{nm}]}. \quad (3)$$

The obtained values are shown in Table 3.

As seen from Table 3, the estimated electron number densities for the two laser systems are similar. However,

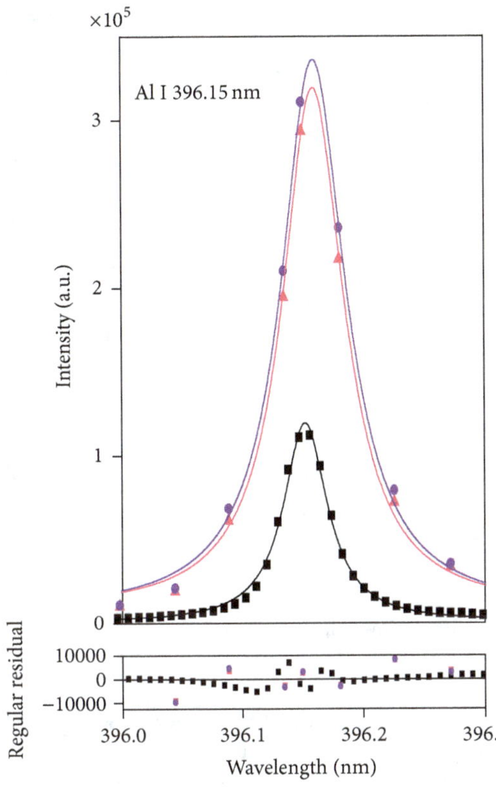


FIGURE 5: Al 396.15 nm line whose Stark width was used for evaluation of electron number density.

TABLE 3: Electron number density calculated from Stark width of Al I 396.15 nm line.

| LIBS system | $w_s$, nm | $n_e$, $10^{16}$ cm$^{-3}$ |
|---|---|---|
| TEA $CO_2$ laser (160 mJ) | 0.033 | 4.3 |
| Nd:YAG laser (50 mJ) | 0.035 | 4.6 |
| Nd:YAG laser (80 mJ) | 0.040 | 5.3 |

these values are not directly comparable, as already discussed for the temperature evaluation. Nevertheless, they provide indication of the excitation conditions in plasma.

For both laser systems, spectral data were processed by calculating the integrated peak area of the emission line, the baseline continuum emission intensity, and the root mean square (RMS) noise of the continuum radiation intensity in regions adjacent to the emission line. The signal-to-noise ratio (SNR) was calculated as the integrated peak area A, divided by the width of the peak area $w$, times the absolute value of the RMS noise: SNR = $A/(w \times \text{rms})$ [30]. Signal-to-noise (SNR) ratio for the elements under study, obtained for TEA $CO_2$ and Nd:YAG laser-induced basalt plasma, is shown in Table 4.

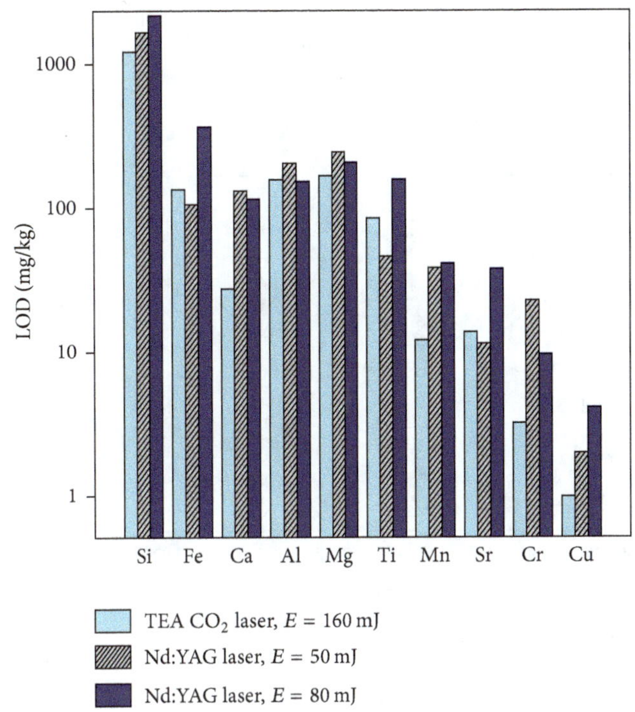


FIGURE 6: Comparison of estimated limits of detection obtained for TEA $CO_2$ and Nd:YAG laser-induced plasma over basalt sample in air at atmospheric pressure.

A simplified single-sample method was applied for estimation of the limit of detection (LOD) [15]. LOD was calculated using the formula LOD = $(3 \times c)/\text{SNR}$, where $c$ is a known analyte concentration obtained by ICP measurements (Table 1). Comparison of estimated limits of detection obtained for the two laser systems is shown in Figure 6. As it can be seen, comparable or even better LODs were obtained for TEA $CO_2$ LIBS system. Also, there is no obvious correlation between the energy of the Nd:YAG laser pulse and the estimated LODs.

## 4. Conclusions

Laser-induced breakdown spectroscopy is based on TEA $CO_2$ laser and has been applied to elemental analysis of basalt sample. The analysis was done in air at atmospheric pressure without any sample treatment. Results of qualitative analysis were compared to those obtained using a commercial LIBS system based on pulsed Nd:YAG laser. It may be concluded that, with regard to detectability and limits of detection, less complex and cost-effective TEA $CO_2$ LIBS system has comparable characteristics to the system based on Nd:YAG laser and time-gated detection.

## Conflicts of Interest

The authors declare that there are no conflicts of interest regarding the publication of this paper.

Table 4: Comparison of signal-to-noise (SNR) ratio for the elements under study, obtained for TEA $CO_2$ and Nd:YAG laser induced plasma over basalt sample in air at atmospheric pressure.

| Element/line (nm) | Conc.* | SNR TEA $CO_2$ (160 mJ) | SNR Nd:YAG (50 mJ) | SNR Nd:YAG (80 mJ) |
|---|---|---|---|---|
| Si I 288.16 | 24.8% | 814 | 526 | 454 |
| Fe I 358.12 | 4.0% | 918 | 342 | 160 |
| Ca II 393.37 | 3.4% | 3860 | 913 | 798 |
| Al I 396.15 | 6.5% | 587 | 456 | 611 |
| Mg I 285.21 | 2.9% | 536 | 368 | 433 |
| Ti II 334.94 | 0.66% | 236 | 437 | 127 |
| Mn I 403.08 | 774 ppm | 199 | 63 | 59 |
| Sr II 407.77 | 740 ppm | 165 | 200 | 61 |
| Cr I 359.35 | 300 ppm | 293 | 41 | 97 |
| Cu I 324.75 | 70 ppm | 218 | 109 | 53 |

*Elemental composition of basalt sample obtained by ICP-OES analysis.

## Acknowledgments

This research was supported by (i) the Ministry of Education, Science and Technological Development of the Republic of Serbia through the project "Effects of Laser Radiation on Novel Materials in Their Synthesis, Modifications, and Analysis" (Project no. 172019) and (ii) COST- MP1208 action ("Developing the Physics and the Scientific Community for Inertial Confinement Fusion at the Time of NIF Ignition"). Andrei Stancalie acknowledges financial support of MCI through NUCLEU Programme (ctr. no. PN 1647/2016) and UEFISCDI (ctr. no. 8PM/2010).

## References

[1] D. W. Hahn and N. Omenetto, "Laser-induced breakdown spectroscopy (LIBS)—II: review of instrumental and methodological approaches to material analysis and applications to different fields," *Applied Spectroscopy*, vol. 66, pp. 347–419, 2012.

[2] R. S. Harmon, R. E. Russo, and R. R. Hark, "Applications of laser-induced breakdown spectroscopy for geochemical and environmental analysis: A comprehensive review," *Spectrochimica Acta - Part B Atomic Spectroscopy*, vol. 87, pp. 11–26, 2013.

[3] A. A. Amosova, S. V. Panteeva, V. M. Chubarov, and A. L. Finkelshtein, "Determination of major elements by wavelength-dispersive X-ray fluorescence spectrometry and trace elements by inductively coupled plasma mass spectrometry in igneous rocks from the same fused sample (110 mg)," *Spectrochimica Acta - Part B Atomic Spectroscopy*, vol. 122, pp. 62–68, 2016.

[4] J. Savovic, M. Stoiljkovic, M. Kuzmanovic et al., "The feasibility of TEA CO2 laser-induced plasma for spectrochemical analysis of geological samples in simulated Martian conditions," *Spectrochimica Acta - Part B Atomic Spectroscopy*, vol. 118, pp. 127–136, 2016.

[5] J. M. Vadillo and J. J. Laserna, "Laser-induced breakdown spectroscopy of silicate, vanadate and sulfide rocks," *Talanta*, vol. 43, no. 7, pp. 1149–1154, 1996.

[6] F. Capitelli, F. Colao, M. R. Provenzano, R. Fantoni, G. Brunetti, and N. Senesi, "Determination of heavy metals in soils by Laser Induced Breakdown Spectroscopy," *Geoderma*, vol. 106, no. 1-2, pp. 45–62, 2002.

[7] R. Barbini, F. Colao, V. Lazic, R. Fantoni, A. Palucci, and M. Angelone, "On board LIBS analysis of marine sediments collected during the XVI Italian campaign in Antarctica," *Spectrochimica Acta-Part B Atomic Spectroscopy*, vol. 57, no. 7, pp. 1203–1218, 2002.

[8] B. Sallé, D. A. Cremers, S. Maurice, R. C. Wiens, and P. Fichet, "Evaluation of a compact spectrograph for in-situ and stand-off Laser-Induced Breakdown Spectroscopy analyses of geological samples on Mars missions," *Spectrochimica Acta - Part B Atomic Spectroscopy*, vol. 60, no. 6, pp. 805–815, 2005.

[9] B. Sallé, J.-L. Lacour, P. Mauchien, P. Fichet, S. Maurice, and G. Manhès, "Comparative study of different methodologies for quantitative rock analysis by Laser-Induced Breakdown Spectroscopy in a simulated Martian atmosphere," *Spectrochimica Acta - Part B Atomic Spectroscopy*, vol. 61, no. 3, pp. 301–313, 2006.

[10] N. Idris, K. Kagawa, F. Sakan, K. Tsuyuki, and S. Miura, "Analysis of heavy metal pollution in soil using transversely excited atmospheric CO2 laser-induced plasma by trapping the soil in microstructured holes on metal subtargets," *Applied Spectroscopy*, vol. 61, no. 12, pp. 1344–1351, 2007.

[11] J. M. Anzano, M. A. Villoria, A. Ruíz-Medina, and R. J. Lasheras, "Laser-induced breakdown spectroscopy for quantitative spectrochemical analysis of geological materials: Effects of the matrix and simultaneous determination," *Analytica Chimica Acta*, vol. 575, no. 2, pp. 230–235, 2006.

[12] M. Momcilovic, M. Kuzmanovic, D. Rankovic et al., "Optical emission studies of copper plasma induced using infrared transversely excited atmospheric (IR TEA) carbon dioxide laser pulses," *Applied Spectroscopy*, vol. 69, no. 4, pp. 419–429, 2015.

[13] A. Ruzicka, G. A. Snyder, and L. A. Taylor, "Comparative geochemistry of basalts from the moon, earth, HED asteroid, and mars: Implications for the origin of the moon," *Geochimica et Cosmochimica Acta*, vol. 65, no. 6, pp. 979–997, 2001.

[14] D. Campos, S. S. Harilal, and A. Hassanein, "The effect of laser wavelength on emission and particle dynamics of Sn plasma," *Journal of Applied Physics*, vol. 108, no. 11, p. 113305, 2010.

[15] G. W. Rieger, M. Taschuk, Y. Y. Tsui, and R. Fedosejevs, "Laser-induced breakdown spectroscopy for microanalysis using sub-millijoule UV laser pulses," *Applied Spectroscopy*, vol. 56, no. 6, pp. 689–698, 2002.

[16] V. Cvetković, D. Prelević, H. Downes, M. Jovanović, O. Vaselli, and Z. Pécskay, "Origin and geodynamic significance of Tertiary postcollisional basaltic magmatism in Serbia (central Balkan Peninsula)," *Lithos*, vol. 73, no. 3-4, pp. 161–186, 2004.

[17] M. Thompson and J. N. Walsh, *Handbook of inductively coupled plasma atomic emission spectrometry*, Viridian Publishing, Springer, Boston, MA, 1989.

[18] S. Zivkovic, J. Savovic, M. Trtica, J. Mutic, and M. Momcilovic, "Elemental analysis of aluminum alloys by Laser Induced Breakdown Spectroscopy based on TEA $CO_2$ laser," *Journal of Alloys and Compounds*, vol. 700, pp. 175–184, 2017.

[19] M. S. Trtica, B. M. Gaković, B. B. Radak, and Š. S. Miljanić, "An efficient small scale TEA CO2 laser for material surface modification," in *Proceedings of the International Conference on Atomic and Molecular Pulsed Lasers IV*, pp. 44–49, rus, September 2001.

[20] A. Khumaeni, M. Ramli, Y. Deguchi et al., "New technique for the direct analysis of food powders confined in a small hole using transversely excited atmospheric CO2 laser-induced gas plasma," *Applied Spectroscopy*, vol. 62, no. 12, pp. 1344–1348, 2008.

[21] A. Matsumoto, A. Tamura, K. Fukami, Y. H. Ogata, and T. Sakka, "Single-pulse underwater laser-induced breakdown spectroscopy with nongated detection scheme," *Analytical Chemistry*, vol. 85, no. 8, pp. 3807–3811, 2013.

[22] M. A. Khater, J. T. Costello, and E. T. Kennedy, "Optimization of the emission characteristics of laser-produced steel plasmas in the vacuum ultraviolet: significant improvements in carbon detection limits," *Applied Spectroscopy*, vol. 56, no. 8, pp. 970–983, 2002.

[23] S. Zivkovic, M. Momcilovic, A. Staicu, J. Mutic, M. Trtica, and J. Savovic, "Spectrochemical analysis of powdered biological samples using TEA $CO_2$ laser plasma excitation," *Spectrochimica Acta Part B: Atomic Spectroscopy*, vol. 128, pp. 22–29, 2017.

[24] R. B. Anderson, R. V. Morris, S. M. Clegg et al., "The influence of multivariate analysis methods and target grain size on the accuracy of remote quantitative chemical analysis of rocks using laser induced breakdown spectroscopy," *Icarus*, vol. 215, no. 2, pp. 608–627, 2011.

[25] R. Epstein, "Principles of Plasma Spectroscopy, by Hans R. Griem. Cambridge University Press, 1997, 366 pages. ISBN 0 521 45504 9. $100.00," *Journal of Plasma Physics*, vol. 60, no. 1, pp. 203–207.

[26] P. J. Linstrom and W. G. Mallard, "The NIST Chemistry WebBook: A chemical data resource on the Internet," *Journal of Chemical and Engineering Data*, vol. 46, no. 5, pp. 1059–1063, 2001.

[27] J. J. Olivero and R. L. Longbothum, "Empirical fits to the Voigt line width: A brief review," *Journal of Quantitative Spectroscopy and Radiative Transfer*, vol. 17, no. 2, pp. 233–236, 1977.

[28] M. Ivković and N. Konjević, "Stark width and shift for electron number density diagnostics of low temperature plasma: application to silicon laser induced breakdown spectroscopy," *Spectrochimica Acta Part B: Atomic Spectroscopy*, vol. 131, pp. 79–92, 2017.

[29] M. S. Dimitrijevic and S. Sahal-Brechot, "Stark Broadening of AI I Spectral Lines," *Physica Scripta*, vol. 49, no. 1, pp. 34–38, 1994.

[30] B. T. Fisher, H. A. Johnsen, S. G. Buckley, and D. W. Hahn, "Temporal gating for the optimization of laser-induced breakdown spectroscopy detection and analysis of toxic metals," *Applied Spectroscopy*, vol. 55, no. 10, pp. 1312–1319, 2001.

# Permissions

All chapters in this book were first published in JC, by Hindawi Publishing Corporation; hereby published with permission under the Creative Commons Attribution License or equivalent. Every chapter published in this book has been scrutinized by our experts. Their significance has been extensively debated. The topics covered herein carry significant findings which will fuel the growth of the discipline. They may even be implemented as practical applications or may be referred to as a beginning point for another development.

The contributors of this book come from diverse backgrounds, making this book a truly international effort. This book will bring forth new frontiers with its revolutionizing research information and detailed analysis of the nascent developments around the world.

We would like to thank all the contributing authors for lending their expertise to make the book truly unique. They have played a crucial role in the development of this book. Without their invaluable contributions this book wouldn't have been possible. They have made vital efforts to compile up to date information on the varied aspects of this subject to make this book a valuable addition to the collection of many professionals and students.

This book was conceptualized with the vision of imparting up-to-date information and advanced data in this field. To ensure the same, a matchless editorial board was set up. Every individual on the board went through rigorous rounds of assessment to prove their worth. After which they invested a large part of their time researching and compiling the most relevant data for our readers.

The editorial board has been involved in producing this book since its inception. They have spent rigorous hours researching and exploring the diverse topics which have resulted in the successful publishing of this book. They have passed on their knowledge of decades through this book. To expedite this challenging task, the publisher supported the team at every step. A small team of assistant editors was also appointed to further simplify the editing procedure and attain best results for the readers.

Apart from the editorial board, the designing team has also invested a significant amount of their time in understanding the subject and creating the most relevant covers. They scrutinized every image to scout for the most suitable representation of the subject and create an appropriate cover for the book.

The publishing team has been an ardent support to the editorial, designing and production team. Their endless efforts to recruit the best for this project, has resulted in the accomplishment of this book. They are a veteran in the field of academics and their pool of knowledge is as vast as their experience in printing. Their expertise and guidance has proved useful at every step. Their uncompromising quality standards have made this book an exceptional effort. Their encouragement from time to time has been an inspiration for everyone.

The publisher and the editorial board hope that this book will prove to be a valuable piece of knowledge for researchers, students, practitioners and scholars across the globe.

# List of Contributors

**Elahe Dadvar**
Department of Environmental Engineering, Faculty of Environment and Energy, Islamic Azad University, Science and Research Branch, Tehran, Iran

**Roshanak Rezaei Kalantary**
Department of Environmental Health Engineering, School of Public Health, Iran University of Medical Sciences, Tehran, Iran

**Homayon Ahmad Panahi**
Chemistry Department, Islamic Azad University, Central Tehran Branch, Tehran, Iran

**Majid Peyravi**
Nanotechnology Research Institute, Chemical Engineering Department, Babol University of Technology, Babol, Iran

**Ashu Chaudhary and Anshul Singh**
Department of Chemistry, Kurukshetra University, Kurukshetra 136 119, India

**Salvatore Barreca, Maddalena Busetto, Matteo Vitelli, Luisa Colzani, Laura Clerici and Pierluisa Dellavedova**
Agenzia Regionale per la Protezione dell'Ambiente della Lombardia (ARPA Lombardia), U.O.C. Laboratorio di Milano, Sede laboratoristica di Monza e Brianza via Solferino N° 16, 20900 Monza, MB, Italy

**Tien Duc Pham, Hoang Hiep Nguyen, Ngoc Viet Nguyen, Thanh Tu Vu, Thi Ngoc Mai Pham, Thi Hai Yen Doan and Manh Ha Nguyen**
Faculty of Chemistry, VNU-University of Science, Vietnam National University-Hanoi, 19 Le Thanh Tong, Hoan Kiem, Hanoi 10000, Vietnam

**Thi Mai Viet Ngo**
Thai Nguyen University of Education, Thai Nguyen University, 20 Luong Ngoc Quyen, Quang Trung, Thai Nguyen, Vietnam

**Elisa Robotti, Federica Campo, Marta Riviello, Marcello Manfredi, Eleonora Mazzucco, Fabio Gosetti and Emilio Marengo**
Department of Sciences and Technological Innovation, University of Piemonte Orientale, Viale Michel 11, 15121 Alessandria, Italy

**Marco Bobba**
Department of Sciences and Technological Innovation, University of Piemonte Orientale, Viale Michel 11, 15121 Alessandria, Italy
Istituto Zooprofilattico Sperimentale della Lombardia e dell'Emilia Romagna, Via Bianchi 9, 25124 Brescia, Italy

**Emanuele Sangiorgi**
Istituto Zooprofilattico Sperimentale della Lombardia e dell'Emilia Romagna, Via Bianchi 9, 25124 Brescia, Italy

**Giorgio Calabrese**
Department of Pharmacy, Università degli Studi di Napoli Federico II, Via Montesano 49, 80131 Napoli, Italy

**Escarlet T. D. Santana and Anizio M. Faria**
Instituto de Ciências Exatas e Naturais do Pontal, Universidade Federal de Uberlândia, Ituiutaba, MG 38304-402, Brazil

**Dayane F. Soares**
Instituto Federal de Educação, Ciência e Tecnologia do Triângulo Mineiro, Ituiutaba, MG 38305-200, Brazil

**Jiaze Liu, Jun Yin, Xiaozheng He, Ting Chen and Dongsheng Shen**
School of Environmental Science and Engineering, Zhejiang Gongshang University, Hangzhou 310012, China
Zhejiang Provincial Key Laboratory of Solid Waste Treatment and Recycling, Hangzhou 310012, China

**Tilal Elsaman**
Department of Pharmaceutical Chemistry, College of Pharmacy, Omdurman Islamic University, 11452 Khartoum, Sudan

**Mohamed H. M. Hamid**
Department of Pharmaceutical Chemistry, College of Pharmacy, Omdurman Islamic University, 11452 Khartoum, Sudan
Department of Pharmaceutical Chemistry, College of Pharmacy, University of Medical Sciences and Technology, Khartoum, Sudan

**Mindaugas Liaudanskas, Kristina Zymon, Almantas Klevinskas and Valdimaras Janulis**
Department of Pharmacognosy, Faculty of Pharmacy, Lithuanian University of Health Sciences, SukilėliJ St. 13, LT-44307 Kaunas, Lithuania

**Jonas Viškelis**
Institute of Horticulture, Lithuanian Research Centre for Agriculture and Forestry, Kauno St. 30, Babtai, LT-54333 Kaunas, Lithuania

**Paulina Wiśniewska, Magdalena Śliwińska, Tomasz Dymerski, Waldemar Wardencki and Jacek Namieśnik**
Department of Analytical Chemistry, Faculty of Chemistry, Gdánsk University of Technology, 11/12 Narutowicza St., 80-233 Gdánsk, Poland

**A. O. Adeniji and O. O. Okoh**
SAMRC Microbial Water Quality Monitoring Centre, University of Fort Hare, Alice 5700, South Africa
Department of Chemistry, University of Fort Hare, Alice 5700, South Africa

**A. I. Okoh**
SAMRC Microbial Water Quality Monitoring Centre, University of Fort Hare, Alice 5700, South Africa
Applied and Environmental Microbiology Research Group, Department of Biochemistry and Microbiology, University of Fort Hare, Alice 5700, South Africa

**Zhi-juan Zhang and Li-li Cui**
Institute of Chinese Materia Medica, Henan University, Kaifeng 475004, China

**Jin-feng Wei and Wen-yi Kang**
Institute of Chinese Materia Medica, Henan University, Kaifeng 475004, China
Kaifeng Key Laboratory of Functional Components in Health Food, Kaifeng 475004, China

**Maths Halstensen, Wathsala N. Jinadasa and Klaus-J. Jens**
Applied Chemometrics Research Group (ACRG), University College of Southeast Norway, Porsgrunn, Norway

**Henrik Jilvero**
Department of Energy and Environment, Chalmers University of Technology, 412 96 Göteborg, Sweden

**H. Ait Ahsaine, Z. Anfar, M. Ezahri and N. El Alem**
Materials and Environment Laboratory, Department of Chemistry, Ibn Zohr University, Agadir, Morocco

**M. Zbair**
Catalysis and Materials Corrosion Laboratory, Department of Chemistry, Chouaib Doukkali University, El Jadida, Morocco

**Xingming Jie and Yiming Cao**
Dalian Nation Library for Clean Energy (DNL), Dalian Institute of Chemical Physics, Chinese Academy of Sciences, Dalian 116023, China

**Haitao Zhu**
University of Chinese Academy of Sciences, Beijing 100049, China

**Ruiqing Ming**
Research Institute of Petroleum Exploration & Development, Beijing 100083, China

**Huiqun He**
CNPC Drilling Research Institute, Beijing 100195, China

**Serena Carlesi and Marilena Ricci**
Department of Chemistry "Ugo Schiff", University of Florence, Via della Lastruccia 3-13, 50019 Sesto Fiorentino, Italy

**Maurizio Becucci**
Department of Chemistry "Ugo Schiff", University of Florence, Via della Lastruccia 3-13, 50019 Sesto Fiorentino, Italy
The European Laboratory for Non-Linear Spectroscopy (LENS), Via N. Carrara 1, 50019 Sesto Fiorentino (FI), Italy

**Dinh Quang Khieu, Bui Hai Dang Son and Nguyen Hai Phong**
College of Science, Hue University, Hue City 530000, Vietnam

**Vo Thi Thanh Chau**
Faculty of Technology, Industrial University of Ho Chi Minh City, Quang Ngai Campus, Quang Ngai City 570000, Vietnam

**Pham Dinh Du**
Faculty of Natural Science, Thu Dau Mot University, Thu Dau Mot City 590000, Vietnam

**Nguyen Thi Diem Chau**
Department of Agriculture and Rural Development, Tam Ky City, Quang Nam Province 560000, Vietnam

**Upenyu Guyo, Lycenter Yard Phiri and Fidelis Chigondo**
Department of Chemical Technology, Midlands State University, Senga, Gweru, Zimbabwe

**Tae Gyu Nam, Hae Won Jang, Yong-Sun Cho, Miyoung Yoo, Dongwon Seo and Jaeho Ha**
Korea Food Research Institute, Seongnam, Gyeonggi 13539, Republic of Korea

**Ahmad Rois Mansur**
Department of Food Biotechnology, Korea University of Science and Technology, Daejeon 34113, Republic of Korea

**Ebru Aktan and Tahsin Uyar**
Faculty of Science, Department of Chemistry, Gazi University, Teknikokullar, 06500 Ankara, Turkey

**Jelena Savovic, Milos Momcilovic, Sanja Zivkovic and Milan Trtica**
Vinĉa Institute of Nuclear Sciences, University of Belgrade, Mike Alasa 12-14, 11001 Belgrade, Serbia

**Andrei Stancalie**
National Institute for Laser, Plasma and Radiation Physics, Center for Advanced Laser Technologies (CETAL), Atomistilor 409, 077125 Magurele, Romania

**Miroslav Kuzmanovic**
Faculty of Physical Chemistry, University of Belgrade, Studentski trg 12-16, 11000 Belgrade, Serbia

# Index


**A**

Acetonitrile, 57-60, 77-79, 87-88, 90, 122-123, 220, 223

Activated Carbon, 2, 4, 8-9, 32, 39, 54, 75, 145-148, 156-158, 198, 205

Activated Carbon Composite, 145, 157-158

Activated Sludge, 75

Adsorption Isotherms, 32-33, 38-40, 152, 156, 203

Adsorption Kinetics, 41, 150, 156-157, 203

Anthocyanins, 87, 94

Atomic Absorption Spectrometry, 32-33, 39, 187

Autoxidation, 217

Azo Dye, 1, 3-4, 9-11, 40, 227-228

**B**

Basalt Plasma, 229, 235

Benzimidazole Ring Systems, 219

Benzophenone, 45, 51-53

Benzothiazole, 219, 221-222, 225-227

Bioactive Compounds, 94, 207, 218

Biosorption, 205

Bovine Serum Albumin, 68

Butylated Hydroxytoluene, 123

**C**

Ca(II) Ions, 197-199, 201-204

Carbon Paste Electrode, 187, 196

Cauda Epididymis, 17, 20

Chemical Hydrolysis, 77-78, 80

Cluster Analysis, 217

Column Chromatography, 78

Critical Liquid-carrying Gas Velocity, 168, 175

**D**

Density Functional Theory, 156, 219-220, 227-228

Dissolved Oxygen, 110

**E**

Energy Dispersive X-ray, 145

Escherichia Coli, 14-16

**F**

Ferric Reducing Antioxidant Power, 88

Flame Ionization Detector, 57, 61, 109, 114, 207

Flavonoids, 53, 93-94, 122, 125, 130

Fluorescence Regional Integration, 66-67

Fourier Transform Infrared Spectroscopy, 32-34, 39, 133, 145, 197

Frontiermolecular Orbitals, 225

Ft-nir Spectroscopy, 177, 185

Fusarium Oxysporum, 14-16

**G**

Gas Chromatography, 11, 54-58, 61, 64-65, 96-98, 107-110, 113-114, 117-118, 120, 130-131, 177, 185, 207, 217

Gas Chromatography Mass Spectrometry, 54

Gas Permeance, 159, 163-165

Gasoline Range Organics, 113-114

Go-tio2, 1-2, 5-8

Graphene Oxide, 1-2, 5-9, 12-13, 156, 196

**H**

Hetarylazopyrazolone Dyes, 219-220, 227-228

Heterobimetallic Complexes of Copper, 14-15, 17, 21

High-performance Liquid Chromatography, 24, 30, 31, 65, 93-94, 123, 130-131

Hydrolysis Kinetics, 10, 77, 84-85

Hydrothermal Treatment, 66-68, 72, 75

**I**

Inductively Coupled Plasma Atomic Emission Spectrometry, 187, 237

Ionic Liquids, 122, 130-131

Isoquercitrin, 87-88, 90-92, 122-129

Isorhamnetin Glucoside, 87-88, 90-92

**K**

Kaempferol, 122-129

Kaolin Clay, 37, 197-198

Kaolinite, 39, 41, 196, 205

**L**

Laser-induced Breakdown Spectroscopy, 229, 235-237

Libs Analysis, 229-231, 236

Lipid Oxidation, 215

Liquid Chromatography, 10, 24, 30-31, 64-65, 93-94, 97, 123, 130-131, 177, 207

Liquid-liquid Extraction, 111, 114, 130

Lysimachia Clethroides Duby, 122, 130

**M**

Mass Spectrometric Detector, 109, 114

Mass Spectrometry, 10, 23-25, 30-33, 42-43, 54-55, 58, 64, 94, 97, 107-108, 117, 130-131, 177, 185, 187, 196, 207-208, 210, 236

Metakaolin Zeolite, 197-200, 204

Metal-organic Framework, 159, 166
Methylene Blue, 1-5, 10, 145, 155-158, 205
Micro-raman Spectra, 179, 182, 185
Mixed Matrix Hollow Fiber Membrane, 159-160, 166
Mixed Matrix Membrane, 159, 166
Molar Conductance, 15, 17, 19
Molecular Weight Fractionation, 66-68, 71
Mptms-diatomite, 187-189, 192-195
Multivariate Analysis, 139, 177, 185, 237

**N**
Naproxen Prodrug, 77-78
Natural Antioxidants, 215
Nh2-mil-53, 159-160

**O**
O-phenylenediamine, 14-15, 17
Organic Pollutants, 1, 6, 8-9, 33, 64, 109, 117, 121
Oxidative Stability, 215, 218
Ozonation, 4, 11

**P**
Perfluorinated Compounds, 23-24, 30
Phenolic Compounds, 53, 87-94, 130, 215
Phytoextraction, 197
Polyphenols, 93-94
Polysaccharides, 181, 183
Principal Component Analysis, 30, 42, 44, 55, 177-179, 185

**S**
Scanning Electron Microscopy, 6, 145
Silicon Dioxide, 89
Sodium Dodecyl Sulfate, 32-33, 35, 39-40
Solid-phase Microextraction, 54-55, 207-208, 217-218
Stable Isotope, 217

**T**
Thiabendazole, 56-64
Time-gated Detection, 229-232, 234-235
Total Petroleum Hydrocarbons, 109-110, 113-114, 116, 118-121
Total Phenolic Content, 88

**U**
Ultem Hfm, 159, 161-165
Ultrasonic-assisted Extraction, 123, 129
Unrefined Sesame Oil, 207-208, 215, 217

**V**
Vibrational Spectroscopies, 177-178

**X**
X-ray Diffraction, 20, 32-34, 39, 133, 145, 161, 187, 189, 197-198
X-ray Fluorescence Spectrometry, 236

**Z**
Zirconium Oxide, 145-146, 157

www.ingramcontent.com/pod-product-compliance
Lightning Source LLC
LaVergne TN
LVHW081014210926
848485LV00004B/861